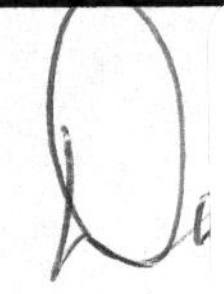

-1

ASSESSMENT METHODOLOGIES FOR PREVENTING FAILURE: DETERMINISTIC AND PROBABILISTIC ASPECTS AND WELD RESIDUAL STRESS

VOLUME 1

presented at
THE 2000 ASME PRESSURE VESSELS AND PIPING CONFERENCE
SEATTLE, WASHINGTON
JULY 23–27, 2000

sponsored by
THE PRESSURE VESSELS AND PIPING DIVISION, ASME

principal editor
R. MOHAN
ROUGE STEEL COMPANY

contributing editors
F. W. BRUST AND P. DONG
BATTELLE

G. WILKOWSKI AND Y.-Y. WANG
ENGINEERING MECHANICS CORPORATION OF COLUMBUS

M. KHALEEL
PACIFIC NORTHWEST NATIONAL LABORATORY

S. RAHMAN
UNIVERSITY OF IOWA

R. W. WARKE
SOUTHWEST RESEARCH INSTITUTE

THE AMERICAN SOCIETY OF MECHANICAL ENGINEERS
Three Park Avenue / New York, N.Y. 10016

ISBN No. 0-7918-1891-8

FOREWORD

The 2000 ASME Pressure Vessels and Piping Conference was held July 23–27, 2000, Seattle, Washington. As part of this conference, several symposia on weld residual stresses, deterministic and probabilistic fracture methods, environmental degradation and assessment and service experience were organized. The current trends in failure assessment, particularly in nuclear power plant applications, were addressed in these symposia. The papers presented in these symposia were gathered in a two-part volume series. Part I of the series contains papers that discuss the deterministic and probabilistic aspects of failure in pressure vessel and piping components and the important role of weld residual stresses in failure assessment and prevention. Many unique and clever ideas that will help broaden our engineering knowledge base have been advanced in these papers. We hope that the pressure vessel and piping community, in general, and the nuclear power plant community, in particular, will benefit from the work reported here.

The Materials and Fabrication Committee of the ASME Pressure Vessels and Piping Division sponsored the symposia. As required by ASME, each paper has been subjected to a formal peer review. All of the reviewers' comments have been addressed prior to the preparation of the final manuscripts.

Raj Mohan
Rouge Steel Company

ACKNOWLEDGMENTS

We would like to thank the authors for their willingness and diligent efforts in preparing the final manuscripts and for sharing their research results. The many reviewers, in spite of their busy schedule, have taken the time to enhance the quality and clarity of the papers. We extend our thanks to them. The support and encouragement provided by our institutions, namely Rouge Steel Company, Battelle, Engineering Mechanics Corporation of Columbus, Pacific Northwest National Laboratory, and Southwest Research Institute.

We extend our gratitude to Dr. Gery Wilkowski, Chairman of the Materials and Fabrication Committee, and to Mr. Joe Sinnappan, Technical Program Chairman, for their patience and guidance. Finally, we would like to thank the ASME editorial staff for their cooperation, assistance, and professional work in finalizing this publication.

F. W. Brust
P. Dong
M. Khaleel
R. Mohan
Y. Y. Wang
R. W. Warke
G. Wilkowski

CONTENTS

CHALLENGES AND RESOLUTIONS IN STRUCTURAL LIFE PREDICTION

Introduction

Raj Mohan
Rouge Steel Company
Dearborn, MI 48121

Yong-Yi Wang
Engineering Mechanics Corporation of Columbus
3518 Riverside Dr., Suite 202
Columbus, OH 43221
Wangemc2@columbus.rr.com

As part of the ASME-PVP conference held in Seattle, WA, during July 23-27, 2000, a major symposium on "Challenges and Resolutions in Structural Life Prediction" was organized. Most of the papers presented in the symposium appear in a separate volume. However, one of the papers authored by Wang et al. fit better in this volume. In this paper entitled "Plastic Collapse Analysis of Pipeline Girth Welds," Wang et al. examine the predictability of the various plastic collapse solutions relevant to pipe girth welds by comparing with the available experimental data. Their paper provides comprehensive guidelines for use of the various solutions. Although the authors have addressed flaws in gas transmission pipelines, the results of the study will be useful for piping used in other applications as well.

PVP-Vol. 410-1, Assessment Methodologies for Preventing Failure:
Deterministic and Probabilistic Aspects and Weld Residual Stress
Volume 1
ASME 2000

PLASTIC COLLAPSE ANALYSIS OF PIPELINE GIRTH WELDS

Yong-Yi Wang
Engineering Mechanics Corporation
Of Columbus
3518 Riverside Dr., Suite 202
Columbus, Ohio 43221
Tel/Fax: 614-850-8792/-8793
wangemc2@columbus.rr.com

Gery M. Wilkowski
Engineering Mechanics Corporation
of Columbus
3518 Riverside Dr., Suite 202
Columbus, OH 43221
Tel/Fax: 614-850-8792/-8793
gwilkows@columbus.rr.com

David J. Horsley
TransCanada Transmission
P.O. Box 2535, Station M
Calgary, Alberta, Canada T2P 2N6
Tel/Fax: 403-948-8159/-8268
david_horsley@transcanada.com

ABSTRACT

The integirty of pipelines under axial tensile stresses (due to bending and axial loads) is often controlled by the quality of the girth welds. Two possible failure modes, brittle fracture and plastic collapse, are often considered for girth welds containing welding defects. Failure by plastic collapse is often viewed more likely than brittle fracture in modern pipelines, where material toughness is typically "high." Selection of an appropriate plastic collapse solution is often difficult partly due to the "over-abundance" of the available solutions, but also because there is often very limited information on the applicable range of these solutions. This paper focuses on plastic collapse solutions relevant to the girth welds of gas transmission pipelines containing surface-breaking circumferential defects. A total of seven plastic collapse solutions are reviewed and compared with available experimental data. These data are from full-scale bend tests under either pure bending or combined internal pressure and bending. For pipes containing defects of depth less than 50% of the wall thickness and length less than 10% of the circumference, the solutions of Miller, Wilkowski, Kastner, and ASME NSC predict similar plastic collapse stresses when pure bending loads are considered. These solutions also agree well with experimental data from pipes with $D/t \geq 28$ and relatively short cracks (up to 10% of the circumference). The Kastner solution provides the best agreement with experimental data for pipes of smaller D/t ratios with longer and deeper cracks and loaded in pure bending. For pipes loaded in a combined internal pressure and bending, the solutions of Miller, Wilkowski, and ASME NSC outperform the Kastner solution. For crack length up to 60% of the circumference, the Miller and the ASME NSC solutions provide almost identical plastic collapse stresses when pure bending is considered. It was found that experimental data relevant to large D/t pipes containing relatively long circumferential defects are few, although there are a large number of experimental data of pipes with smaller D/t ratios. More tests are needed if defect lengths greater than 10% of the circumference were allowed in gas transmission pipelines.

INTRODUCTION AND BACKGROUND

Modern pipeline steels joined using modern welding procedures and consumables usually produce girth welds with good toughness. Consequently, girth weld failure by plastic collapse is often considered more likely than brittle fracture. Plastic collapse has been recognized as one of the possible failure modes in several pipeline codes. Based on a large number of small- and large-scale experimental tests, the EPRG guidelines (Knauf and Hopkins, 1996) stipulate that girth weld integrity be assessed using limit load/net-section-collapse calculation, provided that certain level of toughness is met. The Tier 2 guidelines set the minimum Charpy energy (CVN) at 30 J and the minimum averaged CVN of 40 J. The Tier 3 guidelines add the minimum CTOD toughness value of 0.1 mm and the minimum averaged value of 0.15 mm. The Appendix A of API 1104, "Alternative Girth Weld Acceptance Criteria," has no provision for plastic collapse analysis (API, 1999). The Canadian code, "CSA Z662-99 Oil and Gas Pipeline Systems," requires girth weld integrity to be assessed on both brittle fracture and plastic collapse considerations (CSA, 1999).

One of the major challenges in assessing girth weld plastic collapse is the "over-abundance" of plastic collapse solutions. Selection of an appropriate plastic collapse solution can be a daunting task. The available solutions were, in general, developed to meet a specific application. Naturally therefore, there is a range of pipe sizes and material behavior where each solution was "fit" to experimental data. The successful application of a given solution should therefore take into account the range where it can be considered accurate. There is often very limited information on the applicable range of certain solutions.

The focus of this paper is the plastic collapse solutions relevant to the girth welds of gas transmission pipelines. A total of seven collapse solutions have been reviewed and compared with available experimental data. The differences between the plastic collapse solutions and the experimental data are rationalized whenever possible. This information should prove beneficial to engineers for making more informed decisions.

REVIEW OF PLASTIC COLLAPSE SOLUTIONS

Several in-depth reviews of plastic collapse solutions are available. Miller (1988) conducted extensive review of limit load solutions for a wide range of geometries, from simple laboratory test specimens to complex structural geometries. A review by Rosenfeld (1995) was targeted at pipelines. Wilkowski, et al., (1998) provided a comprehensive review of piping fracture, with a focus on nuclear piping systems. Scott, et al., (1998) gave a more general review of fitness-for-service methodology. Seven plastic collapse solutions pertinent to gas transmission pipelines, are reviewed in this paper.

(1). Miller (1988) presented a limit load solution for pipes having a finite-length part-through wall circumferential crack under bending load,

$$\frac{M_c^{Miller}}{4R^2t\sigma_f} = \cos\left[\frac{\eta\beta\pi}{2}\right] - \frac{\eta\sin(\beta\pi)}{2} \quad if \quad \beta < \frac{1}{2-\eta} \tag{1a}$$

$$\frac{M_c^{Miller}}{4R^2t\sigma_f} = (1-\eta)\sin\left[\frac{(1-\eta\beta)\pi}{2(1-\eta)}\right] + \frac{\eta\sin(\beta\pi)}{2} \quad if \quad \beta > \frac{1}{2-\eta} \tag{1b}$$

where M_c is the bending moment at plastic collapse, or limit bending moment. The other parameters are:

R = pipe radius
t = pipe thickness
a = crack depth
c = crack half length
D = pipe diameter
β = $c/\pi R = 2c/\pi D$, ratio of crack length to circumference
η = a/t = crack-depth ratio
σ_f = material flow stress.

From Eq. (1), the bending moment at plastic collapse for pipes without a crack ($\eta = 0$) is given as:

$$M_c^{Miller} = 4R^2t\sigma_f \tag{2}$$

Therefore, the stress at plastic collapse can be defined as:

$$\sigma_c^{Miller} = \frac{M_c^{Miller}}{4R^2t} \tag{3}$$

or

$$\sigma_c^{Miller} = \sigma_f\left[\cos\left(\frac{\mu\beta\pi}{2}\right) - \frac{\eta\sin(\beta\pi)}{2}\right] \quad if \quad \beta < \frac{1}{2-\eta} \tag{4a}$$

$$\sigma_c^{Miller} = \sigma_f\left[(1-\eta)\sin\left(\frac{(1-\eta\beta)\pi}{2(1-\eta)}\right) + \frac{\eta\sin(\beta\pi)}{2}\right] \quad if \quad \beta > \frac{1}{2-\eta} \tag{4b}$$

(2). Wilkowski and Eiber (1979) give a plastic collapse solution using a surface defect transformation,

$$\frac{M_c^{Wilk}}{\pi R^2 t\sigma_f} = \frac{1-\eta}{1-\eta/N} \tag{5}$$

where

$$N = \left[1 + 0.26\beta + 47\beta^2 - 59\beta^3\right]^{0.5} \tag{6}$$

At $\eta = 0$, Eq. (5) is reduced to

$$M_c^{Wilk} = \pi R^2 t\sigma_f \tag{7}$$

Therefore, the stress at plastic collapse is,

$$\sigma_c^{Wilk} = \frac{M_c^{Wilk}}{\pi R^2 t} = \sigma_f\left[\frac{1-\eta}{1-\eta/N}\right] \tag{8}$$

It should be noted that Eq. (7) corresponds to a bending moment when only the outer-fiber furthest away from the bending neutral axis is plastic, whereas Eq. (2) corresponds to the bending moment when the entire circumference is plastic. The plastic collapse moments (when a crack is infinitesimally small) defined in Eqs. (2) and (7) differ by a factor of $4/\pi$. The definition of plastic collapse stresses, Eqs. (3) and (8), thus reflects this difference.

(3). Willoughby (1982) provides an empirical lower bound solution,

$$M_c^{Will} = \pi R^2 t\sigma_f(1 - 1.6\pi\eta\beta) \tag{9}$$

The stress at plastic collapse is

$$\sigma_c^{Will} = \frac{M_c^{Will}}{\pi R^2 t} = \sigma_f(1 - 1.6\pi\eta\beta) \tag{10}$$

(4). The plastic collapse stress provided by CSA Z662, Appendix K (CSA, 1999) is similar to Eq. (10),

$$\sigma_c^{CSAI} = \sigma_f(1.03 - 18\eta\beta) \tag{11}$$

The second term in the right-hand side of Eqs. (10) and (11) differs by a factor of $1.6\pi/18$. Therefore, the solution in CSA Z662 is expected to be more conservative than that of Willoughby.

(5). Kastner, et al. (1981) proposed a local collapse criterion for internal pressure loading,

$$\frac{PR}{2t} = \sigma_f \frac{(1-\eta)(\pi - \beta\eta)}{\pi(1-\eta) + 2\eta \sin\beta} \quad (12)$$

where P is the internal pressure. The left-hand term in Eq. (12) is the nominal axial stress in a pressure vessel. Based on an observation by Miller (1988), Clyne and Jones (1995) contended that the left-hand term can be replaced by the applied axial stress due to internal pressure, lateral bending, or the combination of two, provided that the failure is not caused by hoop stress. Therefore, in the context of plastic collapse by lateral bending, Eq. (12) is reduced to,

$$\sigma_c^{Kastner} = \sigma_f \frac{(1-\eta)(\pi - \beta\eta)}{\pi(1-\eta) + 2\eta \sin\beta} \quad (13)$$

Miller (1988) stated that Eq. (12) is applicable to any end loading if $\pi R^2 P$ is replaced by total axial force.

(6). ASME Section XI pipe flaw evaluation criteria is based on Net-Section-Collapse (NSC) analysis. Broek developed the original NSC solution (Kanninen, et al., 1976),

$$M_c^{ASMENSC} = 2\sigma_f t R_m^2 (2\sin\zeta - \eta \sin(\beta\pi)) \quad (14a)$$

$$\zeta = 0.5\pi(1-\beta\eta) - \frac{\pi R_i^2 P}{4R_m t \sigma_f} \quad (14b)$$

where R_m=mean radius of the pipe
R_i=inner radius of the pipe.

At $\eta = 0$, Eq. (14) becomes

$$M_c^{ASMENSC} = 4R_m^2 t\, \sigma_f \quad (15)$$

Therefore, the stress at plastic collapse can be defined as:

$$\sigma_c^{ASMENSC} = \frac{M_c^{ASMENSC}}{4R_m^2 t} = \sigma_f [\sin\zeta - 0.5\eta \sin(\beta\pi)] \quad (16a)$$

and

$$\zeta = 0.5\pi(1-\beta\eta) - \frac{\pi R_i^2 P}{4R_m t \sigma_f} \quad (16b)$$

(7). Scott and Wilkowski compared experimental data with the ASME NSC solution. They found that the solution over-predicts collapse loads for large *D/t* pipe (Krishnaswamy, 1995). A *D/t* correction factor was proposed,

$$V_R = 1.257 - 0.01919(R_m / t) \quad (17a)$$

$$\sigma_c^{ASMENSC\,mod} = \sigma_f [\sin\zeta - 0.5\eta \sin(\beta\pi)] \cdot V_R \quad (17b)$$

The corrected solution was validated experimentally for $3.5 < R_m/t < 21$. Scott and Wilkowski cautioned against using the correction factor for very large *D/t* pipes.

COMPARISION WITH EXPERIMENTAL DATA

Overview

The formulation of the plastic collapse solutions reviewed in the previous section varies greatly. Some are from rigorous analytical derivation, others are empirical lower bound fit to experimental data. In this section, the predicted plastic collapse stresses are compared with available experimental data. The database from Rosenfeld (1995) was used extensively in compiling a relevant experimental database for comparison with various solutions. Only surface-breaking defects are included in the experimental database.

In computing the plastic collapse stress of various solutions, flow stress was taken as the averaged value of yield and tensile strengths when both values were available. In the cases when only yield stress was available, the flow was taken as yield stress plus 10 ksi (68.92 MPa)

Experimental Tests Loaded in Bending

There were a total of 132 experimental tests loaded in pure bending (Reynolds, 1970; Kanninen, et. Al., 1982; Wilkowski and Eiber, 1981; Wilkowski, et al., 1989; Joyce, 1983; Erdogan, 1982; Schutze, et al., 1980; Roos, et. al., 1990; Anderson, 1988; Glover, 1984; Worswick, et al, 1985; Glover, 1985; Glover, 1987; Glover and Lazor, 1987; Coote, et al., 1985; Coote, et al., 1986; Hopkins, et al., 1991; Pistone, 1991; Leggat and Challenger, 1993). The pipe diameter ranged from 4 inch to 36 inch (101.6 mm to 914.4 mm). The wall thickness varied from 0.125 inch to 1.115 inch (3.28 mm to 28.32 mm). The *D/t* ratio ranged from 7.5 to 90. The database included 38 tests of stainless steel pipes with yield strength as low as 20 ksi (137.84 MPa). The yield strength of the remaining carbon steels ranged approximately from 50 ksi (344.60 MPa) to 100 ksi (689.20 MPa). The CTOD toughness ranged from 0.02 mm to over 4 mm. The full database is lengthy therefore not listed in this paper. To provide an overview of the database, the median, the standard deviation, and the coefficient of variation (COV) of relevant geometry dimensions and material properties are summarized in Table 1. The use of these statistical variables is to provide a snapshot of the various parameters. A strict statistical analysis usually requires some type of grouping based on certain physical characteristics. No such grouping was attempted in this database.

Table 1 Statistical overview of the full-scale test database loaded in pure bending. The tensile strength was reported in 81 of the 132 tests contained in this database.

	D		*t*		*D/t*	η	β	Yield		Tensile		*Y/T*
	(inch)	(mm)	(inch)	(mm)				(ksi)	(MPa)	(ksi)	(MPa)	
Median	17.00	431.80	0.48	12.21	31.94	0.50	0.10	68.17	469.80	82.90	571.35	0.71
Standard Deviation	13.93	353.82	0.24	6.10	28.52	0.19	0.26	20.27	139.70	10.37	71.44	0.23
COV	0.82		0.49		0.89	0.38	2.48	0.30		0.13		0.32

The median value of the diameters is 17 inch (431.80 mm) with a COV of 0.82. The median value of *D/t* is 31.94 with a COV of 0.89. The large COV values indicate relatively large spread in pipe diameters and *D/t* ratios. A large number of data are from pipes of medium diameter and relatively small *D/t* ratio. The median values of yield stress and *Y/T* ratio are 68 ksi (469.80 MPa) and 0.71, respectively.

The plastic collapse stresses were computed using the listed solutions. The computed values were normalized by the corresponding experimental values. A ratio less than 1 indicates that the predicted value is smaller than the experimental value, and is therefore conservative. Table 2 provides a statistical summary of such comparison. If a solution agrees perfectly with all the experimental data, the median value would be 1.0 and the COV would be zero.

Table 2 Statistical summary of the ratio of the predicted plastic stresses over the experimentally measured plastic collapse stresses using the entire database of tests loaded in pure bending

	Miller	Wilkowski	Willoughby	CSA Z662	Kastner	ASME NSC	Mod. ASME NSC
Median	0.87	0.84	0.74	0.10	0.77	0.89	0.73
Standard Deviation	1.47	0.77	6.08	33.82	0.43	1.47	1.50
COV	1.69	0.92	8.21	332.45	0.56	1.65	2.05

Both the CSA Z662 and Willoughby solutions have very high COV. This indicates that the accuracy of these solutions varies widely as compared to the experimental data. The Kastner solution has the smallest COV, indicating the highest consistence among the nine solutions. The median value of 0.77 indicates the solution is conservative, but not overly conservative. If one were to draw a conclusion based on this current database, the Kastner solution would appear to be the best.

Table 2 shows that the CSA Z662 solution is the most conservative by a wide margin with a median value of 0.1 when the entire database is considered. The CSA Z662 specifies an applicable range of a maximum defect depth of 50% of pipe wall thickness ($\eta \leq 0.5$) and a maximum defect length of 10% of the pipe circumference ($\beta \leq 0.1$). No such restrictions were applied in producing Table 2. In this section, only experimental data conforming to the CSA Z662 defect length and depth requirements are included. This subset of database contains 63 experimental tests. A statistical summary of this subset of database is given in Table 3. As compared to the entire database listed in Table 1, the median value of the diameters is increased to 36 inch (914.40mm). The median value of the defect length is reduced to 5%

Table 3 A statistical overview of the subset of experimental database containing tested pipes with $\beta \leq 0.1$ and $\eta \leq 0.5$

	D		t		D/t	η	β	Yield		Tensile		Y/T
	(inch)	(mm)	(inch)	(mm)				(ksi)	(MPa)	(ksi)	(MPa)	
Median	36.00	914.40	0.44	11.10	71.12	0.46	0.05	71.94	495.78	82.90	571.35	0.89
Standard Deviation	9.50	241.38	0.25	6.46	24.19	0.18	0.03	8.21	56.59	5.95	41.00	0.09
COV	0.26		0.58		0.34	0.38	0.57	0.11		0.07		0.10

The results of the comparison are summarized in Table 4. The CSA Z662 solution is much more reasonable when it is compared with the experimental data within its specified limits. However, it is still much more conservative than the solutions of Miller, Wilkowski, Kastner, ASME NSC, and Willoughby. This is understandable since many code-based solutions are typically set to be quite conservative. The median value of the modified ASME NSC solution is substantially lower than that of the ASME NSC due to the R_m/t "over-correction" for large D/t pipes. Scott and Wilkowski cautioned against using the R_m/t correction solutions for large D/t pipes (Krishnaswamy, 1995).

Table 4 A statistical summary of the ratio of the predicted plastic collapse stresses over the experimentally measured plastic collapse stresses. The subset of the experimental database with $\beta \leq 0.1$ and $\eta \leq 0.5$ is used for comparison.

	Miller	Wilkowski	Willoughby	CSA Z662	Kastner	ASME NSC	Mod. ASME NSC
Median	0.94	0.93	0.85	0.55	0.87	0.94	0.58
Standard Deviation	0.22	0.18	0.47	1.86	0.17	0.22	0.25
COV	0.24	0.20	0.56	3.35	0.20	0.23	0.43

Pipes Loaded by Internal Pressure and Bending

There were a total of 50 experimental tests loaded in combined internal pressure and bending (Reynolds, 1970; Schulze, et al., 1980; Pistone, 1991; Darlaston and Harrison, 1977; Wilkowski, et al., 1989; Bodman and Fuhlrot; 1981; Brenner, et al., 1983; Kim, 1993). As shown in Table 5, this database is quite different from that of pipes loaded in pure bending. The diameters are much smaller with a median value of only 6.63 inch (168.3 mm). The actual diameters varied from 2.95 inch to 48 inch. Therefore the standard deviation is very high. Only 14 of the 50 tests were for carbon steels with diameters of 17-, 36-, and 48-inch (431.8-, 914.4-, and 1219.2 mm). The rest of the tests were conducted on stainless steels with much smaller diameters. The defects in this database are quite long (median β=0.24) and deep (median η= 0.69).

Table 5 A statistical overview of the experimental database containing pipes tested under internal pressure and bending loads.

	D		t		D/t	η	β	Yield		Tensile		Y/T
	(inch)	(mm)	(inch)	(mm)				(ksi)	(MPa)	(ksi)	(MPa)	
Median	6.63	168.28	0.48	12.19	18.75	0.69	0.24	48.59	334.88	71.07	489.81	0.68
Standard Deviation	14.81	376.10	0.39	10.02	25.84	0.14	0.22	17.16	118.25	15.43	106.36	0.15
COV	2.24		0.82		1.38	0.20	0.91	0.35		0.22		0.22

The "predicted" plastic collapse stresses were computed using the listed equations and the geometric parameters and material properties given in the database. Among the seven solutions examined, only the ASME NSC solution explicitly accounts for internal pressure. The "experimental" plastic collapse stresses are the summation of nominal tensile stresses arisen from internal pressure and bending. The predicted values were normalized by the corresponding experimental values. Table 6 provides a statistical summary of such comparison. The long and deep defects prevented us from obtaining meaningful plastic collapse stresses using the CSA Z662 solution. The Willoughby solution has a very high COV, indicating inconsistent accuracy. The solutions of Miller, Wilkowski, Kastner, ASME NSC, and Modified ASME NSC all provide reasonable predictions. Among those solutions, the Kastner solution is the most conservative. The Miller solution provides the smaller COV, indicating good consistence. The overall small D/t ratio did not contribute much to the R_m/t correction in the modified ASME NSC solution. The ASME NSC and the modified ASME NSC predicted similar plastic collapse stresses.

Table 6 Statistical summary of the ratio of the predicted plastic collapse stresses over the experimentally measured plastic collapse stresses for pipes loaded in combined internal pressure and bending

	Miller	Wilkowski	Willoughby	CSA Z662	Kastner	ASME NSC	Mod. ASME NSC
Median	0.72	0.58	0.32	N/A	0.46	0.74	0.74
Standard Deviation	0.31	0.27	1.28	N/A	0.26	0.47	0.51
COV	0.43	0.48	4.04	N/A	0.57	0.63	0.69

DISCUSSION

The previous sections demonstrate how easy it is to reach vastly different conclusions as to the "optimal" plastic collapse solutions. For instance, if one were to select the entire database of pipes tested in pure bending, it may be concluded that the Kastner solution is the best. However, when the database is further divided into subsets using certain physical criteria, different conclusions may be reached. To understand the fundamental differences of these solutions, they are compared with a full range of defect depth and length. The Miller solution is used as the baseline solution because it is at least qualitatively consistent: the plastic collapse stress decreases with increasing defect length up to 100% of the circumference (β=1.0). Some other solutions, such as that of Wilkowski (developed for defect length less than 20% of the circumference), break down for very long defects as evident by the increased plastic collapse stress with increased defect length. The other solutions, when normalized by the Miller solution, are material property independent, see Eqs. (4), (8), (13), and (16). The only significant variables are defect length and depth.

In Figure 1, the ASME NSC solution is normalized by the Miller solution. For defect length up to 60% of the circumference (β = 0.6), the two solutions are identical. At greater defect lengths, the ASME NSC solution provides higher plastic collapse stresses. There is no fundamental basis for the sharp upturn at β = 0.6. It is safe to conclude that the ASME NSC solutions should not be used for defect greater than 60% of the circumference.

The Kastner and Wilkowski solutions are compared with the Miller solution in Figures 2 and 3. Up to a certain defect length, the Kastner and Wilkowski solutions predict progressively lower plastic collapse stresses for longer and deeper defects, as compared to the Miller solution. Beyond a certain defect length and depth, the ratios have an upward turn, indicating the applicable limits of those solutions.

It should be noted that no rigorous screening of the failure mechanism was conducted to separate the failure into brittle fracture, elastic-plastic fracture, or plastic collapse. In the 132 experimental tests included in the pure bending database, 97 tests reported CTOD toughness values. The median value of the CTOD toughness was 0.10 mm, considered quite high. A substantial number of those 97 tests had much higher CTOD toughness. In the 50 experimental tests included in the combined internal pressure and bending database, 38 tests reported CTOD toughness. The median value of the CTOD toughness was 0.37 mm, a very high value. It may be speculated that rigorous screening of the failure mechanism may have excluded some experimental tests. However, the majority of the failure would still be classified as plastic collapse. The median values and COVs of the comparisions may change. However, it is unlikely that the main conclusion of the work would change.

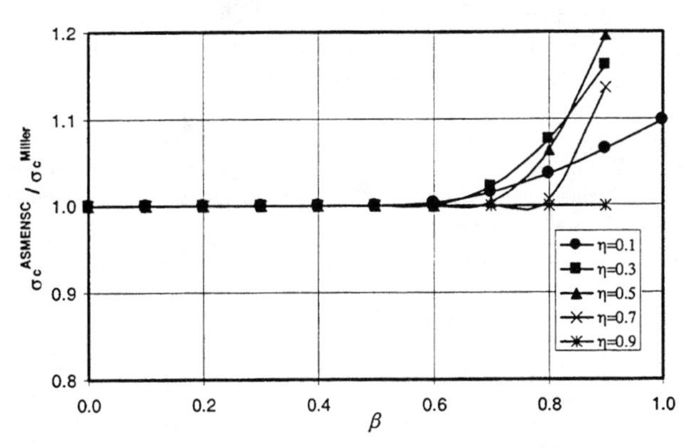

Figure 1. Plastic collapse stress from the ASME NSC solution, normalized by that of the Miller solution, as a function of defect length at various defect depths

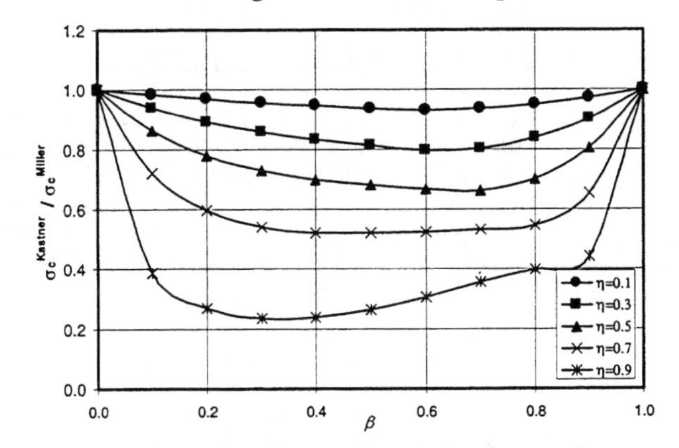

Figure 2. Plastic collapse stress from the Kastner solution, normalized by that of the Miller solution, as a function of defect length at various defect depths

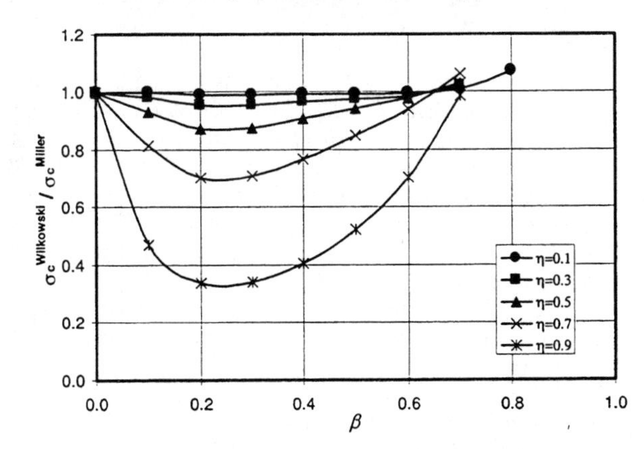

Figure 3. Plastic collapse stress from the Wilkowski solution, normalized by that of the Miller solution, as a function of defect length at various defect depths

CONCLUDING REMARKS

Given the large number of available experimental data, it might be concluded that selecting the optimal plastic collapse solutions should be easy. The reality is more complicated as researchers have recommended seemingly very different plastic collapse solutions, sometimes after analyzing basically the same set of experimental data. Clyne and Jones recommended the Kastner solution (Clyne and Jones, 1995). Wang, et al., recommended the Miller solution based on extensive finite element analysis of large *D/t* pipes with girth weld cracks up to 40% of the circumference and some limited experimental tests (Wang, et al., 1996; Wang and Mohr, 1997). It is illustrative to see how such different conclusions could be drawn if the scope of the experimental data and analysis matrix is not well quantified.

Table 7 Plastic collapse stresses predicted by various solutions as compared to the Miller solution for a defect of 50% wall thickness and 10% circumference. A number smaller than 1 indicates the solution is more conservative than the Miller solution.

ASME NSC	Wilkowski	Kastner	CSA Z662
1.00	0.93	0.86	0.14

The plastic collapse stresses predicted by Miller, Wilkowski, Kastner, and ASME NSC are quite close for pipes with relatively shallow and short cracks. The maximum difference for a defect of 50% of wall thickness and 10% of the circumference is given in Table 7. The difference between the Kastner and Miller solutions is 14%. The differences among the Wilkowski, ASME NSC, and Miller solutions are even smaller. The CSA Z622 predicts a plastic collapse stress only one seventh of that of Miller solution and one sixth of that of Kastner solution. The following conclusions may be drawn:

1. The good candidate plastic collapse solutions for gas transmission pipeline applications are those of Miller, Wilkowski, Kastner, and ASME NSC. The solutions of Willoughby and CSA Z662 may be overly conservative. The modified ASME NSC solution appears over-correcting for the *D/t* effects for large *D/t* pipes.

2. Within the regime of practical interests to the gas transmission industry, i.e., defects with depths less than 50% of wall thickness and lengths less than 10% of the circumference, the plastic collapse solutions of Miller, Wilkowski, Kastner, and ASME NSC should all prove satisfactory. The predicted plastic collapse stresses from these solutions are quite close when the defects are within the above limits. There is no fundamental basis to claim one's superiority over the others. The conflicting claims to the "best" solutions are the manifestation of the characteristics of the experimental database, not the inherent limitations of the solutions.

3. For defects with depths greater than 50% of the pipe wall thickness and lengths greater than 10% of the circumference, significant differences exist among the solutions of Miller, Wilkowski, Kastner, and ASME NSC. It is possible that solutions of Wilkowski and Kastner may be more accurate for larger defects (η>50% and β>10%) in large *D/t* pipes. Judging by the trend exhibited in Figures 2 and 3, the solutions of Wilkowski and Kastner have a defect length upper limit of 20% and 40%, respectively. Unfortunately, there is insufficient experimental data to definitely establish the "best" solution for large defect sizes.

4. Experimental data are lacking for large *D/t* pipes with relatively long defects. In the 71 pipes with $D/t \geq 28$ tested under pure bending, only 7 tests had a β value greater than 0.10 (10% of the circumference). They are at β=0.12, 0.15, 0.15, 0.51. 0.54, and 1.00, respectively. Among the 50 pipes tested under combined internal pressure and bending, only 2 tests were for pipe with $D/t \geq 28$ and a β value greater than 0.10 (both at the 20% of the circumference). If defect lengths greater than 10% of the circumference are to be allowed with confidence, more tests are needed.

5. For crack lengths up to 60% of the circumference, the Miller and the ASME NSC solutions of pure bending provide almost identical plastic collapse stresses.

ACKNOWLEDGEMENTS

The ongoing support of the Welding Supervisory Committee of the Pipeline Research Council International (PRCI) is gratefully acknowledged. The help of Dr. Zhili Feng of Emc^2 in formating and editing this paper is greatly appreciated.

REFERENCES

Anderson, T. L., 1988, "Plastic Collapse of Girth Weld Repair Grooves in Pipe Subjected to Offshore Laying Stresses," Int. J. PVP, Vol. 31.

API, 1999, API Standard 1104, "Welding of Pipelines and Related Facilities," 17th Edition, September.

Bodman, E., and Fuhlrot, H., 1981, "Investigation of Critical Crack Geometries in Pipes," Paper L6/6, SMIRT 6, Paris, August.

Brenner, U., et al., 1983, "Fracture Mechanics Analysis of Stable Crack Growth under Sustained Load and Instability of Circumferential Cracks in Straight Water-Steam Pipes," International Journal of Pressure Vessel and Piping, Vol. 11.

CSA, 1999, Canadian Standards Association, CSA-Z662-99, "Oil and Gas Pipeline Systems."

Clyne, A. J. and Jones, D. G., 1995, "The Integrity of Transmission Pipelines Containing Circumferential Girth Weld Defects," Pipeline Technology, Eds. R. Denys, Vol. II, pp. 299-314.

Coote, R. I., et al, 1985, "Girth Weld Acceptance Standards Based on Engineering Critical Assessment - A Canadian Viewpoint," Paper No. 24, 6th EPRG/NG-18 Joint Technical Meeting.

Coote, R. I., et al, 1986, "Alternative Girth Weld Acceptance Standards in the Canadian Gas Pipeline Code," Paper 21, Welding and Performance of Pipelines Conference, London, Nov. 18-21.

Darlaston, B. J., and Harrison, R. P., 1977, "Ductile Failure of Thin Walled Pipes with Defects under Combinations of Internal Pressure and Bending," 3rd International Conference on Pressure Vessel Technology, Tokyo, April.

Erdogan, F., 1982, "Theoretical and Experimental Study of Fracture in Pipelines Containing Circumferential Flaws," DOT-RSPA-DMA-50/83/3, Contract DOT-RC-82007 Final Report to USDOT, September.

Glover, A. G., 1984, "The Effects of Real Defects on Girth Weld Fracture Behavior," A.G.A. Report PR-140-38.

Glover, A. G., 1985, "The Effect of Stress Relief Due to Hydrostatic Testing on Girth Weld Failure," A.G.A. Report PR-140-413.

Glover, A. G., 1987, "Small and Full Scale Fracture of Thick Section Girth Weldments," A.G.A. Report PR-140-512.

Glover, A. G., and Lazor, R. B., 1987, "Fracture Behavior of Stress Relieved Girth Welds," A.G.A. Report PR-140-626

Hopkins, P. et al, 1991, "An Experimental Appraisal of the Significance of Defects in Pipeline Girth Welds, " Paper No. 20, 8th EPRG/NG-18 Joint Technical Meeting.

Joyce, J. A., 1983, "Instability Testing of Compact and Pipe Specimens Utilizing a Test System Made Compliant by Computer Control," ASMT STP 803.

Kanninen, M. F., and others, 1976, "Mechanical Fracture Predictions for Sensitized Stainless Steel Piping with Circumferential Cracks," EPRI Report NP-192, September.

Kanninen, M. F., et al, 1982, "Instability Predictions for Circumferentially Cracked Type 304 Stainless Steel Pipes Under Dynamic Loading," EPRI NP-2347, Research Project T118-2 Final Report, April.

Kastner, W., Roehrich, E., Schmitt, W., and Steinbuch, R., 1981, "Critical Crack Sizes in Ductile Piping," International Journal of Pressure Vessel and Piping, Vol. 9, pp.197-219.

Kim, H. O., 1983, "Model Simplifies Estimate of Bending Strength in Corroded Pipe," Oil & Gas Journal, April 19.

Knauf, G. and Hopkins, P., 1996, "EPRG Guidelines on the Assessment of Defects in Transmission Pipeline Girth Welds," Sonderdruck aus 3R International, 35, Jahrgang, Heft, s. 620-624.

Krishnaswamy, P., 1995, "Fracture Behavior of Short Circumferentially Surface-Cracked Pipe," NEREG/CR-6298, November.

Leggat,, R. H., and Challenger, N. V., 1993, "Investigation of Validity of BSI PD6493:1991 Defect Assessment Procedures by Analysis of Full Scale Pipe Bend Tests," Proc. 8th Symp. on Line Pipe Research.

Miller, A. G., 1988, "Review of Limit Codes of Structure Containing Defects," *International Journal of Pressure Vessels and Piping,* Vol. 32, pp. 191-327.

Pistone, V., 1991, "Fitness-for-Purpose Assessment of Defects in Pipeline Girth Welds," Paper No. 21, 8th EPRG/NG-18 Joint Technical Meeting.

Reynolds, M. B. 1970, "Failure Behavior of Flawed Carbon Steel Pipes and Fittings," GEAP-10236, GE report to USAEC, Contract AT(04-3)-189, Project Agreement 37, October.

Rosenfeld, M. J., 1995, "Serviceability of Corroded Girth Welds," Draft Final Report, PRI Contract No. PR 218-9438, March 31.

Roos, E., et al, 1990, "Fracture Mechanical Assessment of Pipes Under Quasistatic and Cyclic Loading," Steel Research, No. 4.

Schulze, H. D., et al, 1980, "Fracture Mechanics Analysis on the Initiation and Propagation of Circumferential and Longitudinal Cracks in Straight Pipe and Bends," Nuc. Eng. and Des., v. 58.

Scott, P. M., Anderson, T. L., Osage, D. A., Wilkowski, G. M., 1998, "Review of Existing Fitness-for-Service Criteria for Crack-Like Flaws," Welding Research Council Bulletin 430, April.

Wang, Y.-Y., Sun, X., and Horsley, D. J., 1996, "Plastic Collapse Analysis of Girth-Welded Pipes with Circumferential Cracks," Proceedings of the 9th Symposium on Pipeline Research, Paper 41, Houston, TX, September 30 - October 2.

Wang, Y.-Y. and Mohr, W., 1997, "Plastic Collapse Loads of Girth-Welded and Sleeved Pipes: Analyses and Experiments," *Proceedings of the PRC/EPRG 11th Biennial Joint Technical Meeting,* Paper 18, Arlington, VA, April 7-11.

Wilkowski, G. M. and Eiber, R. J., 1979, "Determination of the Maximum Size of Girth Weld Repair on Offshore Pipelines," A.G.A report PR-3-100.

Wilkowski, G. M., and Eiber, R. J., 1981, "Evaluation of Tensile Failure of Girth Weld Repair Grooves in Pipe Subject to Offshore Laying Stresses," Journal of Energy Resources Technology, v. 103, March.

Wilkowski, G. M, et al., 1989, "Degraded Piping Program – Phase II," NUREG/CR-4082, MBI-2120, Vol. 8, Battelle Summary Report to USNRC, March.

Wilkowski, G. M., Olson, R. J., and Scott, P. M., 1998, "State-of-the-Art Report on Piping Fracture Mechanics," U.S. Nuclear Regulatory Commission, NUREG/CR-6540, BMI-2196, January.

Willoughby, A. A., 1982, "A Survey of Plastic Collapse Solutions Used in Failure Assessment," TWI Report 191/1982.

Worswick, M. J., et al, 1985, "The Effect of Misalignment on the Fracture Behavior of Girth Welds," A.G.A. Report PR-140-406.

WELD RESIDUAL STRESSES AND FRACTURE

Introduction

Frederick W. Brust and Pingsha Dong
Battelle Memorial Institute
Columbus, OH

Welding remains the key process in connecting metallic components to fabricate complex structural systems. The heavy fabrication, automotive, power generation, chemical process, building and bridge construction, and aerospace, are some of the industry's that can benefit from better control of the weld fabrication process. Unfortunately, many service problems occur in or near the welded regions of the components. Corrosion, fatigue, and fracture control of welds' remains a key goal of the structural designer.

The residual stresses and distortions caused by the weld fabrication process have long been known to have an important impact on service performance and fabrication costs. Weld process models have been used in attempts to control residual stresses and distortions in welded fabrications since the late 1930's. The early models were analytically based, and could provide insight for single pass welds in simple components. In the middle to late 1970's, two-dimensional nonlinear finite element models were developed and used to model the weld process in an effort to control the process. In particular, the nuclear industry used such models to develop weld techniques in order to control inter-granular stress corrosion cracking (IGSCC) in power plants. Today, it is possible to perform fully three dimensional weld analyses on very large structural components. These new models are beginning to revolutionize the weld fabrication industry by permitting fabrication techniques such as pre-machining and weld sequence design to become a reality. Great cost savings are being realized by some farsighted industries, as these models become a part of the design process. This volume discusses these models and their use in increasing service performance and decreasing fabrication costs in a number of industries. Keep in mind that, because of the peculiarities of the weld process, where material is melted and re-melted as various passes are deposited, modeling the weld process is difficult for the non-expert.

In the first paper, Keim, Fricke, and Schmidt illustrate the use of weld process models to control weld residual stresses in pipelines to reduce corrosion-related problems. It is well known that altering the residual stress state in corrosion prone regions, such as along the inside diameter of a pipe, which carries corrosive fluids, from tension to compression, can reduce the propensity for corrosion damage. This paper uses numerically based weld process models to investigate a method called last pass heat sink welding to control these residual stresses.

Next Goldak and colleagues illustrate the use of an advanced weld process model to investigate burn-through in pressurized natural gas pipelines. n-service weld repair of damaged natural gas transmission pipelines is desirable from a cost standpoint. However, it is dangerous if the weld repair burns through the pipe wall. Goldak, et al. Illustrate the use of the weld process model to gain a complete understanding of the process. The unique aspects associated with welding on material, which is actively being pressurized, are accounted for.

Mochizuki and Toyoda, in the following paper, investigate the use of a weld process model to control distortions in fillet weld fabrication. In particular, they investigate a reverse side heating process for this purpose. Reverse side heating involves using a traveling heat source, in addition to the weld torch, which is strategically placed as the weld fabrication process proceeds. The method is shown to be quite promising.

The fourth paper, by Cao et. al., overviews a high speed thermal solution procedure which models the temperature versus time histories produced during welding. This solution procedure is based on the enhancement of some closed form solutions, along with the development of new solutions for modeling fully three dimensional weld processes. The speed and accuracy of the technique makes it ideal for performing analyses of complex structures.

Zhang, Dong, and Bouchard next illustrate the use of weld process models to investigate the weld repair process in pipes. With aging of power plants, and the corresponding desire to extend life via repair, makes this study particularly useful. In particular, they illustrate that the length of the weld repair can have a marked effect on the final residual stresses produced. The usefulness' of using a weld process model to design weld repairs is clearly shown.

The sixth paper by Zhang, et al. overviews the development of a weld repair model. The model is first illustrated, and then validation of the model is provided via comparing predictions of weld repair cases to experimental measurements. It is important to note that active design of weld repairs is not only desirable but also even necessary, especially in older structures.

Next, George, Kingston, and Smith discuss an advanced hole drilling technique for measuring residual stresses in thick components. The methodology produces perhaps the most accurate through-wall residual stress determinations yet available. This is an important paper because there continues to be concern regarding the accuracy of residual stress measurements made by other methods.

Leggett and Stacey, in the eighth paper, overview a methodology for including the effects of weld residual stresses on fracture performance. This is an important paper especially for the practitioner since procedures are illustrated for accounting for residual stresses in a service environment. The importance of properly accounting for residual stresses in fracture assessment are also pointed our.

Smith then discusses the effect of residual stresses on crack opening areas. For leak-before-break (LBB) considerations, the residual stresses induced by welding can have a marked effect on the crack opening area, which in turn can greatly affect leak rate predictions. For certain size pipes, especially small diameter pipes, the need to account for residual stresses is pointed out and procedures for including such stresses are shown.

Next, Dong and Brust provide an overview paper, which provides insights regarding the effects of residual stresses on fracture of various components. Guidelines regarding the expected residual stress states for different types of components are discussed. In addition, this paper adds to the previous paper by Smith by also discussing the effect of residual stresses on crack opening area and thus LBB.

The eleventh paper within this section, by Hong, et. al. investigates the effect of weld parameters on residual stresses. A numerical model is used for this purpose and the effect of varying the numerous weld process parameters on the resulting residual stress state is discussed.

Finally, Yang, et. al., illustrate some of the unique features associated with welding of thin stock plate materials to fabricate components. In particular, the effect of plate buckling caused by residual stresses, and more importantly, techniques for properly modeling buckling in thin plate welding are discussed. This application is particularly useful for the automotive and aerospace industries that typically use thin gage metallic materials for fabrications.

PVP-Vol. 410-1, Assessment Methodologies for Preventing Failure:
Deterministic and Probabilistic Aspects and Weld Residual Stress
Volume 1
ASME 2000

WELD OPTIMIZATION BY MODELLING OF LAST PASS HEAT SINK WELDING AND EXPERIMENTAL VERIFICATION

E. Keim/Siemens KWU

J. Schmidt/Siemens KWU

S. Fricke/Siemens KWU

ABSTRACT

Residual stresses play an important role for crack initiation during operation of pipes in nuclear power plants. Formation, magnitude and distribution of the residual stresses contribute significantly to the integrity of components, particularly if corrosion mechanisms are of importance. The origin of the residual stresses is at first the heat, which causes the melting pool, in the following the cooling down, which causes local deformations in the welding region.

The aim of this paper is to demonstrate the effect of "Last Pass Heat Sink Welding" (LPHSW) on the residual stresses of an austenitic pipe by means of three dimensional finite element calculations and experimental measurements. During LPHSW the high tensile residual stresses in the heat affected zones (HAZ) near the weld are redistributed and result in compression stresses. With the aid of numerical simulations and experimental methods the distributions of the residual stresses in and near the weld can be calculated and measured.

After the welding process the pipes undergo the normal operation of the plants. Because there was no evidence up to now how the residual stresses behave during service transients, a test pipe was welded with LPHSW and loaded by combined thermal/mechanical cycles. The residual stresses were calculated and measured before and after the test. Because of the small diameter of the pipe (100 mm), the residual stress distribution is not constant along the pipe circumference, therefore only 3D finite element calculations will lead to reasonable results. In addition to the calculations the residual stresses have been measured.

It could be shown that the residual stresses will be influenced by the operational load, but the compression stresses in the HAZ still remain; this means the protection against any corrosion attack is maintained.

INTRODUCTION

Intergranular corrosion attack of SS piping in BWR plants is concentrated in an area immediately adjacent to the fusion line of the weld root. A remarkable feature is the synergistic occurrence of such defects in association with contraction folds.

Residual stresses are thought to trigger IGSCC. Many different measurement methods have therefore been used to determine residual stresses both in representative new welds and in weldments (circumferential pipe welds) which are already in service. However, a disadvantage common to all of these measurement methods to a greater or lesser degree, is that they integrate the results over a specific measurement length and therefore yield only an approximation of the actual value measured in the component only a short distance away from the fusion line.

Tolerances, whether dimensional (root dimensions) or those associated with welding parameters, which unavoidably affect the component during fabrication, and tolerances inherent to the measurement technique tend to produce results with wide-ranging tolerance limits. Numerous measurements must be performed to express these limits in a tangible and acceptable form, and this, in turn, demands complex and costly experimental procedures. This latter consideration is an important starting point for applying numerical simulation to obtain quasi-neutral results with defined boundary conditions while also allowing parameters to be varied without the need for expensive experimental work.

DESCRIPTION OF THE EXPERIMENT

The experimental part of the work consisted mainly of two steps:

- welding of an austenitic pipe by manual arc weld (TIG process) with an additional last pass heat sink weld

(LPHSW = welding of a last pass on the outer surface during simultaneous cooling of the inner surface with water) to improve the stresses at the inner surface
- after welding loading of pipe by pressure and temperature cycles.

The according residual stresses were measured after the normal welding, the LPHSW and after operational load.

Welding of the pipe

The pipe with the following geometry
outer Diameter: 114.3 mm
wall thickness: 6.3 mm
pipe length 700 mm
has been welded together by a 6 pass weld and an additional LPHSW in 2G - welding position (horizontal pipe axis).

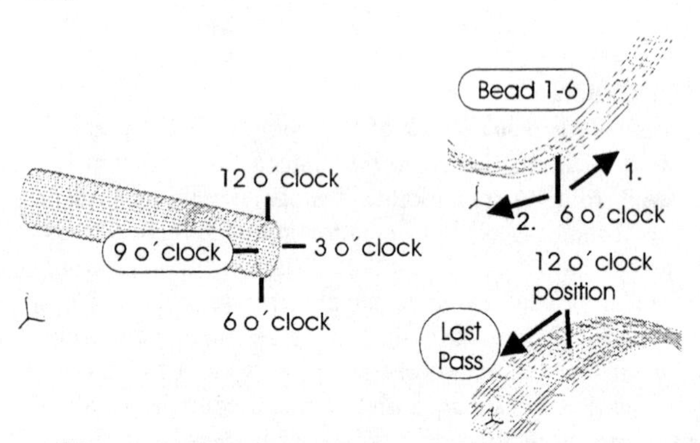

Figure 1: Welding sequence of the single passes

In the case of the thin pipe, the welding sequence is given in Figure 1. Passes 1 to 6 are welded in two steps, whereas the LPHSW has been performed in one step.

In Figure 2 a cross section over the weld is given. During welding the temperatures were recorded at different positions at the inner and outer surface of the pipe, see Figure 3, for comparison with the numerical data.

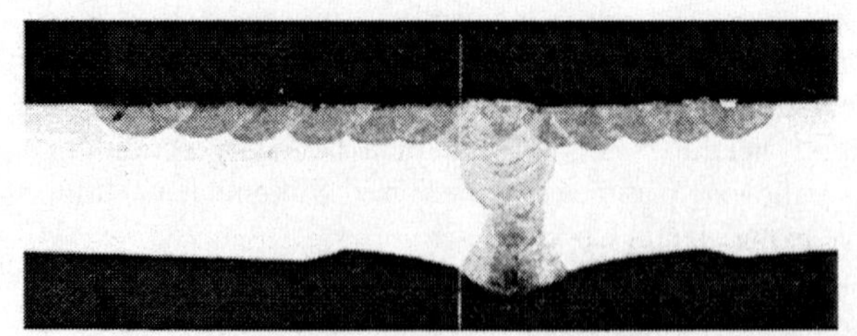

Figure 2: Cross section of the manual TIG weld including the LPHSW

All other necessary input data for the numerical investigations have also been recorded and fixed in the welding procedures (number and sequence of passes, heat input for each pass, velocity).

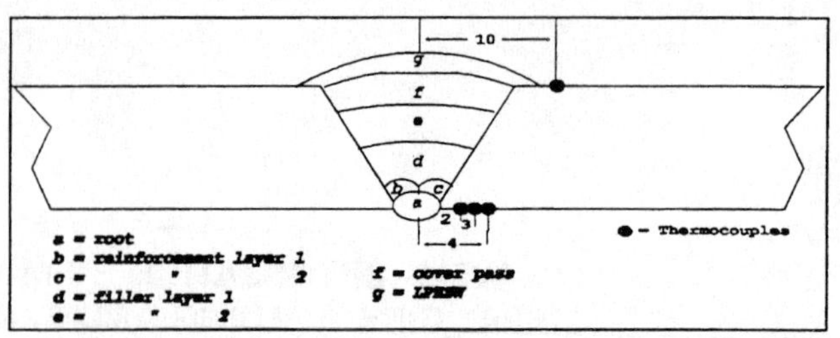

Figure 3: Position of thermo- couples during welding

In Figure 4 the results of the temperature measurement with the thermo-couples is shown during welding of the 2nd reinforcement layer as an example.

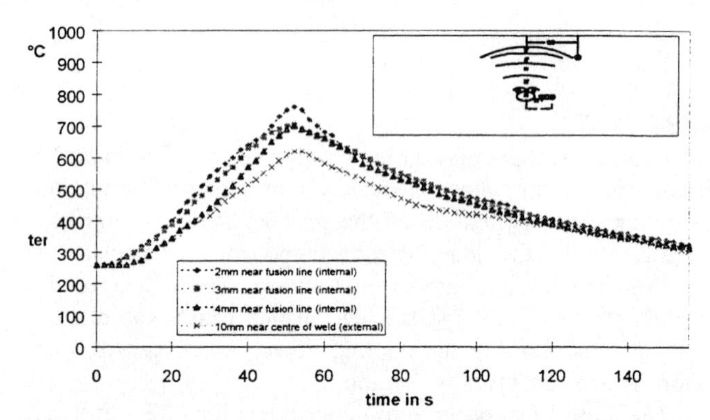

Figure 4: Measured temperatures during welding of the 2nd reinforcement layer

Ageing test

The test pipe has been welded in the Benson test facility and thermo- couples have been mounted along the circumference of the welds and in the base metal to get an exact input for the finite element calculations. In addition an extensometer was fixed on one end of the pipe.

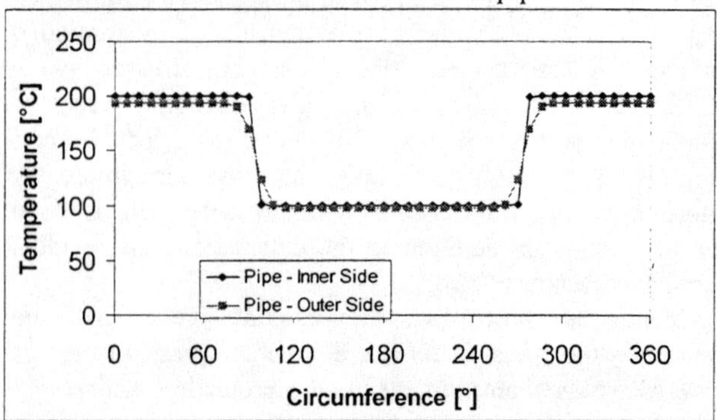

Figure 5: Circumferential temperature during cold / warm water injection

The welded pipe has been loaded by 77 cycles of thermal stratification (Figure 5) under full operation pressure cycles (Figure 6).

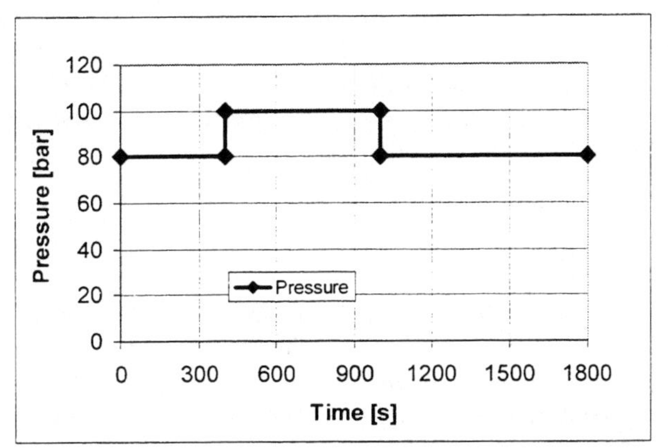

Figure 6: Pressure distribution during ageing cycle

A plate was brought into the pipe in such a way that the two inlet mass flows were separated into cold and hot water regions, to obtain the desired stratification as can be seen from the initial conditions of the finite element analysis, Figure 7. The partition with this method was carried out so that the plate inside had a distance of 2 mm to the welding joint. Therefore no additional stresses were introduced by any restraining effects. A small amount of mass flow mixing was accepted. By controlling the valves, it was possible to hold one inlet temperature constant at 100°C and to vary the other inlet temperature between 200°C and 100°C.

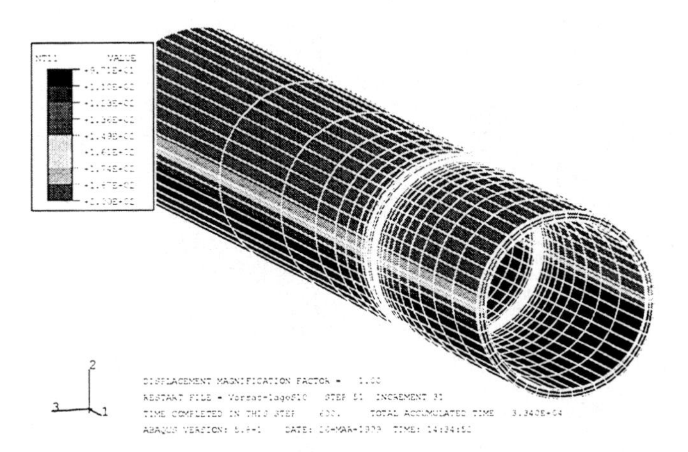

Figure 7: Temperature distribution in the FE model for simulation of temperature stratification

Residual stress measurement

The residual stresses were measured through the thickness by applying the ring-core-method. It is a well known procedure for residual stress measurement and is a purely mechanical technique which always integrates residual stresses over a certain area.

On the inner surface of the pipe in the heat affected zone (HAZ) of the weld a three-directional strain gage rosette is attached to the surface. Then a circular annulus with a diameter of 5 mm is eroded step by step around the rosette. The remaining cylindrical core becomes more and more free from the surrounding stress field. The resulting relaxation strains are measured with the aid of the attached strain gauges. From these data, the residual stresses can be calculated. With increasing depth of the eroded annulus, the reaction of the strain gauge decreases for constant stress in the material. This must be taken under consideration by an experimental declining function.

In the actual case, strain gauge rosettes with an active length of the strain gauges of about 1,6 mm were used. The strain gauge rosettes were attached in the HAZ of the welds in a distance of 1 mm from the fusion line (=middle of the active length). Using this rosette the maximum penetration depth is 2 mm. After reaching this depth, the remaining cylindrical core is removed completely by electric erosion and a new strain gauge rosette is attached at the ground of the cylindrical bore. Then again a circular annulus is eroded step by step until a penetration depth of 4 mm is reached. At this depth, the remaining cylindrical core is removed again and the whole procedure is repeated. The three measurements (0-2 mm, 2-4 mm and 4-6 mm) are afterwards mathematically connected (by using calibration curves).

WELD SIMULATION

The aim is to perform appropriate numerical analyses to assist the welding processes. The process of welding has therefore been simulated under the most realistic conditions possible using the three dimensional finite element method. The mesh configuration is shown in Fig. 8. For appropriate calculation with the FE-code ABAQUS, Hibbitt [1], three dimensional 8 node bricks are used with reduced integration order.

The problem is treated as an uncoupled thermal and mechanical problem, first the temperature field is evaluated and from these results stresses and displacements are calculated. An elastic-plastic mixed hardening material behavior is supposed for austenitic material. All material data are introduced in the model with their temperature dependency.

The model size could best be described by the number of nodes and elements:

Number of nodes:	8370 nodes full geometry pipe
Number of elements:	6570

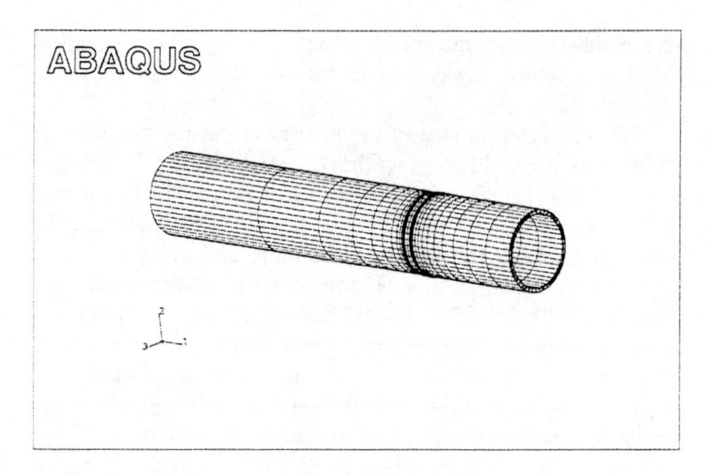

Figure 8: Finite Element model of the welded pipe

To get valuable information about the influence of the LPHSW on the residual stresses, it was impossible to use a symmetrical model. Each individual bead was located in a conventional weld geometry using realistic welding parameters. The element density was increased considerably in the region of the heat-affected zone (elements width approx. 0.1 to 0.2 mm), so that a realistic temperature and stress evaluation could be performed. Mesh details including the last pass are shown in Figure 9.

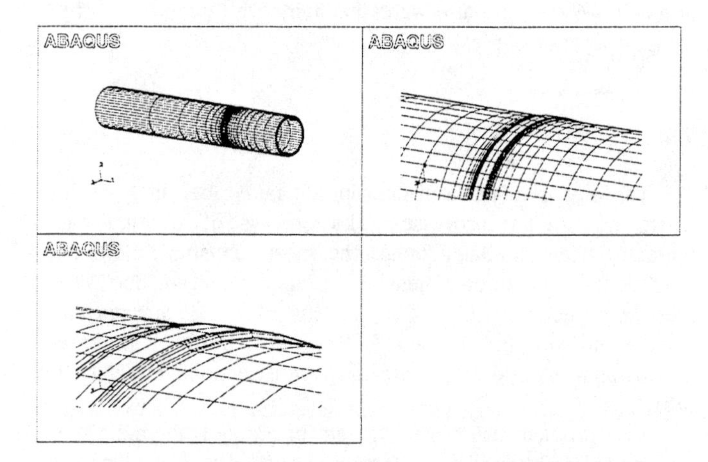

Figure 9: Detailed mesh in the vicinity of the weld

The welding process was simulated as a freely moving heat source controlled by corresponding time functions of the elements. Transition from liquid to solid phase was allowed for on the basis of latent heat; parameters for surface radiation and convection were also defined. The preheat and inter pass temperatures specified in the welding plans were observed for each bead. According to available information from numerous numerical parameter calculations, these parameters significantly affects the formation of the 3 D temperature and stress fields and thus, ultimately, post-weld deformation. The LPHSW was performed after the last bead of the joint weld has been welded. During welding of the LPHSW the pipe has been cooled at the inner surface with cold water.

The aging cycles after welding were performed as a restart run of the weld simulation. This means, the residual stress and deformation field after last pass heat sink welding was the basis for this analysis. The aging cycles were again performed as an uncoupled thermal and mechanical problem. After 10 cycles the analysis was interrupted, because no further effect on the stress and strain field could be observed.

RESULTS

The results of the simulation are mainly presented as residual stress fields during the welding process, after welding all passes, after LPHSW and finally after the operation cycles. A comparison is made between numerical and experimental results.

The axial stresses are the most relevant stress components, because possible flaws are supposed to originate in the HAZ of this circumferential weld in circumferential direction. Therefore the Figures are focused on the axial stresses.

In Figures 10 and 11 these stresses are shown during the welding process for each welded pass at the inner and outer surface of the pipe (I1 and O1 in the legend means first pass, I6 and O6 means 6th pass) and the LPHSW, which is denoted as IL and OL, depending on which surface is regarded. The stresses are plotted along the axis of the pipe, where the middle of the weld is located at a distance of 500 mm, this means the stresses are given over a distance of 100 mm (50 mm left and right to the weld). The condition is that each single pass is welded and cooled down to interpass or room temperature.

The circumferential position is 12 o'clock which is opposite to the starting point of the weld.

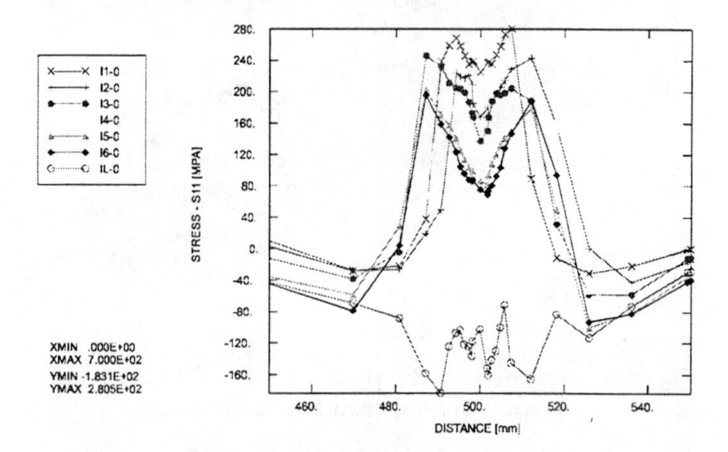

Figure 10: Distribution of axial stresses along the pipe axis at 12 o'clock position on the inner pipe surface

In Figure 10 it can be seen that the axial stresses at the inner surface of the pipe become smaller during the progress of the welding process. The maximum stresses are obtained after welding of the root pass. In the weld middle a local minimum forms and the stresses are highest in the HAZ.

The LPHSW is applied to improve the stresses at the inner surface for prevention of the IGSCC attack. In Figure 11 it can be clearly seen that this measure yields the desired effect. The tensile stresses after the 6th pass change to compression stresses over the whole weld region and also for all positions along the circumference at the inner surface, which can be seen later.

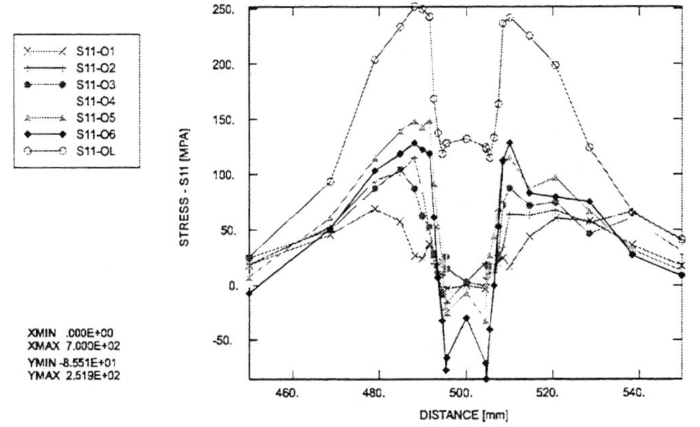

Figure 11: Distribution of axial stresses along the pipe axis at 12 o'clock position on the outer pipe surface

Because the compression axial stresses need to be balanced across thickness, the stress distribution at the outer surface of this thin pipe, Figure 11, looks contrary to the inner surface. The stresses being set up during welding of each pass show compression over the weld after the 6th pass has been welded. The LPHSW turns back the compression stresses to tensile stresses.

In Figures 12 and 13 a cross section of the weld is shown also at the 12 o'clock position of the circumference before (Figure 10) and after LPHSW (Figure 13). The axial stresses S11 are in tension over a long distance across wall thickness, only near the outer surface they are in compression (Figure 10). As mentioned before this situation turns after LPHSW (Figure 13). The significant change into compression stresses in the HAZ of the ID is obvious.

From the 3D analysis not only information about the stress distribution in thickness direction is available, but also along the circumference of the pipe. The welding sequence of the single passes is important for the development of the residual stresses.

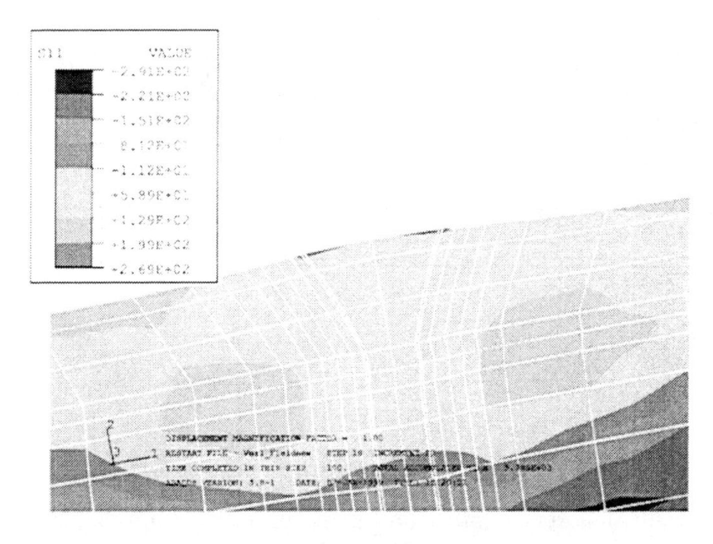

Figure 12: Distribution of axial stresses through the thickness in the vicinity of the weld at 12 o'clock position after welding of pass 1 to 6 and cooling down to room temperature

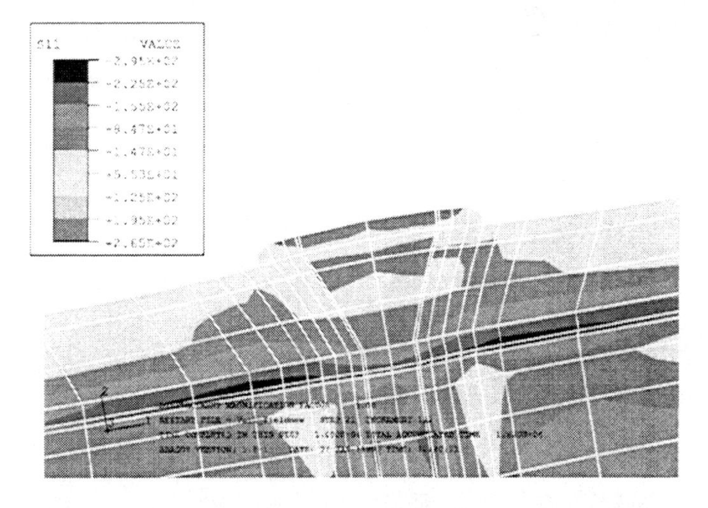

Figure 13: Distribution of axial stresses through the thickness in the vicinity of the weld at 12 o'clock position after welding of LPHSW and cooling down to room temperature

The distribution of the axial stresses along the inner circumference is shown in Figure 14 for the most interesting region, the HAZ, being relevant for an IGSCC attack. As mentioned before, the distribution is quite irregular along the circumference. Because passes 1 to 6 are welded in two steps, a minimum is obtained in 6 o'clock (begin of welding) and 12 o'clock position (stop of first half of welding). After LPHSW all axial stresses at the inner surface are in compression mode not only in the position 0.2 mm from fusion line, but up to an axial distance of 5 mm from weld middle.

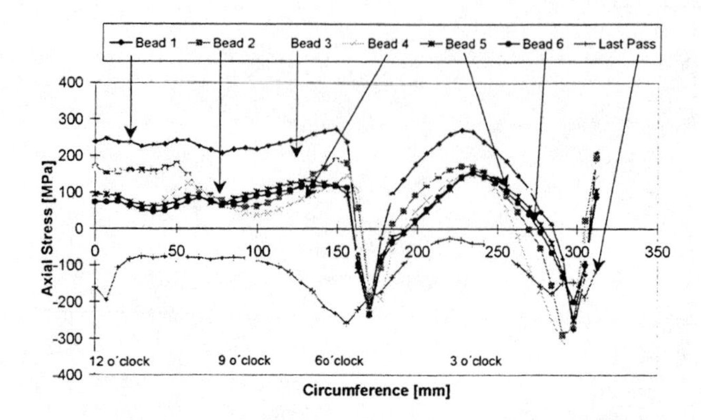

Figure 14: Axial stresses at the inner surface (0.2 mm pos. direction from fusion line) - after cooling down to interpass temperature (1-6 pass) or room temperature (last pass)

One of the most important question is whether this improved compression stresses in the HAZ will alter during normal operation. For this purpose the test pipe was subjected to thermal stratification and pressure cycles measured in plants by monitoring systems and described before. Simultaneously the numerical simulation was also continued, based on the residual stress state after LPHSW.

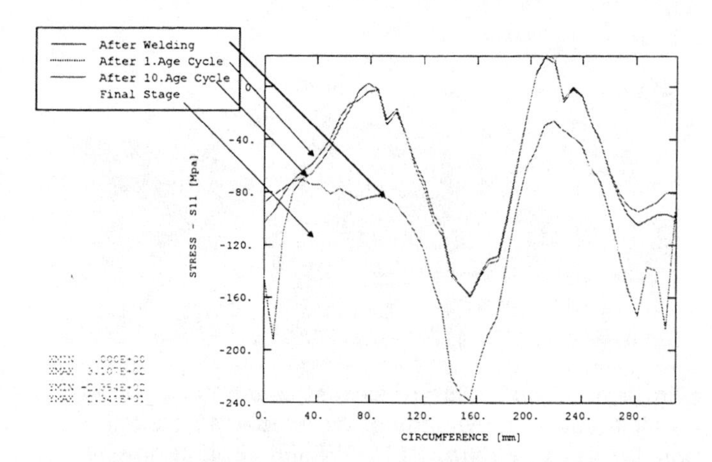

Figure 15: Axial stress (inner surface - 0.2 mm from fusion line) after welding sequence and ageing 1st and 10th cycle

77 pressure and temperature cycles have been conducted in the experiment. In the analysis the effects seemed to vanish after 10 cycles, therefore it has been interrupted after this time. The influence of operation cycles can be seen in Figure 15 where again the results of the calculated axial stresses are given along the inner circumference in the HAZ starting at 12 o'clock position. The condition after welding represents the residual stresses after LPHSW and cooled down to room temperature. Following the pipe has been loaded by internal pressure and operation temperature and afterwards loaded by cycles. It is important to note that during operation cycles (condition 1st and 10th age cycle) the compression residual stresses reduce, but still remain in the compression stage. At the final stage, which means cooled down to room temperature, a certain relaxation took place compared to the stage after welding, but all stresses are still in compression. This means the beneficial effect of LPHSW has not been compensated by loading cycles.

After the experiment has been conducted, the residual stresses have been measured by ring core method. In Figure 16 the numerical and experimental (gained on 4 different welds) results are compared at four different stages (measured at room temperature):

- pass 6 (without LPHSW) before and after operation cycles
- LPHSW before and after operation cycles.

The agreement between measurement and calculation is quite good, but the difficulties are also obvious. Because the stress distribution is not homogeneous around the circumference, it is very important to measure at a location where the stresses are not influenced by start or end point of the welding.

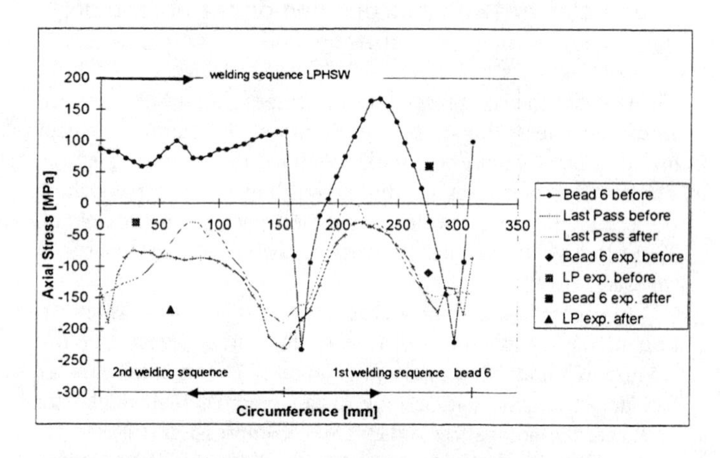

Figure 16: Comparison of experimental and numerical results: Axial stresses at the inner surface (0.2 mm neg. direction from Fusion line) - after cooling down to interpass temperature (1-6 pass) or room temperature (last pass), each before and after operation cycles

SUMMARY

The aim of this paper was to demonstrate the effect of "Last Pass Heat Sink Welding" (LPHSW) on the residual stresses of an austenitic pipe by means of three dimensional finite element calculations and experimental measurements. In addition it should be demonstrated whether the beneficial effect of the LPHSW remains during operation cycles. For this purpose a test pipe has been welded and after welding it has been tested in a special test facility by measured operation temperature and pressure cycles.

It could be shown that

- the residual axial stresses are not homogeneous around the circumference, this means only a 3D simulation is adequate for an according stress simulation
- the LPHSW alters the stresses in the HAZ from tensile to compression
- the residual stresses will be influenced by the operational load, but the compression stresses in the HAZ still remain; this means the protection against any corrosion attack is maintained.

REFERENCES

1. HIBBITT, KARLSSON & SORENSEN, Inc.: ABAQUS USERS MANUAL, Version 5.8, 1998

PVP-Vol. 410-1, Assessment Methodologies for Preventing Failure:
Deterministic and Probabilistic Aspects and Weld Residual Stress
Volume 1
ASME 2000

PREDICTING BURN-THROUGH WHEN WELDING ON PRESSURIZED NATURAL GAS PIPELINES

John Goldak[1] **, Mihaela Mocanita, Victor Aldea, Jianguo Zhou, Dan Downey**
Department of Mechanical and Aeronautical Engineering
Carleton University
Ottawa, Ontario, K1S 5B6
Canada
Tel: 613-520-2600 x5688, Fax: 613-520-5715
Email: jgoldak@mrco2.carleton.ca

David Dorling
Engineering and Maintenance Services
TransCanada Pipelines
1401 Irricana Road,
Airdrie, Alberta, T4B 2B8
Tel: 403 948-8147 Fax: 403 948-8268
Email: david_dorling@transcanada.com

ABSTRACT

Welding on a pressurized natural gas pipeline is simulated using Computational Weld Mechanics (CWM) to determine if CWM can provide useful estimates of the risk of burn-through. The critical input data in addition to the internal pressure in the pipe, the geometry of the pipe, the size and shape of the weld pool including weld reinforcement, are the convection coefficient on the internal pipe surface and the temperature dependence of the viscosity of the pipe metal near the melting point. The CWM analysis shows that above a critical state creep under the weld pool thins the pipe wall and forms a groove. When the pipe wall is thinned by the groove during welding, the internal pipe wall temperature increases under the weld pool during welding. This nonlinear interaction further increases the temperature on the internal pipe wall under the weld pool and further accelerates creep and the actual burn-through. The analyses shows significant thinning or groove formation exactly in those welds that burned-through or were at high risk of burn-through. No significant thinning and no significant groove formation is predicted in those welds for which no significant thinning was reported experiments. We conclude that CWM can compute useful estimates of the risk of burn-through when welding on pressurized pipelines.

NOMENCLATURE

b Body force at a point (x,y,z).
B Matrix that maps FEM nodal displacement vector to strain tensor at a point.
dU FEM nodal displacement increment in a time step.
du Correction to the FEM nodal displacement vector in a time step.
D Elasticity tensor at a point (x,y,z).
F Total Deformation gradient at a point (x,y,z).
F_{El} Elastic Deformation gradient.
F_{Pl} Plastic Deformation gradient.
F_{Init} Initial Deformation gradient due to thermal expansion and phase changes.
$\dot{h}$ Time rate of specific enthalpy at a point (x,y,z).
$\mathcal{K}$ FEM global stiffness matrix.
$\mathcal{P}$ FEM nodal external load vector.
q Thermal flux at a point (x,y,z).
Q Density of energy production at a point (x,y,z).
$\mathcal{R}$ FEM global residual vector.
T Temperature at a point (x,y,z).
κ Thermal conductivity.
ρ Density at a point (x,y,z).
σ Cauchy stress at a point (x,y,z).
σ_{2PK} 2nd Piola-Kirchhoff stress at a point (x,y,z).
$d\varepsilon$ Green-Lagrange finite strain tensor at a point (x,y,z).

INTRODUCTION

It is desirable to weld natural gas pipelines under pressure both for making new connections to the pipeline, called hot-taps, and to repair damage to the pipeline. Welding on the pressurized pipeline avoids the costs of shutting down the pipeline and depressurizing. In thinner walled pipes,

[1]Address all correspondence to this author.

there is a risk of burn-through when welding while the pipeline is pressurized. Should burn-through occur, the safety of the welding personnel are at risk and the pipe will have to be shut down to repair the burn-through. The objective of this research was to determine if CWM could simulate the process of welding on a pressurized pipeline and provides useful estimates of the risk of burn-through.

Kiefner (2)(3) studied the problem in some detail. He recognized that the risk of burn-through was greatest in thin wall pipes. He also proposed a model to predict the risk of burn-through. The model used a rather simple 2D finite difference method with forward Euler time integration for the thermal analysis. If the maximum internal wall temperature exceeded 1800 F, then the risk of burn-through was considered high. Kiefner included the effect of flowing the gas or liquid in the thermal analysis through the convection coefficient. However, he did not consider the effect of stress and deformation.

The CWM methodology used in this paper is simplicity itself. Simply compute the 3D transient temperature field, microstructure evolution, displacement or distortion, strain and stress during the welding process. The CWM analysis of the process of welding on pressurized pipelines involves several difficult problems that make this a particularly challenging problem. The greatest challenge is to develop the capability to deal with creep at temperatures in the range 1652 F (900 C, 1173 K) to 2732 F (1500 C, 1773 K). For sufficiently short times at temperatures below 1652 F (900 C, 1173 K), the creep rate is sufficiently slow that during the rather short times at high temperatures during welding creep can be ignored in most CWM problems and rate independent plasticity can be an appropriate material model. However, it is exactly at temperatures above 1652 F (900 C, 1173 K) that internal pressure can cause burn-through. It is exactly in this temperature range that the steel deforms primarily by creep, which is rate-dependent plastic or viscoplastic or viscous flow. In this temperature range, a realistic analysis must deal with viscous flow and use realistic values of the viscosity, the material parameter that relates the deviatoric strain-rate to stress.

A second challenge is the need to couple the thermal analysis with the stress analysis more strongly than is usually done in CWM. In particular, the thermal analysis must be done on the geometry deformed by the stress analysis. A third challenge is to deal with the length scales that range from a fraction of a mm near the weld pool to a 12 ft. (3.66 m) long pressure vessel length. This large range of length scales can make the FEM problem numerically ill-conditioned. Special solvers that were sufficiently robust are needed to deal with this problem. A fourth challenge is to manage the complexity of the software and the complexity of input data required for analyses. The addition of filler metal, while not critical to this project, is desirable to provide a more complete and realistic simulation.

The paper is organized as follows: First the mechanics theory and numerical methods used are summarized. Then the analysis procedure and results are described. Finally, the results are described and conclusions stated.

THEORETICAL MECHANICS AND NUMERICAL METHODS

Thermal Analysis

The thermal analysis computes an FEM approximate solution to the transient 3D energy equation

$$\dot{h} + \nabla \cdot q + Q = 0 \tag{1}$$

where $\dot{h}$ is the time rate of the specific enthalpy, $q = -\kappa \nabla T$ is the thermal flux, T is the temperature, κ is the thermal conductivity, Q is the density of heat generation. The energy equation is solved using 8-node bricks with backward Euler time integration. It is solved on the current deformed geometry for each time step. Temperature dependent thermal conductivity and specific enthalpy, including the effect of latent heats of phase transformations were used.

The heating affect of the arc is described by a Dirichlet boundary condition on nodes in the weld pool. The weld pool size, shape and position is taken as data determined from experiment for each weld. The temperature in a weld pool is assumed to vary parabolically from the melting point at the boundary of the weld pool to a maximum temperature of the melting point plus 400 K at the weld pool centroid. Temperatures in the weld pool liquid/solid boundary were prescribed to 2780 F (1526 C, 1800 K). The maximum temperature in the weld pool was prescribed to 3500 F (1927 C, 2200 K). The size and shape of the weld pool was estimated from macrographs contained in (1).

External boundaries of the structure which are assumed to be in still air, have a convection boundary condition with ambient temperature of 300 K and convection coefficient of 10 $W/m^2\ K$. The pipe is filled with water and tilted slightly to avoid trapped air under the weld. On the internal surface of the water-filled pipe a coefficient in the range 500 to 5000 $W/m^2\ K$ and ambient temperature of 80 F (27 C, 300 K) was applied. For each weld for a fixed weld pool length, the thermal flux from the weld pool was computed for this range of convection coefficients on the internal pipe surface. Assuming a weld efficiency factor of 0.65, the computed net heat input from the arc was compared to the total heat input for the weld estimated from data taken from (1). The convection coefficient that provided best agreement between weld nugget geometry and heat input data taken from (1) was 1500 $W/m^2\ K$. Convections coefficients in the range 1000 to 2000 $W/m^2\ K$ did substantially change the results.

Microstructure Analysis

The evolution of microstructure is computed using the methodology described by Watt and Henwood in (6) (7).

Thermal Stress Analysis

The thermal stress analysis computes an FEM approximate solution to the conservation of momentum and mass. The loads, material properties and geometry are time dependent.

$$\nabla \cdot \sigma + b = 0$$
$$\frac{D\rho}{Dt} + \rho \nabla \cdot v = \rho_{,t} + \nabla \cdot (\rho v) = 0 \quad (2)$$

where σ is the Cauchy stress, b is the body force and ρ is the density. Inertial forces are ignored.

For each time step, a displacement increment dU is computed using a Newton-Raphson iterative method in an updated Lagrangian formulation. The deformation gradient for a time step is $F = I + \nabla dU$. The Green-Lagrange strain increment for a time step is $d\varepsilon = (F^T F - 1)/2$. For details describing the finite strain theory, see (4). The deformation gradient is decomposed into elastic, initial and plastic deformation gradients, $F = F_{El} F_{Init} F_{Pl}$, where F_{Init} includes deformation due to thermal expansion and phase transformations. For stress analysis, F_{Init} is made piecewise constant in an element. For small strains, the elastic strain increment is the linear part of $d\varepsilon_{El} = (F_{El}^T F_{El} - 1)/2$. The increment in the 2nd Piola-Kirchoff stress is $d\sigma_{2PK} = D\, d\varepsilon_{El}$ where D is the elasticity tensor. At temperatures below 1100 K rate independent plasticity is used. At temperatures above 1100 K a linear viscous constitutive model is used. See (4) for details on the plasticity and viscoplasticity formulations.

The residual for the Newton-Raphson iteration is

$$\mathcal{R} = P - \int B^T \sigma_{2PK} \quad (3)$$

where B is the usual FEM matrix that maps nodal displacements to the strain at a point and P is the nodal vector of external loads. Each Newton-Raphson iteration computes a correction du to the current trial displacement dU by solving

$$\mathcal{K} du = -\mathcal{R} \quad (4)$$

where $\mathcal{K}$ is the global stiffness matrix. For each Newton-Raphson iteration, the linear equation 4 is solved to a tolerance of 10^{-9} in the energy norm. When the 1-norm of the residual $\mathcal{R}$ is less than 10^{-4}, the time step is considered to have converged.

The 3D transient nonlinear thermal stress analysis used a viscoplastic material model and 8-node bricks. Temperature dependent Young's modulus, Poisson's ratio, Yield strength, hardening modulus and viscosity are used. Rigid body modes in the pipe were constrained. The internal surface of the pipe was pressurized to 900 psig (6.2 MPa). The transient temperature field of the thermal analysis was applied to each time step. The displacement, strain and stress were computed for 40 time steps for each weld, i.e., the weld moved approximately 0.1 in (2.5 mm) in each time step. The stress analysis was done on the full vessel.

Coupling Thermal and Stress Analysis

The first time step computed the displacement, strain and stress due to applying the internal pressure. Then the weld pool was positioned in the deformed geometry for each subsequent time step. If a groove formed under the weld, the thermal analysis took into account this thinning of the pipe wall. When the arc was extinguished, the weld was allowed to cool to room temperature. The final time step returned the internal pressure to 1 atm.

METHODOLOGY

Task Sequence:

1. Build an FEM mesh of the pressure vessel including the slots machined for welds and filler metal to simulate the welds described in the (1).
2. For each repair weld, specify data of position, current, voltage, arc efficiency, welding speed, weld metal cross-section geometry and weld pool semi-axes lengths.
3. For each weld compute the 3D transient temperature, microstructure, displacement, strain and stress.
4. Post-process results of each weld analysis to visualize the temperature, displacement, strain and stress. The deformation of the internal pipe wall under the weld was of particular interest.

5. Compare the temperature and deformation computed with CWM with experimental temperature and deformation as estimated from macrographs in (1).
6. Interpret the results of the analyses.

Structure and Materials

The pipe was 20-in. (508 mm) diameter and 0.312-in. (7.9 mm) wall API 5LX-65 ERW. See Figure 1. Table 1 taken from (1) specifies the properties of the steel. The welding consumables were Type E7018 low hydrogen shielded metal arc electrodes.

Table 1. PIPE MATERIAL PROPERTIES.

Designation	API-SLX65
Yield Strength (MPa)	71.3 ksi (483 MPa)
Tensile Strength	89.8 ksi (621 MPa)
Elongation %	29.4

Deposition of Welds

At each test weld location a flat grove was machined into the pipe and a single weld pass was deposited along the centerline of the machined area. All welds were made at an internal pressure of 900 psig (6.2 MPa). Tables 1 and 2 which are taken from (1) specifies the data for each weld.

In (1) welds A, B, D, G, H and I were rated as "safe" welds; C, J and L were rated as marginal and weld E and F were rated as burnthrough. In this paper, we only consider one weld from each class, i.e., welds H, J and E. For these three welds, the welding current was 80 amps, the voltage 21 volts and the electrode size 3/32 in.

Table 2. SUMMARY OF WELD DATA.

Test ID	Wall Thickness (in)	Heat Input (KJ/in)	Travel Speed (in/min)
H	0.156	20	5.0
J	0.156	25	4.0
E	0.125	20	5.0

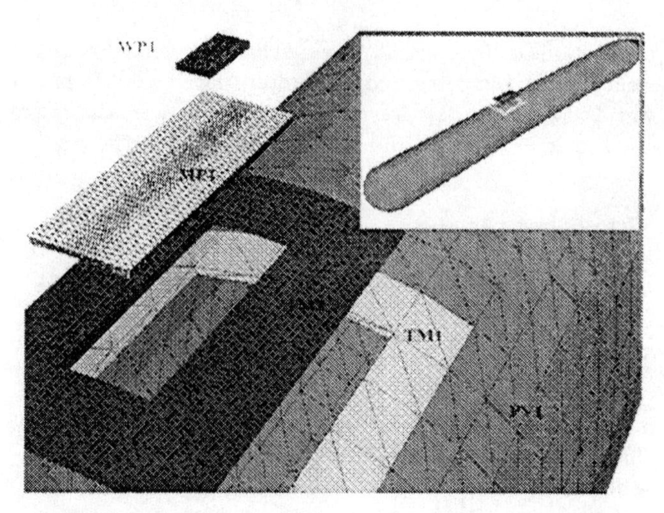

Figure 1. THE VARIOUS PARTS OF THE FEM MESH ARE SHOWN. THE SMALLEST PART, WP1, AT THE TOP OF THE FIGURE, CONTAINS THE WELD POOL.

FEM Mesh

The mesh for the pressure vessel, called PV1, had 3006 8-node brick elements. For each weld, a fine mesh with 3456 elements, called MP1, was created for the region around the machined slot. Outside of MP1, a transition mesh with 768 elements, called TM2, and coarser transition mesh called TM1 with 140 elements was made. Inside MP1 an even finer mesh was created, called WP1, with 3840 elements. The mesh WP1 was moved with the arc during the welding process.

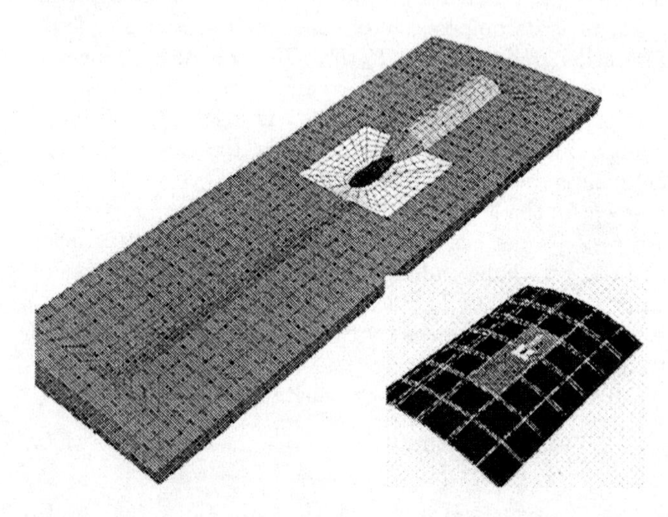
Figure 2. FILLER METAL BEING ADDED DURING THE WELDING PROCESS IS SHOWN.

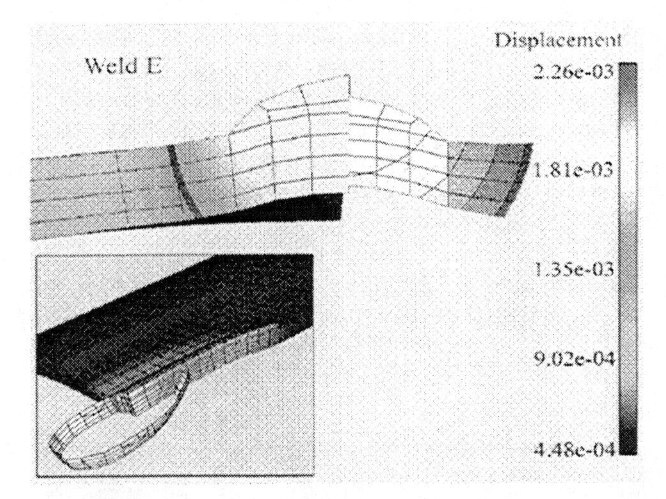

Figure 3. THE 1255 K TEMPERATURE ISO-SURFACE IS SHOWN ON DISPLACED GEOMETRY JUST BEFORE THE ARC WAS EXTINGUISHED.

RESULTS AND DISCUSSION

Some typical results of the thermal analysis are shown in Figures 3 and 4.

The accuracy of the thermal analyses are considered to be limited primarily by the values chosen for the convection coefficient on the internal surface of the pipe. By setting the weld pool cross-section size and power input from data given in (1) and choosing an arc efficiency factor of 0.65, we find the best agreement with data from (1) with the convection coefficient in the range 1000 to 2000. We consider 1500 $W/m^2\ K$ to be the best estimate of the value of the convection coefficient.

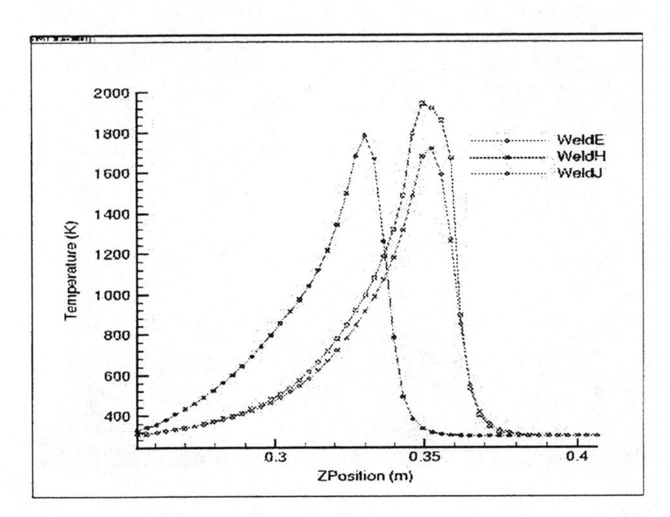

Figure 4. TEMPERATURE VS Z, WHERE TIME =Z/SPEED, ON THE INTERNAL SURFACE DIRECTLY BELOW THE WELD PATH JUST BEFORE THE ARC WAS EXTINGUISHED.

We first note that the stress state in the machined slot is rather different from that of a thin walled pipe with internal pressure. The thin walled slot actually deforms more like a bubble than a thin walled pipe. See Figure 5, curve A would be constant in a thin-walled pipe. The bending changes the stress distribution from that of a thin-walled pipe. Instead of a constant hoop stress, the hoop stress is higher on the outer surface where bending adds a tensile component and lower on the internal surface where bending adds a compressive component. Some of the hoop stress load is shunted around the slot as it would be shunted around a crack.

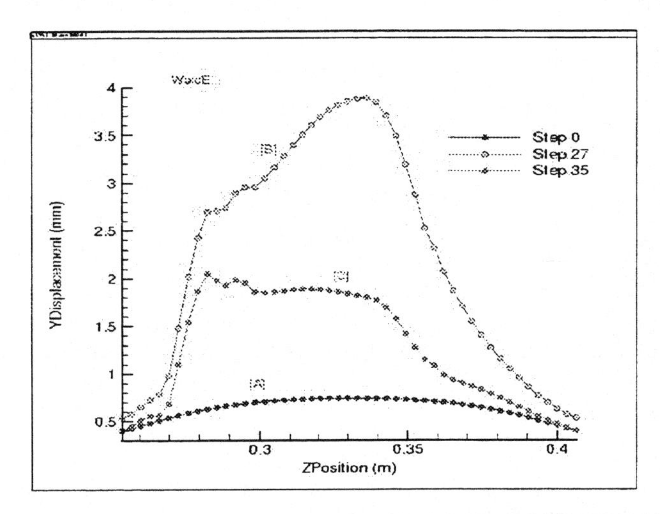

Figure 5. DISPLACEMENT OF INTERNAL SURFACE DIRECTLY BELOW THE WELD PATH. A) AFTER PRESSURIZING THE VESSEL BUT BEFORE STARTING WELD (Step 0). B) JUST AFTER THE ARC WAS EXTINGUISHED BUT BEFORE COOLING TO ROOM TEMPERATURE (Step 27). C) AFTER THE WELD HAS COOLED TO ROOM TEMPERATURE (Step 35).

Figures 7 to 12 show that the agreement between experiment and FEM analysis for the total displacement of the deformed cross-sections, i.e., groove formation, in welds H, J and E is quite good. In welds J and E the FEM results predict slightly deeper groove formation than experiment. The slightly deeper groove formed in the FEM analyses of welds J and E could be caused by a convection coefficient value that was too low and/or a high temperature viscosity that was too high. This suggests the FEM model can predict burnthrough reasonably accurately. Although the FEM analyses were done on a total twelve welds, only details of FEM analyses on welds H, J and E are given here because of space limitations. In the “safe” welds A, B, D, G, H and I, both experiment and FEM showed no significant groove formation. In the “marginal” welds C, J and L and "burnthrough" welds E and F the FEM analyses predicted somewhat deeper groove formation. From an engineering viewpoint, this is desirable because it is preferable for the prediction to err on the safe side.

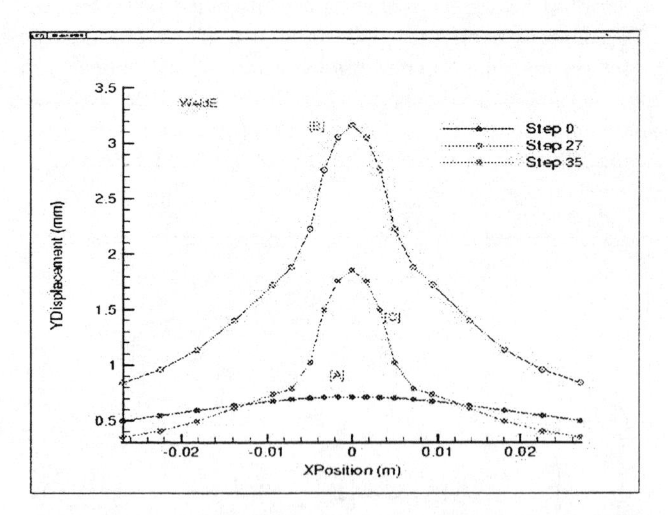

Figure 6. DISPLACEMENT OF INTERNAL SURFACE VS. TRANVERSE DISTANCE FROM THE WELD. THE DATA IS TAKEN APPROXIMATELY ONE WELD POOL LENGTH BEHIND THE TRAILING EDGE OF THE WELD POOL JUST BEFORE THE WELD WAS COMPLETED. A) AFTER PRESSURIZING THE VESSEL BUT BEFORE STARTING WELD (Step 0). B) AFTER COMPLETING THE WELD BUT BEFORE COOLING TO ROOM TEMPERATURE (Step 27). C) AFTER THE WELD HAS COOLED TO ROOM TEMPERATURE (Step 35).

Kiefner (2)(3) states that the Battelle model , "It has been shown that for certain welding process, the level of 1800 F (980 C, 1255 K) at the inside surface is a safe upper limit for avoiding burnthrough." Figure 4 clearly shows that the difference in peak temperature on the internal wall of the pipe is not a sensitive measure of risk of burnthrough. This is not surprising because the peak temperature criterion ignores time and burnthrough is a time-dependent phenomena. We argue that a much better model is obtained by CWM because, in addition to the peak temperature, it includes several additional parameters that are critical to the burnthrough phenomena. These include the value of the internal pressure, the pipe diameter and wall thickness, the weld pool geometry and speed, convection coefficient, the area of the internal surface at temperatures above temperatures of 900 C, and the time spent at high temperatures. The CWM FEM analyses of burnthrough presented in this paper do consider all of these factors. We argue that this is a significant advance in modeling burnthrough when welding on pressurized pipelines.

It should be noted that almost no use has been made of adjustable or tuning parameters in the FEM analysis described in this report. Certainly there is some uncertainty in the convection coefficient on the internal surface of the pipe. However, the values used are in the range expected for a water filled pipe. There is also some uncertainty in the value of the high temperature viscosity for steel. Again the values used are in the range expected for thermal activated dislocation motion, i.e., high temperature creep in steels.

To simulate the actual burn-through, i.e., the formation of a hole, we conjecture that a dynamic analysis including inertial forces would be required.

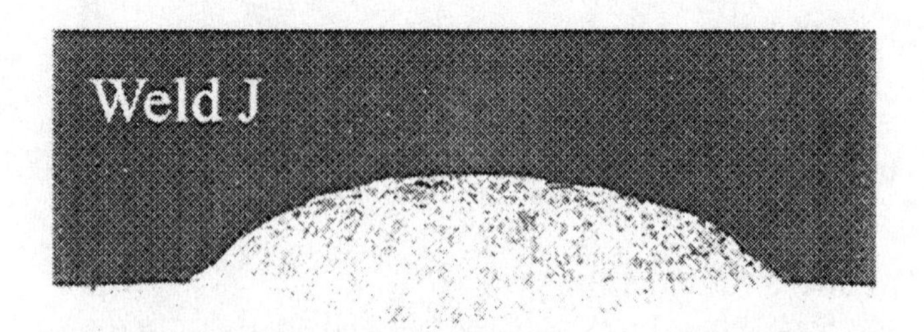

Figure 7. EXPERIMENTAL DATA FOR A MARGINAL WELD.

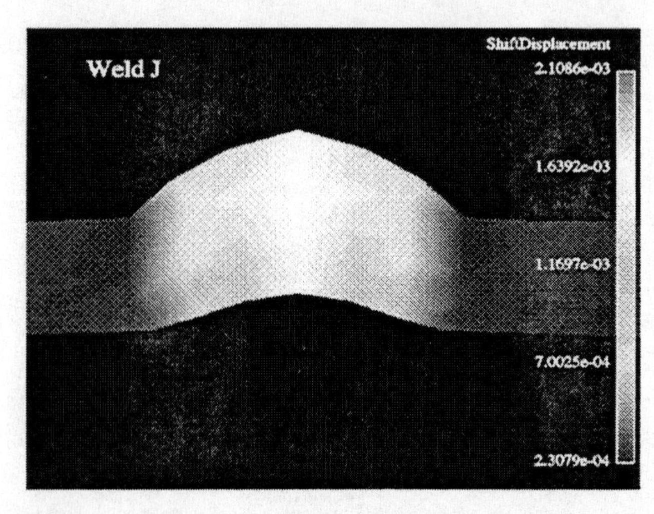

Figure 8. COMPUTED DEFORMED CROSS-SECTION FOR A MARGINAL WELD. CROSS-SECTIONS SHOW REASONABLE AGREEMENT BETWEEN THE PREDICTED AND EXPERIMENTAL DEFORMATION OF A MARGINAL WELD.

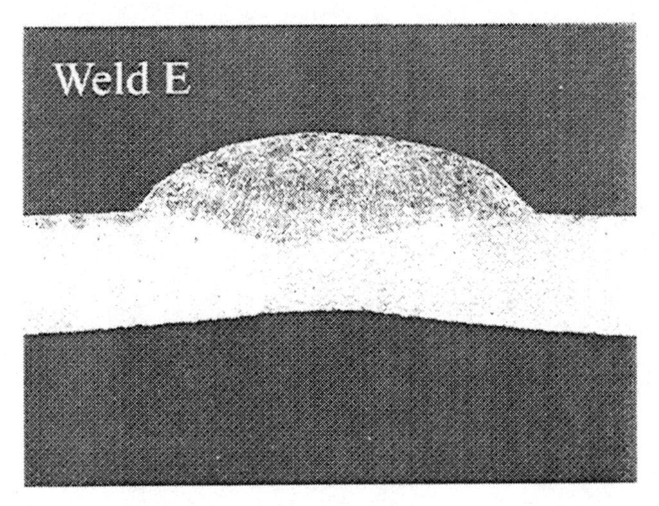

Figure 9. EXPERIMENTAL DATA FOR WELD E, A BURNTHRUOGH WELD.

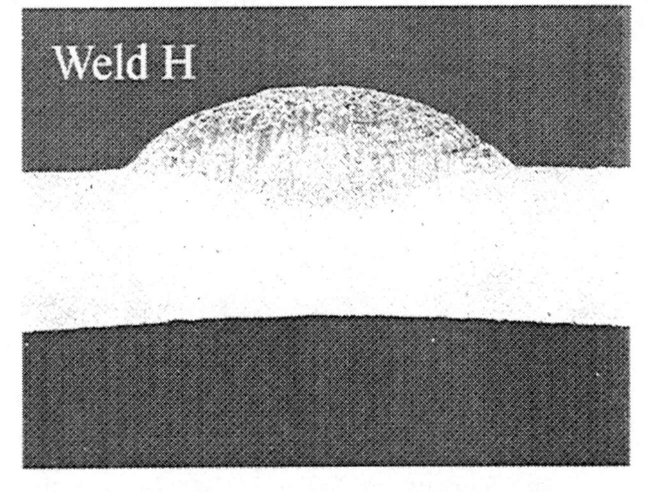

Figure 11. EXPERIMENTAL DATA FOR WELD H WHICH IS A SAFE WELD.

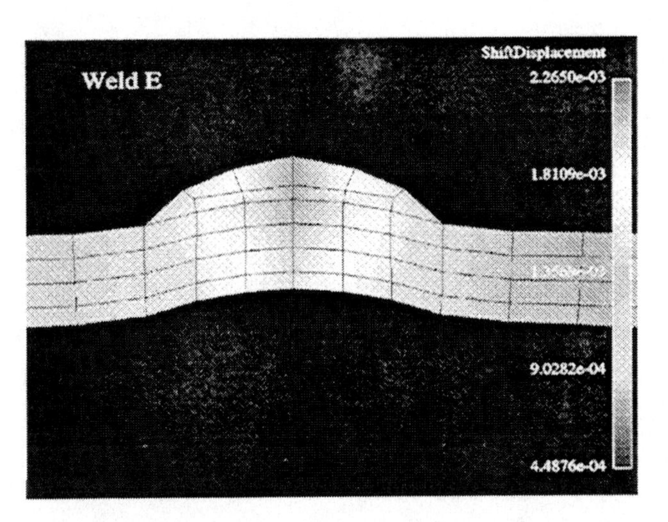

Figure 10. COMPUTED CROSS-SEECTION FOR WELD E. CROSS-SECTIONS SHOW EXCELLENT AGREEMENT BETWEEN THE PREDICTED AND EXPERIMENTAL DEFORMATION OF WELD E.

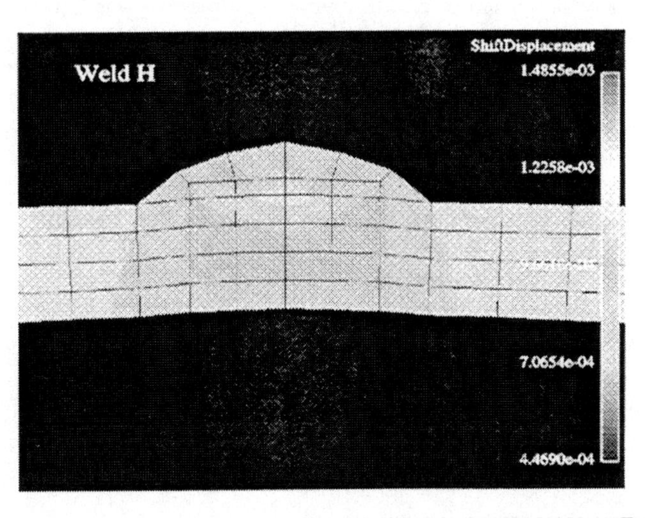

Figure 12. COMPUTATIONAL DATA FOR WELD H. COMPARISON WITH FIGURE 11 SHOWS EXCELLENT AGREEMENT.

All FEM analyses were performed on a Pentium III 600 MHz CPU with 512 Mb RAM. The operating system was Redhat Linux 6.1. Computing times did not exceed 10 hours for any weld simulation.

If internal surface temperatures could be measured experimentally, the convection coefficient could be estimated more accurately. If deformation of the internal surface could be measured more accurately, then even higher accuracy comparisons could be made with FEM simulations.

CONCLUSIONS

We conclude that useful estimates of the risk of burn-through when welding on pressurized thin walled steel pipes can be computed using the FEM methodologies described in this report. Burn-through is primarily a high temperature creep phenomena and requires modeling viscous flow. Solving the energy equation on the current deformed geometry captures an important nonlinear coupling between stress analysis and energy analysis.

ACKNOWLEDGMENT

We thank the American Gas Association (Contract No. PR-206-013) and the National Science and Engineering Research Council of Canada for financial support.

REFERENCES

1. PRC-185-9515, PRC Project No. August 3, 1995. *Repair of Pipelines by Direct Deposition of Weld Metal - Further Studies.*

2. John Kiefner, *Effects of flowing products on line weldability*, Oil and Gas Journal, pp. 49-54, July 18, 1988.

3. John Kiefner, Robert Fischer, *Models aid in pipeline repair welding procedure*, Oil and Gas Journal, pp. 41-547, March 7, 1988.

4. J. Goldak, V. Breiguine, N. Dai, J. Zhou, *Thermal stress analysis in welds for hot cracking, Welding-induced residual stresses in pressure vessels and piping*, 1996 ASME PVP Conference, July 21-26, Montreal, Quebec, Canada, to be published in ASME Journal of Pressure Vessel Technology

5. J. Goldak, V. Breiguine, N. Dai, J. Zhou, *Thermal Stress Analysis in Welds for Hot Cracking*, Editor H. Cerjak, 3rd Seminar, Numerical Analysis of Weldability, Graz, Austria, Sept. 25-26, 1995.

6. Watt, D.F., Coon, L., Bibby, M.J., Goldak, J.A. and Henwood, C., *Modelling Microstructural Development in Weld Heat-Affected Zones*, (Part A), Acta Met., vol. 36, No. 11, pp. 3029-3035, 1988.

7. Henwood, C., Bibby, M.J., Goldak, J.A. and Watt, D.F., *Coupled Transient Heat Transfer-Microstructure Weld Computations (Part B)*, Acta Met., vol. 36, No. 11, pp. 3037-3046, 1988.

PVP-Vol. 410-1, Assessment Methodologies for Preventing Failure:
Deterministic and Probabilistic Aspects and Weld Residual Stress
Volume 1
ASME 2000

IN-PROCESS CONTROL OF WELDING DISTORTION BY REVERSE-SIDE HEATING IN FILLET WELDS

Masahito MOCHIZUKI and Masao TOYODA
Department of Manufacturing Science
Graduate School of Engineering
Osaka University
Suita, Osaka 565-0871
JAPAN
Phone: +81-6-6879-7561
FAX: +81-6-6879-7561
E-mail: mmochi@mapse.eng.osaka-u.ac.jp

ABSTRACT

An in-process controlling method of welding distortion in fillet welds enables angular distortion to be reduced by performing reverse-side TIG heating while at a fixed distance ahead of MIG welding. Various heat conditions were examined by experiment and by finite element analysis to choose appropriate executions and to study the mechanism for reducing welding distortion. The mechanism of reducing angular distortion results from two main effects: TIG heating on the reverse side which provides the opposite angular distortion, and preheating effect.

INTRODUCTION

It is well known that evaluating welding distortion and residual stress is important for material weldability, structural fabricability, and structural integrity. In particular, prediction of welding distortion and its reduction are essential in order to control dimensions and reduce the fabrication cost of welded structures. Therefore, in-process control of welding distortion is more desirable than post-welding rectification method or other methods, considering manufacturing efficiency.

There are many predicting methods for welding distortion and residual stress, and these are useful for developing an in-process control method of welding distortion and residual stress. Thermal elastic-plastic analysis is one of the most powerful tools for predicting the mechanical behavior of welding [1-9], and three-dimensional calculations have recently been performed on the latest high-speed computer [10-14]. Welding distortion and residual stress can be calculated under various welding conditions and for various configurations of welded structures by using a long computing time. On the other hand, a simplified method using the principle of eigen strain can rapidly estimate residual stress and distortion, once welding conditions have been determined [15-19]. The most suitable analytical method can be applied for studying in-process control of welding distortion and residual stress.

Methods using mechanical prestraining, cooling [20], optimizing the welding sequence [21, 22], or using vibration [23] have been developed for residual stress control in welding. The reduction mechanism of residual stress was verified and validated mainly by numerical analysis. On the other hand, in-process control of welding distortion is performed by using restraint [24-26], elastic prestraining [27], preheating [28], preheating or/and stretching [29], or heat control by cooling [30-34]. These methods have been applied to actual welded structures, but the reduction mechanism of welding distortion has not been completely studied because most of these techniques were experientially established for specific conditions.

The easiest method of reducing welding distortion is in-process heat control considering equipment needed to prevent distortion and to the general purpose of the reduction method. Therefore, in-process control of welding distortion by using a heating or cooling source in addition to the welding torch is considered to be very useful for actual use.

In this paper, an in-process controlling method of welding distortion in fillet welds was studied. Angular distortion can be reduced by conducting reverse-side TIG heating at a fixed distance ahead of MIG welding.
The effectiveness of this method was verified by experiment and by three-dimensional thermal elastic-plastic finite element analysis in a fillet welded joint made of type A5083 aluminum alloy. Various heat conditions were examined to choose

appropriate ones and to study the mechanism for reducing welding distortion, and the effectiveness of the proposed in-process method during welding was validated.

EXPERIMENTAL

The effectiveness of the method for in-process control of welding distortion was investigated by an actual experiment using a fillet-welded joint made of aluminum alloy. Several parameters related to welding and heating conditions were varied to study their effect on welding distortion. The two main types of welding distortion were considered to be angular distortion and longitudinal shrinkage. This is because that angular distortion is much larger than other distortion modes and that longitudinal shrinkage may eventually cause buckling deformation for thin plate structures.

Principle of Distortion Control

The configuration of a fillet-welded plate joint, which is made of type A5083-O aluminum alloy, is shown in Fig. 1. The plate thickness of the flange was varied from 3 to 6 mm, and a 3 mm-thick web was fillet-welded to the flange by MIG welding. The two sides were fillet-welded sequentially with 30 mm/s constant speed of MIG welding with type 5183 filler metal (ϕ1.2 mm). No constraint was done during heating and welding processes. The chemical composition and mechanical properties of the material are shown in Table 1. The reverse-side TIG heating ahead of the MIG welding was performed mainly to prevent angular distortion of the flange.

The in-process welding distortion control by reverse-side heating method is schematically illustrated in Fig. 2. The reverse-side TIG heating is conducted under keeping a fixed distance ahead of MIG welding. It is considered that the distance D between the MIG and TIG torches and the heat input of TIG heating Q_{TIG} are important parameters for controlling welding angular distortion.

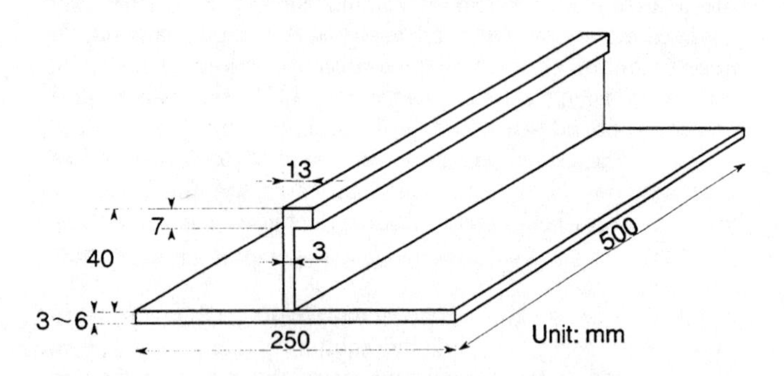

Fig. 1 Configuration of a fillet-welded joint.

Table 1 Chemical composition and mechanical properties of material used.

Chemical composition (%)									Mechanical properties		
Si	Fe	Cu	Mn	Mg	Cr	Zn	Ti	Al	σ_Y (MPa)	σ_T (MPa)	El.(%)
0.14	0.21	0.04	0.70	4.61	0.11	0.01	0.02	Bal.	163	315	23

σ_Y : Yield stress (0.2% proof stress), σ_T : Tensile strength, El. : Elongation (G.L. = 50 mm).

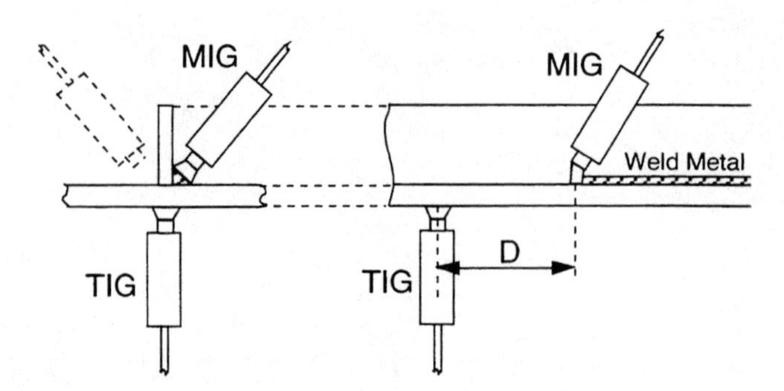

Fig. 2 Schematic illustration of welding distortion control by reverse-side heating.

Results of Experiment

The relationship between the fixed preceding distance D of TIG torch ahead of MIG torch and the residual angular distortion is shown in Fig. 3. Positive distortion represents a concave upper surface. Residual angular distortion was small when the distance of preceding TIG torch D was about from 50 to 300 mm. These angular distortions are smaller than that when MIG welding was performed after cooling the temperature rise caused by TIG heating (TIG + MIG). And the angular distortion caused by only MIG welding was nearly the same as that when TIG heating was conducted just below MIG welding (D = 0 mm). These results suggest that the reduction in distortion caused by the reverse-side heating method is due not only to the opposite angular distortion caused by TIG heating, but also to some factor such as preheating effect.

The effect of TIG heat input Q_{TIG} on angular distortion is shown in Fig. 4. Heat input from the reverse-side TIG heating

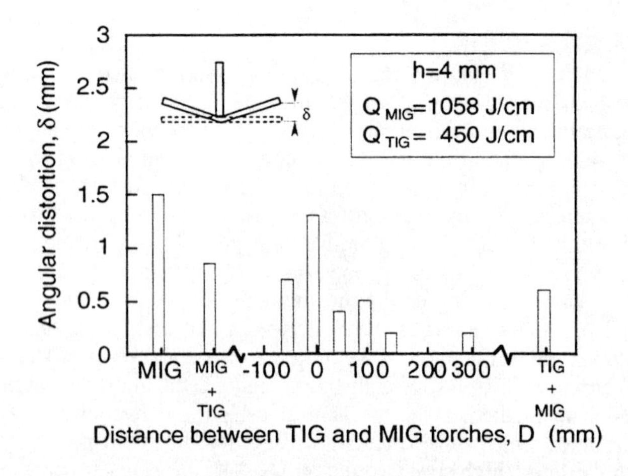

Fig. 3 Effects of distance from MIG torch to TIG torch on angular distortion obtained by experiment.

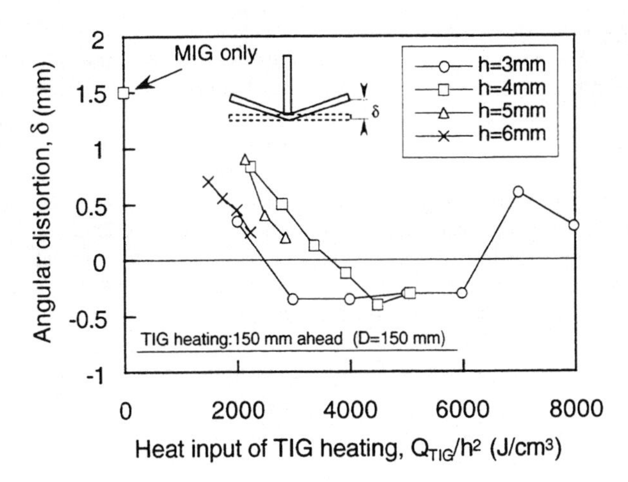

Fig. 4 Experimental effects of TIG heat input on angular distortion.

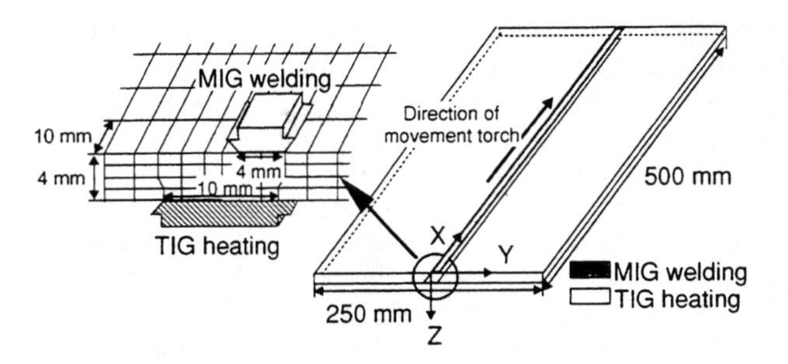

Fig. 5 Finite element analytical model for studying production mechanism of welding distortion by reverse-side heating method.

of about 3000 to 4000 J/cm^3 is appropriate for reducing angular distortion. It is considered that the effect of the opposite angular distortion by TIG heating may saturate at a certain TIG heat input, even though the preheating effect becomes larger with increasing TIG heat input Q_{TIG}.

The reverse-side heating method can reduce angular distortion in welded joints and to obtain an effective result it is important to choose a suitable distance *D* from MIG to TIG torches and adequate heat input Q_{TIG} for TIG heating.

As for the quality of welded joint, total heat input including reverse-side TIG heating is important. However, it is considered that the joint was sufficiently welded because the welding and heating speed was very high and therefore the total heat input did not become higher level which might affect to the quality of welded joint.

MECHANISM OF REDUCING WELDING DISTORTION

The effectiveness of the in-process welding distortion controlling method was confirmed by experiment. The mechanism of reducing welding distortion should be studied in detail for choosing appropriate executions and for applying to other welded components. Various heat conditions, which changes the distance between MIG and TIG torches, TIG heat input, and width of heating area, were examined by three-dimensional thermal elastic-plastic finite element analysis.

Analytical Model

The analytical model for studying the producing mechanism of welding distortion in reverse-side heating is shown in Fig. 5. The web is omitted from the model for simplification. This is because the main purpose of the numerical analysis is for investigation of the tendency and the production mechanism of welding distortion. MIG welding from the upper surface of the plate and TIG heating from the lower surface are simultaneously performed under keeping a

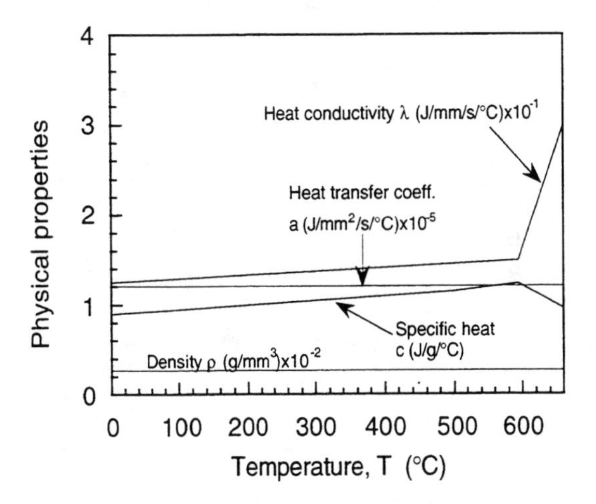

(a) Physical properties for heat conduction analysis

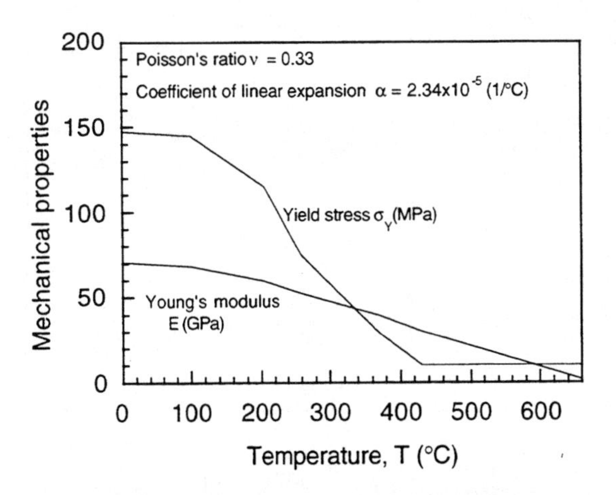

(b) Mechanical properties for thermal elastic-plastic stress analysis

Fig. 6 Physical and mechanical properties for finite element analysis.

fixed distance D. The welding speed v is 30 mm/s for both the MIG welding and TIG heating and simulated by moving the elements which are heat inputted. The condition of the bead-on-plate without filler metal was simulated in the analysis. The welding width for MIG was set to 4 mm and the heating width of TIG to 10 mm based on the results of experiment.

Physical and mechanical properties for heat conduction and thermal stress analyses are shown in Fig. 6. Melting effect are treated by lowering young's modulus and yield stress at high temperature. These properties were determined considering the characteristics of temperature dependence. Heat transfer from the plate surface was considered by heat flux in the temperature analysis for both MIG welding and TIG heating. Heat efficiency η = 0.6 for both MIG and TIG was determined by comparing the measured temperature distributions and histories with the computed velues, as shown in Fig. 7.

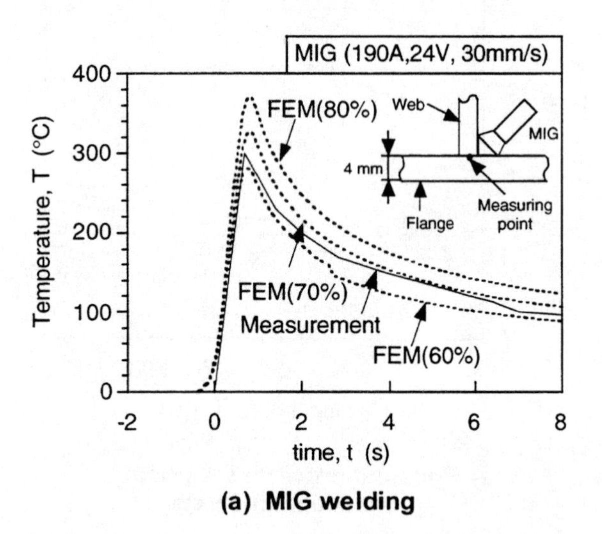

(a) MIG welding

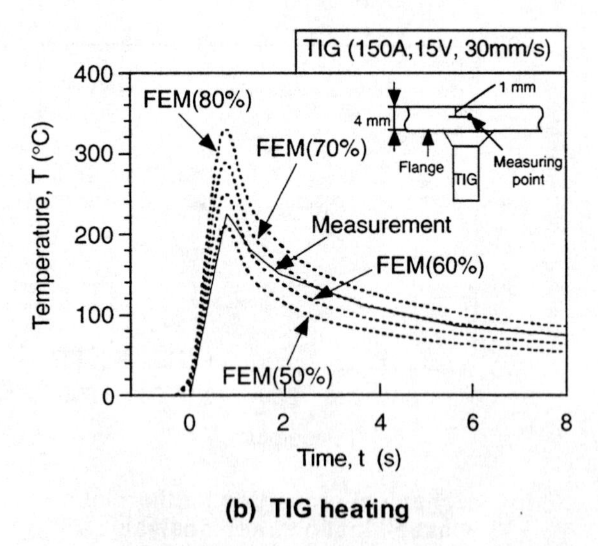

(b) TIG heating

Fig. 7 Comparison of temperature history for determining heat efficiency in heat conduction analysis.

Analytical Results

The angular distortion changes during welding at any position. Figure 8 shows the history of longitudinal distribution of angular distortion by only MIG welding. Angular distortion generates immediately when MIG torch reaches evaluation line and saturates soon after passing MIG torch. Residual angular distortion becomes larger at the end side of welding line than at the start side, but the tendency of generating distortion is the same at any position. Therefore, the angular distortion at the center transverse line (x = 250 mm) is used for the following study.

The history of angular distortion at the cross-section of the plate center is shown in Fig. 9. It is found that large angular

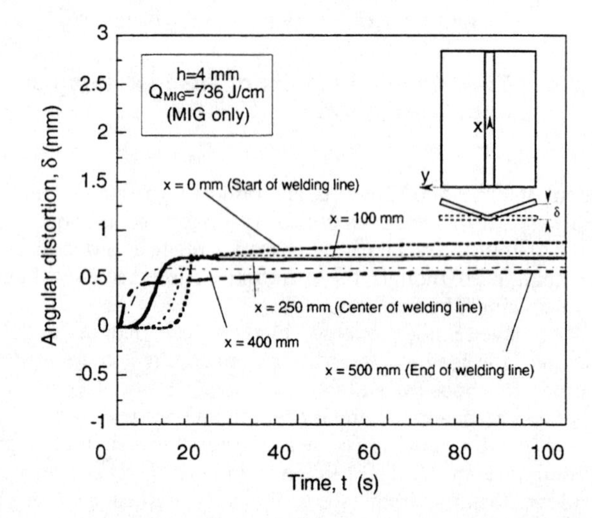

Fig. 8 History of longitudinal distribution of angular distortion by only MIG welding obtained by thermal elastic-plastic analysis.

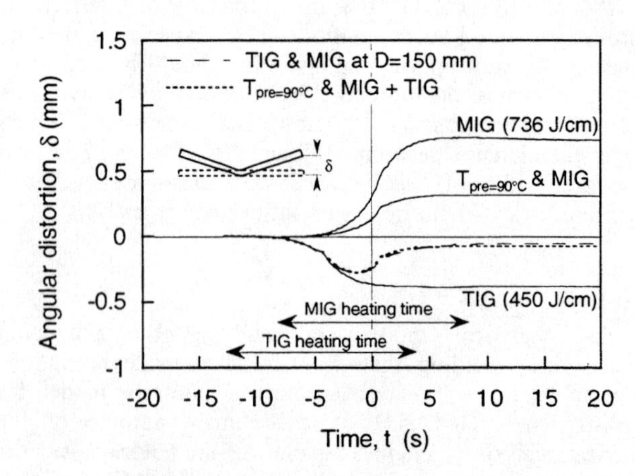

Fig. 9 History and production mechanism of angular distortion obtained by thermal elastic-plastic analysis.

distortion is produced by only MIG welding, and the opposite large distortion is generated by only reverse-side TIG heating. By performing reverse-side TIG heating under keeping 150 mm ahead of MIG welding, residual angular distortion can be reduced. Opposite distortion is produced by the TIG heating until the MIG torch arrives at the center of the plate, and then the distortion finally becomes almost zero due to MIG welding.

Preheating effect is considered to be also important, therefore the phenomena of the MIG welding after preheating at 90 ℃ was calculated. This is because that the plate just before MIG welding was heated by TIG heating and temperature distribution near the weldment was almost a flat 90 ℃. This distortion behavior and the behavior by only TIG heating are elastically superposed. The sum is nearly equal to the welding distortion behavior caused by the reverse-side heating method.

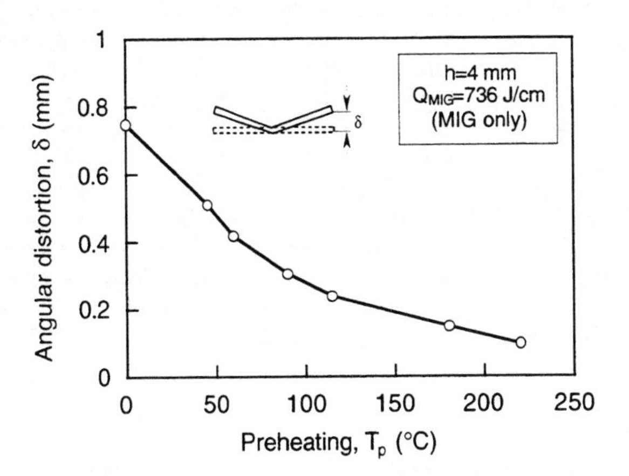

(a) Effect of preheating on angular distortion

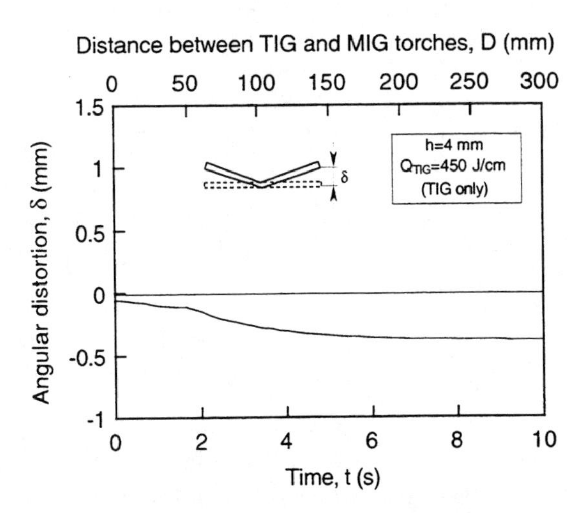

(b) History of angular distortion by only TIG heating

Fig. 10 Two reduction mechanism from results of angular distortion obtained by thermal elastic-plastic analysis.

This result suggests that the mechanism of distortion reduction in reverse-side heating method comes from both the effect of preheating by TIG heating and the effect of the opposite-side angular distortion by TIG heating.

It is considered that the mechanism for reducing angular distortion results from two main effects: the preheating effect, and the TIG heating on the reverse side, which provides the opposite angular distortion by thermal strain. The effect of preheating on angular distortion is shown in Fig. 10 (a). MIG welding was conducted after preheating. Angular distortion decreases with increasing preheating temperature. The temperature field near the weldment just before MIG welding becomes higher as the preceding distance *D* of TIG heating ahead of MIG welding decreases, which means that the preheating effect increases as the distance *D* decreases. On the other hand, the history of angular distortion at the center of the plate causes by only TIG heating is shown in Fig. 10 (b). The horizontal axis, time *t*, means the elapsed time until the start of MIG welding, and the equivalent distance *D* between TIG and MIG torches can be calculated from the conversion of welding speed. This figure shows that the effect of the opposite-side angular distortion can't be well obtained when the distance *D* is small. It is necessary to choose appropriate conditions considering both the preheating effect and the opposite-side distortion effect, though they have effects on the TIG heat input, as shown in Fig. 10.

The comparison of residual stress between only MIG welding and reverse-side heating method is shown in Fig. 11. Longitudinal stress is dominant in the model because the plate is only 4 mm thick and transverse stress hardly generates. Longitudinal stress distribution by only MIG welding is nearly the same as those by any conditions of TIG heating ahead of MIG welding. This is because the magnitude and distribution of residual stress are mainly determined by total heat input, and added TIG heat input affects the width of the tensile stress area a little larger, comparing the distribution by only MIG welding, but it does not affect to the magnitude of residual stress.

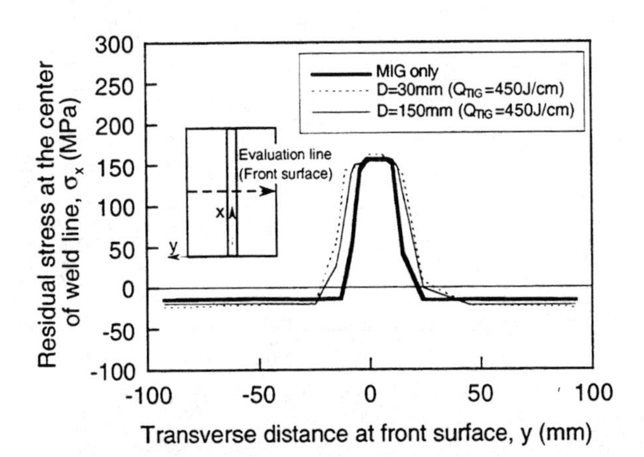

Fig. 11 Comparison of longitudinal residual stress at the center of weld line on front surface obtained by thermal elastic-plastic analysis.

Appropriate Conditions for Reverse-Side Heating

The effect of distance D between TIG and MIG torches on angular distortion is shown in Fig. 12. For Q_{TIG}= 450 J/cm, residual angular distortion becomes small when the preceding distance D is 100 to 300 mm, and the tendency matches the experimental results shown in Fig. 3. The angular distortion for MIG welding performed after cooling the temperature rise caused by TIG heating (TIG + MIG) is larger than that caused by reverse-side heating. The reverse-side heating method is more effective than the current method of bending to produce opposite distortion prior to welding. This is because it not only produces opposite distortion but also produces a preheating effect that reduces welding distortion. On the other hand, the effectiveness of the preceding TIG heating are not so large when TIG heat input Q_{TIG}= 225 J/cm. It is important that the appropriate TIG heat input should be chosen by comparing with MIG heat input.

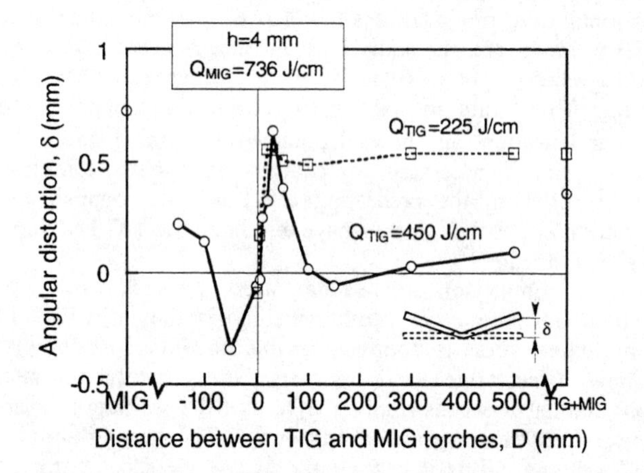

Fig. 12 Effects of distance between TIG and MIG torches on angular distortion obtained by thermal elastic-plastic analysis.

The effect of TIG heat input Q_{TIG} on angular distortion is shown in Fig. 13. The relationships between longitudinal shrinkage and TIG heat input are also represented. Residual angular distortion becomes smaller when an appropriate TIG heat input is used, and the tendency of the results agree with measured values shown in Fig. 4. Residual longitudinal shrinkage increases in proportion to TIG heat input Q_{TIG}. Therefore, it is suitable for controlling both angular distortion and longitudinal shrinkage to choose a condition of smaller heat input in the range of TIG heating that allows angular distortion to be reduced.

Other heating sources should also be investigated for usage of distortion control. Gas heating or induction heating generally offers a larger preheating area than TIG heating even for the same heat input. The effects of preheating area on angular distortion and longitudinal shrinkage in the reverse-side heating method are shown in Fig. 14, for two types of preheating area having the same heat input. The heat source with high energy density (type W) can effectively reduce the residual angular distortion with lower heat input. This result shows that a heat source with high energy density of reverse-side heating is effective for preventing angular distortion.

These theoretical results show that the in-process reverse-side heating method during welding effectively reduces angular distortion in fillet welds when suitable conditions are chosen. This method is readily applicable to the fabrication of welded structures with complicated shapes [35].

SUMMARY

An in-process control method can reduce welding angular distortion in fillet welds by performing reverse-side TIG heating at a fixed distance ahead of MIG welding. The effectiveness of this method was verified by experiment and by three-dimensional thermal elastic-plastic finite element analysis

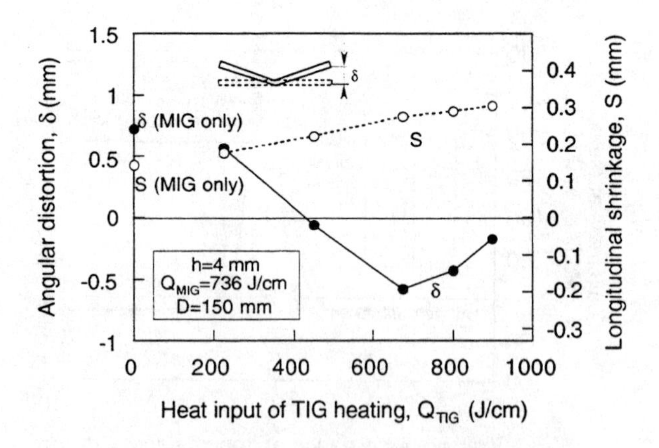

Fig. 13 Effects of TIG heat input on angular distortion and longitudinal shrinkage obtained by thermal elastic-plastic analysis.

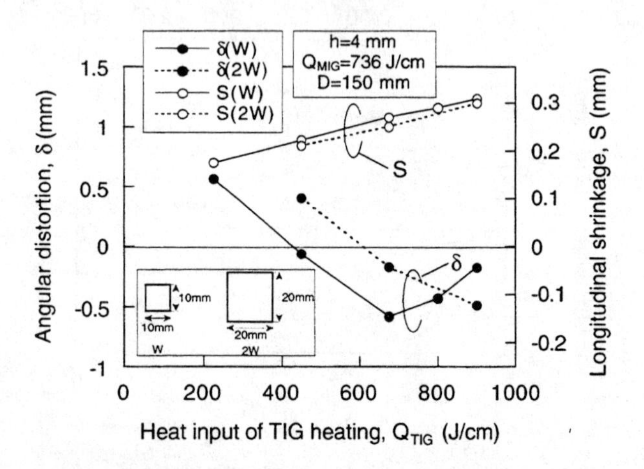

Fig. 14 Effects of preheating area on angular distortion and longitudinal shrinkage obtained by thermal elastic-plastic analysis.

in a fillet welded joint. Various heat conditions were examined to choose appropriate conditions and to study the mechanism for reducing welding distortion. The reducing mechanism for angular distortion results from two main effects: one is that TIG heating on reverse side provides the opposite angular distortion by thermal strain, and the other is the preheating effect by TIG heating. This method is effective at reducing angular distortion.

ACKNOWLEDGMENTS

The authors would like to gratefully acknowledge Mr. Shinji Takeno of Sky Aluminum, Ltd., and Mr. Yoshiki Toda and Mr. Nobuki Takahashi, graduate student of the Department of Manufacturing Science, Osaka University, for their kind help and cooperation in this study.

REFERENCES

1. E. Friedman, "Thermomechanical Analysis of the Welding Process Using the Finite Element Method," ASME *Journal of Pressure Vessel Technology*, Vol. 97, pp. 206-212, 1975.
2. E. F. Rybicki and R. B. Stonesifer, "Computation of Residual Stress due to Multipass Welds in Piping Systems," ASME *Journal of Pressure Vessel Technology*, Vol. 101, pp. 149-154, 1979.
3. J. H. Argyris and J. St. Doltsinis, "On the Natural Formulation and Analysis of Large Deformation Coupled Thermomechanical Problems," *Computer Methods in Applied Mechanics and Engineering*, Vol. 25, pp. 195-253, 1981.
4. J. B. Leblond, G. Mottet and J. C. Devaux, "A Theoretical and Numerical Approach to the Plastic Behavior of Steels During Phase Transformation," *Journal of Mechanical Physics of Solid*, Vol. 34, pp. 395-432, 1986.
5. J. A. Goldak, "Modelling Thermal Stresses and Distortion in Welds," *Proceedings of the Conference on Recent Trends in Welding Science and Technology*, Gatlinburg, Tennessee, ASM, pp. 71-82, 1989.
6. P. Tekrewal and J. Mazumder, "Transient and Residual Thermal Strain-Stress Analysis of GMAW," ASME *Journal of Engineering Materials and Technology*, Vol. 113, pp. 336-343, 1991.
7. M. Mochizuki, K. Enomoto, N. Okamoto, H. Saito and E. Hayashi, "Welding Residual Stresses at the Intersection of a Small Diameter Pipe Penetrating a Thick Plate," *Nuclear Engineering and Design*, Vol. 144, No. 3, pp. 439-447, 1993.
8. J. J. Janosch and M. Clerg , "Numerical Welding Simulation of Two Pipes — Determination of the Evolution of Residual Stresses During Proof Test Pressure," ASME-PVP, Orlando, Vol. 347, pp. 103-113, 1997.
9. P. Dong, J. Zhang and M. V. Li, "Computational Modeling of Weld Residual Stresses and Distortions — An Integrated Framework and Industrial Applications," ASME-PVP, San Diego, Vol. 373, pp. 311-335, 1998.
10. C. T. Karlsson and B. L. Josefson, "Three-Dimensional Finite Element Analysis of Temperatures and Stresses in a Single-Pass Butt-Welded Pipe," ASME *Journal of Pressure Vessel Technology*, Vol. 112, pp. 76-84, 1990.
11. S. Brown and H. Song, "Finite Element Simulation of Welding of Large Structures," ASME *Journal of Engineering for Industry*, Vol. 114, pp. 441-451, 1992.
12. L. Karlsson, L-E. Lindgren, M. Jonsson, L. Josefson and A. Oddy, "Modeling of Residual Stresses and Distortion Development," *Mathematical Modelling of Weld Phenomena 3*, H. Cerjak Ed., The Institute of Materials, London, pp. 571-589, 1997.
13. P. Michaleris, Z. Feng and G. Campbell, "Evaluation of 2D and 3D FEA Models for Predicting Residual Stress and Distortion," ASME-PVP, Orlando, Vol. 347, pp. 91-102, 1997.
14. P. Dong, J. K. Hong, J. Zhang, P. Rogers, J. Bynum and S. Shah, "Effects of Repair Weld Residual Stresses on Wide-Panel Specimens Loaded in Tension," ASME *Journal of Pressure Vessel Technology*, Vol. 120, pp. 122-128, 1998.
15. Y. Ueda and K. Fukuda, "New Measuring Method of Three-Dimensional Residual Stresses in Long Welded Joints Using Inherent Strains as Parameters — L_z Method," ASME *Journal of Engineering Materials and Technology*, Vol. 111, pp. 1-8, 1989.
16. M. Mochizuki, N. Saito, K. Enomoto, S. Sakata and H. Saito, "A Study on Residual Stress of Butt-Welded Plate Joint Using Inherent Strain Analysis," *Transactions of the 13th International Conference on Structural Mechanics in Reactor Technology*, Porto Alegre, Brazil, Vol. 2, pp. 243-248, 1995.
17. M. Mochizuki, M. Hayashi, M. Nakagawa, N. Tada and S. Shimizu, "A Simplified Analysis of Residual Stress at Welded Joints Between Plate and Penetrating Pipe," *JSME International Journal (Series A)*, Vol. 40, No. 1, pp. 8-14, 1997.
18. H. Murakawa, Y. Luo and Y. Ueda, "Inherent Strain as an Interface between Computational Welding Mechanics and Its Industrial Application," *Mathematical Modelling of Weld Phenomena 4*, H. Cerjak Ed., The Institute of Materials, London, pp. 597-619, 1998.
19. M. R. Hill and D. V. Nelson, "The Localized Eigenstrain Method for Determination of Triaxial Residual Stress in Welds," ASME-PVP, San Diego, Vol. 373, pp. 397-403, 1998.
20. Y. Ueda, K. Nakacho and T. Shimizu, "Improvement of Residual Stress of Circumferential Joint of Pipe by Heat-Sink Welding," ASME *Journal of Pressure Vessel Technology*, Vol. 108, pp. 14-23, 1986.
21. M. Mochizuki, M. Hayashi and T. Hattori, "Effect of Welding Sequence on Residual Stress in Multi-Pass Butt-Welded Pipe Joints," *Transactions of the Japan Society of Mechanical Engineers*, Vol. 62A, No. 604, pp. 2719-2725, 1996 (in Japanese).
22. C. L. Tsai, S. C. Park and W. T. Cheng, "Welding Distortion of a Thin-Plate Panel Structure — The Effect of Welding Sequence on Panel Distortion is Evaluated," *Welding Journal*, Welding Research Supplement, Vol. 78, No. 5, pp. 156s-165s, 1999.
23. S. Aoki, T. Nishimura, T. Hiroi and Y. Amano, "Reduction of Residual Stress of Welded Joint Using

Vibration," *Transactions of the Japan Society of Mechanical Engineers*, Vol. 61C, pp. 4800-4804, 1995 (in Japanese).
24. Ya. I. Burak, L. P. Besedina, Ya. P. Romanchuk, A. A. Kazimirov and V. P. Morgun, "Controlling the Longitudinal Plastic Shrinkage of Metal During Welding," *Automatic Welding*, No. 3, pp. 21-24, 1977.
25. Ya. I. Burak, Ya. P. Romanchuk, A. A. Kazimirov and V. P. Morgun, "Selection of the Optimum Fields for Preheating Plates Before Welding," *Automatic Welding*, No. 5, pp. 5-9, 1979.
26. Q. Guan, D. L. Guo and R. H. Reggatt, "Low Stress Non-Distortion (LSND) Welding — A New Technique for Thin Materials," *Welding in the World*, Vol. 33, No. 3, pp. 160-167, 1994.
27. T. Kumose, T. Yoshida, T. Abe and H. Onoue, "Prediction of Angular Distortion Caused by One-Pass Fillet Welding," *Welding Journal*, Vol. 33, pp. 945-956, 1954.
28. K. Masubuchi, *Analysis of Welded Structures — Residual Stresses, Distortion, and Their Consequences*, Pergamon Press, Oxford, pp. 320-322, 1980.
29. K. Terai, S. Matsui, T. Kinoshita, S. Yamashita, T. Tomoto, H. Horiuchi, K. Tsujimoto and K. Nishio, "Study on Prevention of Welding Deformation in Thin-Skin Plate Structures," *Kawasaki Technical Review*, No. 61, pp. 61-66, 1978 (in Japanese).
30. D. Q. Cole, *Development of Techniques for Controlling Warpage and Residual Stresses in Welded Structures*, Final Report, NAS8-21174, Harvey Engineering Laboratories, 1968.
31. M. Watanabe, Z. Murakami and M. Nishida, "Residual Stress Reduction in Welded Joint by Local Low-Temperature Cooling," *Preprints of the National Meeting of the Japan Welding Society*, Vol. 33, pp. 308-309, 1983 (in Japanese).
32. Q. Guan, C. X. Zhang and D. L. Guo, "Dynamic Control of Welding Distortion by Moving Spot Heat Sink," *Welding in the World*, Vol. 33, No. 4, pp. 308-312, 1994.
33. Q. Guan, "A Survey of Development in Welding Stress and Distortion Controlling in Aerospace Manufacturing Engineering in China," IIW Doc. X-1413-98, 1998.
34. M. Ohata, Y. Toda, M. Toyoda and S. Takeno, "Control of Welding Distortion in Fillet Welds of Aluminum Alloy," *Quarterly Journal of the Japan Welding Society*, Vol. 17, No. 2, pp. 301-309, 1999 (in Japanese).
35. M. Mochizuki and M. Toyoda, "Numerical Analysis of Residual Stress in Welded Structures Using Inherent Strain," *Document for the 5th International Seminar on Numerical Analysis of Weldability*, Graz-Seggau, Austria, No. S-V3, 1999.

PVP-Vol. 410-1, Assessment Methodologies for Preventing Failure:
Deterministic and Probabilistic Aspects and Weld Residual Stress
Volume 1
ASME 2000

A NEW COMPREHENSIVE THERMAL SOLUTION PROCEDURE FOR MULTIPLE PASS AND CURVED WELDS

Z. Cao and F. Brust
Battelle

A. Nanjundan, Y. Dong, and T. Julta
Caterpillar

ABSTRACT

In this paper, a new transient, explicit, and closed-form comprehensive thermal solution procedure (CTSP) is developed to simulate the thermal process of welding in a complex welded structure. The solution can deal with not only multiple-pass and curved welds, but also different types of complex welded structures and their boundary conditions. A series of examples has been calculated for the practical welded structures. Also, the predicted results have been compared with the experimental measurements. A good agreement has been achieved between prediction and measurement. It has been demonstrated that tremendous computational time has been saved using the thermal solution than using FEM. Meanwhile, the accuracy from the solution is very competitive to that from the FEM and usually better when very coarse meshes are used because of its mesh independence.

INTRODUCTION

Over recent years, numerical thermal methods have received increasing attention in the welding community. However, one of the major impediments using numerical methods is that the numerical thermal solutions can be extremely cumbersome to implement, and costly, especially when curved and multiple-passe welds are involved because a more refined mesh is required. Meanwhile, analytical thermal solutions are also of interests for many other applications, such as welding metallurgy prediction, and welding process control.

Many investigations have been studied on analytical thermal modeling and analyses of welding [1-7]. The majority of the studies were concerned with the quasi-steady state (QSS). Rosenthal [1,2] proposed an analytical QQS temperature solution using a point or line heat sources on a semi-infinite body. Rykalin [3] developed a similar solution including transient effects. Eagar and Tsai [4] modified Rosenthal's theory to include a two-dimensional surface Gaussian distributed heat source and obtained an analytical solution on a semi-infinite body with this moving sources. In order to simulate welds with deeper penetration, Kasuya [5] derived a comprehensive solution using a group of point heat sources through thickness direction to predict thermal history and HAZ shapes. However, location and intensity of each heat source are depended upon the weld profiles. Using the conformal mapping technique, Jeong and Cho [6] introduced an transient analytical solution for a fillet weld joint based on a 2D Gaussian heat source. Recently, Nguyen [7] developed a 3D transient solution using Goldak's [8] 3D double ellipsoidal moving heat source. Though the solution can match weld geometry on top surface by adjusting distributed parameters in different directions, it fails to match weld profiles.

All above solutions are for straight line and single pass welds. Also, all the transient solutions have to numerically solve an integral, which can be troublesome and slow down the computation and reduce accuracy. In this paper, a closed-form transient solution is derived for the first time. The solution can be used for curved and multiple-pass welds. Experimental validation is performed. A good agreement between measurement and prediction has been achieved.

THEORY

Constant thermal properties and adiabatic boundary conditions are assumed in the present study. The coordinate system used in the study is shown in Figure 1. The heat transfer equation, which can be used to describe the welding

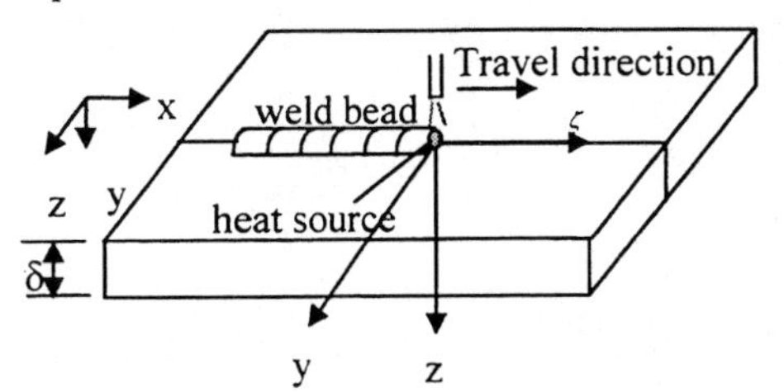

Fig. 1 Coordinate system used in the study

process, is written with respect to Cartesian coordinate system as follows:

$$\frac{\partial}{\partial x}\left(\lambda\frac{\partial T}{\partial x}\right)+\frac{\partial}{\partial y}\left(\lambda\frac{\partial T}{\partial y}\right)+\frac{\partial}{\partial z}\left(\lambda\frac{\partial T}{\partial z}\right)-\rho C_p v\frac{\partial T}{\partial x}=\rho C_p\frac{\partial T}{\partial t} \quad (2.1)$$

where λ is conductivity, ρ is density, C_p is specific heat, v is welding speed, and T is temperature.

If the heat source can be assumed as a point-source, the heat flux through the surface of sphere drawn around the source must tend to the value of the total heat Q delivered to a workpiece, as the radius of the sphere tends to zero.

$$-4\pi R^2 \lambda \frac{\partial T}{\partial R} \rightarrow Q, \text{ as } R = \sqrt{\zeta^2 + y^2 + z^2} \rightarrow 0 \quad (2.2)$$

where $\zeta = x - vt$, with the heat source traveling in the x direction. Initial Conditions in the workpiece are:

$$T = T_\infty, \text{ at } t = t_0 \quad (2.3)$$

where T_∞ is the ambient temperature. If an infinite-size workpiece is assumed, it then follows:

$$T = T_\infty, \text{ at } R \rightarrow \infty \quad (2.4)$$

Therefore, the analytical solution for equation (2.1) together with its initial condition and boundary conditions from equation (2.2) to (2.4) can be obtained:

$$T(x,y,z,t) = \frac{2q}{C_p \rho (4\pi\alpha)^{3/2}} \exp\left(-\frac{v\zeta}{2\alpha}\right) \int_0^t \frac{dt''}{t''^{3/2}} \exp\left(-\frac{v^2 t''}{4\alpha} - \frac{R^2}{4\alpha t''}\right) dt \quad (2.5)$$

where $\zeta = x - vt$ and $R^2 = \zeta^2 + y^2 + z^2$.
After a detailed derivation, the equation (2.5) can be rewritten as:

$$T(\zeta,y,z,t) = T_{SS}(\zeta,y,z) \cdot F(\zeta,y,z,t) \quad (2.6)$$

where $T(\zeta,y,z,t)$ is the transient temperature field, $T_{ss}(\zeta,y,z,t)$ is the quasi-static state temperature field, $F(\zeta,y,z,t)$ is the transient transform function. These functions can be written as:

$$T_{SS}(\zeta,y,z,t) = \frac{Q}{4\lambda\pi R} \exp\left(-\frac{v\zeta + vR}{2\alpha}\right) \quad (2.7)$$

$$F(\zeta,y,z,t) = \frac{1}{2}\left[\exp\left(\frac{vR}{\alpha}\right)\left(1 + erf\left(\frac{vt+R}{2\sqrt{\alpha t}}\right)\right) + \left(1 + erf\left(\frac{vt-R}{2\sqrt{\alpha t}}\right)\right)\right] \quad (2.8)$$

where $\alpha = \frac{\lambda}{\rho C_p}$ and erf is error function.

Equation (2.6) together with equations (2.7) and (2.8) can be used to obtain the transient temperature field for a moving heat source in an infinite body.

Boundary Conditions

Equation (2.6) is obtained by assuming an infinite-size workpiece. In order to consider boundary effects at the top and bottom surfaces for a finite thickness plate, it is assumed that the heat loss from the surfaces is negligible in comparison with that due to conduction in an infinite body, i.e., $-\lambda \partial T / \partial z = 0$. As shown in Figure 2, to prevent heat loss from the top surface, an auxiliary heat source (referred as 2)

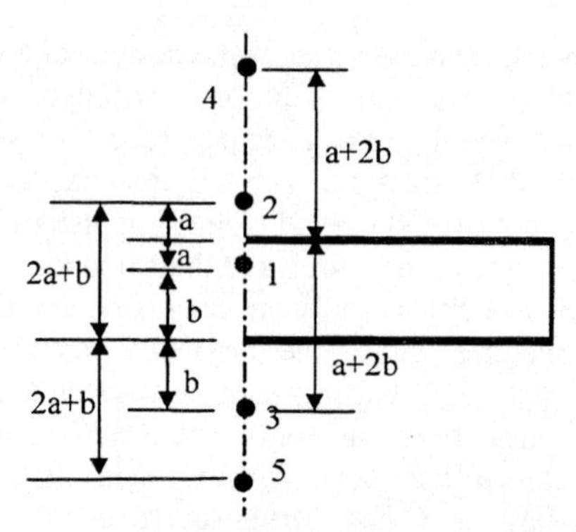

Fig. 2 Heat source reflection to maintain boundary conditions

with the same heat input and travel speed as the original heat source is placed at the reflected mirror position. Similarly, the auxiliary heat source 3 is placed for the bottom surface. Once the auxiliary source 3 is imposed, another auxiliary source 4 needs to be placed above the top surface so that the top surface is insulated. As such, a number of the heat sources are required to obtain a reasonable temperature solution to satisfy the top and bottom boundary conditions. The temperature solution then becomes:

$$T(\zeta,y,z,t) = \sum_{i=1}^{i=m} \left(\sum_{n=0}^{n=\infty} \left(f(\zeta,y,z-z_{n,i}^t,t) \cdot T_{SS}(\zeta,y,z-z_{n,i}^t)\right) + \sum_{n=1}^{n=\infty} \left(f(\zeta,y,z-z_{n,i}^b,t) \cdot T_{SS}(\zeta,y,z-z_{n,i}^b)\right) \right) \quad (2.9)$$

where

$$T_{SS}(\zeta,y,z-z_{n,i}^t,t) = \frac{Q_i}{4\lambda\pi R_{n,i}^t} \exp\left(-\frac{v\zeta + vR_{n,i}^t}{2\alpha}\right)$$

$$F(\zeta,y,z-z_{n,i}^t,t) = \frac{1}{2}\left[\exp\left(\frac{vR_{n,i}^t}{\alpha}\right)\left(1 + erf\left(\frac{vt+R_{n,i}^t}{2\sqrt{\alpha t}}\right)\right) + \left(1 + erf\left(\frac{vt-R_{n,i}^t}{2\sqrt{\alpha t}}\right)\right)\right]$$

$$T_{ss}(\zeta, y, z - z^b_{n,i}, t) = \frac{Q_i}{4\lambda\pi R^b_{n,i}} \exp(-\frac{v\zeta + vR^b_{n,i}}{2\alpha})$$

$$F\left(\zeta, y, z - z^b_{n,i}, t\right) = \frac{1}{2}\left[\exp\left(\frac{vR^b_{n,i}}{\alpha}\right)\left(1 + erf\left(\frac{vt + R^b_{n,i}}{2\sqrt{\alpha t}}\right)\right) + \left(1 + erf\left(\frac{vt - R^b_{n,i}}{2\sqrt{\alpha t}}\right)\right)\right]$$

$$z^t_{n,i} = \left[\text{truncate}\left|\frac{n}{2}\right| \cdot 2\delta + z_i\right](-1)^n, \quad n = 0,1,...,\infty \quad \text{(on the top)}$$

$$z^b_{n,i} = \left[\text{truncate}\left|\frac{n+1}{2}\right| \cdot 2\delta - z_i\right](-1)^{n+1}, \quad n = 1,2,...,\infty \quad \text{(on the bottom surface)}$$

where n is the number of auxiliary heat sources and i is the number of distributed heat sources through thickness δ.

Transient Heating and Cooling

Transient heating can be simulated by using the transient solution in Equation (2.6). To obtain transient cooling solutions (i.e., after heat source is turned off), the following solution scheme can be devised as shown in Figure 3. An imaginary heat source and an imaginary heat sink are activated once the actual heat source is extinguished (see Figure 3 (b)). The resultant temperature is contributed by both the heat source and the heat sink (see Figure 3 (c)):

$$T(\zeta, y, z, t) = T_{SS}(\zeta, y, z) \cdot \left(F(\zeta, y, z, t) - F(\zeta, y, z, t - t_0)\right) \quad (2.10)$$

where t_0 is welding duration time.

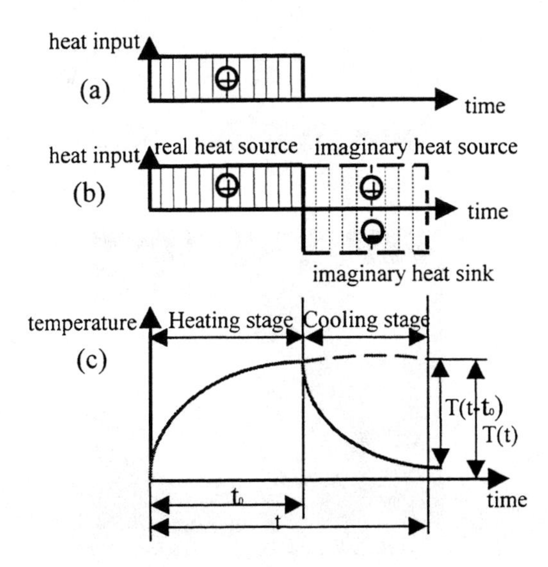

Fig.3 Sketch of temperature calculation during cooling stage (a) Actual heat source; (b) Imaginary heat source and imaginary heat sink; (c) Resultant temperature history

Curved Welds

Any curves can be approached by a series of small straight-line segments. Temperature distributions by each individual segment can be calculated by the equation (2.10). Therefore, the overall temperature distribution of the curve is contributed by every segment on the curve:

$$T(\zeta, y, z, t) = \sum_{i=1}^{i=n} T_{ss}(\zeta_i, y_i, z_i) \cdot \left(F(\zeta_i, y_i, z_i, t) - F(\zeta_i, y_i, z_i, t - t_i)\right) \quad (2.11)$$

where n is the total number of segments for the curve, ζ_i, y_i, z_i are the moving coordinates of the ith segment, and t_i is the welding duration time of the ith segment.

Multiple Passes

Each pass can consist of a set of straight-line segments. The overall temperature of the multiple-pass weld is contributed by every pass at the weld:

$$T(\zeta, y, z, t) = \sum_{j=1}^{j=m}\left[\sum_{i=1}^{i=n} T_{SS}\left(\zeta_{i,j}, y_{i,j}, z_{i,j}\right) \cdot \left(F\left(\zeta_{i,j}, y_{i,j}, z_{i,j}, t\right) - F\left(\zeta_{i,j}, y_{i,j}, z_{i,j}, t - t_{i,j}\right)\right)\right] \quad (2.12)$$

where m is the total number of passes on the weld.

RESULTS AND DISCUSSIONS

The CTSP has been programmed as a software package written in FORTRAN 77, which is used to predict temperature history. The package has been directly interfaced with finite element codes, such as ABAQUS for weld residual stress/distortion predictions. It has the following primary features: (1) simple user input format, (2) both shell and solid model, (3) automatic weld path tracing and time step determination, and (4) minimized output file size.

Figure 4 shows the temperature distribution comparison between the transient and quasi-steady state (QSS) solutions. (a1) and (b1) are the top view, and (a2) and (b2) are the longitudinal cross section. It indicates that overall temperature from the transient solution is lower than that from the QSS solution since arc from the transient solution is started at x=20mm while arc from the QSS is assumed to start from infinite farther location. Therefore, for a limited size of the weld workpiece, the transient solution has to be used in order to obtain a more realistic temperature solution. Figure 5 shows temperature distribution during the cooling stage using the transient solution of equation (2.10). Here arc starts at x=20mm and ends at x=80mm. In this case, if the

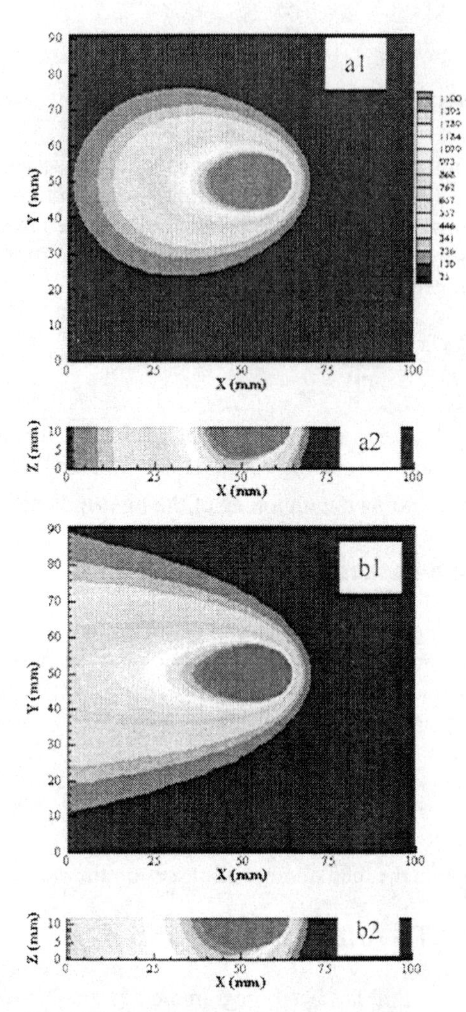

Fig. 4 Temperature comparison between transient and quasi-steady state solutions: (a1) and (a2): transient (start at x=20mm) (b1) and (b2) quasi-steady state

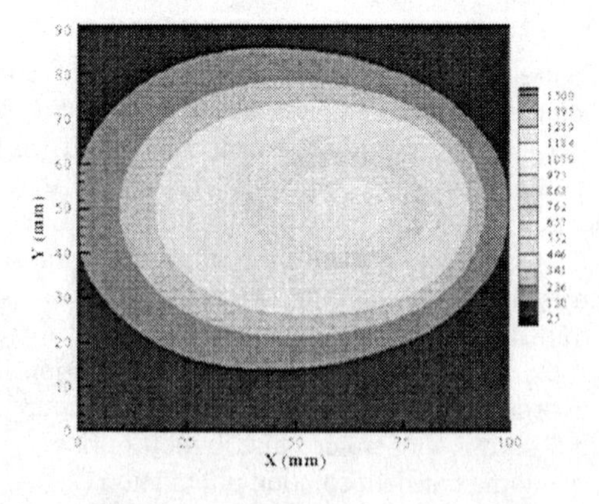

Fig. 5 Temperature distribution during cooling: start at x=20mm and end at x=80mm

QSS solution is used, the overall heat input Q in the equation (2.7) is equal to zero. So temperature solution is always zero during the cooling stage if the equation (2.10) is used, which means that it will be room temperature everywhere once the cooling stage starts. Obviously, this is not true, which demonstrated that the transient solution has to be used for the entire welding process, i.e., both heating and cooling stages.

Figure 6 shows temperature distribution of an arbitrary-shaped weld using the equation (2.11). The real weld path is marked by the solid white line. The arc starts at x=25mm and y=20mm. It clearly captured temperature accumulation from the entire weld path. Figure 7 shows temperature distribution during the first pass and the third pass using the equation (2.12). The first pass starts at y=50 and x=20 and ends at x=80, and the second comes back to the starting point. The third pass starts and ends at the same locations as the first pass. From the Figure 7(b), it can be seen that there are overall three temperature regions contributed by the three passes, respectively. Therefore, the final temperature field is accumulated from the all three passes.

To consider boundary conditions, a group of reflected heat sources is used to balance the heat loss from the top and bottom surfaces (see Figure 2). Effects of reflected heat

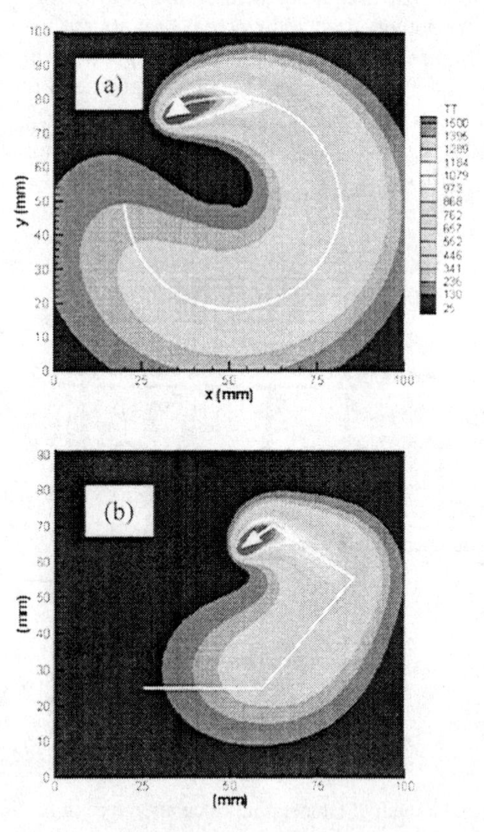

Fig. 6 Temperature distribution of curved welds (a) circular welds; (b) arbitrary-shaped welds

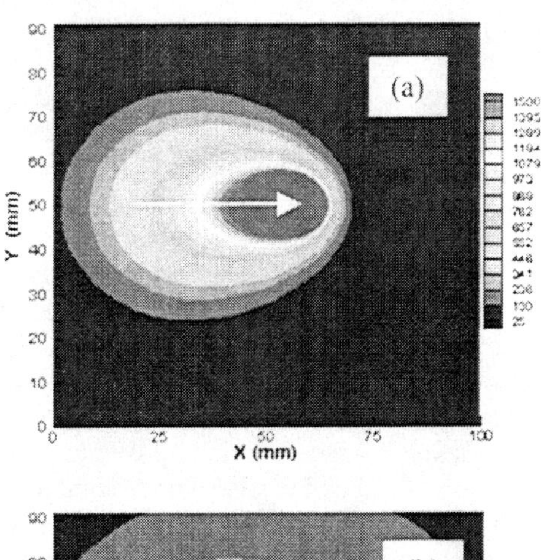

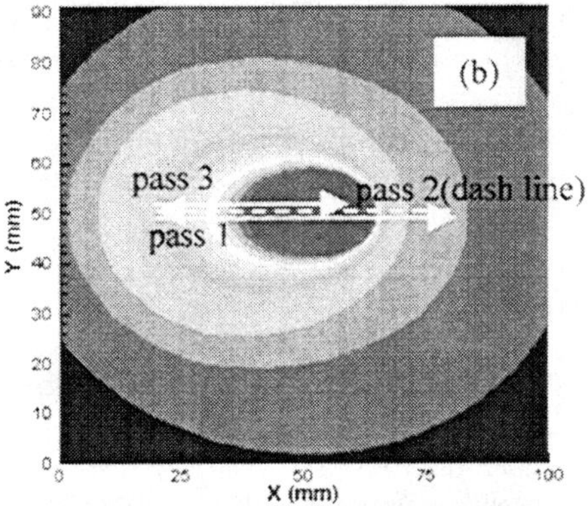

Fig. 7 temperature distribution during multiple pass welds: (a) the first pass; (b) the thirst pass

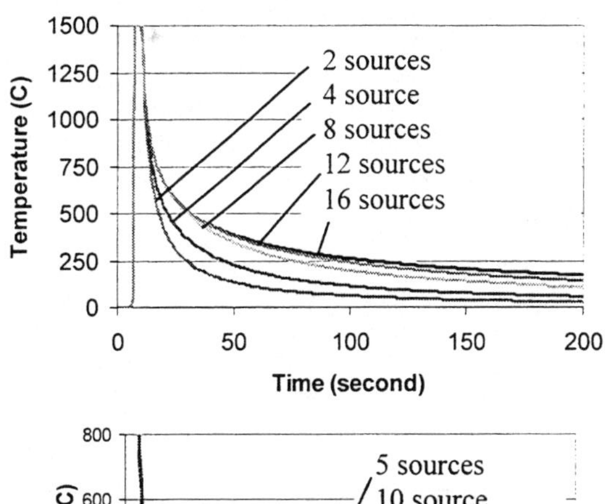

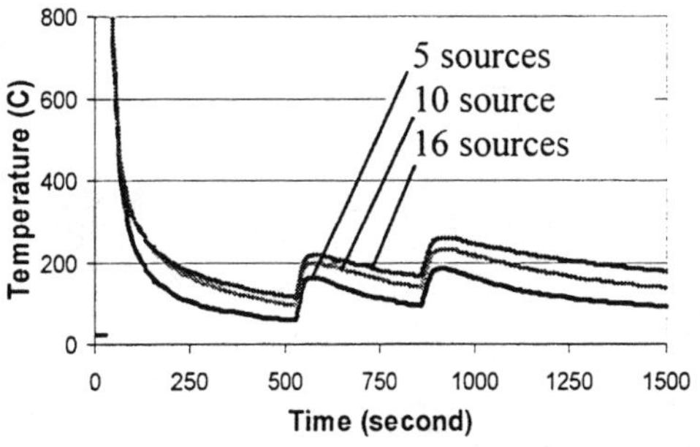

Fig. 8 Effect of heat source numbers on temperature history: (a) single pass; (b) multiple pass (low carbon steel, 11200W,6.8mm/s, 5mm from the weld center line, Bead on plate)

source numbers on the temperature history for a single and multiple passes is showed in Figure 8, respectively.

For the single pass weld, the cooling speed decreases and difference between them becomes smaller as the number of heat sources increase. Obviously, the more the number of heat sources, the better the adiabatic boundary condition is served. On the other hand, the computational speed will be slow down when the number of heat sources is increased. Therefore, as long as the computational accuracy is satisfactory, the smaller number of heat sources is desired in order to speed up the calculation. Compared with the single pass case, the temperature history difference between using 10 and 16 sources for the multiple pass case becomes bigger as the number of passes increases since low temperature accumulation from different passes becomes more and more significant. Therefore, for the multiple pass applications, the more heat sources are needed in order to achieve the same accuracy as the single pass case.

An example of a real welded structure is simulated using the CTSP. Temperature distribution on the complex welded structure during welding process is showed in the Figure 9. Obviously, the weld paths are curved arbitrarily. The calculation is carried out on SUN ULTRA2 CREATOR3D. Also, an ABAQUS thermal analysis is

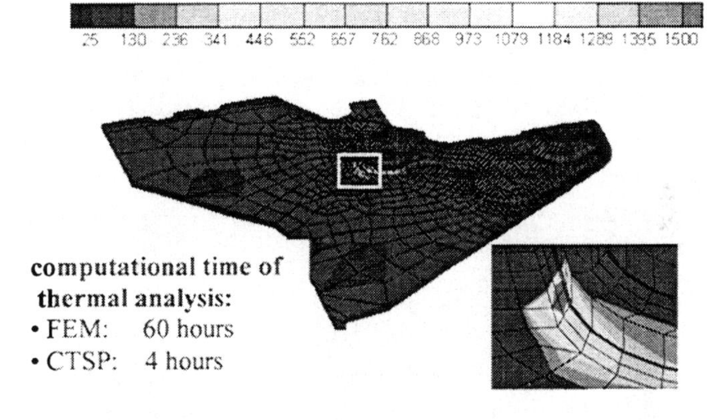

Fig. 9 Temperature distribution on a complex welded structure during welding

performed on the same model. To complete the thermal analysis, it takes 60 hours for FEM and 4 hours for the CTSP. Thus, the computational time has been saved tremendously using the new solution. More importantly, unlike the FEM analysis, the analytical solution does not have convergent problems because of its mesh independence.

In order to validate the thermal solution, the analytical results are compared with the FEM results and the experimental measurement. Temperature history at three nodal points are compared between FEM and the analytical solution (as shown in Figure 10). The nodal 25 is located at the weld root, and the nodal 31 and 45 are located farther from the weld. The temperature differences on these three points between FEM and the analytical solution are all small, which means that the analytical thermal solution is as good as FEM analysis.

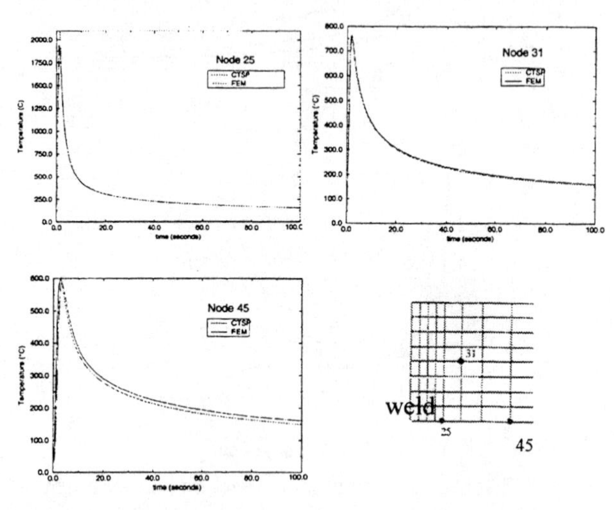

Fig. 10 Predicted temperature history between CTSP and FEM

Temperature comparison between the analytical solution and thermocouple measurements is showed in the Figure 11. The thick lines represent the prediction and the thin lines indicate the experiments. It can be seen that the temperature comparison at these two points is reasonably good, especially the peak temperature matches very well.

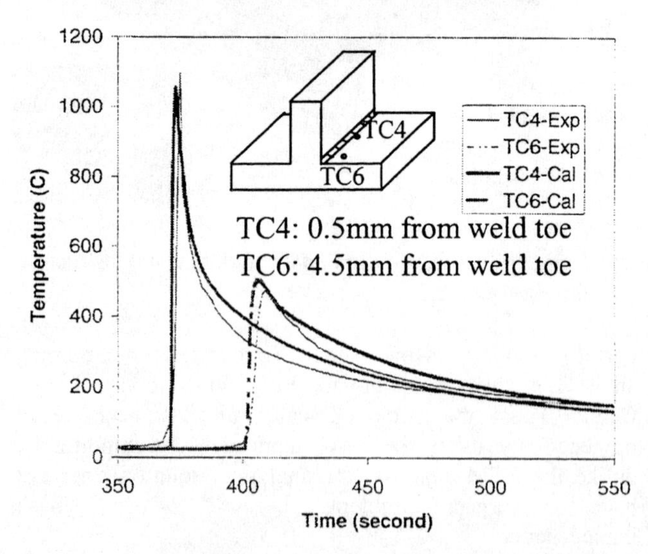

Fig. 11 Temperature history comparison between prediction and measurement

CONCLUSIONS

A new transient, explicit, and closed-form comprehensive thermal solution procedure has been developed to simulate temperature history for the curved and multiple pass welds in complex welded structures. The technique which applies one imaginary heat source and one imaginary heat sink has been successful to technique has also been adopted to treat multiple pass and curved welds. Boundary conditions have been served using imaginary reflected-heat sources. The more the sources, the better boundary conditions are satisfied. Compared with FEM thermal analysis, the solution provides not only high computational speed, but also good accuracy. The overall solution scheme has been programmed in a software package, which has direct interfaces with commercial FEM software for weld residual stress/distortion predictions. Comparisons between the solution and FEM and experiments have been made and good agreements have been achieved.

REFERENCES

1. D. Rosenthal, "Mathematical theory of heat distribution during welding and cutting", Welding Journal, Vol.20 (5), 1941, p.220-234s.
2. D. Rosenthal and M. Cambridge, "The theory of moving source of heat and its application to metal treatments", Trans. ASME, Vol.68 (11), 1946, p.849-866.
3. N. N. Rykalin, "Calculations of Thermal Processes in Welding", Mashgiz, Moscow, 1951.
4. T. W. Eager and N. S. Tsai, "Temperature fields produced by traveling distributed heat sources", Welding Journal, Vol. 62 (12), 1983, p.346s-355s.
5. T. Kasuya and N. Yurioka, "Prediction of welding thermal history by a comprehensive solution", Welding Journal, Vol.72 (3), 1993, p.107-115s.
6. S. K. Jeong and H. S. Cho, "An analytical solution to predict the transient temperature distribution in fillet arc welds", Welding Journal, 76 (6), 1997, p.223-232s.
7. N. T. Nguyen, A. Ohta, K. Matsuoka, N. Suzuki and Y. Maeda, "Analytical solutions for transient temperature of semi-infinite body subjected to 3-D moving heat sources", Welding Journal, Vol. 78 (8). 1999.

ACKNOWLEDGMENT

This work is sponsored by FASIP program in the Advanced Technology Program (ATP) funded by NIST, of the Department of Commerce.

PVP-Vol. 410-1, Assessment Methodologies for Preventing Failure:
Deterministic and Probabilistic Aspects and Weld Residual Stress
Volume 1
ASME 2000

EFFECTS OF REPAIR WELD LENGTH ON RESIDUAL STRESS DISTRIBUTION

J. ZHANG, P. DONG
Battelle Memorial Institute, Columbus, Ohio

P. J. BOUCHARD
British Energy, Gloucester, UK

ABSTRACT

This paper discusses residual stress distributions induced by repairing a stainless steel girth weld in a 19mm thick pipe of outer diameter 541mm. In particular, the effects of repair weld circumferential length are examined using finite element modeling. Results for three different repair lengths are presented having circumferential angular spans of 20 degrees (short repair), 57 degrees (medium repair) and 114 degrees (long repair). A special 3-D shell element model is used which facilitates the simulation of multi-pass welds in 3-D piping components. The results shed light on a number of important 3-D residual stress features associated with repairs. Outer surface axial residual stresses in the weld and adjacent base material are tensile along the length of the repair, reach maxima values near the arc start/stop positions and then drop into compression beyond the repair ends. The short repair develops the highest axial tensile stresses due to the overlay of start/stop effects. The circumferentially remote residual stresses are unaffected by the repairs. At mid-length of the repair, profiles of axial stress along the pipe show tensile peaks at ≈40 mm away from the weld centre-line; these peaks decrease in magnitude with increasing repair length. However, the medium repair axial stresses show the greatest range of influence along the pipe. The pre-existing original girth weld residual stresses have very little effect on the repair residual stress characteristics. Finally, residual stress measurements on mock-up components are discussed which confirm the validity of the finite element methods used.

1.0 INTRODUCTION

Repair weld induced residual stresses can be a significant structural integrity concern for the safe operation of nuclear piping and vessel components; for example, they can cause localized creep damage [1] and stress corrosion cracking. The characteristics of residual stress fields associated with repair welds are significantly different from those in the original girth weld, as illustrated by numerical predictions and recent measurements [2, 3 and 4]. Factors influencing the stress field include the repair geometry (length and depth), the weld repair procedure (heat input, pass sequence etc.), the component wall thickness and the global stiffness of the structure. For accurate structural integrity assessments of safety-related plant, a thorough understanding of the residual stress field and how it affects component life is required. When the role of residual stress has been quantified, strategies can be developed to improve component life and thereby reduce design and operating costs, or to underwrite safe operating conditions and inspection regimes for degraded plant. Weld residual stresses in real structures have complex distributions and exhibit significant variability. Finite element modeling has become a very effective tool for quantifying residual stress levels and understanding factors affecting the spatial distribution. However, residual stress distributions associated with limited-length repair welds in pipes often exhibit strong 3-D characteristics that cannot be effectively captured by traditional 2-D finite element weld simulation models, for example:

- Circumferential variations in stresses
- Repair start/stop effects
- Repair length effects
- Peak stress locations along the circumference
- Axial region of influence of residual stresses
- Repair effects on stresses circumferentially remote from the repair area
- Effects of weld sequence and arc traveling directions
- Non-axisymmetric local deformations

In this paper, a series of finite element analyses are reported which investigate the 3-D characteristics of residual stress fields induced by part-circumferential repairs to a pipe girth weld. In particular, the effects of repair weld circumferential length are examined, covering angular spans of 20 degrees (short repair), 57 degrees (medium repair) and 114 degrees (long repair). A special 3-D shell finite element modeling approach [5] is used which facilitates the simulation of multi-pass welds in 3-D piping components. For completeness, the basic concept and implementation of the special 3-D composite shell element model is summarized in section 2. This is followed by a description of the pipe weld repair finite element analyses in section 3 and presentation of the results with detailed discussion in section 4. Finally, in section 5, the validity of the weld simulation approach is discussed in the context of experimental measurements on mock-up components.

2.0 SPECIAL 3-D COMPOSITE SHELL ELEMENT MODEL

A special 3-D composite shell element approach has been used in the present analyses. It was initially developed in earlier work [5] to overcome the computational difficulty of performing a full 3-D multi-pass weld simulation with solid elements. The new approach reduces the solution time, simplifies the simulation procedure, and facilitates both model generation and post-processing. It is particularly suited to analysis of repair welds in pipe/vessel components.

2.1 Concept

The basic concept of the composite shell element model is illustrated in Figure 1. A multi-pass weld in a thick-shell structure is idealized as a laminated composite shell composed of multiple weld layers that are sequentially deposited. Each weld layer can be regarded as a lumped weld pass (even though multiple passes in one layer can also be represented and simulated). The weld groove geometry and the material mismatch between the weld metal and base material are modeled by assigning weld material properties to the appropriate weld layers of the elements within the weld area. To do so, a refined finite element mesh is required in the weld region. The resulting weld groove profile in the cross-section of the model has a step-wise geometry, as shown in Figure 1.

2.2 Implementation

The special 3-D composite shell element approach has been successfully implemented by the combination of commercial finite element codes, e.g., ABAQUS [6], with purpose written routines and interfaces. A brief description of the finite element formulation for the composite shell element and a specific weld material constitutive model can be found in [5, 7]. To overcome some of the limitations in the commercial codes, further development in both thermal and mechanical models have been made as described below.

a) Thermal Model. Welding heat flow simulation for the 3-D composite shell model is based on analytical solutions by Rosenthal [8] and Kasuya [9] (both steady-state and transient solutions) with modifications to model specific welding processes. The reason for choosing analytical solutions rather than numerical ones is to obtain realistic temperature gradients through the shell thickness. In a multi-pass repair weld, the temperature gradient through the section thickness is highly non-linear as each weld pass is deposited. Numerical thermal solutions based on shell theory tend to smear out these non-linear temperature gradients because they are typically obtained by applying body or surface heat fluxes on the shell element. It is well known that residual stresses are strongly dependent on the temperature gradients. Analytical thermal solutions can provide much more realistic temperature gradients through the shell thickness. To account for boundary effects on the analytical solutions, a series of reflecting heat sources are employed. Contributions by Rykalin [10] for start/stop effects of the weld torch are also incorporated. This analytical thermal model not only permits accurate temperature predictions but also provides much faster solution speeds than finite element ones. The resulting temperature solutions are three dimensional and independent of the finite element mesh. To feed the temperature solutions into the shell finite element model for stress analysis, a unique automatic mapping procedure has been developed. The output is a file containing the temperature history associated with the deposition of each weld layer at each sectional integration point of every element.

b) Mechanical Model. For the mechanical analysis, ABAQUS's built-in solution procedure cannot directly model the process of sequential weld layer deposition as demanded by the composite shell model. A special ABAQUS user material subroutine has been developed to cope with this difficulty. It uses a unified weld constitutive model [7] and incorporates a weld layer deposition scheme. The mathematical modular structure of the developed weld material routine is shown in Figure 2.

In the special user subroutine, the process of sequential weld layer deposition is simulated using a consistent layer activation/deactivation scheme. This scheme keeps track of the movement of the weld torch and updates the status of weld element and layer (deposited or not). For a weld layer that is not yet deposited at a given time, a very small material stiffness (Jacobian matrix) is assigned so that it would not affect any deformation mechanism on the shell section. This process is called layer deactivation. The deactivated layers can be regarded as virtual layers that are not actually present. After the weld layer is deposited, it is reactivated and temperature-dependent material properties are assigned. Through such a consistent layer activation/deactivation process, the effects of sequential weld metal deposition in a multi-pass weld can be properly modeled. For a repair weld, this process can also be used to simulate groove excavation before repair welding.

3.0 PIPE REPAIR WELD ANALYSIS

This section describes the series of pipe repair weld analyses performed using the special 3-D composite shell element modeling approach described above. The analyses are focused on assessing the effects of repair weld length on the residual stress field. The significance of the original weld residual stresses on the repair stress fields are also investigated.

3.1 Analysis Model.

The 3-D composite shell finite element model used for the pipe repair welds is shown in Figure 3. The pipe is made of AISI Type 316H stainless steel, and has an initial J-preparation manual metal arc girth weld at mid-length [11]. The modeled axial half-length, outer diameter, and thickness of the pipe are 700 mm, 541 mm, and 19 mm, respectively. The repair weld is made by excavating original weld metal and filling using a manual metal arc welding procedure. The width of the repair weld, symmetrically embedded within the original weld, is 22.5 mm at the outer surface. The repair depth is 75% of the pipe shell thickness. As shown in Table 1, three different repair lengths are considered, with the angular span ranging from 20 degrees to 114 degrees in pipe circumference. Note that only half of the welded pipe is modeled because of structural symmetry along the weld centre-line.

Table 1. Definition of Repair Lengths and Angles

Repair Case	Mid-Surface Length (mm)	Angle (degree)
Short	91	20
Medium	260	57
Long	520	114

Figure 4 shows the definition of the composite shell section for the pipe repair welds. It is composed of 4 weld layers, each layer having 3 sectional integration points making a total of 12 for the entire section. The original and repair weld grooves through the shell section thickness are indicated in the shaded areas. Three layers are used to represent three lumped repair weld passes. Note that a typical repair of this size would use about 12 weld beads.

Temperature dependent material properties for the base and weld materials up to melting point were used. The room temperature 1% proof stress values for the base and weld materials were 280 MPa and 476 MPa respectively. An isotropic hardening model was employed for the base material with a saturation level defined by the 10% proof stress, and the weld metal was treated as elastic-perfectly-plastic at the 1% proof stress level.

3.2 Modeling Procedure

The modeling procedure for the pipe repair weld consisted of three main steps: simulation of the original girth weld, excavation of the repair weld area and deposition of three repair weld passes. The original pipe girth weld residual stress field was simulated separately [11] using a 2-D cross-section model and a traditional weld modeling procedure [e.g. see 11-13]. This stress field was mapped onto the 3-D composite shell element model giving an initial axi-symmetric distribution of as-welded residual stress.

Excavation of the repair groove process was represented using the layer deactivation scheme discussed in section 2.2. This resulted in a re-distribution of the original weld residual stresses around the repair to accommodate the stress-free conditions in the groove. The preparation run-out at the ends of the repair cavity was not modeled in the present work, although this refinement is feasible using the composite shell approach. The repair weld was then simulated by three lumped passes using the thermal and mechanical models described in the section 2.2. Note that the direction of welding was alternated between passes.

The significance of the original girth weld residual stress distribution was also investigated by repeating the above modeling procedure but leaving out the step mapping the initial axi-symmetric stress field onto the 3-D model.

4.0 RESULTS AND DISCUSSION

Transient temperature distributions associated with the three-pass repair weld are shown in Figure 5 for the medium-length repair weld (55 degrees). The short and long repairs share the same temperature characteristics. Note that the temperatures are plotted on different surfaces for each repair pass, surface A-A (integration point #5) for pass 1, surface B-B for pass 2 (integration point #8), and surface C-C (integration point #12, outer surface) for pass 3. The red and orange areas in the plots indicate the approximate melt zone (>1400°C) of each repair pass. The start/stop positions for each pass are also annotated.

The overall axial (transverse) residual stress distributions on the pipe outer surface are shown in Figure 6 for the three different repair welds. Axial residual stresses are tensile along the length of all three repairs, but become compressive beyond the ends. The maximum tensile stresses are located near the arc start/stop positions. The stress distributions for the short and long repairs show marked differences. The short repair clearly exhibits an overlay of the arc start/stop effects. As a result, the tensile residual stresses are higher in the middle portion of the short repair weld. For the long repair, however, the start/stop effects are isolated and the axial residual stress distributions are more uniform in the middle portion of the repair. Note that a low stress region can also be identified within the weld itself. The characteristics of the medium-length repair lie between the short and long repairs. However, it appears that the axial stresses of the medium repair have the largest axial range of influence.

The circumferential variations in axial residual stresses are shown in Figure 7. The stresses are plotted along the weld centre-line on the pipe outer surface. Again, it is shown that the maximum tensile axial stresses are located near the start/stop positions. The short repair start/stop maxima have merged giving significantly greater stresses at mid-length than the medium and long repairs, although of similar magnitude to the start/stop peaks of the medium and long repairs. Two compressive stress zones are present in the region immediately outside the repair area (adjacent to the start/stop ends). Stresses circumferentially remote from the repair area are almost unaffected by the repair weld and therefore are primarily due to the original weld. Note that the axial tensile stresses are higher at the last pass stop position than the start position.

Variations in axial surface residual stresses along the length of the pipe at the repair centre are shown in Figure 8. All the repair stress profiles show a tensile peak about 40 mm away from the weld centre-line and die away with axial distance from the weld area. However, the medium repair residual stresses have by far the greatest range of influence. The short repair exhibits higher peak axial stresses than the medium repair which are similarly greater than the long repairs.

The effect of the original girth weld residual stresses on the repair residual stress field characteristics are shown in Figure 9 for the medium repair. It is seen that the overall characteristics of the axial repair residual stresses are very similar for these two cases. The same observation was made for the short and long repairs. This implies that the repair process itself dominates the induced residual stress field, for the specific deep repairs (75% section thickness) and structural geometry examined.

The 3-D deformation characteristics for the three different repair welds are shown in Figure 10. The deformations shown in the figure are magnified by a factor of 10. The deformation results indicate that the short repair has the smallest radial shrinkage in the region of repair and axial range of influence.

5.0 EXPERIMENTAL MEASUREMENTS

A mock-up of the original 19mm thick pipe girth weld has been made and neutron diffraction measurements performed to characterize the through-wall distribution of residual stresses in the base material heat affected zone [11]. The measurements support the overall residual stress distribution predicted by the 2-D finite element prediction (thick bead model), which has been used for the original girth weld in the present repair weld studies. However, when comparing the measurements and predictions it was observed that the pattern and sizes of weld beads, and particularly the final capping pass, have a strong influence on the detailed stress profile. Further residual stress measurements using neutron diffraction, the deep hole drilling method and sectioning are being performed on the mock-up after the introduction of short and medium length, 75% depth repairs.

Residual stresses in a short (20° arc span), 75% thickness repair to a 35mm thick pipe girth weld (of similar diameter to the 19mm pipe) have been studied using a virtually identical finite element simulation approach; the main difference being the use of four lumped repair weld passes rather than three. A mock-up of the 35mm repaired pipe has also been made and through-wall stresses in the heat affected zone at mid-length of the repair measured using the deep hole drilling technique [4]. The measured results are compared with the shell element prediction in Figures 11 and 12. A good general correlation is observed. This result confirms that the composite shell element approach has accurately captured the global response of the structure to the repair weld. It should be noted that the deep hole method tends to measure the general residual stress distribution rather than detailed stress variations in thinner section welds.

6.0 CONCLUDING REMARKS

A series of finite element analyses have been performed to investigate the 3-D characteristics of residual stresses induced by repairs to pipe girth welds. The repair length significantly affects both the magnitudes and overall spatial distributions of residual stresses.

a) Circumferential Effects. Maximum tensile axial residual stresses occur on the outer surface near the arc start/stop positions. Due to the overlay of start/stop effects, the short repair develops higher tensile residual stresses. Residual stresses for the long repair are quite uniform in the middle portion of the repair region. The distributions of stress are fairly symmetric with respect to the centre of the repair weld, despite the fact that the moving arc was simulated in all three repairs. Compressive axial residual stresses are present beyond the repair start/stop ends. The repair welds appear to have a negligible effect on circumferentially remote stresses (i.e. 180 degrees away from the repair center). These remote stresses are primarily due to the original girth weld.

b) Axial Effects at the Weld Centre. The axial stress profiles show a tensile peak at about 40 mm away from the weld centre-line and die away with distance from the weld area; these peaks decrease in magnitude with repair length. However, the medium repair residual stresses have by far the greatest range of influence.

c) Effect of Original Weld. For the 75% through-thickness repair welds studied, the pre-existing original weld residual stresses have very little effect on the repair residual stress characteristics. The situation might be different if the repair depth is smaller, or if the repair weld is located near the original weld start/stop positions (no original weld start/stop effects were included in this study since the original weld residual stresses were mapped from a 2-D axi-symmetric model).

d) Experimental Validation. Deep hole residual stress measurements on a mock-up pipe with repairs have demonstrated that the special composite shell element weld modeling approach can be used to accurately capture the global response of a thin structure to repair welds. Neutron diffraction measurements have been performed that support the residual stress distribution predicted by the 2-D finite element simulation of the original girth weld; this distribution has been used as a starting condition in the present repair weld studies. Further residual stress measurements using neutron diffraction, the deep hole drilling method and sectioning are being performed on the mock-up after the introduction of short and medium length, 75% depth repairs.

7.0 ACKNOWLEDGEMENTS

This work is published with permission of British Energy Generation Ltd. The deep hole residual stress measurements were performed by the University of Bristol, UK.

8.0 REFERENCES

[1] J. Dunn, J. MacGuigan, R. J. McLean, L. Miles, and R. A. Stevens, Investigation and repair of a leak at a high temperature stainless steel butt weld, *Proc. Int. Conf. Integrity of High Temperature Welds*, pp. 241-258, Prof. Eng. Publishing, London (1998).

[2] P. Dong, J. K. Hong, J. Zhang, P. Roger, J. Bynum, and S. Shah, Effects of Repair Weld Residual Stresses on Wide-Panel Specimens Loaded in Tension. *ASME Journal of Pressure Vessel Technology*, Vol. 120, No. 2, pp. 122-128 (1998).

[3] L. Edwards, P. J. Bouchard, M. Dutta, M. E. Fitzpatrick, Direct measurement of residual stresses at a repair weld in an austenitic steel tube, *Proc. Conf. on Integrity of High Temperature Welds,* I. Mech. E., Nottingham, pp. 181-191 (1998).

[4] D. George, D. J. Smith and P. J. Bouchard, Evaluation of through wall residual stresses in stainless steel repair welds, *Proc. Fifth European Conf. On Residual Stresses (ECRS5)*, (Sept. 1999, Delft-Noordwijkerhout, The Netherlands), to be published.

[5] J. Zhang, P. Dong, and F. W. Brust, A 3-D Composite Shell Element Model for Residual Stress Analysis in Multi-Pass Welds, *Trans. of the 14th Int. Conf. on Structural Mechanics in Reactor Technology (SMiRT 14),* Lyon, France,

Vol. 1, pp. 335-344 (1997).

[6] ABAQUS/Standard User Manual (Version 5.8), Vol. I, II, III, Hibbit, Karlsson & Sorensen, Inc. (1998).

[7] F. W. Brust, P. Dong, and J. Zhang, A Constitutive Model for Welding Process Simulation using Finite Element Methods, *Advances in Computational Engineering Science*, S. N. Atluri and G. Yagawa, Eds., pp. 51-56 (1997).

[8] D. Rosenthal, Mathematical Theory of Heat Distribution During Welding and Cutting. *Welding Journal Research Supplement*: 220-234 (1941).

[9] T. Kasuya, and N. Yurioka, Prediction of Welding Thermal History by Comprehensive Solution. *Welding Journal* 72(3): 107-115S (1993).

[10] N. N. Rykalin and A. V. Nikolaev, Welding Arc Heat Flow. *Welding in the World* (1971).

[11] L. Edwards, G. Bruno, M. Dutta, P. J. Bouchard, K. L. Abbott, R. and Lin Peng, Validation of Residual Stress Predictions for a 19mm thick J-Preparation Stainless Steel Pipe Girth Weld using Neutron Diffraction, *Proc. Sixth Int. Conf. on Residual Stresses (ICRS6)*, (July 2000, Oxford, UK), to be published.

[12] P. Dong, J. Zhang, and F. W. Brust, Residual Stresses in Strength-Mismatched Welds and Implications on Fracture Behavior. To appear in *Engineering Fracture Mechanics*.

[13] J. Zhang, P. Dong, and F. W. Brust, Analysis of Residual Stresses in a Girth Weld of a BWR Core Shroud, *Approximate Methods in the Design and Analysis of Pressure Vessels and Piping Components*, W. J. Bees, Ed., PVP-Vol. 347, pp. 141-156 (1997).

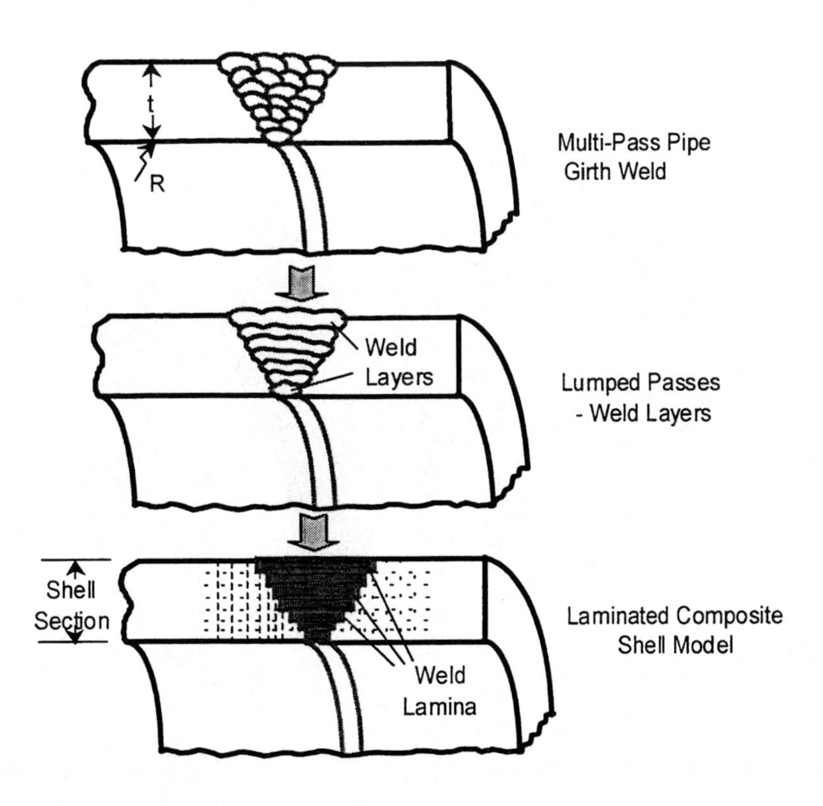

Figure 1. Illustration of the basic concept of a special 3-D composite shell model

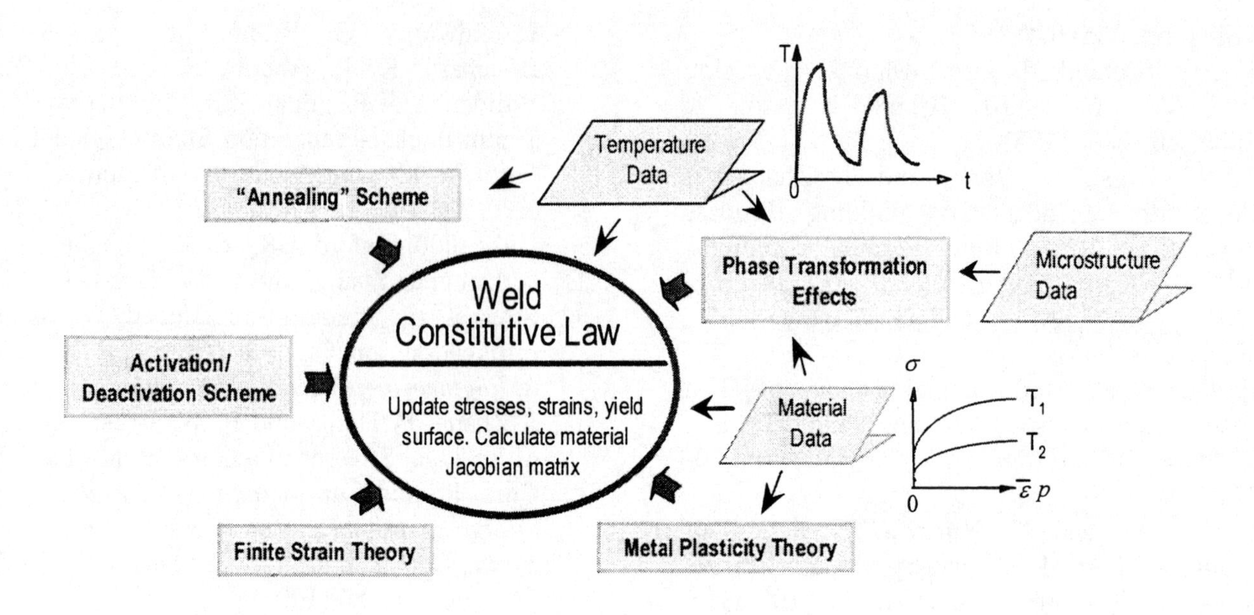

Figure 2. Mathematical structure of a weld material routine

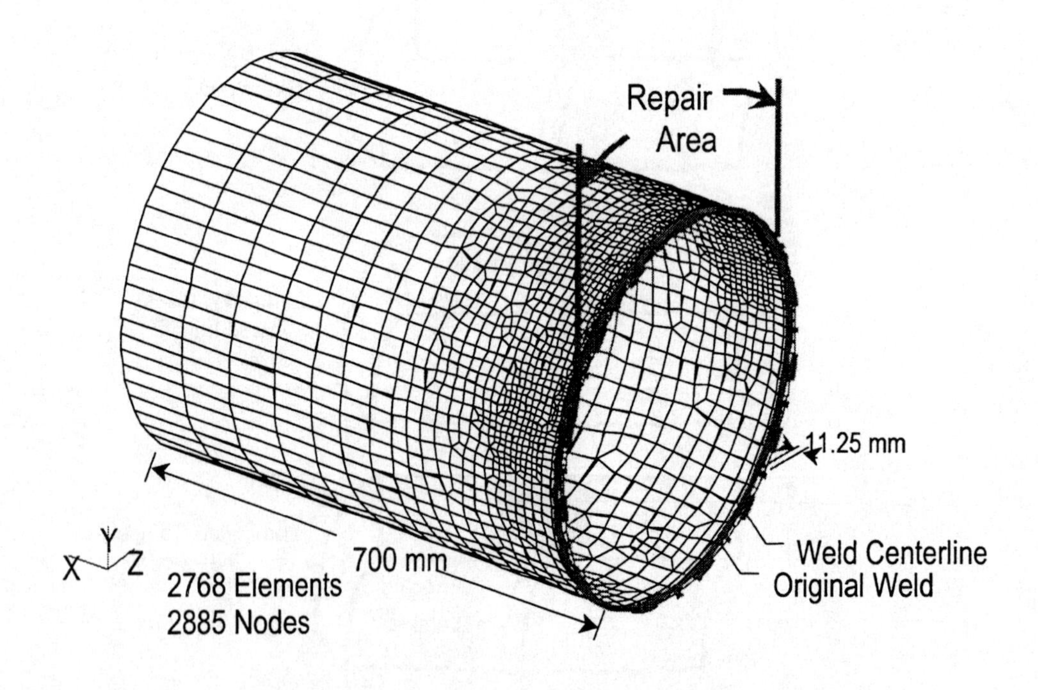

Figure 3. 3-D composite shell element model for pipe repair welds

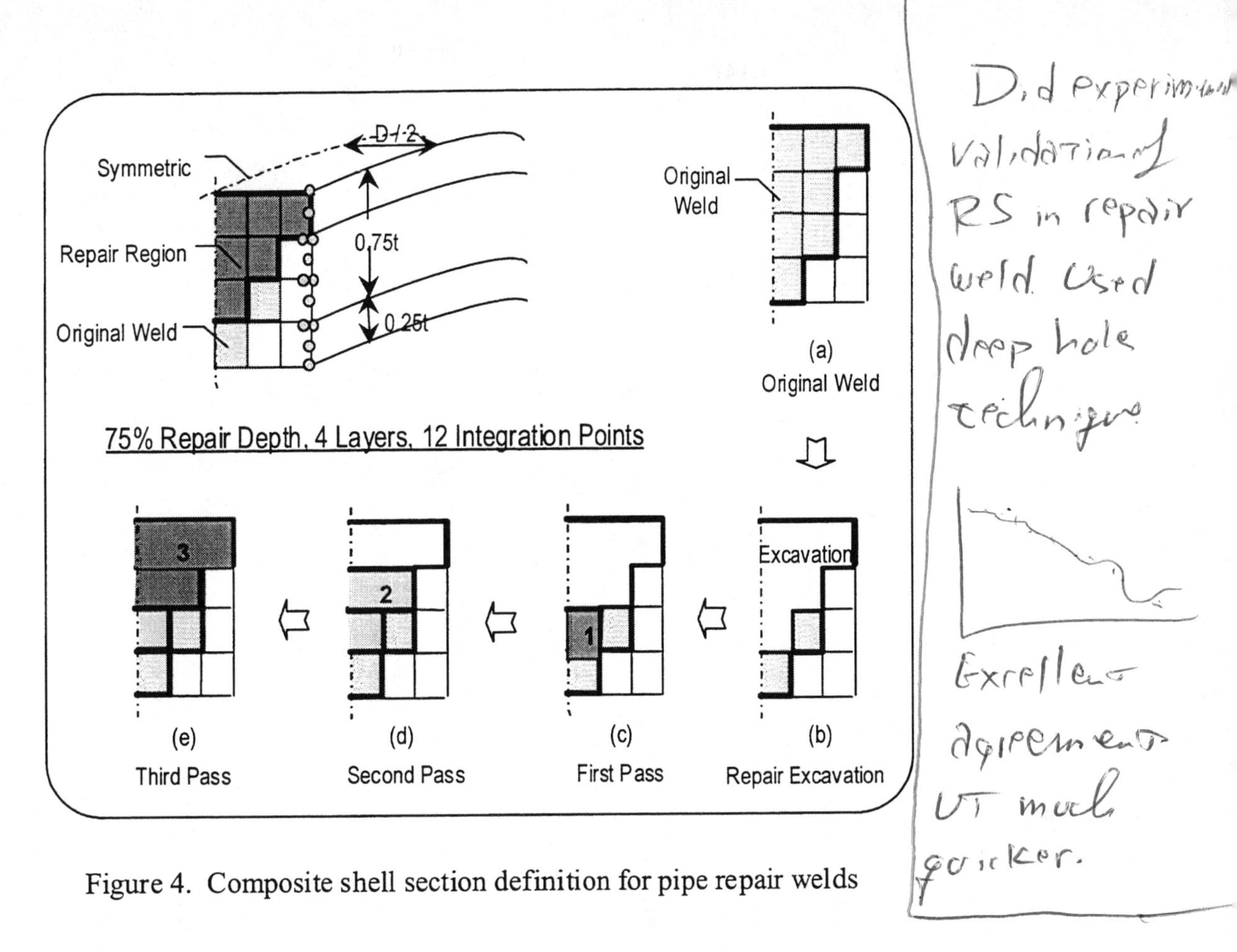

Figure 4. Composite shell section definition for pipe repair welds

Figure 5. Temperature distributions during pipe repair welding (medium repair length)

PVP-Vol. 410-1, Assessment Methodologies for Preventing Failure:
Deterministic and Probabilistic Aspects and Weld Residual Stress
Volume 1
ASME 2000

ANALYTICAL AND EXPERIMENTAL STUDY OF RESIDUAL STRESSES IN A MULTI-PASS REPAIR WELD

J. Zhang, P. Dong, J. K. Hong
Battelle Memorial Institute,
Columbus, Ohio

W. Bell
Mitsui Babcock, Renfrew, UK

E. J. McDonald
BNFL Magnox Generation, Berkeley, UK

ABSTRACT

This paper summarizes the study of residual stresses induced in a multi-pass repair weld in a thick section test plate. The geometry of the test plate consisted of a butt-welded base plate (60mm thick) restrained by seven transverse strong back stiffeners. The repair excavation is centered upon the original weld and is wide enough to encompass the weld and a minimum of 10 mm beyond each fusion boundary. A V-shaped flat-bottomed repair weld preparation is used with a depth of 35mm. The excavation of the repair weld is formed by machining and hand grinding. The excavation is then welded using a four-layer temper bead technique. A total of 135 repair passes are used to fill the cavity.

A global-to-local computational model is used to predict the residual stresses in the repair weld. The global 3-D model is used to capture the overall structural deformation during repair. With the input of restraints from the global model, a local 2-D cross section model is used to capture residual stress details at the mid-section of the repair weld. Experimental measurements on the residual stress distributions are also carried out using centre hole drilling, X-ray diffraction, and deep hole techniques. Both surface and through-thickness residual stress distributions show good agreement between the predictions and measurements, particularly in the repair area.

1.0 INTRODUCTION

Following the discovery of a deep flaw in a 6.8 m diameter x 57 mm thick vertical boiler shell, a testing program was set up to establish a repair procedure to successfully repair the defect. One of the major concerns for the repair weld is the residual stress induced by the repair welding process. As many of the recent studies have indicated [1-5], repair weld residual stresses not only possess unique characteristics in distribution but also have higher tensile stresses in and near the repair area. This is mainly caused by a higher degree of restraint imposed on the repair weld. To ensure structural integrity, knowledge of residual stresses in the repair weld is necessary.

To obtain the residual stress information for the repair weld, a combined analytical and experimental study is undertaken to predict and validate the residual stress distributions. A representative test plate is used to validate the finite element predictions by comparing results with experimental measurements taken using various measuring techniques. With this validation, the finite element modeling technique can be applied to the actual boiler shell repair to optimize the welding procedure and minimize the residual stresses. The following sections of the paper describe the details of the analysis, the mock-up test plate, and the results obtained from the predictions and measurements. A concluding remark is provided at the end of the paper regarding the residual stress distributions in the test plate.

2.0 REPAIR WELD IN A TEST PLATE

2.1 Preparation of Test Plate

The geometry of the test plate is shown in Fig. 1. The test plate consists of a butt-welded base plate (60mm thick) restrained by seven transverse strong back stiffeners. The dimension of the base plate is 1050 mm long and 794 mm wide. The butt weld has a double-V type weld groove preparation. Prior to the making of the butt weld, seven transverse strong back stiffeners are welded on the backside (side 2) of the base plate. Then, half of the butt weld is made from the opposite side (side 1) to fill one V-groove. After completion of the weld on side 1, additional backing plates are welded to side 1 and the strong back stiffeners on side 2 are removed to allow access for welding on side 2. The butt weld is then completed on side 2. On completion of the butt weld, the test plate is fully stress relieved and strain aged.

2.2 Repair Weld

After the test plate is stress relieved and strain aged, a flat bottom V-shaped repair cavity is excavated by means of controlled machining followed by hand dressing using a high speed grinderette to a maximum depth of 35 mm. The width of the repair cavity on the surface extends to at least 10mm beyond the fusion line of the original butt weld. The repair welding is then carried out in a very controlled manner. The repair cavity is welded using a "four layer" welding technique. This technique results in a refined HAZ microstructure. On completion of temper bead welding a normal welding procedure is followed to complete the bulk fill of the weld . Fig. 2 illustrates the details of weld groove geometry and deposition of the first two temper bead layers. It is estimated that a total of 135 weld passes are used to fill the cavity, as shown in Figs. 3 and 4. To reheat the underlying weld passes, a capping layer of temper beads is also deposited on top of the repair weld surface. This layer of surface temper beads is subsequently ground off to leave a flat surface.

2.3 Material Properties

The base plate used in the test plate is Ducol, manufactured in 1990. Tensile tests indicated it had an initial yield strength of 636 MPa. The original butt weld and repair weld have similar yield strengths of about 490 MPa. Therefore, the repair weld represents about 23% strength under-match. Details of the temperature-dependent physical and mechanical properties of the base plate and weld materials used for the finite element analysis are listed in Table 1 and 2, respectively.

3.0 RESIDUAL STRESS MODELING

To model the residual stress development during the repair welding of the test plate, a combined global-to-local modeling approach is used. The global model is used to capture the overall deformation characteristics of the test plate during repair. This is accomplished by using a special 3-D shell element model developed by Battelle[6]. With the input of structural restraints from the global model, a 2-D cross section model is used to capture residual stress details in the mid-section of the repair weld. Details of these models are described in the following sections.

3.1 Global Shell Element Model

The concept of using a shell element model for multi-pass welds was originally proposed in earlier work by Battelle [6]. By careful control of material behavior at each sectional integration point through the thickness of the shell elements in the weld area, a multi-pass welding process can be modeled using shell theory. As a full 3-D solid model is computationally expensive, the shell element model is particularly suited for modeling geometries with strong three-dimensional residual stress features, such as repair welds.

The global shell element model used for the test plate analysis is shown in Fig. 5. Only one half of the test plate is modeled due to structural symmetry with respect to the weld centerline. Because the global deformation characteristic of the plate is the primary interest from this model, lumped repair passes are used in this analysis. The resulting deformation characteristics of the test plate after repair is also shown in Fig. 5. Note that the repair welding process causes the base plate to deform downwards in the repair area as the weld metal contracts. However, due to the presence of a transverse strong back stiffener in the middle of the repair area, the test plate is prohibited from deforming, thus resulting in a weaving pattern of deformation in the entire repair area.

3.2 2-D Cross Section Model

The shell element model described above is intended to capture the global deformation of the test plate rather than predict accurate levels of local residual stress. Consequently, local structural details are not reflected in the shell model. To obtain local residual stress details, a 2-D cross section model with local structural details is needed.

Fig. 6 shows the 2-D cross section model for the mid-section of the repair weld. It is a generalized plane strain model whose cross sectional constraints are obtained from the global shell element model. The backing plate of the cross section model is assigned with an equivalent material property calculated by homogenizing the seven backing plates in the entire length of the base plate. To improve the efficiency and time of the finite element analysis, without reducing the accuracy of the predicted residual stress magnitude and distribution, a number of weld beads were combined, effectively lumping them together There are a total of 36 lumped weld passes (from 135 actual weld passes) built in the model, as shown in Fig. 6. Two lumped temper bead layers are present on the bottom and side of the repair groove. One temper bead layer is present on the top of the repair weld. This top capping layer of temper beads is removed after the repair weld is completed.

The welding heat flow and residual stress analyses are carried out using ABAQUS heat transfer and stress analysis procedures along with Battelle's specific weld modules. In particular, a specific welding heat source model is used to simulate heat input from each lumped weld pass. A unified weld material model [7] is used to model the specific material behavior under welding conditions (e.g., melting/re-melting effects). The sequential deposition of weld passes in the repair groove is modeled using ABAQUS element activation /deactivation procedure.

The resulting transverse residual stress distributions are shown in Fig. 7. The as-welded transverse residual stress is shown in Fig. 7a and the transverse stress after the removal of the capping temper bead layer is shown in Fig. 7b. It is clear that the transverse residual stresses are highly tensile on the top surface of the repair weld. The peak tensile stresses are located near the weld fusion line in the base plate (Fig. 7a). While the removal of the temper bead layer largely reduces the peak stress magnitude, the locations of the peak stress stay relatively in the same places. In the bottom part of the base plate under the repair groove, a compressive zone of transverse stress is present.

The corresponding longitudinal residual stress of the repair weld is shown in Fig. 8. The longitudinal stresses are shown after the removal of the capping temper bead layer. Compared to the transverse stress component, the longitudinal residual stresses are predominantly tensile in the entire repair region. The peak longitudinal residual stresses are located near the fusion lines on the side of the repair groove.

4.0 RESIDUAL STRESS MEASUREMENT

To validate the residual stress distributions predicted by the finite element models, experimental studies are carried out to measure the residual stresses on the test plate. Centre hole drilling and X-ray diffraction methods are used to measure the residual stresses on the surface. A deep hole technique is used to measure the residual stress distributions through the thickness of the repair weld at the centerline. Details of the experimental procedure are described in the following sections. Comparisons between the finite element residual stress predictions and measurement values are also given.

4.1 Surface Measurement using Centre Hole Drilling and X-ray Diffraction

Surface residual stress measurements were recorded using the strain gauge rosette centre hole drilling method and X-ray diffraction technique. Although measurements were taken at several areas on the test plate, one area of particular interest was at mid-length of the repair weld. This location was considered directly comparable with the 2-D cross section finite element model. The area of interest covered both base plates, the upper and lower HAZ and the weld metal.

a) Centre Hole Measurement

Strain gauges of Micromeasurement type TEA-06-062RK-120 were attached to the prepared surface following normal residual stress strain gauging methods. The hole, formed at the centre of the strain gauge rosette was made using the Al_2O_3 air abrasion technique. The response from the strain gauges was monitored throughout the drilling process and on reaching a depth of approximately 1.5mm. A final set of strain readings was taken and various hole dimensions noted. The resulting residual stresses were calculated using strain gauge formulae with a correction factor to compensate for the hole at the centre of the rosette.

b) X-ray Diffraction

A dedicated transportable computer controlled X-ray diffractometer system was used which included a goniometer fitted with a linear position sensitive detector in the side inclination orientation. Measurements were made using CrK_{α} X-radiation and a straight flow position sensitive detector filled with 90% argon, 10% methane gas with a pressure of 1 MPa. The measurements were made with an incident beam of 1.5mm diameter on the specimen at $\psi=0$. A detailed description of the instrument and the stress evaluation procedure is given in [8]. Calibration was undertaken using C_eO powder to give an accuracy of ±10MPa for stress measurement.

The surface measurement data obtained from the centre hole drilling and X-ray diffraction are compared with the finite element predictions, as shown in Fig. 9. The comparison shows a good agreement between the finite element predictions and measurements at most of the measurement points. Both centre hole and X-ray data indicate that the peak tensile stresses occur at the weld fusion line, as predicted by the finite element models.

4.2 Through-thickness Measurement using Deep Hole Technique

The through thickness residual stress distribution at the centre of the weld repair was measured using the Deep Hole Method. Full details of the measurement technique are described in [9,10] and a brief description is given below.

The deep hole method begins by drilling a highly uniform, straight axis reference hole at the center of the weld using a gun drill. Prior to drilling the reference hole, a small, fully stress relieved reference block is fitted to the measurement site. This is used to calibrate the drilling process and give an indication of the reliability of the results in the test plate. The diameter of the reference hole is accurately measured around its diameter and along its depth using a compressed air probe. The measurement probe is moved through the reference hole at increments of 0.2 mm using a stepping motor. A column of material, with the reference hole as its axis is trepanned from the test plate by electrochemical machining using a hollow copper electrode. During the machining operation any changes in the axial dimension of the column are monitored. After trepanning, the reference hole diameter is re-measured. The changes in the hole diameter (diametrical strain) and the axial direction changes (axial strain) are used to calculate the residual stresses.

The through-thickness distributions of residual stresses obtained from the deep hole measurement are also compared with the finite element results. The comparison is shown in Fig. 10. While some discrepancy exists, the comparison is still considered relatively good, especially in the top 2/3 part of the plate thickness which covers the repair weld.

5.0 MITIGATION OF RESIDUAL STRESSES

On confirming the predicted residual stress distribution by experimental means concern was raised about the magnitude of the peak tensile surface stress at the weld fusion line. Further studies of the microstructure and hardness in the test plate also indicated that this location could be prone to crack initiation during operation. Additional studies examined a number of methods of managing the magnitude and distribution of peak stresses. Several parameters were investigated, namely increasing preheat temperature, attachment of run-off plates to ensure that all weld beads were tempered by a subsequent weld layer, moving the starting point of the temper bead layer and finally changing the weld sequence.

One of the most successful weld sequences in terms of managing the residual stress distribution is shown in Fig. 11. In this case, the run-off plates were attached by light tack welds at intervals along the length of the weld excavation. The first weld temper bead was deposited at the root of the excavation. Subsequent beads were deposited on the lower face of the excavation before moving to the base and upper face. A similar number of weld beads was used in both test plate models.

The resulting transverse residual stress distribution after removal of the run-off plates and capping beads for this optimized welding procedure is shown in Fig. 12. The tensile stress at the surface of the weld is significantly more uniform compared with that shown in Fig. 7b. A line plot of the transverse residual stress distribution on the top surface is shown in Fig. 13. Also shown in Fig. 13 is a comparison with the distribution obtained from the original procedure. It is clear that the peak stress at the weld fusion line is significantly reduced by using the optimized procedure. The average magnitude of the tensile stresses within the repair region is also reduced.

6.0 CONCLUDING REMARKS

A combined analytical and experimental study is conducted for the residual stress distributions in a repair weld in a thick section test plate. A global-to-local modeling technique is used to predict the deformation of the test plate and residual stress distributions. The finite element predictions are validated by experimental measurements using various measurement techniques.

Based on the results obtained from this study, the following observations can be obtained:

- The global shell model of the repair weld in the test plate shows a significant 3-D deformation in the repair area. Due to the presence of transverse strong back stiffeners on the back side of the test plate, a weaving pattern of deformation occurs as a result of repair weld contraction.
- The finite element models predict that the transverse residual stresses are highly tensile on the surface of the plate. The peak stresses on the surface are located near the weld fusion line in the base plate. X-ray diffraction and centre hole drilling measurements on the surface also confirm this finding.
- The transverse residual stress distribution through the thickness of the plate shows tensile surface residual stress changing to compressive within the depth of the repair at the weld centerline. Both finite element results and deep hole measurements confirm this trend.
- The finite element predictions compare well with the experimental measurements, especially within the repair area. Therefore, the finite element analysis technique can be used to model the actual repair for the vessel component.
- The finite element method can be used to examine the benefits of adding run-off blocks, optimising the preheat temperature and managing the weld sequence in order to reduce peak tensile stresses.

7.0 REFERENCES

1. Zhang, J, Dong, P., and Bouchard, J. P., "Effects of Repair Weld Length on Residual Stress Distributions", to appear in the *Proceedings of ASME PVP 2000*
2. P. Dong, J. K. Hong, J. Zhang, P. Roger, J. Bynum, S. Shah, "Effects of Repair Weld Residual Stresses on Wide-Panel Specimens Loaded in Tension", *Journal of Pressure Vessel Technology*, Vol. 120, No. 2, pp. 122-128 (1998)
3. J. Zhang, P. Dong, "3-D Residual Stress Characteristics in Pipe Repair Welds", *Proceedings of the 5th International Conference on Trends in Welding Research*, Pine Mountain, GA (1998)
4. F. W. Brust, J. Zhang, P. Dong, "Pipe and Pressure Vessel Cracking: The Role of Weld Induced Residual Stresses and Creep Damage during Repair", *Transactions of the 14th International Conference on Structural Mechanics in Reactor Technology (SMiRT 14)*, Lyon, France, Vol. 1, pp. 297-306 (1997)
5. F. W. Brust, P. Dong, J. Zhang, "Influence of Residual Stresses and Weld Repairs on Pipe Fracture", *Approximate Methods in the Design and Analysis of Pressure Vessels and Piping Components*, W. J. Bees, Ed., Proceedings of the 1997 ASME Pressure Vessel & Piping Conference, PVP-Vol. 347, pp. 173-191 (1997)
6. J. Zhang, P. Dong, F. W. Brust, "A 3-D Composite Shell Element Model for Residual Stress Analysis in Multi-Pass Welds", *Transactions of the 14th International Conference on Structural Mechanics in Reactor Technology (SMiRT 14)*, Lyon, France, Vol. 1, pp. 335-344 (1997)
7. F. W. Brust, P. Dong, J. Zhang, "A Constitutive Model for Welding Process Simulation using Finite Element Methods", *Advances in Computational Engineering Science*, S. N. Atluri and G. Yagawa, Eds., pp. 51-56 (1997)
8. P. Doig, D. Lonsdale and P.E.J. Flewitt. Appl. Cryst. 14, 124, 1985
9. J. Bouchard, P. Holt, and D. J. Smith, "Prediction and measurement of residual stresses in a thick section stainless steel weld", in Proceedings of ASME-PVP Conference, 1997, Vol. 347, pp 77-82
10. D. J., Smith, and N. W. Bonner, "Measurement of Residual Stresses Using the Deep Hole Method", in Residual Stresses in Design, Fabrication, Assessment and Repair, Ed R.W. Warke, ASME, PVP Vol 327, pp53-65, 1996.

Table 1. Physical properties of base plate and weld metals

Temp. (°C)	Thermal Conductivity (W/mm/°C)		Specific Heat (J/g/°C)	Density (g/mm³)
	Base Plate, Original Weld	Repair Weld		
20	42.8E-3	51.8E-3	0.463	7.86E-3
100	43.5E-3	50.3E-3	0.503	7.86E-3
150	43.2E-3	49.0E-3	0.512	7.86E-3
200	42.9E-3	47.6E-3	0.532	7.86E-3
250	42.4E-3	46.1E-3	0.544	7.86E-3
300	41.8E-3	44.5E-3	0.555	7.86E-3
350	40.4E-3	43.0E-3	0.570	7.86E-3
400	39.0E-3	41.4E-3	0.589	7.86E-3
1500	39.0E-3	41.4E-3	0.589	7.86E-3

Table 2. Mechanical properties of base plate and weld metals

Temp. (°C)	Young's Modulus (x10³ MPa)	Poisson's Ratio	Thermal Expansion Coefficient (x10⁻⁶ /°C)	Base Plate		Original Weld		Repair Weld	
				0.2% Proof Stress (MPa)	20% Proof Stress (MPa)	0.2% Proof Stress (MPa)	17% Proof Stress (MPa)	0.2% Proof Stress (MPa)	20% Proof Stress (MPa)
20	209	0.29	11.6	636	801	490	592	491	580
100	205	0.29	12.1	624	779	477	604	471	565
150	202	0.29	12.3	617	766	468	611	459	555
200	199	0.29	12.6	609	752	460	619	453	576
250	195	0.29	12.9	602	739	452	626	447	598
300	191	0.29	13.1	594	725	443	634	441	620
350	186	0.29	13.3	587	712	435	641	435	641
400	181	0.29	13.5	579	698	427	648	429	662
1500	1.4	0.29	15.0	2	2	2	2	2	2

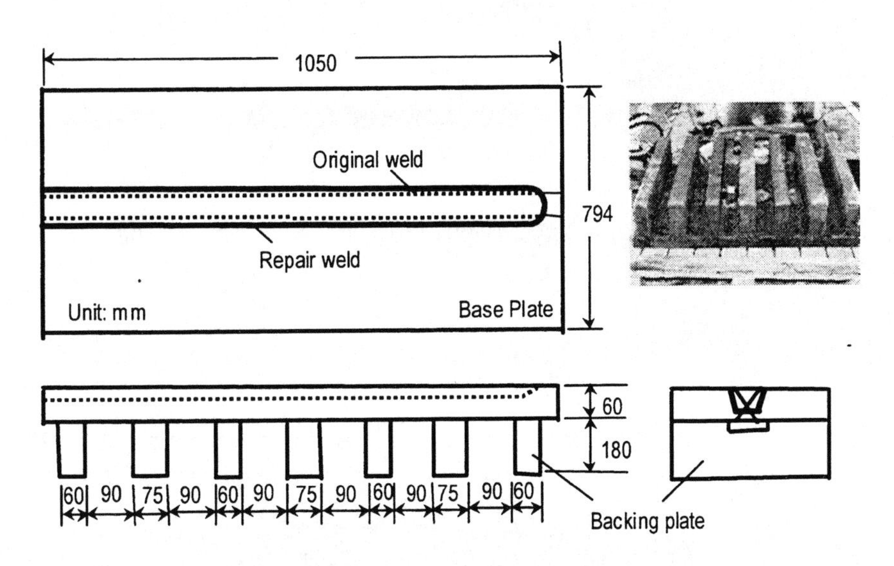

Fig.1 Test Plate Geometry and Repair Weld

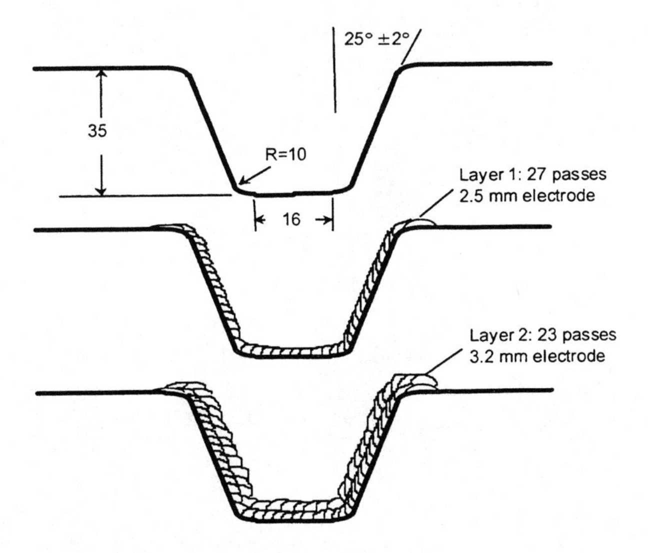

Fig. 2 Repair Weld Profile and Temper Bead Layers

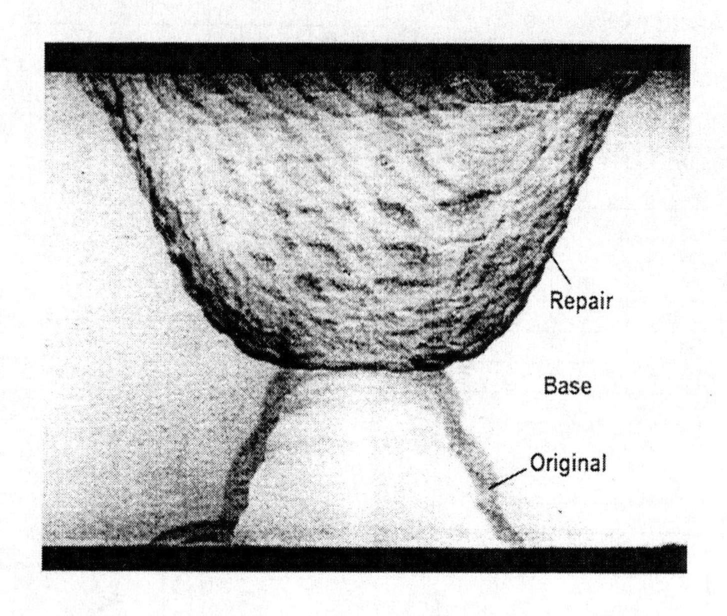

Fig. 3 Macrograph of Repair Weld Cross Section

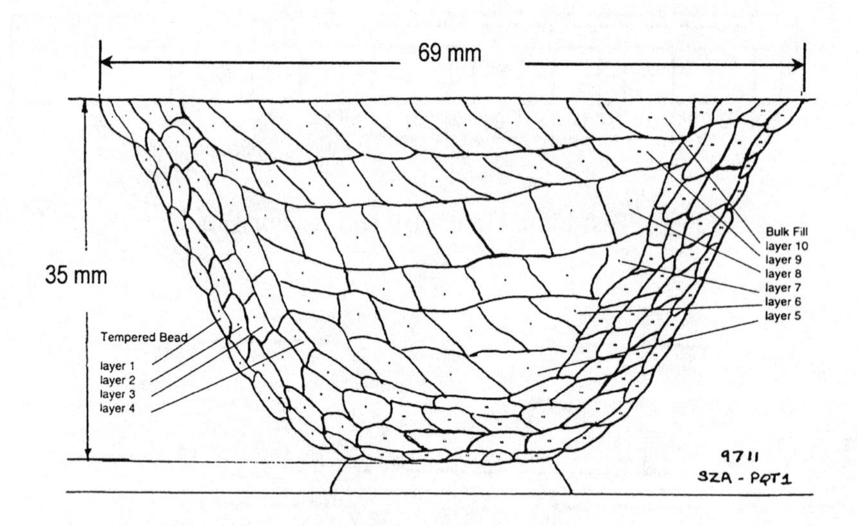

Fig. 4 Repair Weld Passes and Layering Technique

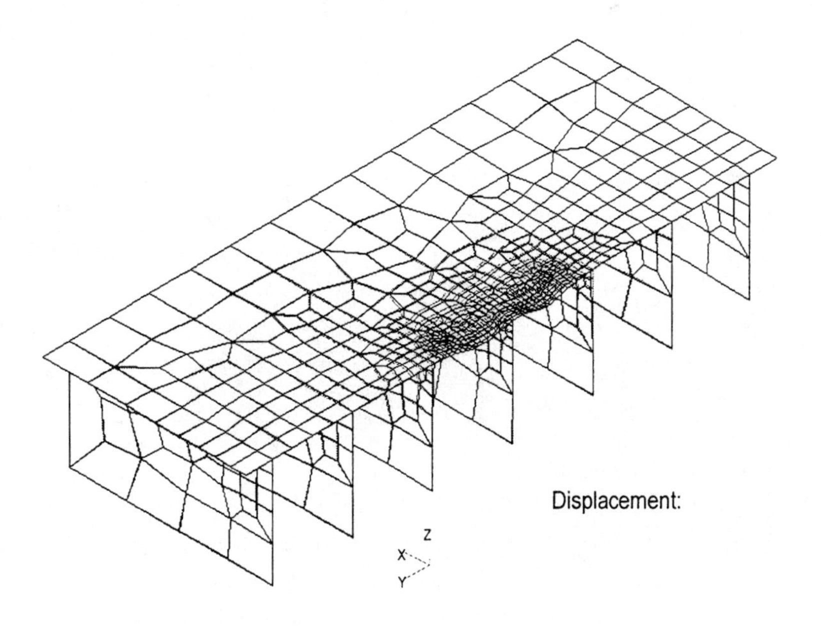

Fig. 5 3-D Deformation Characteristics of the Test Plate after Repair

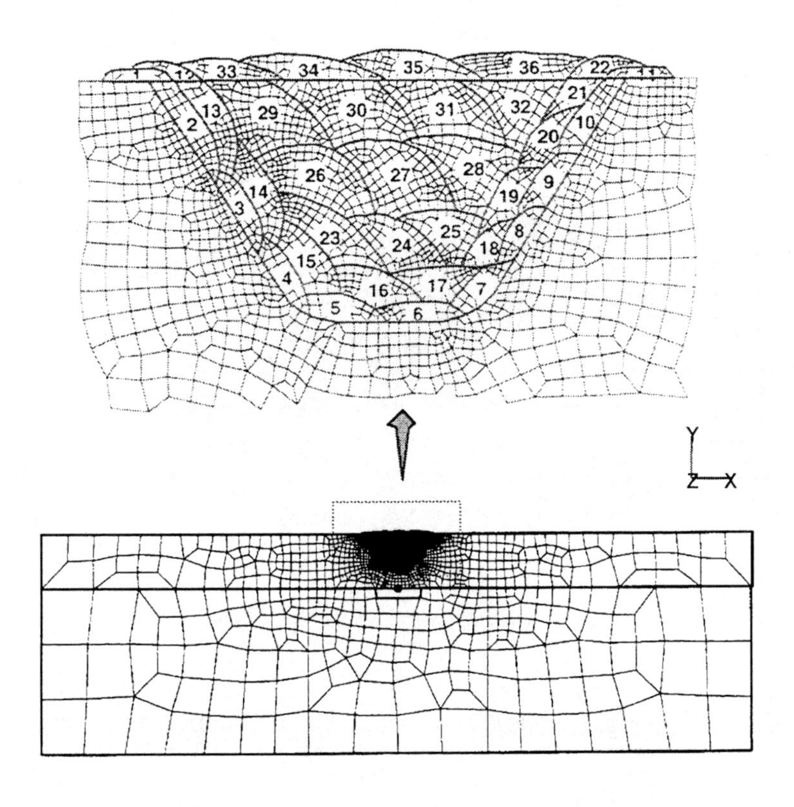

Fig. 6 2-D Cross Sectional Model for Repair Weld

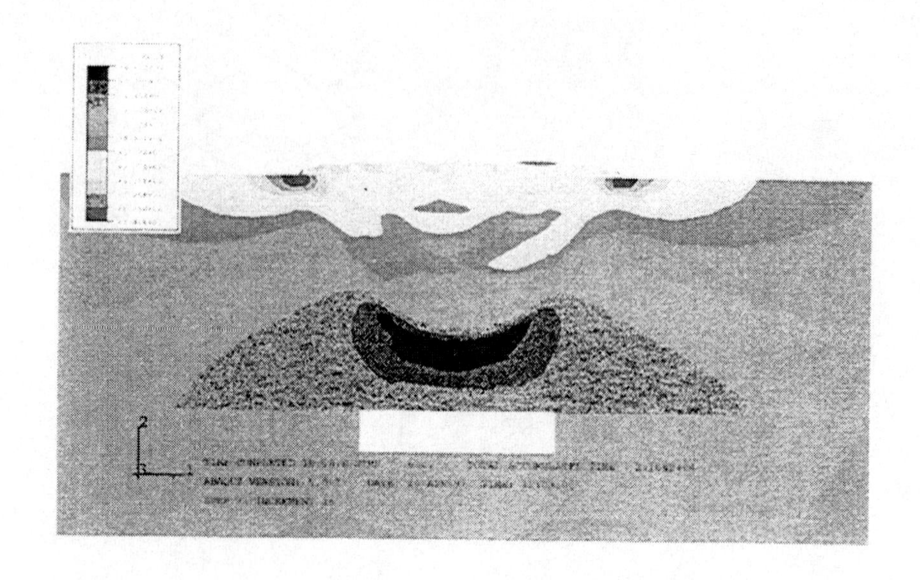

a) As-Welded

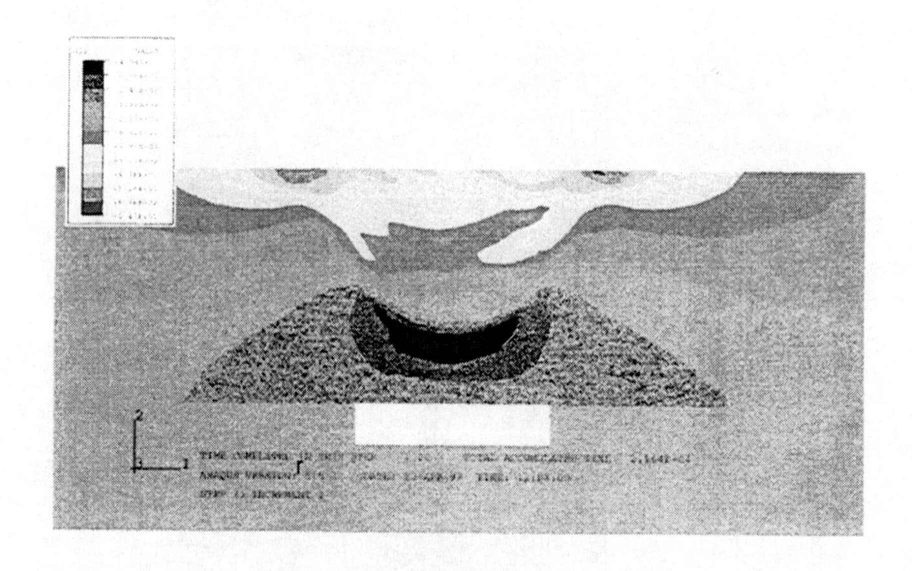

b) After Removal of Temper Bead Layer

Fig. 7 Transverse Residual Stress Distributions

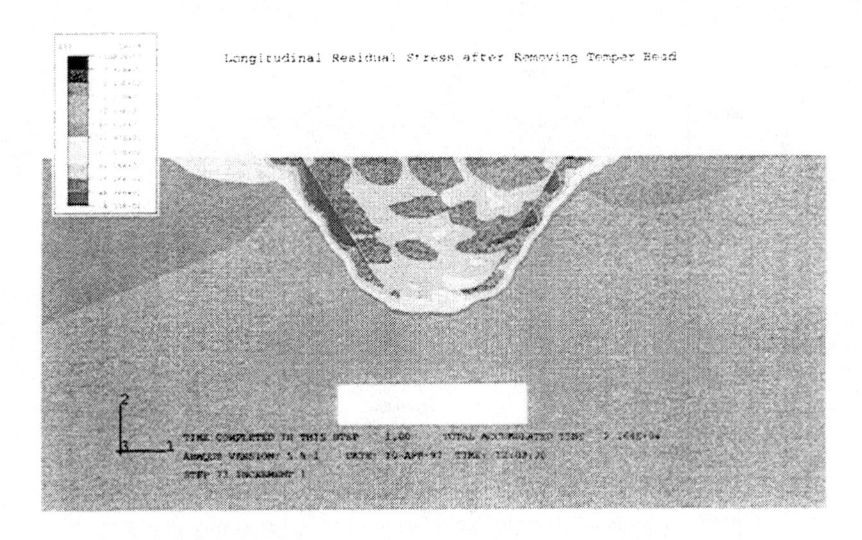

Fig. 8 Longitudinal Residual Stress Distribution

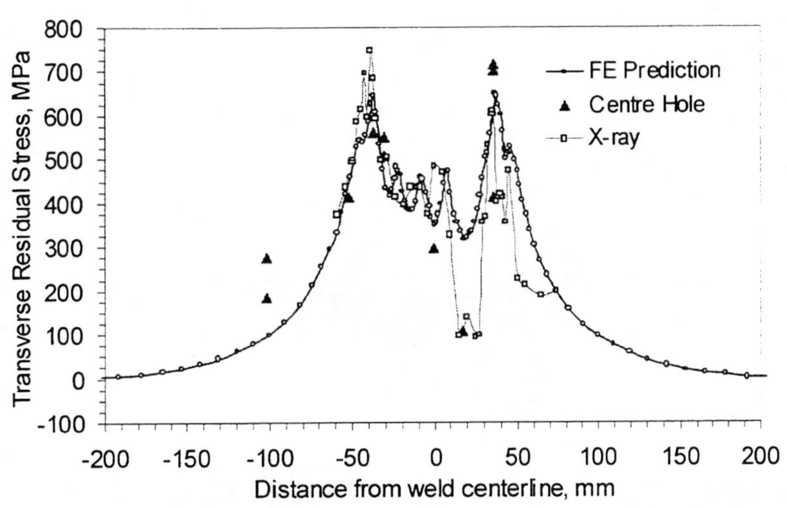

Fig. 9 Comparison of Transverse Residual Stress on Top Surface

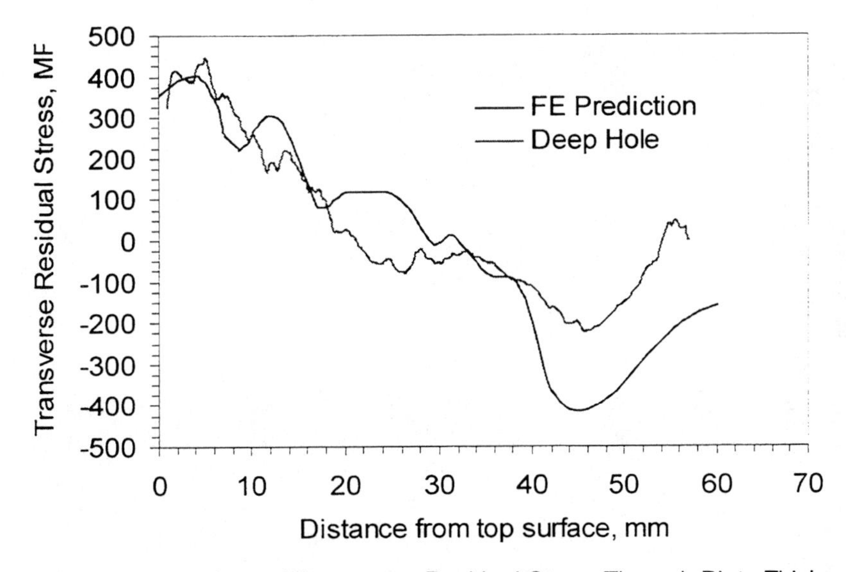

Fig. 10 Comparison of Transverse Residual Stress Through Plate Thickness

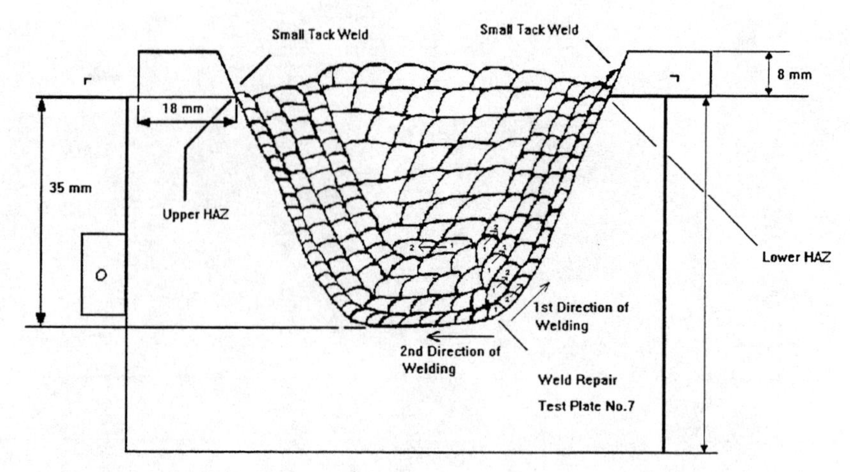

Fig.11 Optimized Repair Weld Passes And Layering Technique

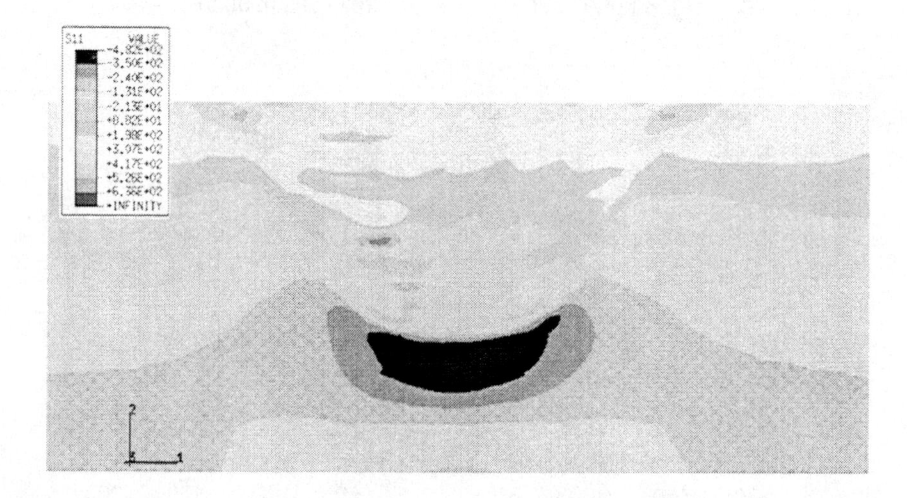

Fig.12 Optimized Residual Stress Distribution after removal of the weld cap and run-off plates

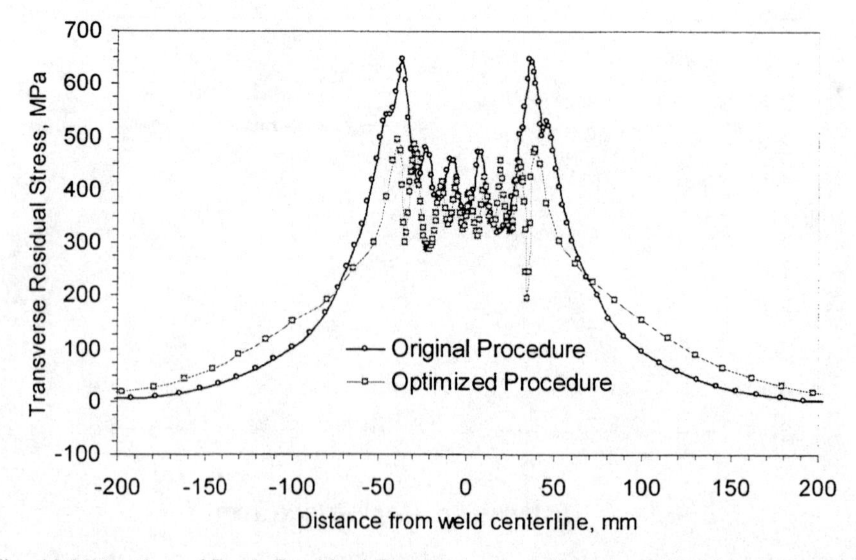

Fig. 13 Mitigation of Peak Residual Stresses using Optimized Welding Procedure

PVP-Vol. 410-1, Assessment Methodologies for Preventing Failure:
Deterministic and Probabilistic Aspects and Weld Residual Stress
Volume 1
ASME 2000

CALCULATION OF STRESS INTENSITY FACTORS CORRESPONDING TO STANDARD RESIDUAL STRESS PROFILES FOR COMMON WELD JOINT GEOMETRIES

R H Leggatt (TWI, Cambridge, UK) A Stacey (Health and Safety Executive, London, UK)

ABSTRACT

Improved methods for including residual stresses in defect assessment procedures are becoming available. The latest generation of guidance documents, including BS 7910, API 579 and reports from the European SINTAP project, provide standard residual stress distributions for common welded joint configurations. Some also give stress intensity factor solutions for the non-linear distributions which are characteristic of residual stresses. In this paper, a standard distribution of residual stresses is referred to as a 'residual stress profile' or RSP, and K_{RSP} is the stress intensity factor at a defect subject to a RSP.

The new procedures for obtaining the RSP and K_{RSP} provide a practical alternative to the traditional assumption that 'residual stresses at welds are equal to yield'. The excessive conservatism which may be associated with that assumption can be reduced or eliminated. However, these new capabilities bring with them the danger of a 'black-box' syndrome. There are many different RSPs for different joint geometries and defect orientations, and the associated stress intensity factor solutions are complex. The fracture analyst may not be aware of the nature of the assumed residual stress distribution, or of the sensitivity of the assessment to the assumed residual stresses.

Examples are given in this paper of the variation of K_{RSP} as a function of crack size for a range of defect and weld configurations including long and short surface defects at butt and T-butt welds in plates and circumferential butt welds in cylinders. In order to demonstrate the benefits of using the standard residual stress profiles, the K_{RSP} values are compared with the stress intensity factor for the same defect subject to uniform yield magnitude residual stresses. A new solution is given for the K_{RSP} at a through-wall transverse crack subject to residual stresses having the trapezoidal shape given in the SINTAP and API 579 compendia for profiles of surface residual stresses.

INTRODUCTION

An interesting new development in residual stress technology is the compilation of standardised residual stress profiles (RSPs) for common types of welded joints. Within the last year, three new compendia of RSPs have been published. These may be found in Appendix Q of British Standard BS 7910:1999[1], in the final report[2] from Task 4 of the European collaborative project SINTAP, and in Appendix E of the draft API 579[3]. Stress intensity factor solutions for the non-linear distributions which are characteristic of RSPs in welded joints are given in Appendix C of API 579[3] and in a report[4] from Task 2 of the SINTAP project.

This paper starts with a summary and comparison of the RSPs in the three documents listed above[1-3]. Examples are given of stress intensity factors for a range of common defect and weld combinations, mostly calculated using the profiles and stress intensity solutions given in reports[2,4] from the SINTAP project. These examples illustrate the variation of stress intensity with crack depth, and the benefits to be obtained by assuming a non-linear RSP, as opposed to the traditional assumption of uniform residual stresses equal to the yield strength. Further examples are given in another SINTAP report[5].

COMPARISON OF PUBLISHED RESIDUAL STRESS PROFILES

Table 1 summarises the RSPs available in BS 7910[1] Appendix Q, the SINTAP compendium[2] and API 579[3] Appendix E, categorised by joint type. The first compendium of RSPs known to the present authors was compiled by Mathieson[6] in 1991 for use in conjunction with the R6 assessment procedure. Very limited guidance on RSPs was given in BS PD 6493:1991[7] (now superseded by BS 7910:1999[1]). Bate, Green and Buttle[8] reviewed the source data and RSPs in References 6 and 7, and considered additional data with particular reference to offshore construction. They provided recommendations for revisions to the R6 compendium[6]. The current British[1] and European[2] compendia are based on further review and refinement of References 6 to 8, and many of the RSPs are common to both documents and are similar or identical to profiles given in the earlier documents. The profiles in API 579[3] Appendix E have some similarities with the BS 7910[1] and the SINTAP compendia[2]: many have similar shapes; like SINTAP, API 579 gives four profiles for each joint type, namely the through-wall and surface profiles of longitudinal and transverse stresses (BS 7910 gives only through-wall profiles); and many of the same source references are quoted. However, the equations of the profiles in API 579 are entirely different from those in References 1 and 2: it is evident that they are based on an independent review of the source data.

BS 7910 and the SINTAP compendium are intended to be generally applicable for the assessment of flaws in welded structures. API 579 is concerned specifically with pressurised equipment used in the refinery and chemical industry. This difference of scope has led to some differences in the categorisation of weld types, as shown in Table 1. It should be noted that different terminologies are used for describing directions in API 579 compared with the other source documents, as summarised in Table 2. At circumferential welds, the term longitudinal refers to the cylinder axial direction in API 579, and to the circumferential direction (along the weld) in the other documents.

Within the general categories of straight and curved full penetration butt welds in Table 1, BS 7910 and SINTAP give one set of RSPs for butt welds in flat plates and pipe axial seam welds, and one set of RSPs for circumferential butt welds in pipes. API 579 does not consider butt welds in flat plates, but gives separate sets of RSPs for welds in (i) piping and pressure vessel cylindrical shells, (ii) storage tanks and (iii) spheres and pressure vessel heads. It gives separate RSPs for longitudinal, circumferential and meridional welds, and for single-V and double-V welds.

For T-butt and T-fillet welds, BS 7910 and SINTAP give separate consideration to welds in plated construction and tubular construction. API 579 does not make this distinction, but does give profiles of stresses in the main plate and the stay plate.

With regard to the actual shapes of the profiles, the through-wall RSPs in BS 7910 and SINTAP include linear, bilinear, polynominal (up to 6th order) and cosine profiles. Those in API 579 are more consistent in shape, all being linear, bilinear or 2nd or 3rd order polynominals. The surface RSPs in SINTAP and API 579 are trapezoidal, i.e. uniform inside the weld and decreasing linearly to zero in parent plate adjacent to the weld.

All the RSPs in API 579 and about half of those in BS 7910 and SINTAP are based on upperbound curves fitted to published residual stress data obtained by measurement or numerical modelling. The remainder of the RSPs in BS 7910 and SINTAP, including the surface profiles of longitudinal stresses and the bilinear through-wall profiles at the toes of T-butt and T-fillet welds, are based on a theoretical upper bound model of the extent of the plastically deformed zone adjacent to a weld developed by Leggatt[9,10]. The dimensions of the deformed zone are given as a function of material properties, weld heat input and plate thickness. The theoretically based profiles have been compared with relevant experimental and numerical solutions in the SINTAP project[2], and found to be satisfactory. The theoretical model assumes a high restraint against thermal contraction after welding, which is an appropriate assumption for longitudinal stresses in nearly all cases, and for transverse stresses at T-joints in thicker materials (i.e. thicker than about 25mm). However, it may be less realistic in thinner materials, such that the RSPs become increasingly conservative as thickness decreases.

The empirically-based RSPs are strictly valid only within the range of plate thickness and welding conditions for which data were available. They have been non-dimensionalised with respect to plate thickness or weld width. This does not reflect the influence of welding and other parameters included in the theoretical model on the width of the plastically deformed zones adjacent to the welds. Hence the empirically-based RSPs may be over-conservative for thicknesses greater than those for which source data was available, and may be non-conservative for thicknesses below the range of the source data.

LONGITUDINAL SURFACE CRACKS AT STRAIGHT BUTT WELD

Longitudinal cracks are subject to transverse residual stresses. Through-wall RSPs for transverse residual stresses at butt welds are shown in Fig.1. The residual stress is normalised with respect to the yield strength, and the depth from the surface, z, is normalised with respect to the wall thickness, t. BS 7910[1] and SINTAP[2] give the same profile, which is a 6th order polynominal and is applicable for plate butt welds and axial seam welds. API 579 gives a 2nd order polynominal (Eq.E.34) for single-V longitudinal welds and uniform stresses (Eq.E.43) for double-V longitudinal welds in piping and cylindrical pressure vessels, and a linear profile (Eq.E.73) for single-V welds and a 2nd order polynominal (Eq.E.79) for double-V welds in storage tanks.

API 579 Eq.E.34 and E.79 are somewhat similar to the BS 7910 and SINTAP profiles, except in the region $0.4<z/t<0.9$ where the API profiles are more tensile. API 579 Eq.E43 adopts the traditional conservative assumption of uniform yield magnitude stresses at double-V longitudinal welds in pipes and cylindrical pressure vessels, but Eq.E73 shows stresses decreasing linearly to zero at the root of single-V welds in storage tanks. The present authors would not be confident of always finding zero stress or less at this location: tensile transverse stresses can develop at the root of the weld as it is stretched by angular distortion during the deposition of the fill passes.

The "normalised K_{RSP}", i.e. the stress intensity factor corresponding to the standard residual stresses profile divided by $\sigma_Y \sqrt{\pi a}$, is plotted versus normalised crack depth for infinite longitudinal surface cracks at straight butt welds in Fig.2. The K_{RSP} was calculated using a weight function solution given as Equation AI.3 in the SINTAP stress intensity factor handbook[4]. The normalised K_{RSP} for the BS 7910 and SINTAP profiles is approximately equal to 1.0 for crack depths 0<a/t<0.4, but then starts to rise, despite the presence of compressive stresses in this region. The normalised K_{RSP} for uniform yield stress $\sigma_R^T = \sigma_Y$ (as in API 579 Eq.E.43) starts at 1.0, but rises faster than the K_{RSP} for a non-linear profile, as would be expected. This demonstrates the benefit to be obtained by using the non-linear RSP, where this is deemed to be applicable. K_{RSP} has not been calculated for the linear profile defined by API 579 Eq.E.73, but this may be expected to lie between the solutions shown in FIg.2.

All the K_{RSP} values discussed above apply to long surface cracks at the side welded last (BS 7910), or at the outside (API 579). For cracks at the opposite face, the normalised K_{RSP} would tend to zero as the crack depth tends to zero for the RSP defined by API Eq.E.73 and to about 1.0 for the other profiles shown in Fig.1. This is clearly a significant difference, and further work to confirm the validity of Eq.E.73 in this region would be desirable.

The normalised K_{RSP} for the deepest and surface intersection points of semi-elliptical surface cracks of aspect ratio, 2c/a, equal to 3.33 and 10 are plotted in Fig.3. K_{RSP} was evaluated using a solution for a 5th order polynominal stress profile[4], and hence it was necessary to fit a 5th order polynominal to the 6th order BS 7910/SINTAP profile shown in Fig.1.

In contrast to the K_{RSP} shown in Fig.2 for infinite surface cracks, the normalised K_{RSP} shown in Fig.3 for the deepest point of finite surface cracks falls continuously with crack depth. The stress intensity factor is always positive, despite the presence of compressive stresses in the range 0.5<z/t<0.8. The normalised K_{RSP} at the surface intersection points of finite surface cracks is approximately constant at about 0.8 for an aspect ratio of 2c/a = 3.33, and is initially lower but rises with crack depth for a crack of aspect ratio 2c/a = 10. The stress intensity at the surface intersection point is greater than that at the deepest point for a/t > 0.1 with 2c/a = 3.33, and for a/t > 0.5 with 2c/a = 10.

TRANSVERSE THROUGH-WALL CRACKS AT BUTT WELD

Transverse cracks at welds are subject to longitudinal residual stresses. The characteristic trapezoidal profile of longitudinal surface residual stresses recommended in BS 7910, SINTAP and API 579 is illustrated in Fig.4. For the purpose of assessing through-wall cracks, the usual practice is to assume that the given cross-weld profile of surface stresses is applicable throughout the thickness of the joint. The issue of interest is how the stress intensity factor changes as the crack length increases beyond the width of the tensile zone.

A solution for a through-wall crack in an infinite flat plate subject to a trapezoidal RSP was given in Ref 5 and is reproduced here. The solution was obtained by integrating the weight function solution given on page 5.11a of Ref 11 for a symmetrical point load, p, at a distance, y, from the centre of a crack length, 2a, in an infinite plate. For a transverse crack subjected to longitudinal stresses, $\sigma_R^L(y)$, the point load, p, is equated with the force $\sigma_R^L(y)dy$, acting on an infinitesimal length, *dy*, of the crack. Then:

$$K_1 = \frac{2}{\sqrt{\pi a}} \int_0^a \frac{\sigma_R^L(y)dy}{\sqrt{1-(y/a)^2}}$$

The solution for the trapezoidal profile shown in Fig.4 is as follows:

For $2a \le W_1$

$$K_1 = \sigma_{YW} \sqrt{\pi a}$$

For $W_1 < 2a \le W_2$

$$K_1 = \sigma_{YW} \sqrt{\pi a}\,(2/\pi)\{(\pi/2) - F_1\}$$

$$F_1 = [(4a^2-W_1^2)^{1/2} - W_1\pi/2 + W_1 \sin^{-1}(W_1/2a)]/(W_2-W_1)$$

For $2a > W_2$

$$K_1 = \sigma_{YW} \sqrt{\pi a}\,(2/\pi)\{\sin^{-1}(W_2/2a) - F_2\}$$

$$F_2 = [(4a^2-W_1^2)^{1/2} - (4a^2-W_2^2)^{1/2} - W_1 \sin^{-1}(W_2/2a) + W_1 \sin^{-1}(W_1/2a)]/(W_2-W_1)$$

The normalised K_{RSP} for through-wall cracks subject to trapezoidal profiles with different widths of tensile zone from W_2/W_1 = 3 to 9 are shown in Fig.5. For $a/W_1 \le 0.5$, the crack is within the weld, and is subject to yield magnitude stresses, giving normalised K_{RSP}=1. For

$a/W_1 > 0.5$, the crack length is greater than the weld width, and the normalised K_{RSP} falls with crack length, at a rate dependent on the total width of the tensile zone. The use of the trapezoidal RSP removes the conservatism inherent in the traditional assumption of residual stresses equal to yield strength for cracks whose width is greater than the weld width.

LONGITUDINAL SURFACE CRACKS AT T-JOINTS

Figure 6 shows the through-wall profiles of transverse residual stresses in the main plate at the toes of T-butt or T-fillet welds. The BS 7910 Eq.Q.13 and SINTAP Fig.4d RSP is a 4th order polynominal with one peak just below the toe and a second peak near the back surface. The other profiles fall from yield stress at the toe to zero at an internal location. For the bilinear profiles given by BS 7910 Eq.Q.9 and SINTAP Fig.2d, the depth of the tensile zone, z_o, is a function of the heat input of the weld pass adjacent to the weld toe and the material properties. These profiles are applicable where $z_o \leq t$. If the validity of the bilinear RSP is accepted, then, for crack depths up to $a/t = 0.5$, the curved profiles of API 579 E.94, BS 7910 Eq.Q.13 and SINTAP Fig.4d would seem to be over-conservative for $z_o/t = 0.25$, approximately correct for $z_o/t = 0.5$, and unconservative (at greater crack depths) for $z_o/t = 0.75$.

Normalised K_{RSP} values for longitudinal surface cracks of aspect ratio $2c/a = 3.33$, 10 and infinity at the toe of a T-butt weld are shown in Fig.7. These were evaluated assuming the bilinear profile shown in Fig.6 with $z_o/t = 0.5$. They were evaluated using a solution from the SINTAP stress intensity factor handbook[4] for a semi-elliptical surface crack in a plate subject to through-wall stress profile expressed as a 5th order polynominal. Hence it was necessary to fit a 5th order polynominal to the bilinear profile, which gave a very good fit. It can be seen that the normalised K_{RSP} rises with crack depth for an infinitely long surface crack, but falls for cracks of finite length.

The normalised K_{RSP} for the deepest and surface intersection points of longitudinal surface cracks of aspect ratio $2c/a = 3.33$ at the toes of T-butt welds subject to uniform residual stress, $\sigma_R^T = \sigma_{YP}$, or the bilinear profile with $z_o/t = 0.5$ are shown in Fig.8. The K_{RSP} values for the bilinear profiles were calculated for the region $0 < a/t < 0.5$, where the profile is linear, and hence could be expressed in terms of a membrane and bending component. All K_{RSP} values in Fig.8 were calculated using the well-known solutions by Newman and Raju[12]. The normalised K_{RSP} values rise with crack depth under uniform yield stresses. They fall with crack depth at the deepest point of a crack subject to the bilinear RSP, but remain approximately constant with crack depth at the surface intersection point, which is the controlling location for crack depths, $a/t > 0.15$. Hence, the benefit of using the bilinear profile is smaller than would be expected on the basis of the stress intensity at the deepest point of the crack.

CIRCUMFERENTIAL SURFACE CRACKS AT CIRCUMFERENTIAL BUTT WELDS

This type of crack is subject to the through-wall profile of transverse residual stresses (i.e. parallel to the cylinder axis) at circumferential butt welds. Published RSPs for these welds are shown in Fig.9. The SINTAP compendium[2], Fig.3d, gives different profiles for values of heat input per unit thickness, q/vt, higher or lower than $60 J/mm^2$. The SINTAP low heat input profile is shown for $q/vt = 60 J/mm^2$: the profile is scaled down at lower values of q/vt, with a limiting scaling factor of 0.5 as q/vt tends to very low value. The SINTAP high heat profile has a cosine distribution and is not subject to scaling. The API 579 "low thickness" profile is applicable for $t^* \leq 20mm$ when $q/v \leq 2kJ/mm$, and for $t^* \leq 50mm$ where $q/v > 2kJ/mm$. The API 579 "high thickness" profile is applicable for $t^* \geq 70mm$ when $q/v \leq 2kJ/mm$, and for $t^* \geq 100mm$ when $q/v > 2kJ/mm$. t^* is a modified thickness parameter dependent on R/t, and is equal to the thickness t for $R/t \leq 10$. The RSP is scaled between the low and high thickness lines at intermediate values of t^*.

Drawings in API 579 and the SINTAP compendium imply that the single sided weld is made from the outside : the profiles would be different if the weld was made from the inside.

API 579 has, in general, adopted a more conservative approach than SINTAP, in as much as the former gives entirely tensile stresses while the latter gives compressive stresses at various locations. The two approaches show some agreement with respect to the stresses at the root of the weld (i.e. at $z/t=1$). Both methods agree that the stress is a function of heat input, q/v, and thickness, t. API gives yield tensile stresses for low thickness conditions, and SINTAP gives yield tensile stresses for high heat input per unit thickness (i.e. for low thickness). Both methods give reduced tensile stresses at the root for high thickness or low q/vt conditions.

Although the two profiles given by SINTAP have undoubtedly been observed in practice, it is difficult to reconcile the abrupt change in profile at the transitional value of $q/vt = 60 J/mm^2$. It is also the case that some observations have been made of zero or low tensile stresses at the outside face, $z/t=0$. Hence, the API profiles may be considered more appropriate for surface defects located at the outside surface. On the other hand, the tensile peak shown in the SINTAP high heat input RSP at $z/t=0.4$ has also been observed in practice, and the API 579 profiles may be unconservative for buried defects at this location.

Normalised K_{RSP} values corresponding to the SINTAP high heat and low heat profiles and to uniform yield stresses acting on a full circumferential defect at the root of a circumferential butt weld are shown in Fig.10. These were calculated using a weight function solution. The normalised K_{RSP} for uniform yield residual stresses rises gradually from 1.2 at a/t=0 to 2.0 at a/t=0.6. The normalised K_{RSP} for the SINTAP high heat RSP falls gradually with crack depth from 1.2 at a/t=0, and hence would provide some benefit in terms of reduced K_{RSP} at crack depths a/t>0.2. The normalised K_{RSP} for API low thickness RSP (not shown) would lie between the uniform yield stress and SINTAP high heat curves and hence would be of limited benefit.

The normalised K_{RSP} for the SINTAP low heat RSP falls rapidly with crack depth, and hence would show significant benefit, even for shallow cracks. The normalised K_{RSP} for the API 579 high thickness RSP (not shown) can be obtained by scaling down the uniform yield stress solution by 0.2, and hence would also show considerable benefit compared with the assumption of uniform yield stresses.

SUMMARY AND CONCLUSIONS

- Three sets of standardised residual stress profiles for common weld joint types have recently become available.
- Many of the profiles given in BS 7910:1999[1] Appendix Q and the SINTAP project residual stress compendium[2] are identical: those in API 579[3] Appendix E show some similarities with the BS 7910 and SINTAP profiles, but have different equations.
- Areas where one or other of the published profiles may be unconservative or over-conservative have been discussed in this paper.
- A SINTAP project report[4] and API 579[3] provide stress intensity factor solutions applicable to non-linear residual stress profiles.
- The reduction in stress intensity factor obtained by using the published profiles instead of the traditional assumption of uniform yield magnitude residual stresses has been quantified for a range of examples.
- A new solution for the stress intensity factor at a through-wall crack subject to a trapezoidal residual stress profile is presented.
- Further investigation of the transverse residual stresses at circumferential butt welds is needed.

NOMENCLATURE

a	half-length of through crack or depth of surface crack
c	half-length of finite surface crack
K_1	stress intensity factor
K_{RSP}	stress intensity factor at crack subject to RSP
q	weld power
R	radius
RS	residual stress
RSP	residual stress profile
t	thickness
t*	modified thickness
v	weld travel speed
W_1	width of weld
W_2	total width of tensile zone
y	transverse direction
z	through-wall direction
z_o	depth of tensile zone
σ_R^L	longitudinal residual stress
σ_R^T	transverse residual stress
σ_Y	yield strength
σ_{YP}	parent metal yield strength
σ_{YW}	weld metal yield strength

ACKNOWLEDGMENTS

This paper was based in part on work done at TWI under sub-contract to the Health and Safety Executive, as part of SINTAP, a collaborative project supported by the European Commission. The authors acknowledge the support of their project partners and sponsors.

REFERENCES

1. BS 7910:1999: 'Guidance on methods for assessing the acceptability of flaws in fusion welded structures'. British Standards Institution, April 1999.
2. Barthelemy J Y: 'SINTAP Task 4. Compendium of residual stress profiles'. Final Report. Institut de Soudure, May 1999.
3. API: 'Recommended practice for fitness-for-service'. Issue 12, September 1999 (Special Release).
4. Al Laham S: 'SINTAP Task 2.6. Stress intensity factor and limit load handbook'. British Energy Generation Ltd, EPD/GEN/REP/0316/98, Issue 2, April 1998.

5. Leggatt R H and Sanderson R M: 'Stress intensity due to residual stresses TWI report SINTAP/TWI/4-6 88269/46/99, September 1999.
6. Mathiesen P A R: 'A compendium of as-welded residual stress profiles'. Nuclear Electric memorandum TD/SIB/MEM/0233, September 1991.
7. BSI: 'Guidance on methods for assessing the acceptability of flaws in fusion welded structures'. BSI Published Document PD 6493:1991.
8. Bate S K, Green D and Buttle D J: 'A review of residual stress distributions in welded joints for the defect assessment of offshore structures'. Offshore Technology Report OTH 95-482. London, The Stationery Office, September 1995.
9. Leggatt R H: 'Welding residual stresses'. In: The 5th International Conference on Residual Stresses (ICRS5), Linköping, Sweden, 16-18 June, 1997. Eds Ericsson T, Oden M, Andersson A, Linköping University, Sweden. Vol.1.pp12-25.
10. Leggatt R H: 'Recommendations for revised surface residual stress profiles'. TWI report SINTAP/TWI/4-3 88269/4-3/98, June 1998.
11. Tada H, Paris P and Irwin G: 'The stress analysis of cracks handbook'. Del Research Corporation, Second Edition, 1985.
12. Newman J C and Raju I S: 'Stress-intensity factor equations for cracks in three-dimensional finite bodies subjected to tension and bending loads'. NASA Technical Memorandum 85793, April 1984.

Table 1: Comparison of residual stress profiles in BS 7910, SINTAP and API 579

	BS 7910[1] Appendix Q	SINTAP Compendium of RS Profiles[2]	API 579[3] Appendix E
Materials	Ferritic steels, austenitic steels	Ferritic steels, austenitic steels, aluminium.	"Currently, a distinction is not made concerning the material of construction"
Distributions	Through-wall profiles	Through-wall and surface profiles.	Through-wall and surface profiles.
Straight full penetration butt welds	Q.1.2 Plate butt welds Q.1.4 Pipe axial seam welds (same as Q.1.2)	Fig.1 Plate butt welds and pipe axial seam welds	E4 Welds in piping and pressure vessel cylindrical shells. Single-V (E4.3) and double-V (E.4.4) longitudinal welds E6 Welds in storage tanks. Single-V (E.6.3) and double-V (E.6.4) longitudinal welds
Curved full penetration butt welds	Q.1.3 Pipe circumferential butt welds	Fig.3 Pipe butt welds	E4 Welds in piping and pressure vessel cylindrical shells. Single-V (E4.1) and double-V (E4.2) circumferential welds. E5 Welds in spheres and pressure vessel heads Single-V (E5.1) and double-V (E5.2) circumferential welds. Single-V (E5.3) and double-V (E5.4) meridional welds E6 Welds in storage tanks. Single-V (E6.1) and double-V (E6.2) circumferential welds
T-butt and T-fillet welds	Q.1.5 T-butt and fillet welds. Stresses in main plate at toe of weld at plate-to-plate, tube-to-plate and tube-to-tube joints. Two alternative profiles are given.	Fig.2 Plate T-butt welds (whose heat input is known) Appendix 3 Plate T-butt welds (where heat input is not known) Fig.4 Pipe T-butt welds	E8 Full penetration and fillet welds at a tee joint E8.1 Main plate E8.2 Stay plate
Nozzle attachment welds		Fig.5 Set-in nozzle Fig.6 Set-on nozzle	E7 Full penetration welds at corner joints (nozzles or piping branch connections) E7.1 Corner joint E7.2 Nozzle fillet weld E7.3 Shell fillet weld at a reinforcing pad
Repair welds	Q.1.6 Repair welds.	Fig.7 Repair welds	E9 Repair welds E9.1 Seam welds E9.2 Nozzle welds

Table 2 Terminology for directions

	This paper/BS 7910/SINTAP	API 579
Direction of stress relative to weld	Longitudinal	Parallel
	Transverse	Perpendicular
Direction of welding in cylinders	Circumferential	Circumferential
	Axial	Longitudinal

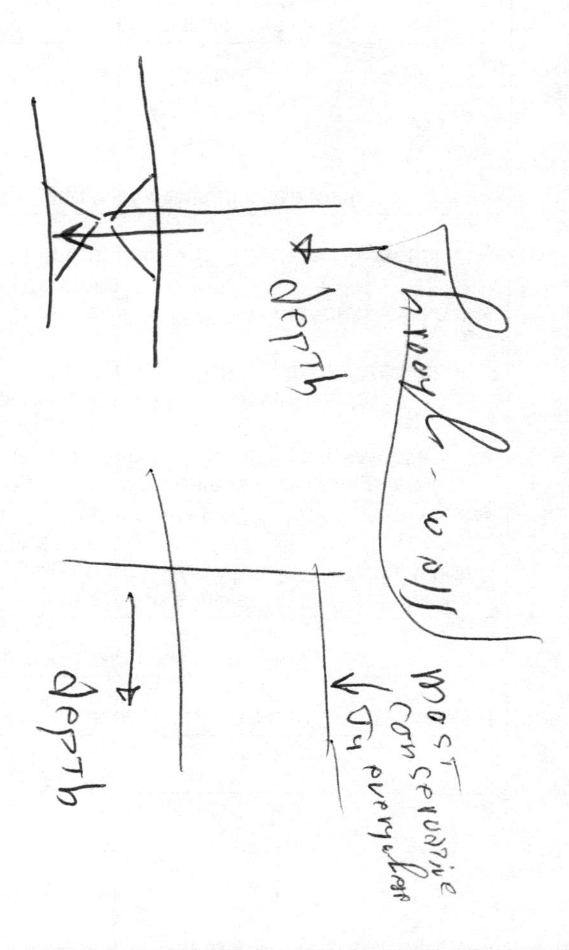

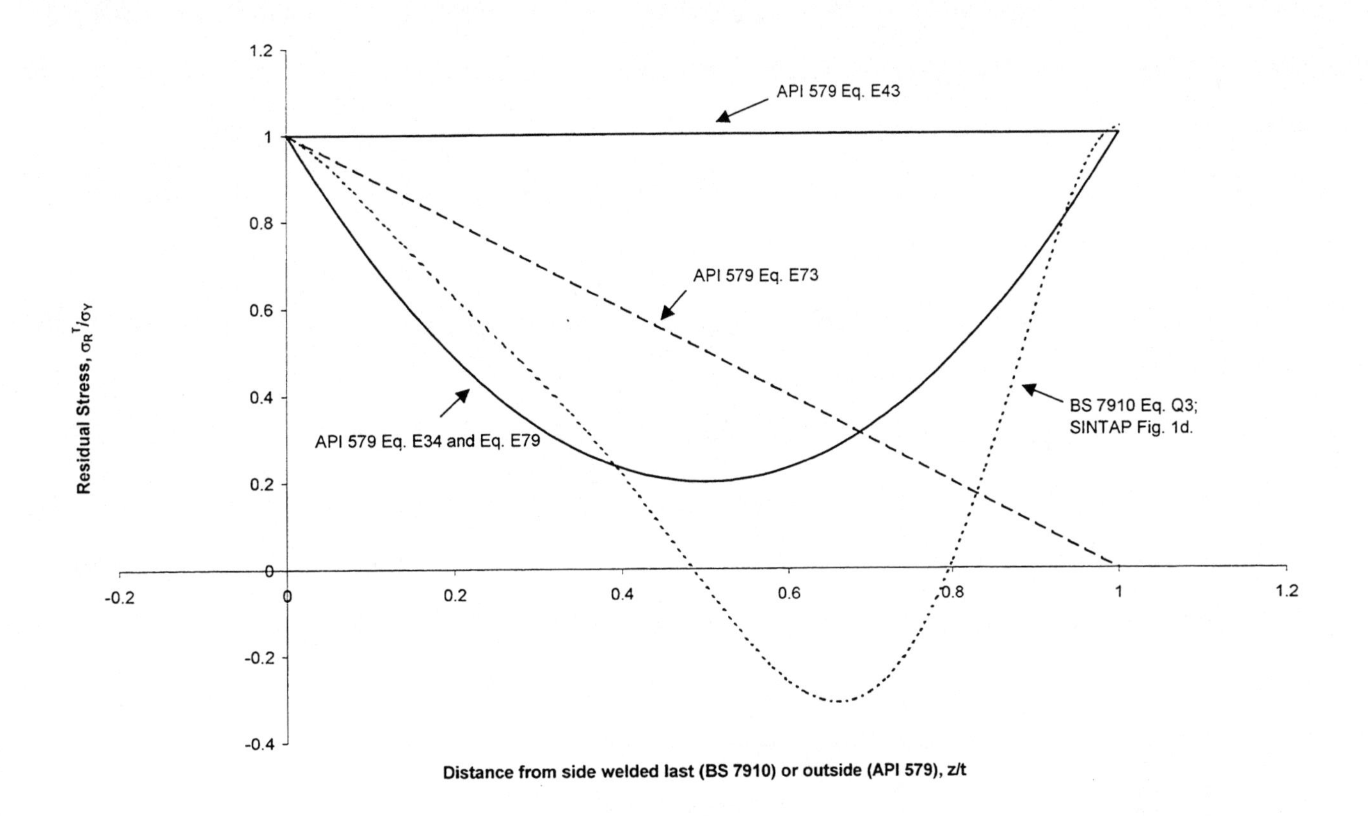

Fig. 1 Through-wall profiles of transverse residual stresses at butt weld.

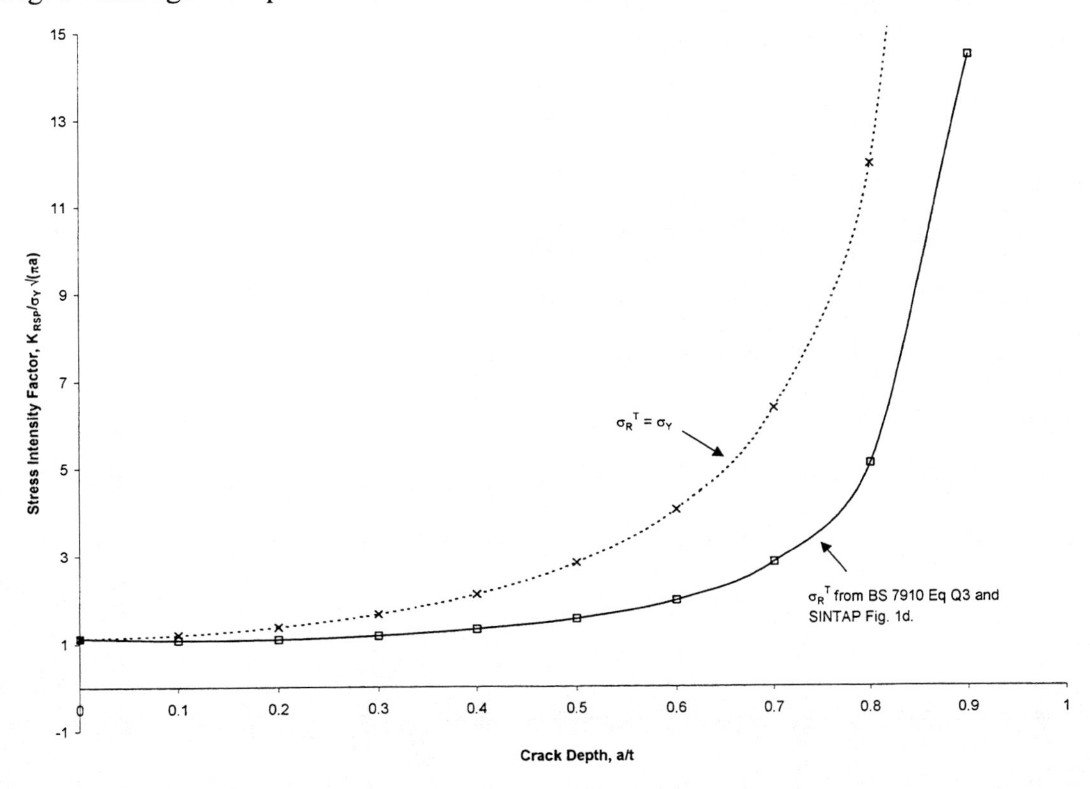

Fig. 2 Normalised K_{RSP} at longitudinal infinite surface crack at straight butt weld. Effect of residual stress profile.

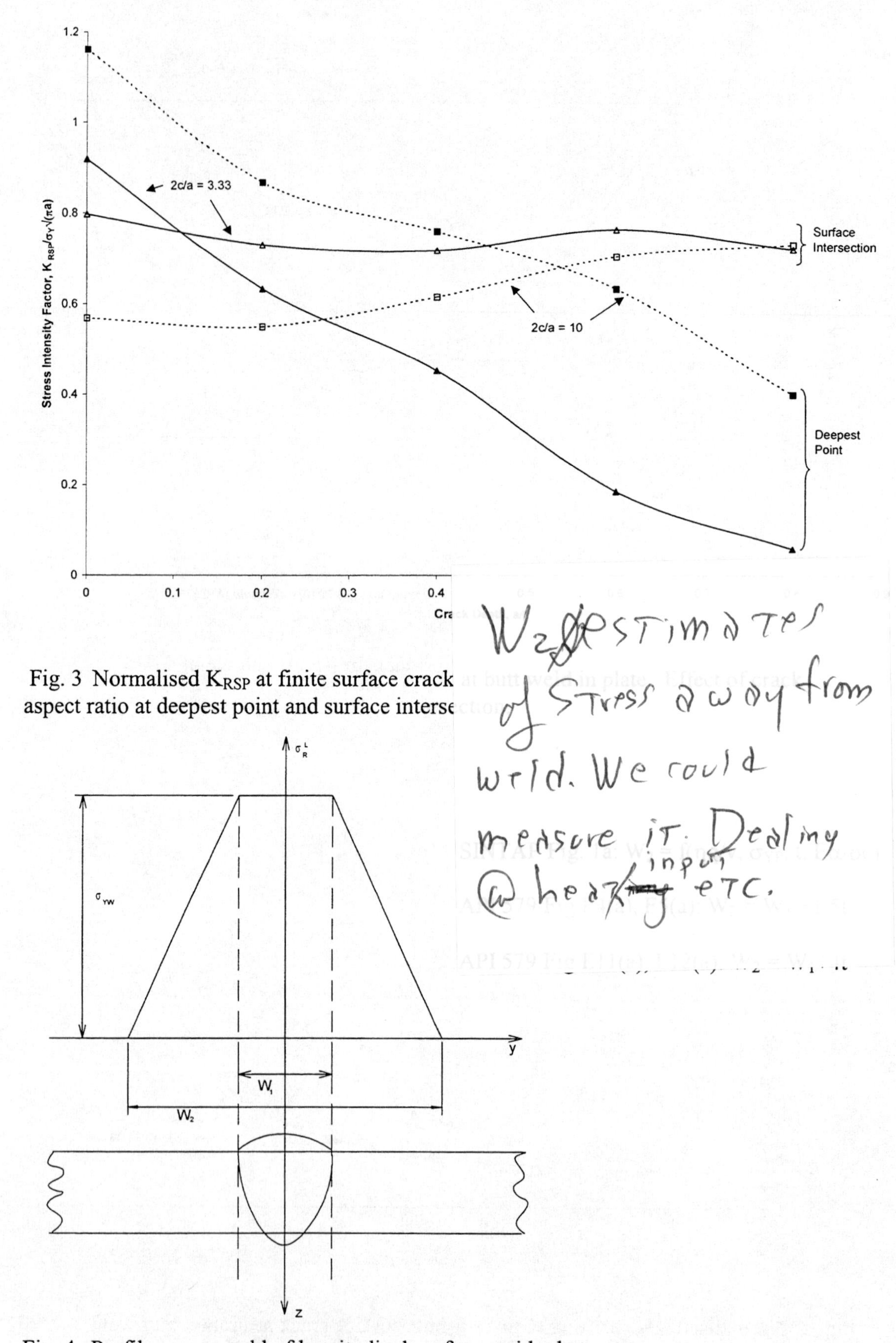

Fig. 3 Normalised K_{RSP} at finite surface crack at buttweld in plate. Effect of crack aspect ratio at deepest point and surface intersection

Fig. 4 Profile across weld of longitudinal surface residual stresses

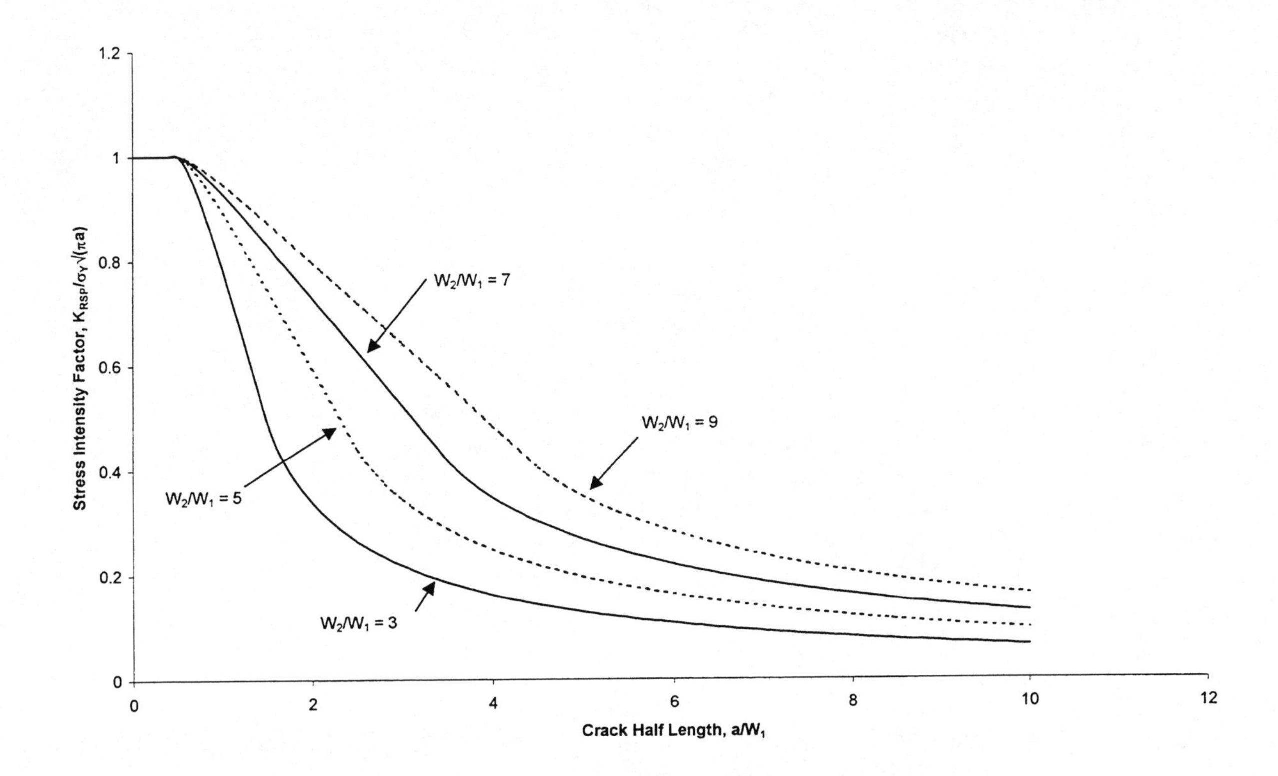

Fig. 5 Normalised K_{RSP} at transverse through thickness crack at butt weld in plate. Effect of residual stress profile.

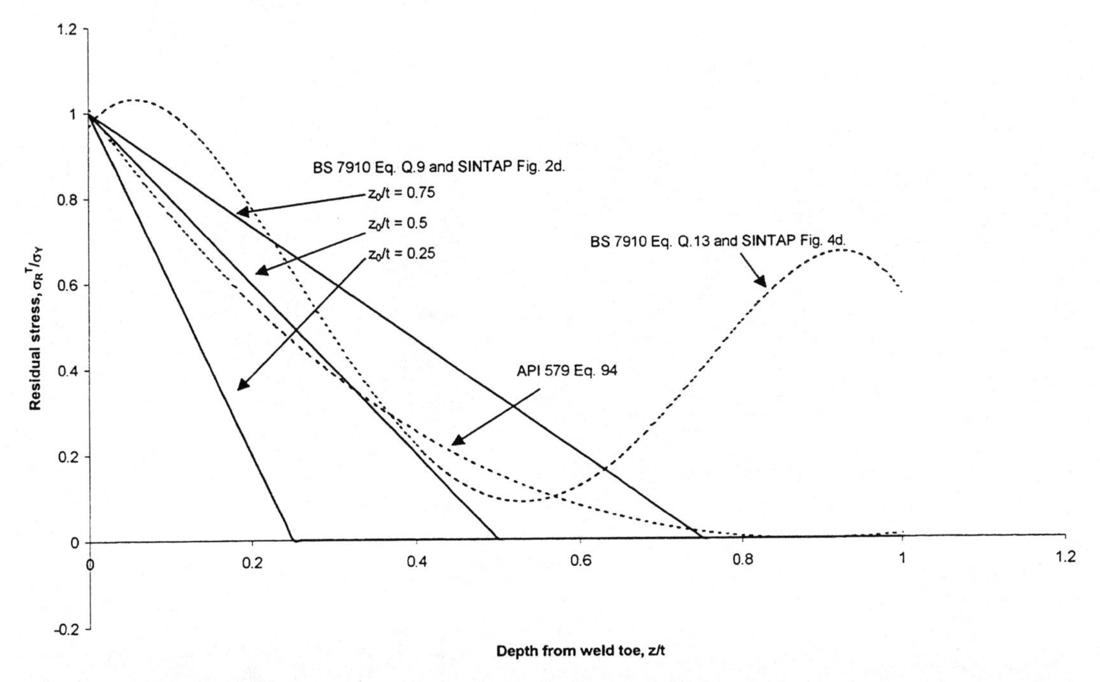

Fig. 6 Through-wall profiles of transverse residual stresses at toe of T-butt or T-fillet weld.

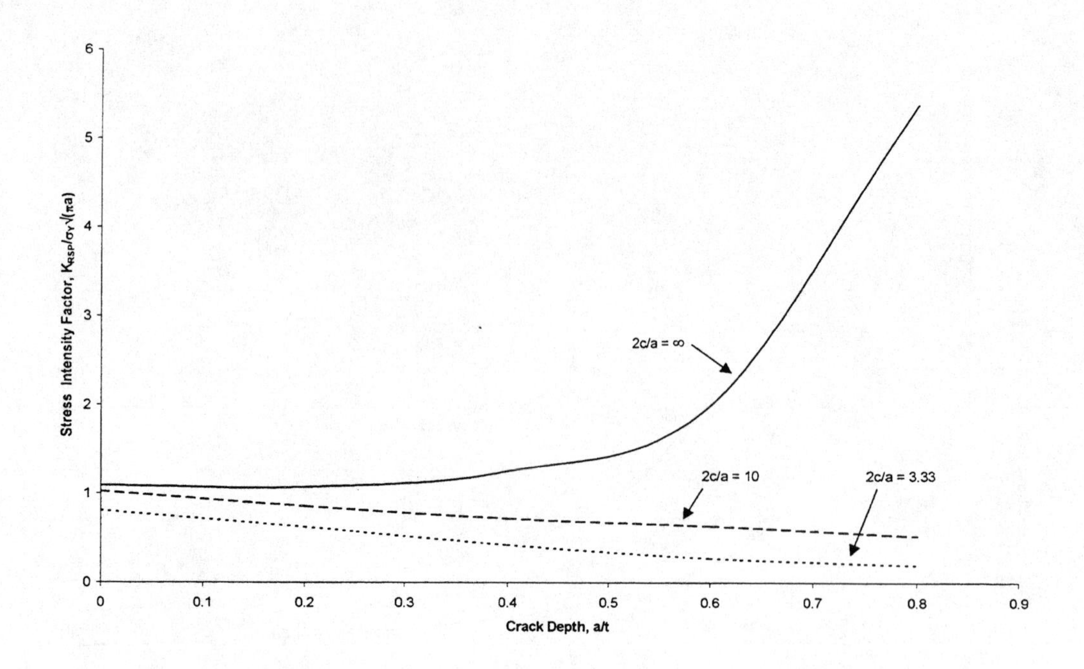

Fig. 7 Normalised K_{RSP} at deepest point of short surface crack at toe of T-butt weld for bilinear RSP with $z_0/t = 0.5$. Effect of aspect ratio.

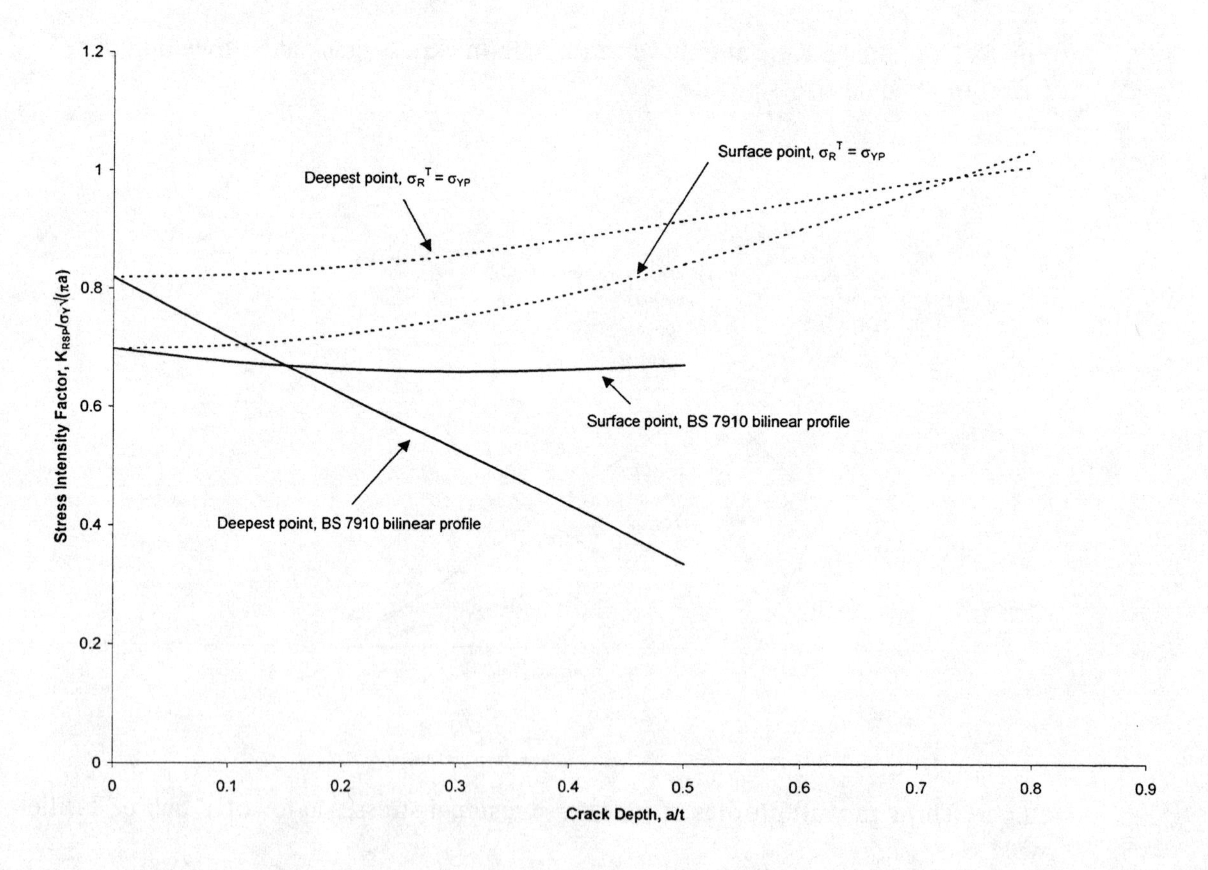

Fig. 8 Normalised K_{RSP} at short surface crack at T-butt weld. Effect of residual stress profiles.

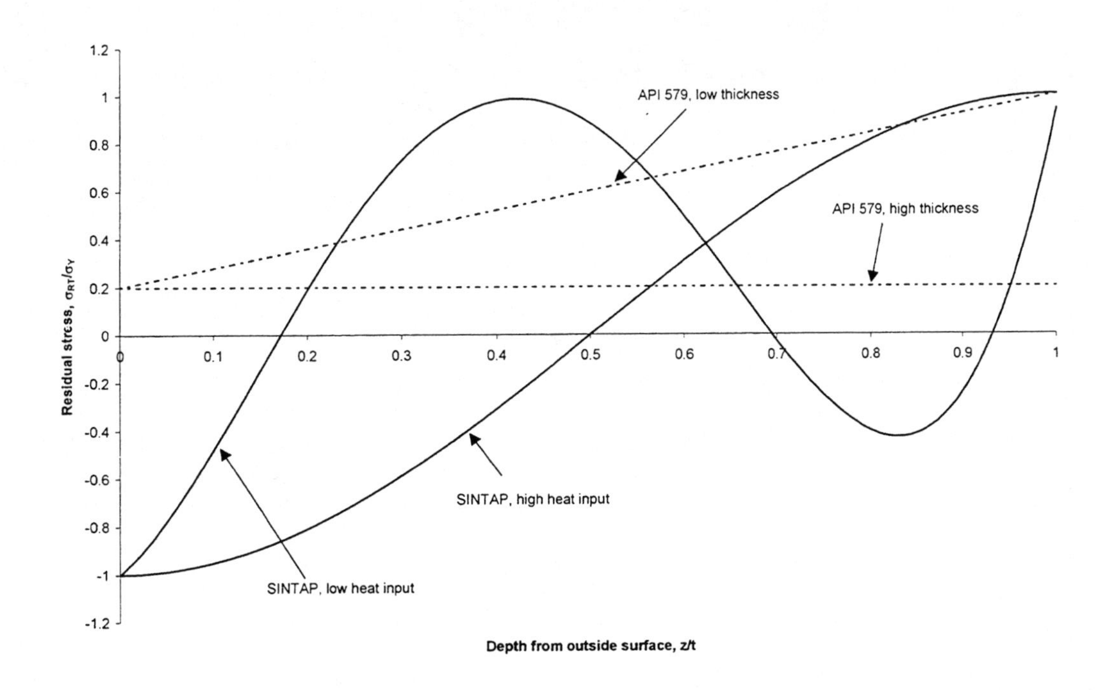

Fig. 9 Through-wall profiles of transverse residual stresses at single-V circumferential butt welds with $q/(vt) = 60 J/mm^2$.

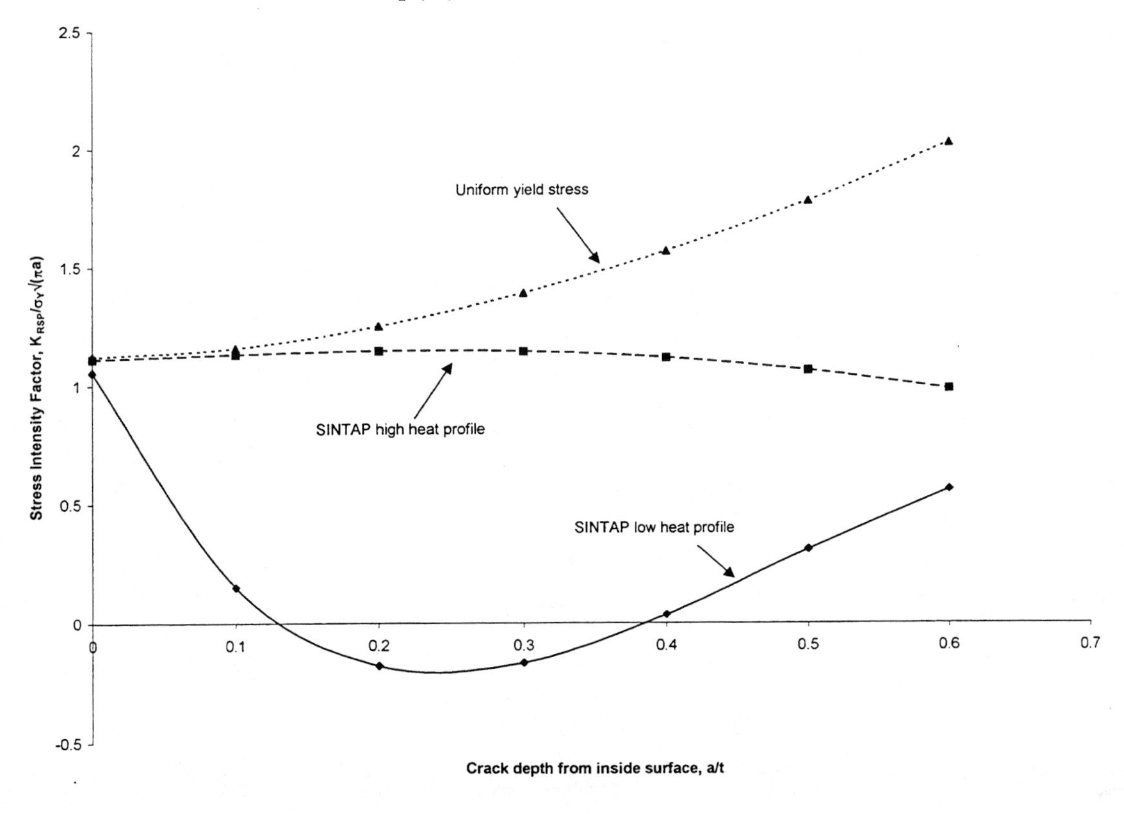

Fig. 10 Normalised K_{RSP} at fully circumferential crack at outside surface of pipe. Effect of residual stress profile.

PVP-Vol. 410-1, Assessment Methodologies for Preventing Failure:
Deterministic and Probabilistic Aspects and Weld Residual Stress
Volume 1
ASME 2000

THE CRACK OPENING AREA DUE TO AN APPLIED AND RESIDUAL STRESS FIELD

E. Smith

**Manchester University/UMIST Materials Science Centre,
Grosvenor Street, Manchester, M1 7HS, United Kingdom**

ABSTRACT

A knowledge of the crack opening area for normal operating conditions is required for the development of a leak-before-break case for a component in a pressurized system. Accordingly the paper provides an exact analytical expression for the opening area associated with a two-dimensional crack which is subjected to a specific tensile stress distribution, that is a realistic combination of applied and weld residual stresses. The analytical results are compared with those determined via a simplified procedure based on the determination of the crack opening displacement at the crack centre coupled with the assumption that the crack opening has an elliptical profile. The paper's analysis builds upon work described at previous PVP Conferences and provides further underpinning for the view that the simplified procedure gives reasonably reliable crack opening area predictions for a wide range of crack sizes and loading conditions.

INTRODUCTION

Formulation of a structural integrity case for a pressurized system often involves the development of a leak-before-break argument which incorporates two key elements. These are: (a) The size of the through-thickness crack that is a stable under accident loading conditions, and (b) The size of through-thickness crack that gives a measurable leakage under normal operating conditions. It is necessary to show that there is an acceptable margin between these two critical crack sizes for a leak-before-break case to be viable.

As regards the second critical crack size, it is important to be able to estimate the crack opening (leakage) area for normal operating conditions, where the opening might be due to a combination of pressure induced and weld residual tensile stresses. An analysis based on the material behaving elastically is probably adequate for many situations, and it is from this basis that this paper derives an exact analytical expression for the opening area associated with a two-dimensional crack that is subjected to a specific tensile stress distribution, which is a realistic combination of applied and weld residual tensile stresses. The analytical results are compared with those obtained via a simplified procedure (Rahman et al, 1995) based on the determination of the crack opening displacement at the crack centre coupled with the assumption that the crack opening has an elliptical profile. The paper's analysis is a further development of work described by the author at previous PVP Conferences (Smith, 1998, 1999) and which involved a power law stress variation. The present paper's results provide further underpinning for the view that the simplified procedure gives reasonably reliable crack opening area predictions for a wide range of crack sizes and loading conditions.

EXACT ANALYSIS FOR CRACK-OPENING AREA

We consider the behaviour of a two-dimensional crack of length 2a which is perpendicular to a longitudinal weld in a large (infinite) plate that is subjected to a uniform applied tensile stress σ_A (Figure 1). The distribution of the tensile residual stress $\sigma_R(x)$ along the line perpendicular to the weld, and in the crack's absence, is represented by the relation

$$\sigma_R(x) = \sigma_R g(x/h) \quad (1)$$

which is shown schematically in Figure 2. The function g(w), with w = x/h, must satisfy certain conditions for it to be compatible with physical requirements (Tada and Paris, 1983). These are:

(a) The residual stress is self-balanced, i.e.

$$\int_{-\infty}^{+\infty} g(w)dw = 0 \quad (2)$$

(b) The residual stress is symmetric with respect to the weld, i.e. g(w) = g(-w).

(c) The residual stress has a maximum at x = 0.

(d) The residual stress diminishes to zero at a great distance from the weld.

(e) The function g(w) decreases from, its (positive) maximum value at w = 0 to negative(minimum) values, being zero at w = ± 1, and then increases and approaches zero monotonically as | w | increases.

The idealised stress distribution shown in Figure 2 exhibits these features.
The specific residual stress distribution (Smith, 1985, 1997) that is described by the function

$$g(w) = \frac{(1-w^2)}{(1+w^2)^2} \quad (3)$$

satisfies the requirements (a) - (e) and is shown in Figure 3. It corresponds to the insertion of edge dislocations with equal (though of opposite sign) Burgers vectors at x = 0, y = ±h with the extra half planes extending from y = h to +∞ and y = -h to y = -∞.

The Mode I stress intensity factor K(a) for a crack of length 2a due to this applied plus residual stress distribution is

$$K(a) = 2\left(\frac{a}{\pi}\right)^{1/2} \int_0^a \frac{(\sigma_A + \sigma_R(x))dx}{(a^2 - x^2)^{1/2}} \quad (4)$$

whereupon relations (1), (3) and (4) give

$$K = \sigma_A(\pi a)^{1/2} + \sigma_R(\pi h)^{1/2}\left[\frac{\lambda}{(1+\lambda^2)^3}\right]^{1/2} \quad (5)$$

with λ = a/h. It follows, by using the arguments developed by Streitenburger and Knott (1995) that the crack opening area A due to this stress distribution, is given by the expression

$$A = \frac{4}{E_o}\int_0^a (\pi s)^{1/2} K(s)ds \quad (6)$$

where E_o = E/(1-v²), with E = Young's modulus, v is Poisson's ratio,and with K(s) being given by relation (5) with a replaced by s. It then follows, after evalutaion of the appropriate integrals, that

$$A = \frac{2\sigma_R \pi a^2}{E_o} + \frac{4\sigma_R \pi a^2}{E_o} \cdot \frac{1}{(1+\lambda^2)^{1/2}\left[1+(1+\lambda^2)^{1/2}\right]} \quad (7)$$

CRACK-OPENING AREA USING APPROXIMATE PROCEDURE

With the elliptical profile assumption, the crack opening area A = $\pi\phi_o$ a/2 where Φ_o is the relative displacement of the crack faces at the crack centre due to the stress distribution $\sigma(x) = \sigma_A + \sigma_R(x)$. Φ_o is given by the expression (Tada et al, 1973)

$$\phi_o = \frac{4}{\pi E_o}\int_o^a \sigma(x)\ell n\left[\frac{a+(a^2-x^2)^{1/2}}{a-(a^2-x^2)^{1/2}}\right]dx \tag{8}$$

Thus with σ_R (x) being given by relations (1) and (3) with w = x/h and λ = a/h, it follows from relation (8) that ϕ_o is given by the relation

$$\phi_o = \frac{4\sigma_A a}{E_o} +$$

$$\frac{4\sigma_R a}{\pi E_o}\int_o^1 \frac{(1-\lambda^2 w^2)}{(1+\lambda^2 w^2)^2}\ell n\left[\frac{1+(1-w^2)^{1/2}}{1-(1-w^2)^{1/2}}\right]dw \tag{9}$$

Substituting w = sin θ and $\mu = \lambda^2$, this relation can be written in the form

$$\phi_o = \frac{4\sigma_A a}{E_o} + \frac{8\sigma_R a}{\pi E_o}[I-2J] \tag{10}$$

where I and J are respectively the integrals

$$I = \int_o^{\pi/2} \frac{\cos\theta}{(1+\mu\sin^2\theta)}\ell n\tan\left(\frac{\theta}{2}\right)d\theta \tag{11}$$

and

$$J = \int_o^{\pi/2} \frac{\cos\theta}{(1+\mu\sin^2\theta)^2}\ell n\tan\left(\frac{\theta}{2}\right)d\theta \tag{12}$$

Differentiation of relation (12) with respect to μ coupled with the use of relation (11) gives

$$J = I + \mu\frac{dI}{d\mu} \tag{13}$$

whereupon expression (10) becomes

$$\Phi_o = \frac{4\sigma_A a}{E_o} - \frac{8\sigma_R a}{\pi E_o}\left[I + 2\mu\frac{dI}{d\mu}\right] \tag{14}$$

With $\beta = (1+\mu)/\mu > 1$, the integral I (see relation (11) is given by the relation

$$2\mu I = \int_o^{\pi/2} \frac{1}{(\beta^{1/2}-\cos\theta)}\ell n\tan\left(\frac{\theta}{2}\right)d\theta - \int_o^{\pi/2} \frac{1}{(\beta^{1/2}+\cos\theta)}\ell n\tan\left(\frac{\theta}{2}\right)d\theta \tag{15}$$

with the right-hand side simplifying to give

$$2\mu I = -\int_o^{\pi} \frac{1}{(\beta^{1/2}+\cos\theta)}\ell n\tan\left(\frac{\theta}{2}\right)d\theta \tag{16}$$

With t = tan (θ/2) and

$\psi = t/[(\beta^{1/2}+1)/(\beta^{1/2}-1)]^{1/2}$, it follows that

$$\mu I = -\frac{1}{(\beta-1)^{1/2}}\int_o^{\infty} \frac{\left\{\ln\psi + \frac{1}{2}\ln\left[\frac{\beta^{1/2}+1}{\beta^{1/2}-1}\right]\right\}d\psi}{[1+\psi^2]} \tag{17}$$

and since

$$\int_o^{\infty}\frac{\ln\psi . d\psi}{[1+\psi^2]} = 0 \tag{18}$$

relation (17) simplifies to

$$I = -\frac{\pi}{2\mu^{1/2}} \ln\left[(1+\mu)^{1/2} + \mu^{1/2}\right] \qquad (19)$$

remembering that $\beta = (1+\mu)/\mu$. Relations (10), (13) and (19) then give, with $\mu = \lambda^2 = (a/h)^2$

$$\phi_o = \frac{4\sigma_A a}{E_o} + \frac{4\sigma_R a}{E_o} \frac{1}{(1+\lambda^2)^{1/2}} \qquad (20)$$

or, since $A = \pi\, \phi_o\, a/2$, we have

$$A = \frac{2\pi a^2}{E_o}\left[\sigma_A + \frac{\sigma_R}{(1+\lambda^2)^{1/2}}\right] \qquad (21)$$

and this is the crack-opening area obtained on the basis of the determination of the relative displacement ϕ_o of the crack faces at the crack centre coupled with the assumption of an elliptical opening profile.

DISCUSSION

The analyses in the preceding sections have given us an exact expression, i.e. (7), for the crack-opening area, and an approximate expression, i.e. (21), based on the determination of the relative displacement at the crack faces at the crack centre coupled with the assumption of an elliptical opening profile. The ratio of the approximate to the exact area is given, from expressions (7) and (21), as

$$\frac{A_{APP}}{A_{EX}} = \frac{\left[1 + \frac{\sigma_R}{\sigma_A} \frac{1}{(1+\lambda^2)^{1/2}}\right]}{\left[1 + \frac{2\sigma_R}{\sigma_A} \frac{1}{(1+\lambda^2)^{1/2}\left\{1+(1+\lambda^2)^{1/2}\right\}}\right]} \qquad (22)$$

Table I shows this ratio for a/h=0.25, 0.50, 0.75 and 1.00, for the three cases where $\sigma_R=0$ (no residual stress), $\sigma_R = \sigma_A$ and $\sigma_A = 0$ (no applied stress). The results in the Table clearly show that the approximate procedure, based on the elliptical opening profile assumption, gives reasonably accurate crack opening area-estimates over a wide range of crack sizes and applied/residual stress combinations, particularly when the normal operational stresses are dominated by the applied stress and with small crack sizes.

However, as noted in the author's earlier paper (Smith, 1999), although the elliptical opening profile procedure gives reasonably accurate crack opening area estimates, these estimates are in fact slightly in excess of the actual values (see Table I). Consequently, for a prescribed crack-opening (leakage) area, the critical crack size associated with a critical leakage will be slightly underestimated with the implication that there is a degree of non-conservatism in the context of a leak-before-break argument. However, this comment is made with the recognition that in this paper the crack-opening areas have been calculated on the basis of elastic analyses. If the areas, and in particular the area based on the elliptical opening profile assumption, are calculated on the basis of elastic-plastic analyses, the areas will be increased (Smith, 1998) and then it is likely that the non-conservatism of the area calculated using the simple procedure will be removed.

REFERENCES

Rahman, S., Ghadiali, N., Wilkowski, G., and Bonora, N., 1995, "The Effects of Off-Centred Crack and Restraint of Induced Bending due to Pressure on the Crack-opening Area Analysis of Pipes", ASME, PVP – Vol. 304, pp 149-162.

Smith, E., 1985, "The Stress Intensity for a Crack Perpendicular to a Longitudinal Weld in a Large Plate", International Journal of Fracture, Vol. 27, pp 75-79.

Smith, E., 1997, "Describing the Effect of Secondary Stresses on Crack Extension in the K_r-L_r Format", International Journal of Pressure Vessel and Piping, Vol. 72, pp 51-55.

Smith, E., 1998, "The Viability of the Elliptical Profile Assumption when Estimating Crack Opening Area", ASME, PVP – Vol. 373, pp 485-488.

Smith, E., 1999, "The Opening Area Associated with a Crack that is Subjected to a General Tensile Stress Distribution", ASME, PVP – Vol. 393, pp 165-168.

Streitenberger, P., Knott, J.F., 1995, "The Calculation of Crack Opening Area and Crack Opening Volume from Stress Intensity Factors", International Journal of Fracture, Vol. 76, pp R49-R54.

Tada, H., Paris, P.C., and Irwin, G.R., 1973, "The Stress Analysis of Cracks Handbook", Del. Research Corporation, Hellertown, Pa, USA.

Tada, H., and Paris, P.C., 1983, International Journal of Fracture, Vol. 21, pp 279.

TABLE 1

The ratio A_{APP}/A_{EX} for various values of a/h and for the three cases were $\sigma_R=0$, $\sigma_R=\sigma_A$ and $\sigma_A=0$.

		$\sigma_R=0$	$\sigma_R=\sigma_A$	$\sigma_A=0$
$\lambda=\frac{a}{h}$	0.25	1.000	1.008	1.015
	0.50	1.000	1.027	1.059
	0.75	1.000	1.052	1.125
	1.00	1.000	1.076	1.208

FIGURE CAPTIONS

FIGURE 1

A crack of length 2a which is perpendicular to a longitudinal weld in an infinite plate that is subjected to a tensile stress σ_A, coupled with a residual tensile stress.

FIGURE 2

The schematic residual tensile stress pattern associated with the weld.

FIGURE 3

The specific residual stress distribution analysed in the paper.

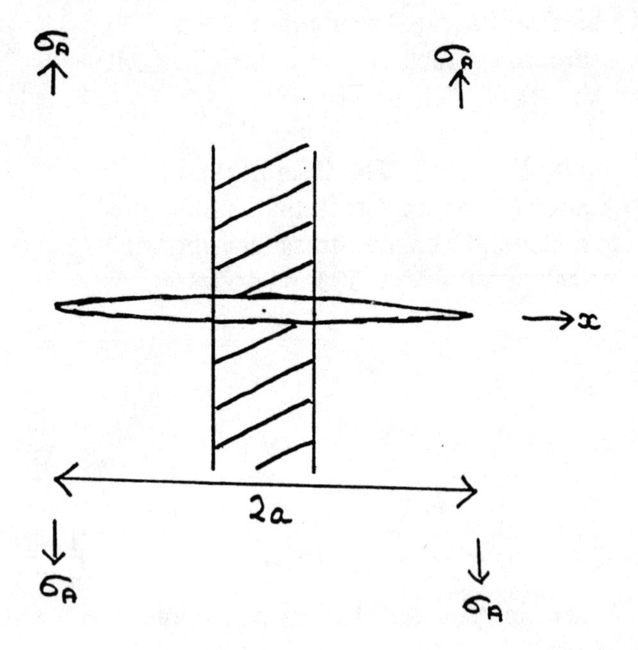

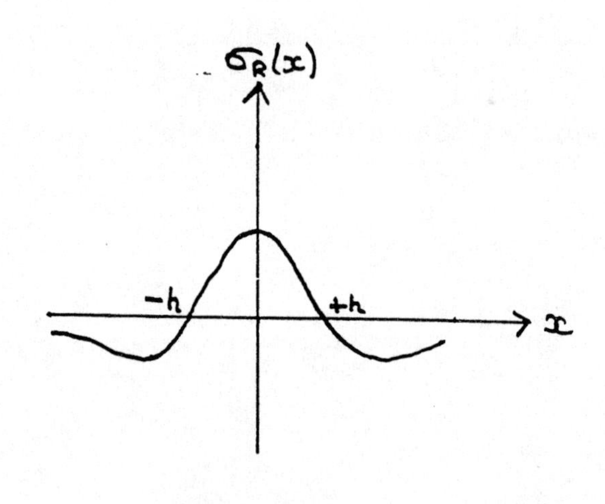

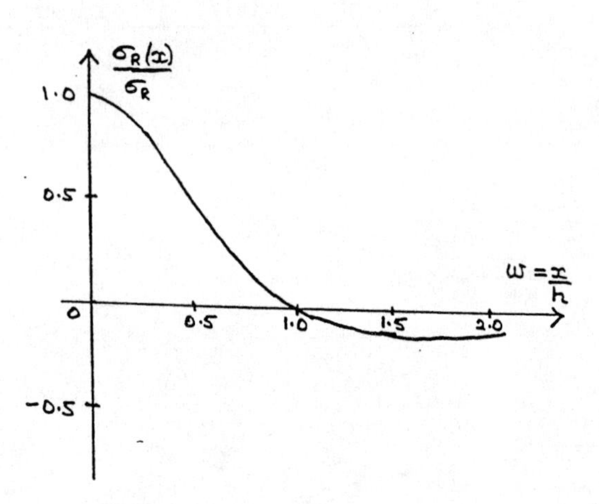

PVP-Vol. 410-1, Assessment Methodologies for Preventing Failure:
Deterministic and Probabilistic Aspects and Weld Residual Stress
Volume 1
ASME 2000

Characteristic Residual Stress Distributions in Pressure Vessels and Piping Components

P. Dong and F. W. Brust
Battelle

ABSTRACT

Typical residual stress distributions as occur in pressure vessel and piping components are discussed. The importance of these stresses in accounting for fracture behavior is pointed out. Several example problems are discussed which illustrate key points.

INTRODUCTION

Over the last decade, welding-induced residual stresses have received an increasing attention in the pressure vessel and piping research community. The primary driving force can be attributed to the fact that recent advances (e.g., References [1,2]) in structural integrity assessment of welded components demand more accurate information on the weld residual stress state. Regardless of the specific details involved, these structural integrity assessment procedures typically assume upper bound residual stress solutions. However, in recent years, there has been major progress in better understanding of weld residual stresses, in part, due to the availability of advanced weld residual stress modeling tools, e.g., Brust, et al [3-6], Dong, et al [7-11], and Zhang et al [12-14]. It has been demonstrated that the current structural integrity assessment procedures can significantly over-estimate the residual stress effects in most cases (e.g., Dong et al, [15] and Brust et al, [16,17]) and under-estimate their effects in others (e.g., Dong, [18]).

The needs become more evident as welding repairs are increasingly used in aging pressure vessel and piping components. As such, both repair procedure development and the subsequent safety assessment require a better understanding of the welding effects on structural components (Zhang et al, [14]). This is because repair weld residual stress distributions can be drastically different from an original weld. As some of the advanced computational modeling techniques as well as new and improved experimental methods have become available over the recent years, more accurate residual stress information can now be obtained for various structural integrity assessment applications.

In this paper, recent advances in weld residual stress analysis techniques are reviewed. A typical girth weld example is then used to demonstrate both the detailed weld residual stress solutions and their effects on fracture behavior from residual stress dominated cracking to leak-before-break assessment. Finally, discussions on some of the recent residual stress solutions for typical pressure vessel and piping components will be given in view of their effects on structural integrity assessment considerations. Future challenges in effectively characterizing residual stress distributions and incorporating them into fracture assessment procedures will be highlighted in view of the recent progress in this area.

RESIDUAL STRESS MODELING

The thermal and thermal-mechanical processes associated with weld residual stress evolution during welding can be extremely complex. Rapid arc heating during welding produces a molten weld pool. The weld pool shape can be largely influenced by the weld metal transfer mode and corresponding fluid-flow characteristics. On cooling, both rapid solidification within the weld pool and solid-state phase transformation in heat affected zones (HAZ) occur, depending on both peak temperature and cooling rate. The process becomes even more complicated in multi-pass welds. There are significant research activities on-going currently to address both the fundamental physics and numerical methods associated with these phenomena. However, as far as weld residual stress modeling is concerned, numerical procedures can be significantly simplified. For residual stress predictions, localized temperature distributions ranging from room temperature to above melting and structural restraint are of primary importance in determining the final residual stress state in a weldment. Considering an equivalent fusion profile in conduction-based heat flow solution techniques can ignore detailed weld pool phenomena. The material-melting behavior within the weld pool can be treated using phenomenological thermal-plasticity model within the continuum mechanics context (Brust et al, [3]) and as discussed by Dong et al, [8]).

TYPICAL GIRTH WELDS

A typical pipe girth weld model is shown in Fig. 1. The girth weld mock-up was made of 316L stainless steel. The idealized weld pass profiles (Figs. 1a and 1b) were derived for computational model generation based on detailed examination of the macrographs of the weld cross section. The specific details can be found in Dong et al [8]. Both axis-symmetric and special shell element models were used and are shown in Fig. 1. The special shell element model is shown in Fig. 1c and assumes symmetry conditions with respect to the weld centerline. The welding arc travel direction and start/stop position (at $\theta = 0$) are shown in Fig. 1c. An overlap of 10 mm for the start and stop was assumed in the analysis. The special shell element model was intended to capture some of the three-dimensional residual stress characteristics and start/stop position effects, while the axis-symmetric model was used to resolve the local residual stress details.

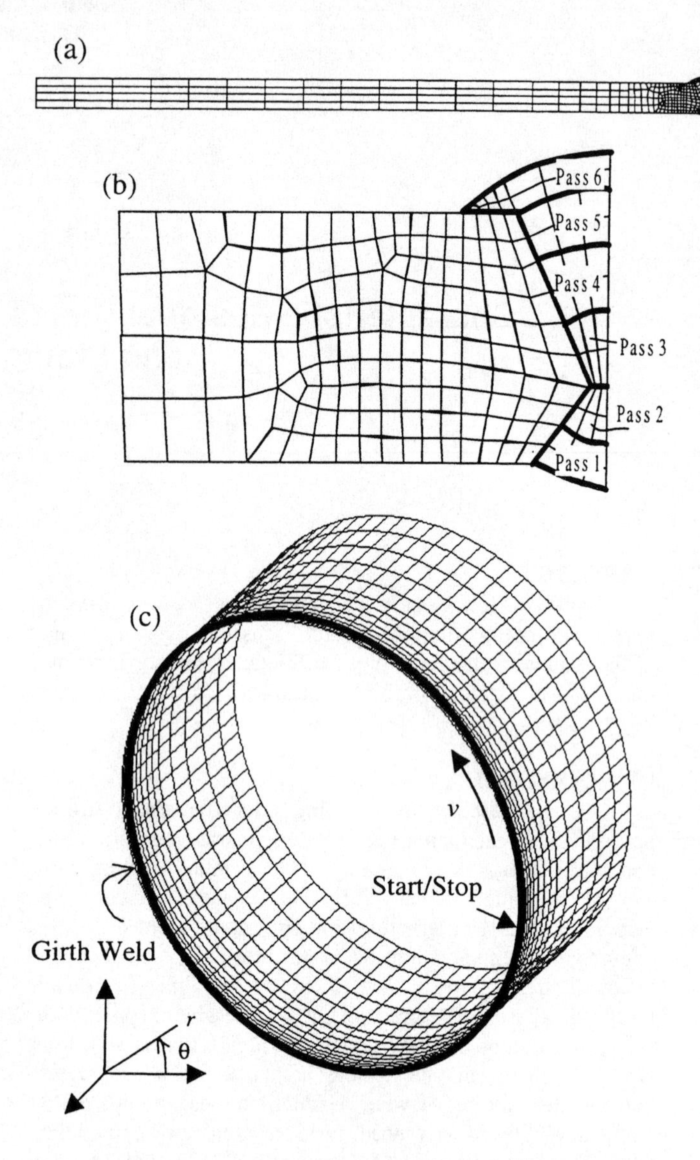

Fig.1 A combined 3D shell element and 2D cross-section model for residual stress analysis of a multi-pass girth weld: (a) Axisymmetric model; (b) Weld area; (c) 3D shell element model

Residual Stresses – Axis-symmetric Model

The final residual stress distributions based on the axis-symmetric model are given in Fig. 2, with respect to axial distance from the weld centerline on both outer and inner surfaces, respectively. Two sets of finite element results are presented in this figure. One was from the 6-pass model and the other from a 5-pass model in which Pass 5 and Pass 6 (Figure 1b) were deliberately lumped together. This was done to investigate the effects of using lumped passes to simplify the computational procedures. Hole-drilling measurements performed on the weld mock-up specimen are also given for comparison purposes. It can be seen that the overall trend between the finite element predictions and the experimental measurements agrees reasonably well.

Of a particular interest is the axial residual stress component that has a direct relevance to fracture assessment of the girth-welded pipe as reported in Dong, et al [18]. The axial residual stresses exhibit a bending type distribution through the pipe wall thickness. Within the weld area, the predicted axial residual stresses are tensile at the inner surface and compressive with a relative low magnitude on the outer surface. The hole drilling measurements showed an excellent agreement, particularly away from the weld. The implication of such a residual stress distribution on fracture behavior was discussed in Brust, et al [4-6] for a thick girth weld and under repair conditions.

Residual Stresses - Special Shell Element Model

The results presented earlier were based on axis-symmetric assumptions. Before the residual stress results can be generalized for the girth-welded pipe components of concern, the limitations of the axis-symmetric assumptions must be examined. There have been numerous publications where the detailed residual stress measurements collected on similar girth-welded pipes showed the residual stresses were not symmetric, for example, the work by Shack, et al [19]. The deviation from axis-symmetry is believed to be dependent on welding conditions and pipe Radius to thickness (R/t) ratio, etc., as discussed in Dong, et al [11]. To quantify some of the 3D effects, a special 3D shell element model recently developed for multi-weld simulation (Zhang, et al, [12]) was used in this investigation. The shell element is shown in Fig. 1c. To simplify the computational effort, it was further assumed that the last weld plays a dominant role in the final residual stress distribution on the outer surface. This proves to be reasonable assumption (Zhang, et al, [12-14]) for assessing 3D global residual stress distributions in many circumstances.

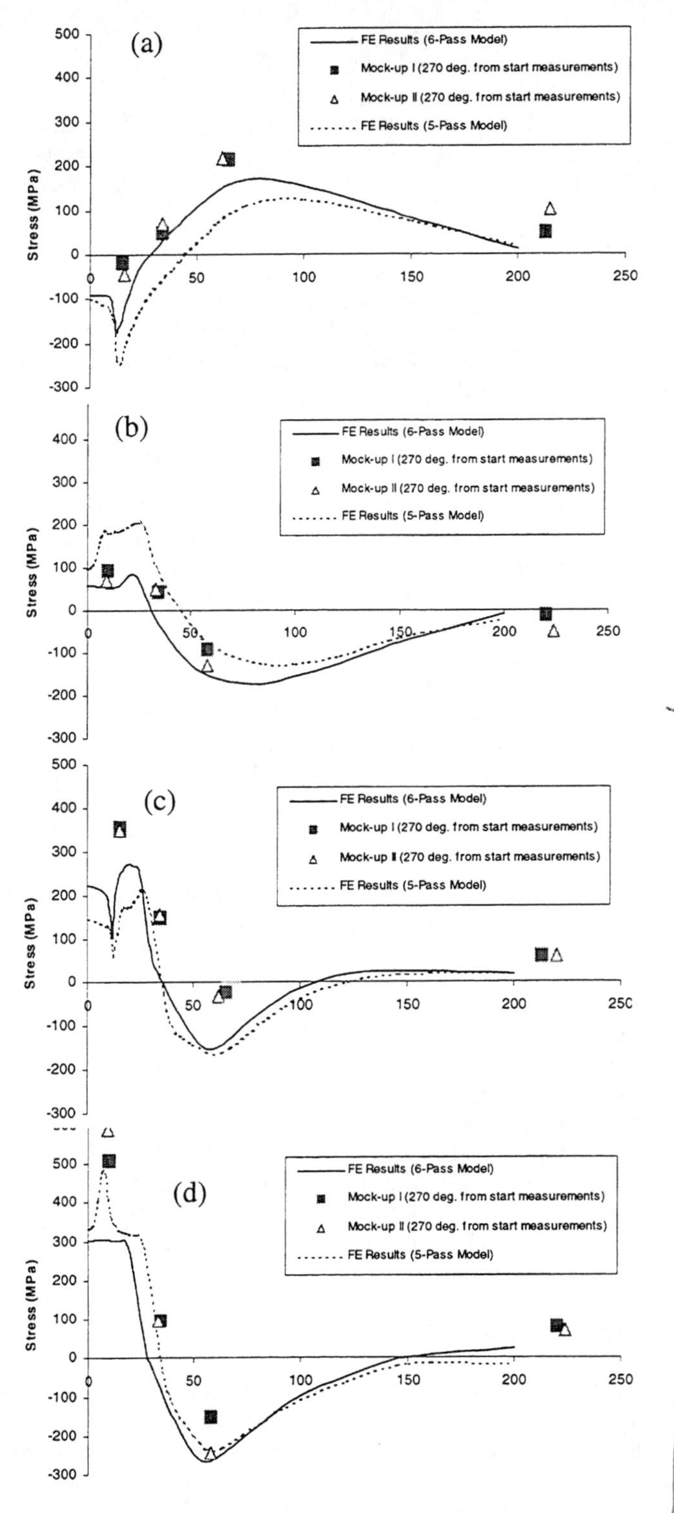

Fig. 2: Axisymmetric residual stress modeling results: (a) Axial residual stress - outer surface; (b) Axial residual stress - inner surface; (c) Hoop residual stress - outer surface; (d) Hoop residual stress - inner.

The lumped heat input for the moving arc heat flow solution in the shell element model was based on the combined effects of Passes 5 and 6 in the 5-pass axis-symmetric model. As shown in Fig. 3, the axial residual stresses within the weld region were compressive on the outer surface. This feature is consistent with the axis-symmetric results (Fig. 2a). Some periodic variations are present along the circumference. Such a variation was more pronounced for the axial component. A rapid change in residual stresses was seen at the start/stop positions. Similar residual stress characteristics were also discussed in (Dong et al, [8,11]).

In this pipe girth weld, it can be seen that the axis-

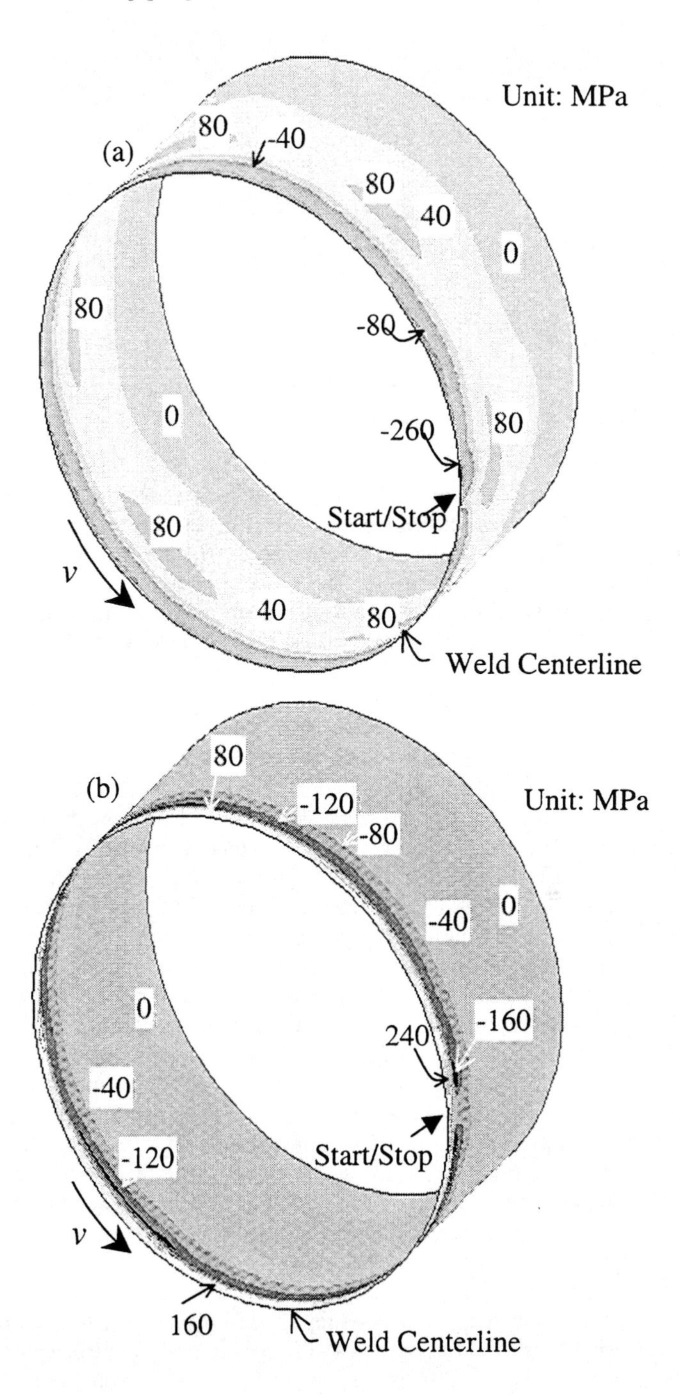

Fig. 3 Predicted Residual Stress Distributions on the Outer Surface: (a) Axial Residual Stress; (b) Hoop Residual Stress

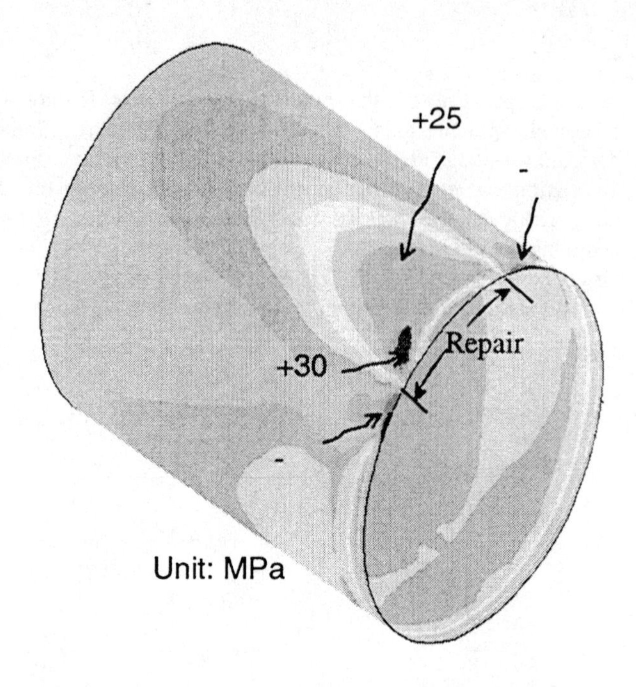

Fig. 4 Predicted final transverse residual stress distribution on the outer surface

symmetric model provided a reasonable prediction of the residual stress distributions. However, in general, a careful interpretation of the axis-symmetric results is required in view of its inherent limitation. The underlying assumptions strictly imply that the entire girth weld was formed in a simultaneous manner. As such, the results obtained tend to reflect the residual stress distribution in a girth weld in an average sense and cannot be interpreted, strictly speaking, as a representation of a cross section away from the start/stop positions. Furthermore, an axis-symmetric weld formation almost never occurs in practice as far as arc-welding processes are concerned. As demonstrated in this investigation, both the traveling arc and start/stop effects tend to violate the axis-symmetric assumptions by introducing circumferential variations of residual stresses. Such effects can be further enhanced in multi-pass welds if travel directions or start/stop positions are changed from pass to pass, which often occur in practice. Perhaps for this reason, past comparisons between axis-symmetric finite element analysis and experimental results often showed significant discrepancy (e.g., Rybicki and Stonsifer, [20] and Brust and Kanninen, [21]). To incorporate some of the three-dimensional effects, the special shell element model offers a computationally efficient approach. By combining a 2D cross-section model with the special shell model, both local residual stress details and global characteristics can be captured (Zhang et al, [12]).

Repair Weld Residual Stresses

In analyzing multi-pass repair weld residual stresses in the same girth weld as in Fig. 1, the special shell element model was used as shown in Fig. 4. The detailed discussions on using the 3D-shell element procedure for simulating multi-pass welds can be found in Zhang et al [12]. In this approach, the analytical based thermal solution technique was used (see Cao [22]) in conjunction with the weld constitutive model, as discussed earlier. A pass-by-pass metal deposition and corresponding temperature history can be directly simulated as shown in Fig.4 for an intermediate repair pass. The final axial residual stress distribution on the outer surface is shown in Fig. 4 after 3-pass repair welding. It can be seen that once the repair weld is introduced, the axial residual stresses are significantly elevated within the entire length of the repair, with the maximum tensile stresses occurring near the stop position. These high tensile stresses occur through the thickness and, as seen next, can result in a significantly different crack growth response. High compression zones are present outside of the repair length.

Stress Intensity Factor Solutions – Surface Cracks

In some piping components, stress corrosion cracking due to weld residual stresses can be a major concern. As discussed in Brust et al [4-6,16,17], Dong and Zhang [15], and Hou, Pan, and Brust [23], the conventional J-Integral approach can no longer be conveniently used in extracting stress intensity factors due to its path dependence in residual stress fields. The finite element alternating method (FEAM) has demonstrated to be an extremely efficient method for computing stress intensity factors for a variety of weld configurations (see for instance Brust et al. [4-6]). With this procedure, a 3D FEAM model is used with mapped residual stress fields from either 2D cross-section or 3D shell element model, or a full 3D model can be used directly. By considering a partial elliptical surface crack in both weld residual stress fields with and without repair, the stress intensity factor solutions were shown in Brust et al [4-6]). Without repair weld, the stress intensity factors along the elliptical crack front are high at sub-surface and rapidly reduced to negative values as the crack depth approaches the mid-thickness of the pipe wall. For any potential crack growth from this point on in this situation, the elliptical crack will be more likely to grow longer along the circumferential direction. With the presence of the repair weld, the overall elevation of the stress intensity factors can be clearly observed. The elliptical crack will be more likely to grow deeper first, and possibly break through the pipe wall, before growing longer. This is extremely important as weld repair continues to be used in aging plants.

Crack Opening Displacement for LBB Assessment

An important consideration in leak-before-break (LBB) analysis for nuclear piping system is an accurate prediction of crack-opening area (COA). A comprehensive discussion on this subject can be found in Rahman et al [24]. Among some of the important issues, effects of weld residual stresses on crack-opening behavior are not well understood. A comparative study using the special shell element discussed earlier and an elastic superposition method was reported in Dong et al [18], where it was demonstrated that a full-field residual stress field can introduce more significant crack closure than predicted using the elastic super-position method.

As the through-wall crack becomes longer, present LBB procedures (Rahman et al, [24]) assumes the crack opening area becomes proportionally larger, according to linear elastic analysis results. As shown in Fig. 5, the actual crack opening behavior becomes rather complicated once the full field residual stresses are considered. As the crack length increases from 2c = 150mm to 300mm and 500mm, the crack opening profile starts to deviate from the typical elliptical shape, particularly at ID and crack closure effects become more significant. Such non-elliptical opening behavior can be attributed to the increasing effects of the hoop residual stresses

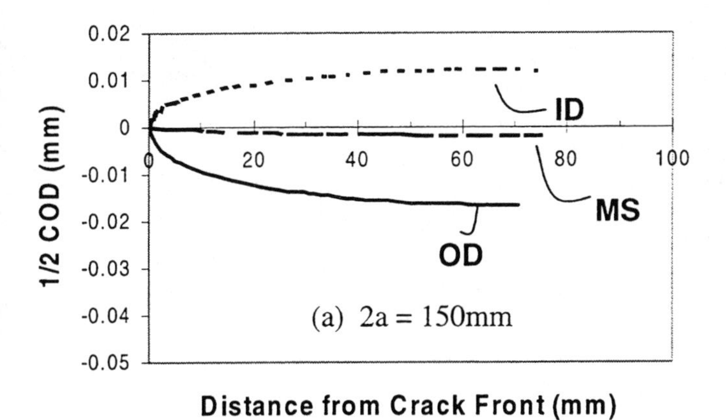

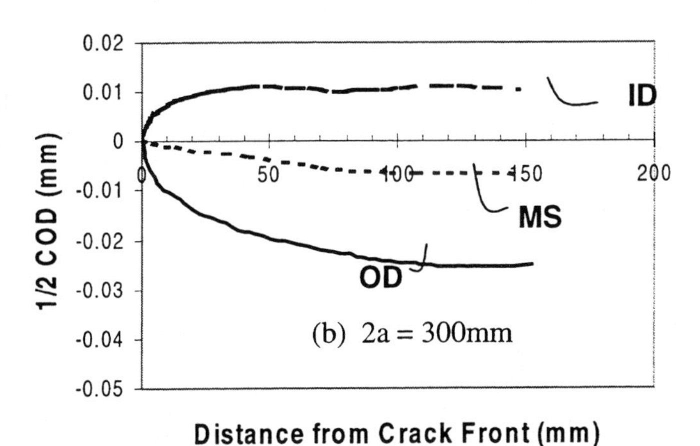

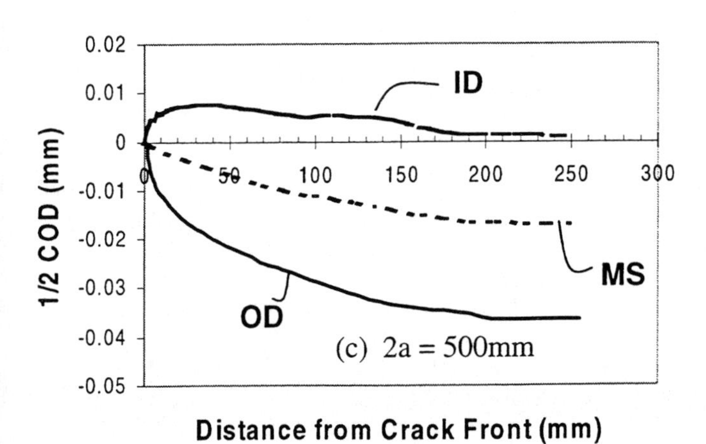

Fig. 5 Crack-opening displacement versus crack length (2c) due to weld residual stresses only: (a) 2c = 150mm, (b) 2c = 300mm; (c) 2c = 500mm.

along the crack face as the crack increases in length. Accordingly, crack opening area calculations must take such effects into account for LBB assessment. Further study revealed that the hoop stresses redistribution, which occurs as a crack grows, contributes to the non-elliptical shape demonstrated by introducing buckling-type deformation mode along the crack face.

Generalization of Residual Stress Distributions

Based on the detailed discussions in the previous section in the context of a typical girth weld, one can readily conclude that as more accurate weld residual stress information becomes available, not only can welding procedures be optimized to mitigate weld residual stress effects, but also to improve structural integrity assessment procedures. In most cases, excessive conservatism can be eliminated from the current code of practice (e.g., code-specified residual stress field), and in others, residual stress effects can be adequately taken into account to ensure code-specified safety margin to be met (e.g., LBB assessment). To achieve this, typical residual stress distributions must be characterized and generalized for fracture assessment purposes.

The particular girth weld discussed earlier represents only one type of residual stress distribution in pressure vessel and piping components. In theory, many parameters can affect the detailed residual stress development, e.g., welding procedures, weld geometry, and material etc. However, an in-depth review of the recent research results for various applications suggests that two types of residual stress distributions dominate the majority of pressure vessel and piping applications, i.e. "bending" and "self-equilibrating" types. Girth welds with drastically different pipe R/t ratio and pass profile still possess essentially the same residual stress distributions, i.e., "bending" type with compressive axial stresses near outer surface and tensile stresses at the inner surface. The "self-equilibrating" type can be inferred for the girth welds with significantly different geometry and weld profiles.

As demonstrated earlier, knowing the residual stress type for a particular application is essential before any structural integrity assessments can be accurately performed. However, although some of the controlling parameters governing each of the two types of residual stress distributions can be qualitatively identified based on some of the existing residual stress results to date, the demarcation line separating the two types of residual stresses in terms of component geometry and welding procedures has not been established. Consequently, there exists an urgent need in the pressure vessel and piping community to develop the required knowledge base for quantitative definition of the residual stress types as well as the corresponding parameterized distributions. Typical residual stress distributions for other weld configurations, such as nozzles, repair welds etc. should also be addressed in a similar manner. For instance, recent results (Dong et al, [10]) showed that residual stresses due to weld repair tend to show essentially the same characteristic distributions, regardless of girth weld or large vessel.

Along this same line, some noted collaborative research efforts are currently under planning, for instance, by ASME Pressure Vessel Research Council (PVRC) and Materials Property Council (MPC) to be focused on improved fitness-for-service assessment procedures for petrochemical applications (Osage et al, [25]). For nuclear utilities, one additional area of concern should be the incorporation of the detailed weld residual stress effects on crack-opening area calculations (e.g., crack length and through-wall residual stress types) for LBB assessment purposes. Furthermore, existing fracture mechanics analysis procedures incorporating weld residual stresses assume that a specified type of residual stress distribution is stationary with respect to crack locations. Such an assumption can lead to underestimation of the residual stress effects on fracture driving force, as shown in Fig. 6. Moreover, as also seen in Figure 6, the response of the crack driving force can significantly change as the crack grows, especially in large vessels.

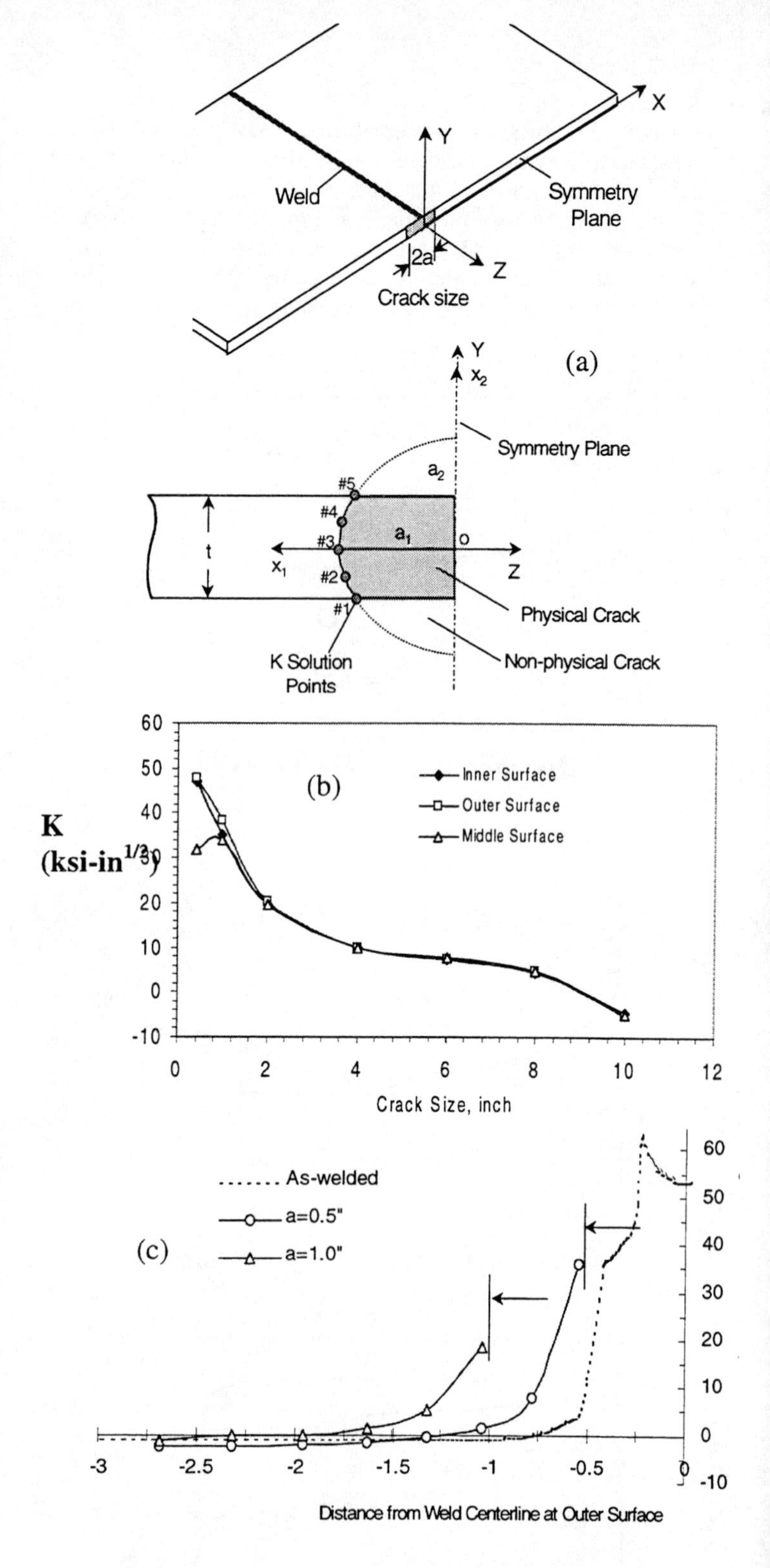

Fig. 6. Stress intensity factors for a through-wall crack orientation in a butt joint: (a) Crack definition; (b) stress intensity factors for growing crack; (c) Residual stress re-distributions versus crack length

CONCLUSIONS

Over the last decade, significant progress has been made towards a better understanding of weld residual stresses and their effects on fatigue, corrosion, fracture, and high temperature damage behavior in pressure vessel and piping components. As a result, novel-welding procedures can be developed to mitigate weld residual stress effects on structural integrity. In addition, fracture mechanics assessment procedures can be refined to provide improved life predictions. Some of the major advances worth noting are:

(1) Residual stresses in typical multi-pass welds are reasonably well understood with the emergence of some commercially available finite element codes in conjunction with appropriate assumptions.
(2) Residual stresses in complicated structures can be analyzed with some of the advanced computational residual stress analysis procedures for case by case applications.
(3) The beneficial effects of incorporating realistic weld residual stresses in fracture mechanics models for structural integrity assessment are being increasingly recognized by industry due to growing concerns in safety and maintenance costs in pressure equipment.

Some of the major challenges are:

(1) Generalized residual stress distributions that cover a wide range of applications remain to be established for fracture mechanics assessment purposes. Such residual stress distributions must be parameterized with respect to weld geometry, material and important welding procedure variables.
(2) Fracture mechanics frame work for effectively incorporating full-field residual stress information (e.g., residual stress induced stress tri-axiality) should be developed for fracture driving force characterization, particularly in the nonlinear response regime.
(3) The present leak-before-break (LBB) assessment procedures can severely under-estimate crack size based on crack opening area without taking into account crack rotation due to full field residual stresses. Improved procedures for crack opening area calculation remain to be developed for various through-wall residual stress

types categorized using today's advanced residual stress analysis techniques.

REFERENCES

1. API RP-579, 2000 First Edition, "Recommended Practices for Fitness-for-Service".
2. BS7910, 1999, British Standards Institution, London.
3. Brust, F.W., Dong, P., and Zhang, J., 1997, "A Constitutive Model for Welding Process Simulation Using Finite Element Methods," *Advances in Computational Engineering Science,* Atluri, S.N., and Yagawa, G., eds., pp 51-56.
4. Brust, F.W., Dong, P., and Zhang, J., 1997, "Influence of Residual Stresses and Weld Repairs on Pipe Fracture," ASME Pressure Vessel and Piping Conference Proceedings, PVP-Vol. 347, Approximate Methods in the Design and Analysis of Pressure Vessel and Piping Components, pp. 173-191, 1997.
5. Brust, F.W.; Zhang, J.; Dong, P., 1997, "Pipe and Pressure Vessel Cracking: The Role of Weld Induced Residual Stresses and Creep Damage during Repair", Transactions of the 14th International Conference on Structural Mechanics in Reactor Technology (SMiRT 14), Lyon, France, Vol. 1, pp. 297-306.
6. F. W. Brust, P. Dong, J. Zhang,, 1997, *Crack Growth Behavior in Residual Stress Fields of a Core Shroud Girth Weld.* Fracture and Fatigue, H. S. Mehta, Ed., PVP-Vol. 350, pp. 391-406.
7. P. Dong:, 1995, "Modeling and Analysis of Alloy 2195 Welds in Super Lightweight Tank", Battelle Project Report to Marshall Space Flight Center, Huntsville, Alabama, July.
8. Dong, P., Ghadiali, and Brust, F.W., 1998 "Residual Stress Analysis of A Multi-Pass Girth Weld," ASME Pressure Vessel and Piping Conference Proceedings, PVP-Vol. 373, *Fatigue, Fracture, and Residual Stresses,* pp. 421-431.
9. Dong, P., Hong, J.K., Zhang, J., Rogers, P., Bynum, J., and Shah, S., 1998, "Effects of Repair Weld Residual Stresses on Wide-Panel Specimens Loaded in Tension", *ASME Journal of Pressure Vessel Technology*, Vol. 120, No. 2, 1998, pp 122-128.
10. Dong, P., Zhang, J., and Li, M.V., 1998, "Computational Modeling of Weld Residual Stresses and Distortions – An Integrated Framework and Industrial Applications, ASME Pressure Vessel and Piping Conference Proceedings, PVP-Vol. 373, *Fatigue, Fracture, and Residual Stresses,* pp. 311-335, 1998.
11. Dong, Y., Hong, J.K., Tsai, C.L., and Dong, P., 1997, "Finite Element Modeling of Residual Stresses in Austenitic Stainless Steel Pipe Girth Welds," *Welding Journal,* Vol. 10, No.10, pp. 442s-449s.
12. Zhang, J., Dong, P., and Brust, F.W., 1997, "A 3-D Composite Shell Element Model for Residual Stress Analysis of Multi-Pass Welds," Transactions of the 14th International Conference on Structural Mechanics in Reactor Technology (SMiRT 14), Lyon, France, Vol. 1, pp. 335-344.
13. Zhang, J., Dong, P., Brust, F. W., William J. Shack, Michael E. Mayfield, Michael McNeil, 2000. "Modeling Weld Residual Stresses in Core Shroud Structures", Nuclear Engineering and Design, 195, pp. 171-187.
14. Zhang, J. and Dong, 2000. "Residual Stresses in Welded Moment Frames and Effects on Fracture," *ASCE Journal of Structural Engineering,* March, No. 3.
15. Dong, P., Zhang, J, 1999. "Residual Stresses in Strength-Mismatched Welds and Implications on Fracture Behavior," *Engineering Fracture Mechanics,* 64, pp. 485-505.
16. Brust, F.W., 1999, "Classical and Emerging Fracture Mechanics Parameters for History Dependent Fracture with Application to Weld Fracture", PVP-Vol. 393, Fracture, Fatigue and Weld Residual Stress ASME
17. F. W. Brust, P. Dong, 2000, "Welding Residual Stresses and Effects on Fracture on Pressure Vessel and Piping Components: A Millennium Review and Beyond", To Appear in the ASME Journal of Pressure Vessel and Piping, Special Millennium Edition, 2000.
18. Dong, P., Rahman, S., Wilkowski, G., Brickstad, Bergman, M., Bouchard, J., and Chivers, T., 1996, "Effect of Weld Residual Stresses on Crack-Opening Area Analysis of Pipes for LBB Applications," ASME Pressure Vessel and Piping Conference Proceedings, PVP-Vol. 324, pp. 47-64.
19. Shack, W.J., Ellington, W.A., and Pahis, L.E., 1980, "Measurement of Residual Stresses in Type 304 Stainless Steel Butt Weldments, EPRI NP-1413.
20. Rybicki, E.F., Stonesifer, R.B., 1979, "Computation of Residual Stresses Due to Multi-Pass Welds in Piping Systems," Journal of Pressure Vessel Technology, Vol. 101, pp. 149-154.
21. F. W. Brust and M. F. Kanninen, 1981b, "Analysis of Residual Stresses in Girth Welded Type 304-Stainless Pipes", *ASME Journal of Materials in Energy Systems,* Vol. 3, No. 3.
22. Cao, Z, Brust, F. W., Dong, Y., Nanjundan, A., and Jutla, T., 2000, "A New Comprehensive Thermal Solution Procedure For Multi-pass and Curved Welds", To be presented at and appear in Proceedings of the 2000 ASME Pressure Vessel and Piping Conference Seattle, Washington, July 23-27.
23. Y.-C. Hou, J. Pan and F. W. Brust, 1998, "A Fracture Analysis of Welded Pipes with Consideration of Residual Stresses," in PVP-Vol. 373, Fatigue, Fracture, and Residual Stresses -1998-, ASME PVP Conference, San Diego, July 26-30, pp. 433-437.
24. Rahman, S., et al., 1995, "Refinement and Evaluation of Crack-Opening Area Analyses for Circumferential Through-Wall Cracks in Pipes," NUREC/CR-6300, BMI-2184, U.S. Nuclear Regulatory Commission, Washington, D.C.
25. Osage, D., Prager, M, and Dong, P., 1999, PVRC/MPC draft long range plan for weld residual stress characterization and local post-weld heat treatment.

ACKNOWLEDGMENT

This work is sponsored in part by the Department of Energy, Office of Basic Engineering Sciences, under Grant No. DE-FG02-90ER14135. The authors thank Dr. Robert Price for his support.

PVP-Vol. 410-1, Assessment Methodologies for Preventing Failure:
Deterministic and Probabilistic Aspects and Weld Residual Stress
Volume 1
ASME 2000

Investigation for Effect of GTA Welding Parameters on Residual Stress by Finite Element Analysis

Seok Joo Hong, Jong Sung Kim, Sung Ho Hong
Structural Integrity & Materials Department,
Korea Power Engineering Company,
Republic of Korea

Abstract

In this paper, residual stress distributions are analytically determined for GTA spot weld plate of structural steel with cladding material. The finite element model is constructed by considering phase transformations. The effects of variations in material properties due to phase transformation and change on residual stresses in welded part are investigated. The effect of GTA welding parameter on residual stress is also investigated.

1. Introduction

One of the major problems in welded structures is residual stress. Residual stresses around the weld joint are detrimental to the integrity and service behavior of the welded part. Residual tensile stresses in the near of weld may promote brittle fracture, stress corrosion cracking, and reducing fatigue life.

A weldment is locally heated by various heat sources, and temperature in the region near the weldment is not uniform. Because of localized heating, thermal stresses are generated during welding. As a result of phase transformation and non-uniform cooling, stresses are remain in the material. In this paper, phase transformation means liquid – solid change in welding, and we consider physical properties variation.

Several factors may effect on the formation of residual stress [1-3]. The structural factors include the geometry, thickness, and joint design. The material factors include metallurgical condition of base metal and weld metal. The fabrication factors include welding process, procedure, welding parameter. In many weld process, various studies of residual stress analysis are accomplished [1-5]. but study of gas tungsten arc (GTA) spot weld is a few. Some study was accomplished in residual stress of same materials [6]. But it is difficult to find that residual stress analysis on dissimilar metal welding.

In this paper, the residual stress states are analytically determined for GTA spot-weld plate of structural steel with nickel-alloy clad material. The effects of GTA weld parameters on residual stresses in welded part are investigated.

2. EXPERIMENTAL PROCEDURE

2.1 Geometry of model and material properties

The geometry of model is shown in Fig. 1. The chemical composition of base metal (SS-400) and clad metal (C-276) is shown in Table 1. The liquidus and solidus temperatures calculated using the formula[8-9] ; liquidus and solidus temperature are 1517.396℃ and 1478.157℃ respectively. Fig. 2 shows density, thermal conductivity, and specific heat [10-12] change with temperature in base metal. The elastic modulus and thermal expansion of base metal vary with temperature change. The Poisson ratio of base metal is determined 0.275 [13].

According to Kozloswki [14], creep behavior can be expressed as ;

$$\dot{\varepsilon} = 0.344e^{-17160/T}\sigma^{n}t^{m} \quad (1)$$

$$n = 6.365 - 4.521\times10^{-3}T + 1.439\times10^{-6}T^{2} \quad (2)$$

$$m = -1.362 + 5.761\times10^{-4}T + 1.982\times10^{-8}T^{2} \quad (3)$$

where $\dot{\varepsilon}$ is creep strain rate (sec^{-1}), σ is stress (MPa), n is strain rate exponent, T is temperature (K), t is time (sec).

Density of C-276 is determined 8.82×10^{-6} kg/mm^3 [15]. Fig. 4 and Fig. 5 show thermal conductivity and specific heat, elastic modulus, thermal expansion change with temperature [15, 16]. The Poisson ratio of C-276 is determined 0.307 [15].

Creep behavior of C-276 is determined as follow ;

$$\dot{\varepsilon} = A\sigma^{n} \quad (4)$$

$$A = 10^{-154.095+0.286749T-0.00018334T^{2}+3.91865\times10^{-8}T^{3}} \quad (5)$$

$$n = 2.7743\times10^{6}T^{-1.92517} \quad (6)$$

where ε is creep strain rate (sec^{-1}), A is constant, σ is stress (kg/mm^2), n is strain rate exponent.

2.2 Boundary condition and FE modeling

In thermal analysis, conductive and convective heat transfer in weldment and ambient temperature are considered.

Fig. 6 shows thermal boundary condition of GTA spot welding. Tungsten electrode with radius r_w is applied heat flux to weldment, and heat convection happen around the air, but radioactive heat transfer is not considered.

Heat flux and heat convection are determined as follows;

$$Q = \eta\frac{I\times V}{\pi\times r_{w}^{2}} \quad (7)$$

$$q_{a} = h_{a}(T - T_{a}) \quad (8)$$

where Q is heat flux (W/m^2), q_a is heat convection (W/m^2), η is heat efficiency, I is current (ampere), V is voltage (volt), h_a is heat convection coefficient with air ($10W/m^2$℃), T_a is temperature of air (25℃).

Table 2. shows welding parameters [19, 20]. Fig. 7 shows mechanical condition. The thermal analysis and residual stress distribution was computed using ABAQUS [7].

2.3 Thermal analysis

Thermal analysis is accomplished as follows; heat efficiency is 0.3, current is 150A, voltage is 15V, and welding time varies from 7 sec to 10 sec. Fig. 9 shows thermal distributions of 10 second welding time. Fig. 10 shows nugget size variation. Nugget width and depth are linearly increased with time. In particular, nugget width is more rapidly increased than nugget depth.

2.4 Creep analysis

It is accomplished similar condition of thermal analysis. Fig. 11 shows stress distribution change with time. In heating process stress around weldment and HAZ is higher than other part, but in cooling process stress in the vicinity of HAZ is high. The reason is why steep temperature gradient at the boundary of weldment and HAZ in cooling process. Fig. 12 shows residual stress distribution change with welding time in HAZ, but residual stress change is small.

3. RESULTS AND DISCUSSION

An FE analysis for residual stress in dissimilar material spot weld was developed. The computations were carried out using ABAQUS code. The results show as follow; (1) nugget width and depth are linearly increased with time. In particular, nugget width is more rapidly increased than nugget depth. (2) Residual stress is the highest at the surface of HAZ. (3) Residual stress shows stress distribution change with welding time, but residual stress change is small.

4. REFERENCES

[1] N.J. Smith, J.T. McGrath, J.A. Gianetto, R.F. Orr, Weld. J. Res. Suppl., Mar., 1989, 112

[2] B. Taljat, T. Zacharia, X.L. Wang, J.R. Keiser, Z. Feng, M.J. Jirinec, Approximate methods in the design and analysis of pressure vessels and piping components, PVP ASME 347, 1997, 83

[3] A. Jovanovic, A.C. Lucia, Int. J. Press. Vessels Pip. 22, 1986, 111

[4] J.K. Hong, C.L.Tsai, P. Dong, Weld. Res. Suppl., September, 1998, 372

[5] P. Dong, N. Ghadiali, F.W. Brust, PVP ASME 373, 1998, 421

[6] B. Taljat, B. Radhakrishnan, T. Zacharia, Mat. Sci. and Eng., A246, 1998, pp.45

[7] ABAQUS manual, ABAQUS/Standard User's Ⅱ, Hibbitt, Karlsson & Sorensen, Inc., 1997

[8] A. Kagawa and T. Okamoto, Mater. Sci. Technol. 1986. 2., pp.997

[9] O. Kubachewski, Iron binary phase diagrams, Berlin Springer Verlog, 1982

[10] A. Jablonka, K. Harse and K. Schwerdtfeger, "Thermomechanical properties of iron and iron-carbon alloys : density and thermal contraction" Steel Rearch. Vol. 62, No.1, 1991, pp.24-33

[11] H. Jacobi, G. Kaestle and K. Wunnenberg, "Heat transfer in cyclic secondary cooling during solidification of steel", Iron making and Steel making, Vol. 11, No. 3, 1984, pp.132-145

[12] B. G. Thomas, "Stress modeling of casting processes : an overview", Modeling of Casting, Welding and Advanced Solidification Process Ⅵ, The Minerals, Metals & Materials Society, 1993, pp.519-533

[13] P. J. Wray, "Mechanical, physical and thermal data for modeling : the solidification processing of steels", Modeling of Casting and Modeling Process, edited by H. D. Brody and D. Apelian, A Publication of the Metallurgical Society of AIME, 1980, August, 3-8

[14] P. F. Kozlowski, B. G. Thomas, J.A.Azzi, Hao Wang, "Simple Constitutive Equations for Steel at High Temperature", Metallurgical Transactions A, Vol. 23A, Mar.,1992, pp.903-918

[15] Inco alloys C-276 Brochure

[16] Haynes International web site (www.haynesintl.com/c276site)

[17] D.R.Gaskell, "Introduction to Metallurgical Thermodynamics", 2nd ed.

[18] INCONEL alloy 690 Brochure

[19] NiDI (Nickel Development Institute), "Guidelines for wallpapering sheet lining with Nickel/Chrome/Molybdenum alloys"

[20] Welding Handbook AWS, 8th ed. Vol. 1, "Welding Technology"

Table 1. Chemical composition of SS400 and C-276.

Material	C	Ni	Mo	Cr	Ti	Co	Fe
SS400	0.15	-	-	-	-	-	Bal
C-276	0.01 max	Bal	15.0 17.0	14.5 16.5	3.0 4.5	2.5 max	4.0 7.0

Table 2. GTA spot welding parameters of SS400 and C-276.

Thermal Efficiency	Current (A)	Voltage (V)	Electrode dia. (mm)	Welding time (sec)
0.2-0.5	60-200	8-15	1.6	5.0-10.0

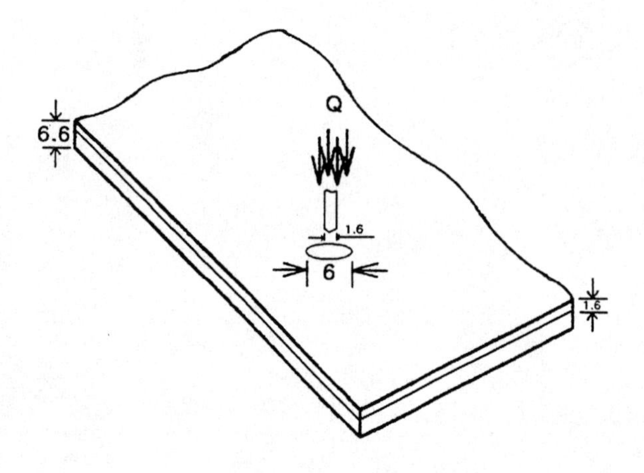

Fig. 1. Geometry of model.

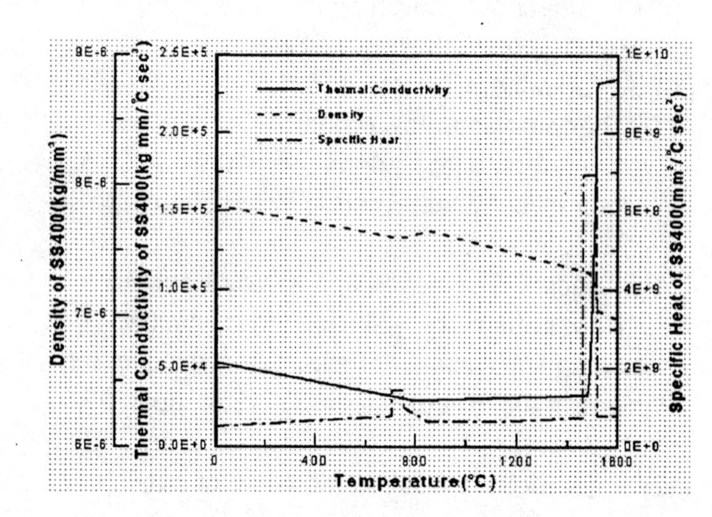

Fig. 2. Thermal conductivity / density / specific heat change vs. temperature

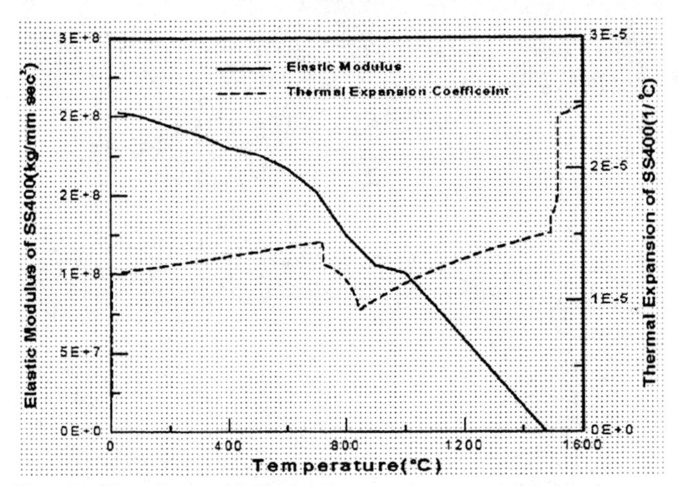

Fig. 3. Elastic modulus / thermal expansion vs. temperature

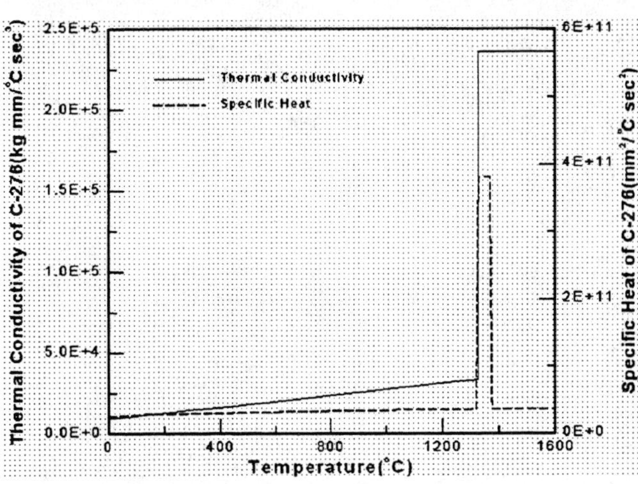

Fig. 4. Thermal conductivity / specific heat. vs. temperature

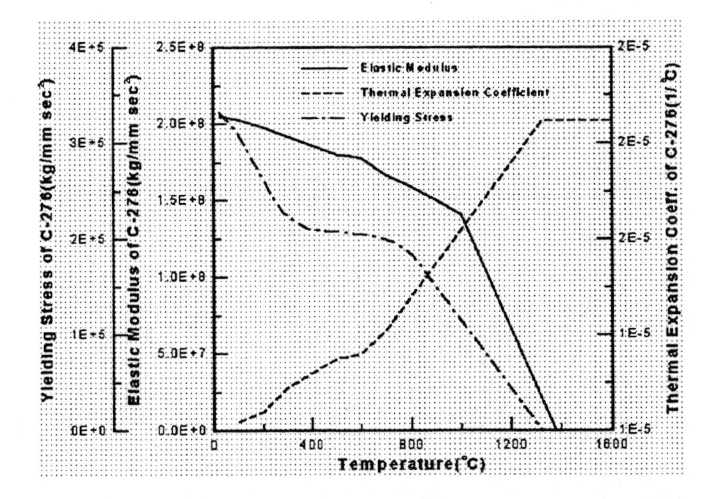

Fig. 5. Elastic modulus / thermal expansion / yield stress vs. temperature

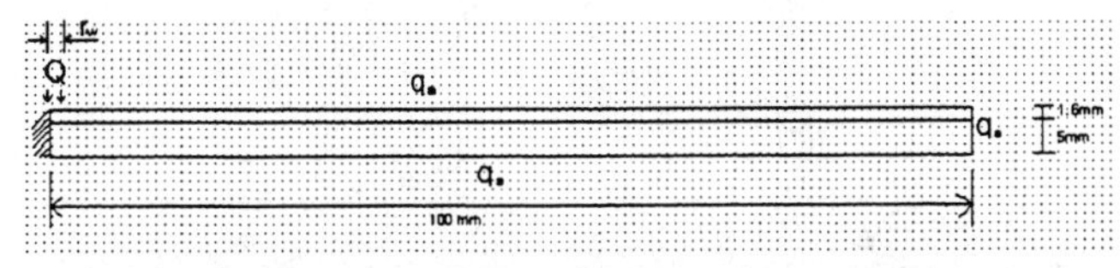

Fig. 6. Thermal boundary condition

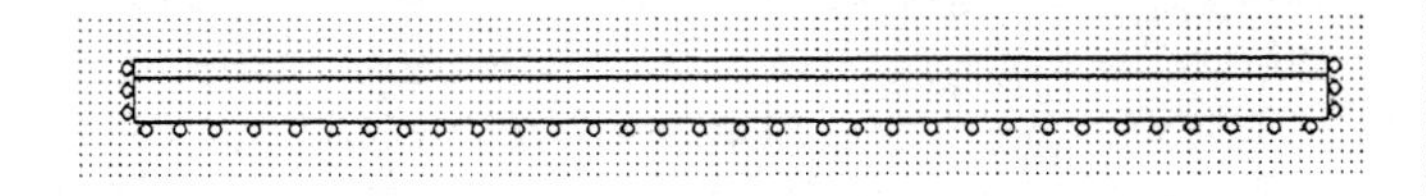

Fig. 7. Mechanical boundary condition

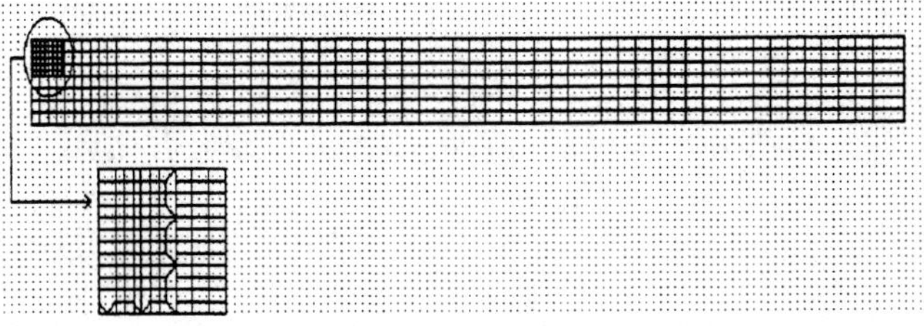

Fig. 8. FE modeling

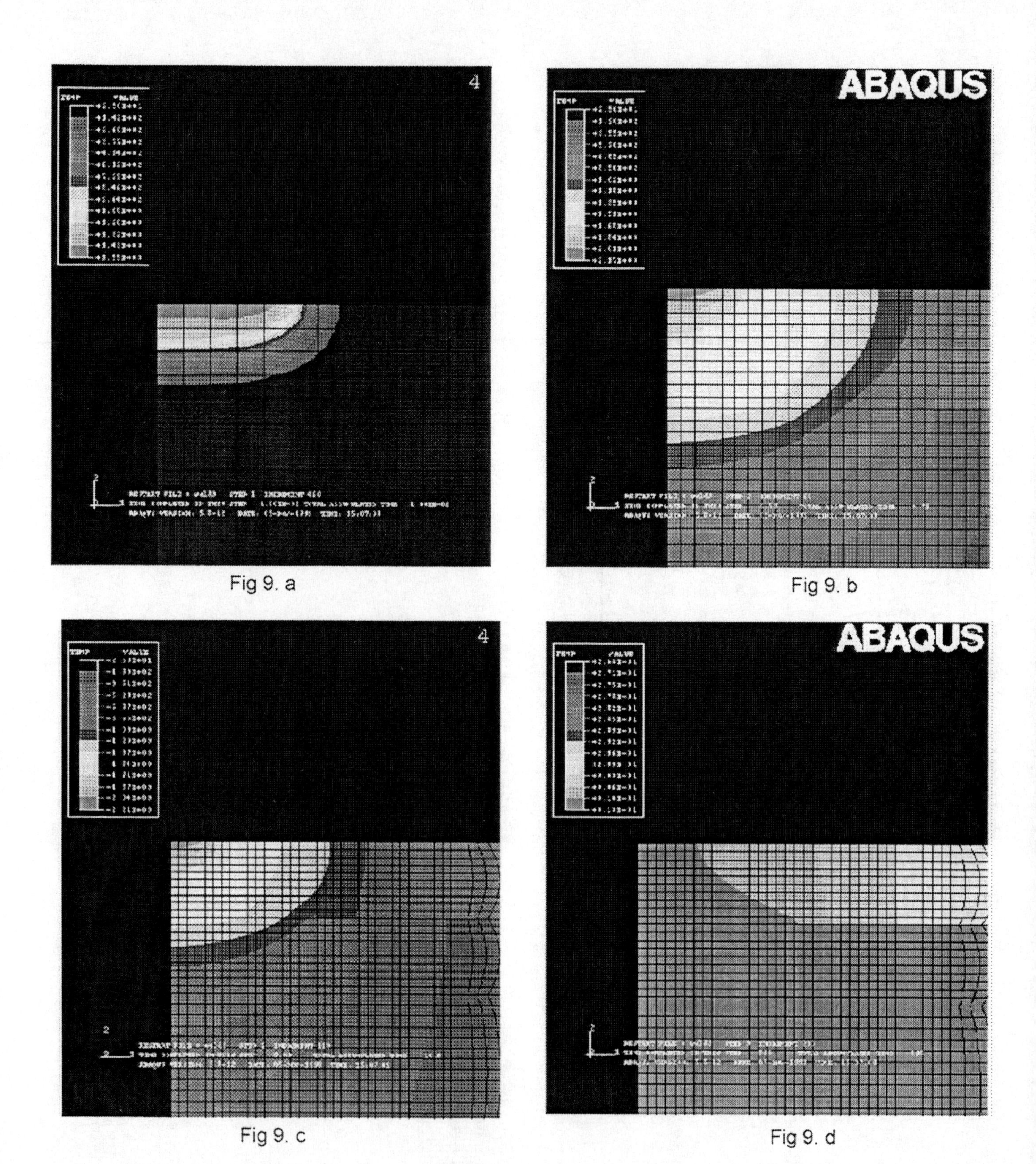

Fig 9. a

Fig 9. b

Fig 9. c

Fig 9. d

Fig 9. Temperature state(a:0.01sec, b:4.8sec, c:10sec, d:595sec)

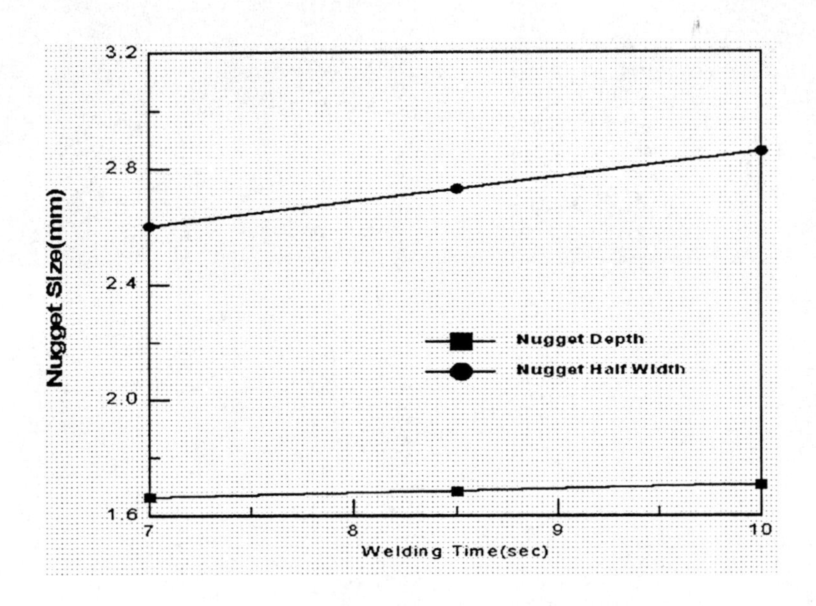

Fig 10. Nugget size vs. welding time

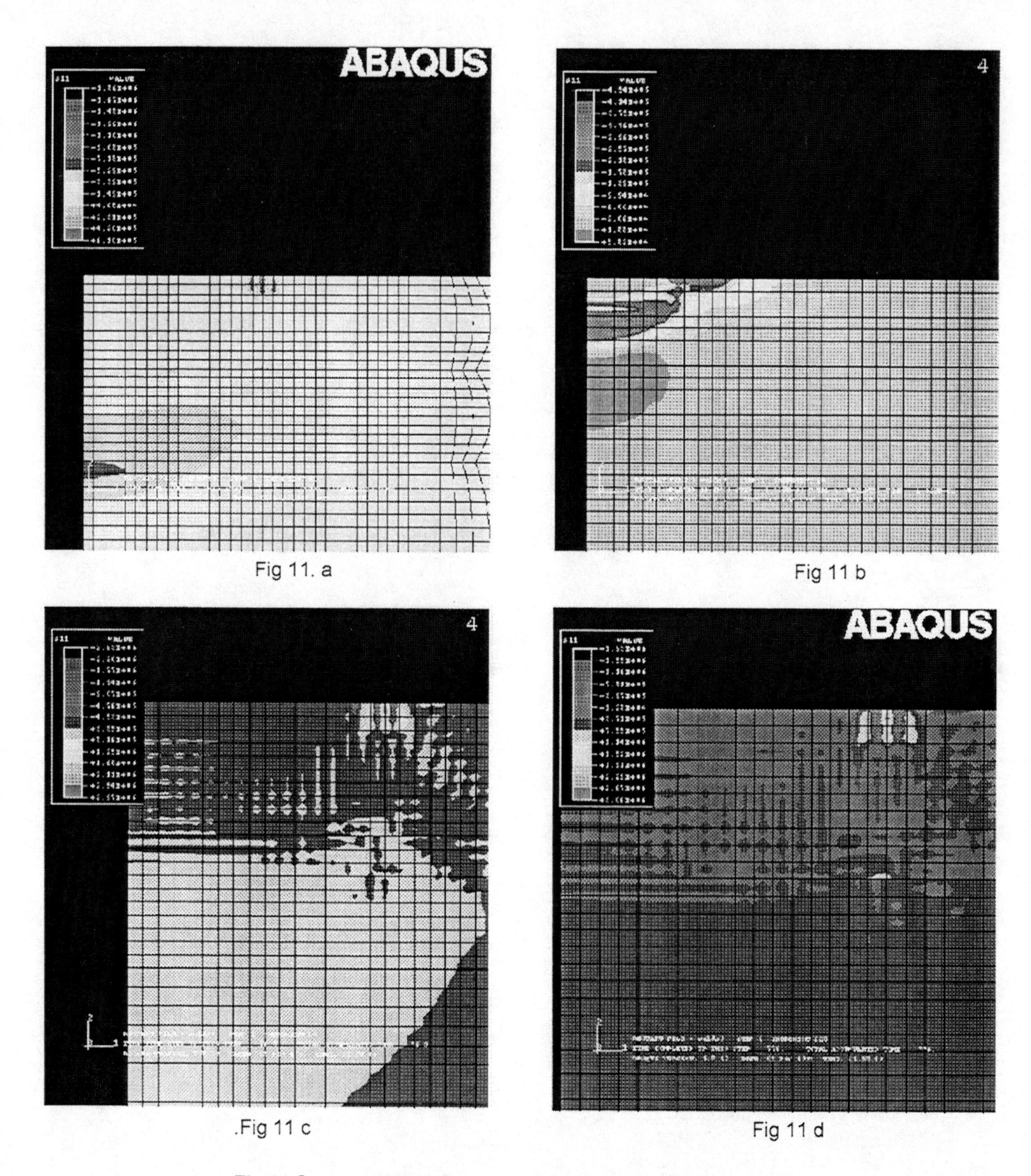

Fig 11. a

Fig 11 b

.Fig 11 c

Fig 11 d

Fig 11 Stress state(a:0.01sec, b:10sec, c:75sec, d:780sec)

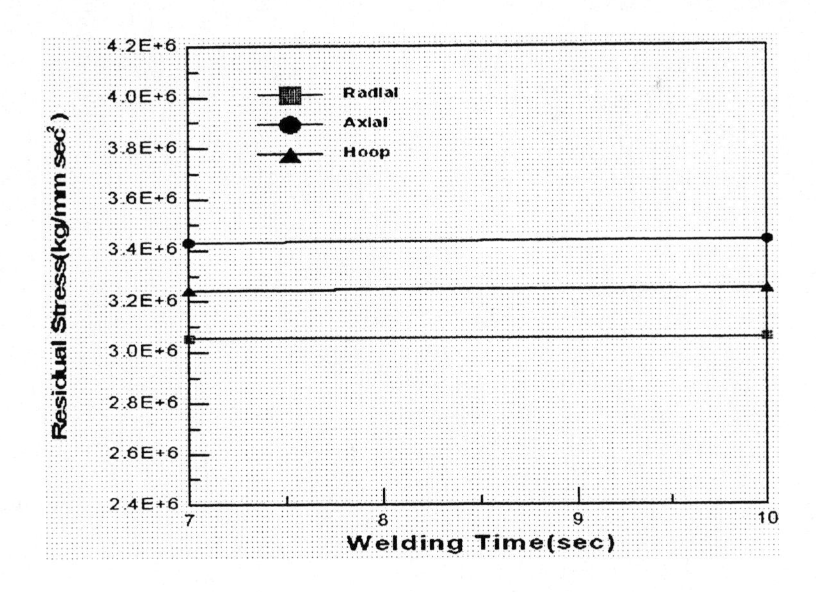

Fig 12. Residual stress vs. welding time

PVP-Vol. 410-1, Assessment Methodologies for Preventing Failure: Deterministic and Probabilistic Aspects and Weld Residual Stress
Volume 1
ASME 2000

MODELING PROCEDURE DEVELOPMENT OF BUCKLING DISTORTION IN THIN PLATE WELDING

Y.P. Yang, F.W. Brust, P. Dong, J. Zhang and Z. Cao
Battelle, 505 King Avenue, Columbus, Ohio, 43201, USA

ABSTRACT

A general numerical procedure was developed to understand buckling mechanism and be capable of predicting buckling distortion based on a developed rapid thermal solution model and material model implemented into ABAQUS commercial codes. In this procedure, a three-dimensional shell model approach was adopted for the simulations to permit practical solution time for weld analysis of complex, thin structures. Weld bead configuration, fixture and initial precambering were properly considered in the model generation, which is a quite challenge for including such detailed information in the three-dimensional shell model. Initial precamber was superposed on three-dimensional shell model by a developed modeling procedure before performing thermal-mechanical analysis. By given different initial precambering factors, the effect of plate initial precamber on buckling distortion and residual stresses was investigated. The simulation results confirmed that initial precambering had very strong effect on buckling distortion.

INTRODUCTION

In the production of the structural components of automobiles, aircraft and shipbuilding, most of the parts are fabricated from thin gage steel. The parts are then welded together to create finished components. However buckling distortion is usually induced in the welding fabrication of thin section large structure, which causes additional time-consuming and costly operations to remove distortions or to relieve residual stress after welding.

With the development of advanced welding modeling techniques [1-4], it is possible to predict welding distortion and residual stress in a thick structure by the finite element method. But the welding modeling procedures developed for a thick structure can not directly used to predict buckling distortion in a thin structure. A further development is presented in this paper to better understand the buckling mechanism and predict buckling distortion.

BUCKLING DISTORTION

A simple example of buckling distortion is a stiff column under axial compressive loading. When the load reaches a certain limit (critical buckling load P_{cr}), the column bends suddenly and exhibits a much lower stiffness. The distortion direction and shape becomes unstable. This phenomenon is known as buckling. When thin plates are welded, welding residual stresses occur in the plates. Longitudinal (parallel to welding direction) tensile residual stresses occur in the weld and longitudinal residual compressive stresses occur in areas away from the weld. If the compressive residual stress exceeds the critical buckling load (P_{cr}) of weldment, buckling distortion can happen.

The critical buckling load (P_{cr}) can be expressed as:

$$P_{cr} = P + \lambda Q \quad (1)$$

Where Q is incremental loading, λ eigenvalues and P preloads. The preloads are often zero in classical buckling problems, but in welding buckling analysis, the preloads should include initial distortion (plate precamber) and welding residual stress. Initial distortion and residual stress can significantly reduced buckling strength [5].

WELDING EXPERIMENT

Fig. 1a shows the experimental setup. A plate with dimensions 10" long and 12" wide sits on three points (two rest pins and one roller bearing) as shown in Fig. 1b. A robotic MIG welding process was used to perform bead-on-plate experiments with the weld placed along the middle of the plate. Plate thickness was 3mm. The welding parameters used were current 261 amps, voltage 22.4 volts, and travel speed 16.93 mm/s. Displacement monitoring sensors (10 LVDT) were set up on the plate to measure out-of-plate and in-plane distortions.

Different distortion patterns were observed during welding experiment. Fig. 2a shows displacement histories recorded by LVDT26 with the plate edges moving down in three bead-on-plate coupons and with the plate edges moving up in two bead-on-plate coupons. The same welding conditions (fixture and parameters) were used for all the coupons. This was indeed quite puzzling when first observed. Experimental pattern inconsistency becomes less in the "T" joint, but the final distortion values have some differences (see marking down

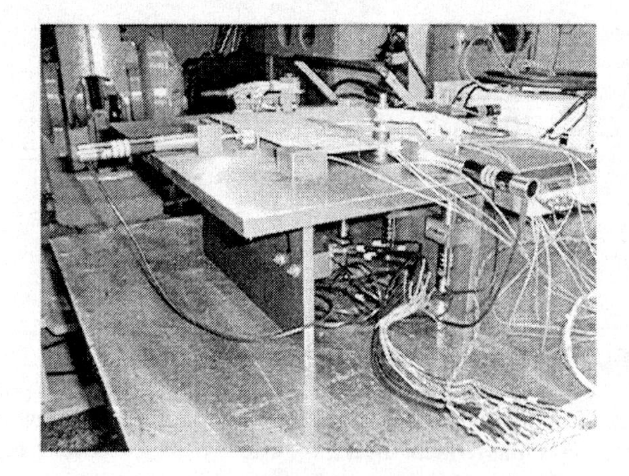

(a) Welding Setup

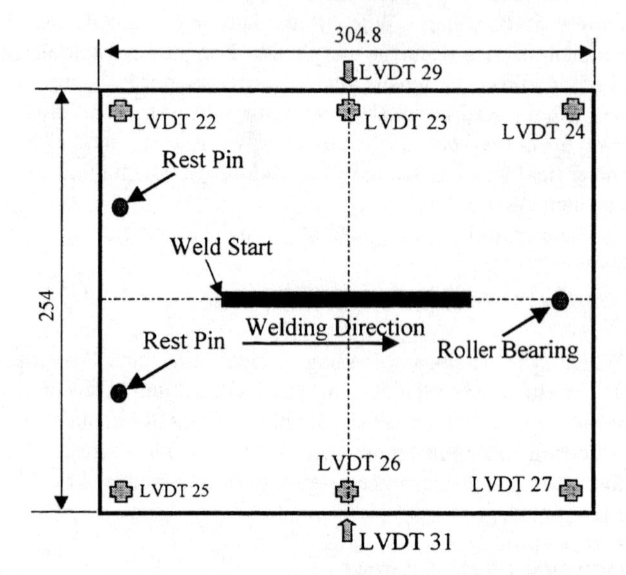

(b) LVDT and Fixture Locations

Fig. 1 Welding Setup and LVDT Locations

values) as shown in Fig. 1b

What causes the distortion pattern differences and final distortion value scatter in the welding experimental results that were discussed above? The possible reasons include the variations of the plate initial shape (see Fig. 3) and the fact that the weld beads are not perfectly centered for all plates. As a result, arc length (arc voltage) could be changed during welding to reduce the weld bead penetration. Finite element modeling was used to address these issues and answer these questions.

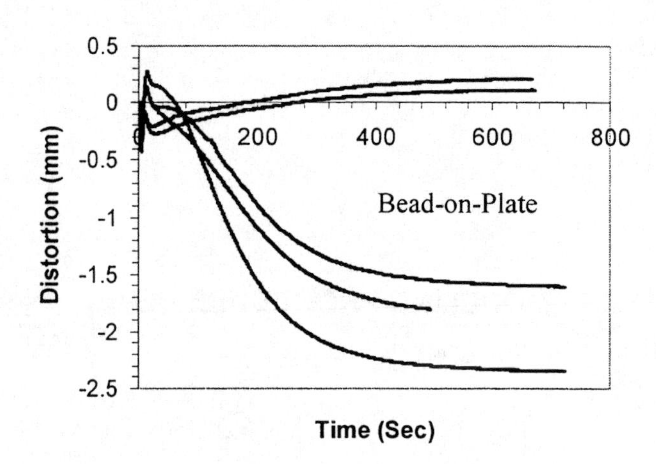

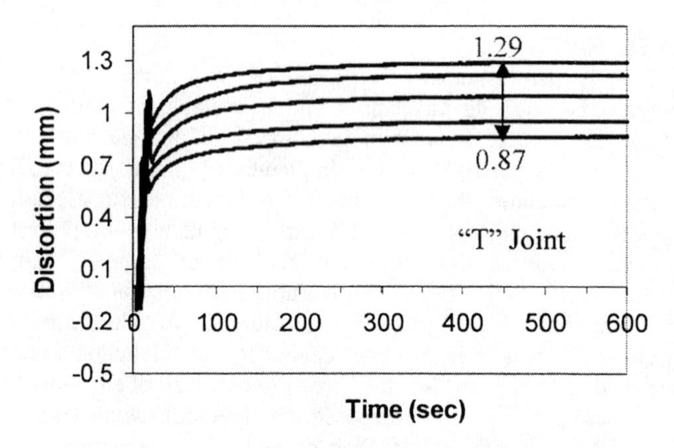

Fig. 2 Distortion Histories at LVDT 26

NUMERICAL SIMULATION

Finite Element Model

As shown in Fig. 3, half of the plate was used to generate the finite model because of the symmetry of welding conditions. The 6" long weld is located in the middle of the plate with a very fine mesh. Three-dimensional (3D) shell models were used to simulate the welding distortion with consideration of weld bead thickness. In the shell model, the plate thickness is 3mm and the thickness of bead position is 5.2 mm which is more than 70% larger than that of plate thickness. Two different kinds of boundary conditions, spring boundary conditions and displacement boundary conditions, were tested to simulate the fixture used in the experiment. Very similar results are obtained although different boundary conditions are used. For simplicity, displacement boundary conditions were used in the following analysis.

Initial Plate Pre-camber

The plates originally cut from coil inherited pre-camber (initial imperfection) distortions before welding as shown in Fig. 4. The numbers shown on the 'plate' outlines in Fig. 4

represent the measured pre-camber in two representative plates. The model needs to include this initial imperfection since these plates are so thin. In order to study the effect of the initial imperfection on the distortion evolution, many cases were

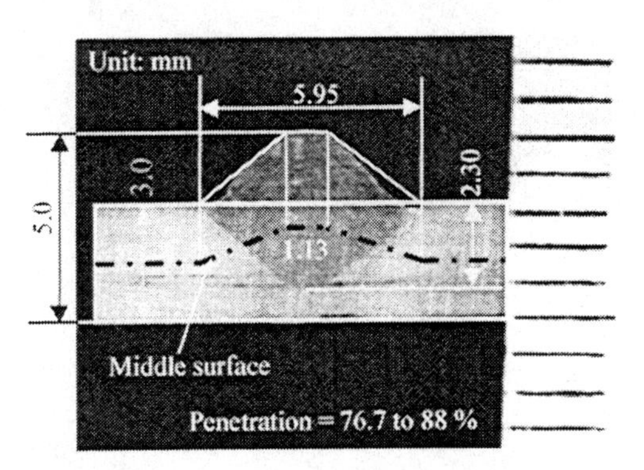

(a) weld cross section

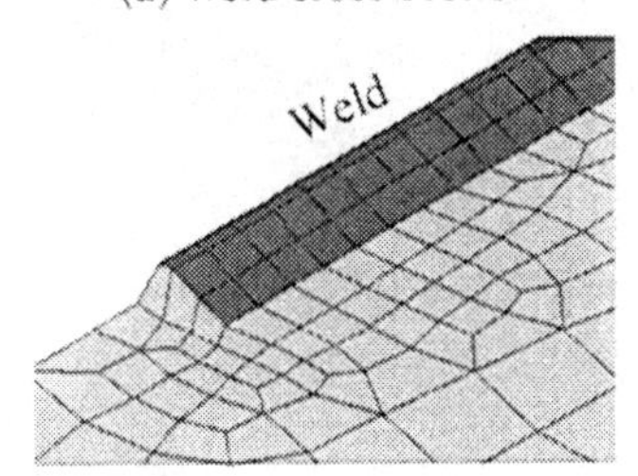

(b) Weld local view

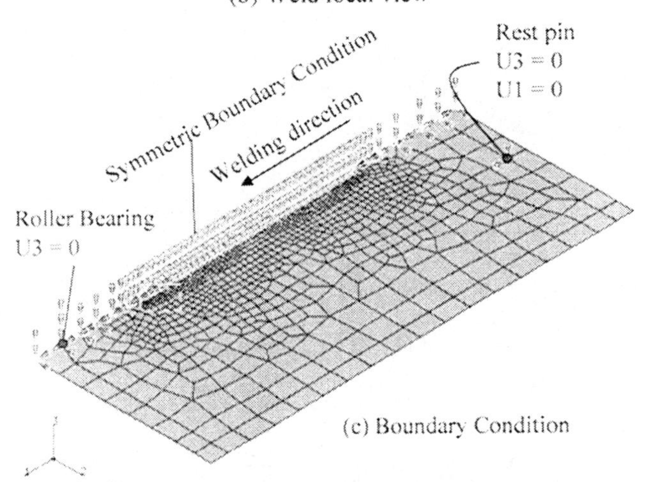

Fig. 3 Shell Model Generation and Boundary Conditions

considered and analyzed. Initial imperfections of 0.0 mm (perfect flat plate), 0.1 mm and 1 mm were considered. For the imperfect plates, two situations were modeled with the bead-on-plate analyses: the weld is deposited on the concave surface or the convex surface.

Pre-camber Shape Generation

Fig. 5a illustrates that a buckling mode analysis was used to generate the initial plate shape. Fig. 5b shows the initial plate shape that was input in the 3D-shell model when performing the mechanical analysis for the concave case.

Unit: inch ($x10^{-3}$)

16		18		13
	15		20	
0		24		0
	26		30	
4		?		11

(a) Plate 1: Max Diff. = 0.024'' (0.61mm)

0		17		2
	14		18	
10		17		8
	6		13	
5		?		5

(b) Plate 2: Max Diff. = 0.017'' (0.43mm)

Fig. 4 Initial Precamber Measurement

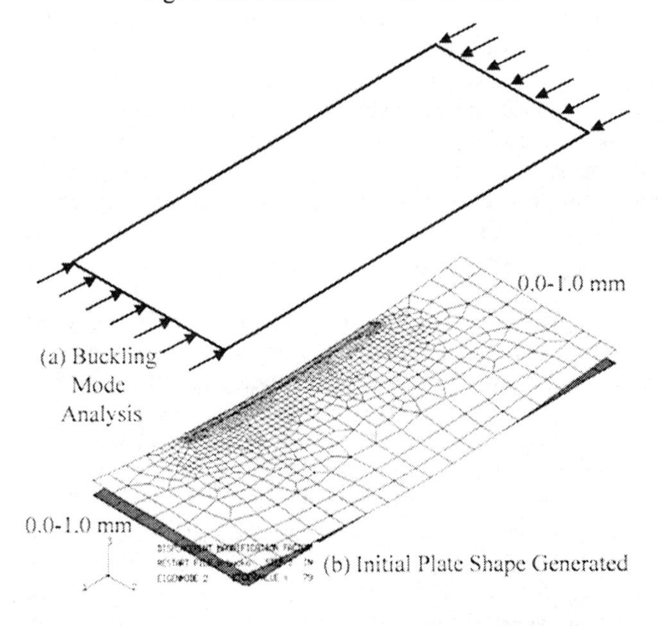

Fig. 5 Initial Plate Shape Generated

Heat Source Model and Material Model

The thermal solution was performed using the rapid thermal solution code [6]. The comparison between experiment and prediction was shown in Fig. 6. The mechanical analyses were conducted by ABAQUS with a material subroutine [7]. The first step in the analysis accounted for plate heating during welding, the second step permitted cooling and the third step was to release clamping. It should be pointed out that the third step could be a static analysis with or without using a RIKs method [8]. If very strong clamping system is used, the third step should use the RIKs method when using ABAQUS.

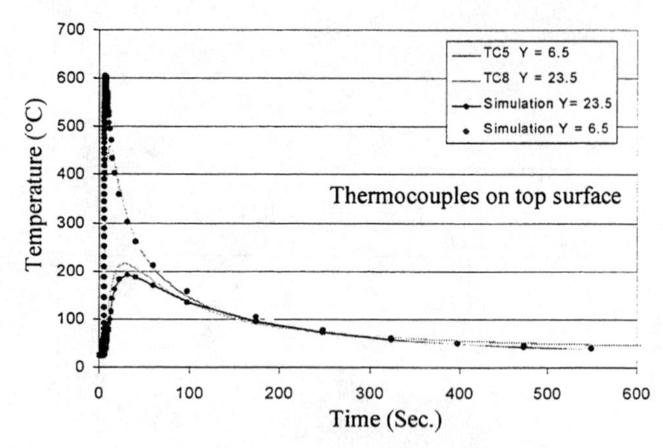

Fig. 6 Temperature Comparison between Experiment and Prediction

Effect of Plate Pre-Camber on Final Distortion Shape

Fig. 7b shows the final distortion shape when the weld was deposited on the concave surface, which was done by inputting an initial imperfection of the plate to the shell model. If the weld was deposited on the convex surface, which was done by inputting a negative initial imperfection into the model (see Fig. 7c), the opposite distortion shape was obtained compared with the weld deposited on the concave surface. These results illustrate the importance of initial pre-camber in welding of thin plates. A good comparison was achieved between experiment and numerical simulation when weld deposited on the concave surface, as shown in Fig. 8.

Effect of Plate Pre-Camber on Residual Stress Distribution

Fig. 9 shows that the transverse residual stress on a 3-mm plate depends strongly on the plate bending direction. Without considering plate pre-camber, the plate edges were moving up during welding as shown in Fig. 7a, which produced tension on the bottom surface of the plate (Fig. 9a) and compression around the weld on the top surface of the plate (Fig. 9b). When welding on the concave surface, the plate bends downward as shown in Fig. 7c. This produces tension on the top surface of the plate (Fig. 9d) and compression around the weld on the bottom surface of the plate (Fig. 9c).

Fig. 10 shows the longitudinal residual stress distribution in the 3-mm plate without and with including plate pre-camber in the model, respectively. It was found that the initial imperfection has a very small effect on the longitudinal residual stress distribution in the weld area, but it has a very strong effect on the overall stress distribution. The stress distribution

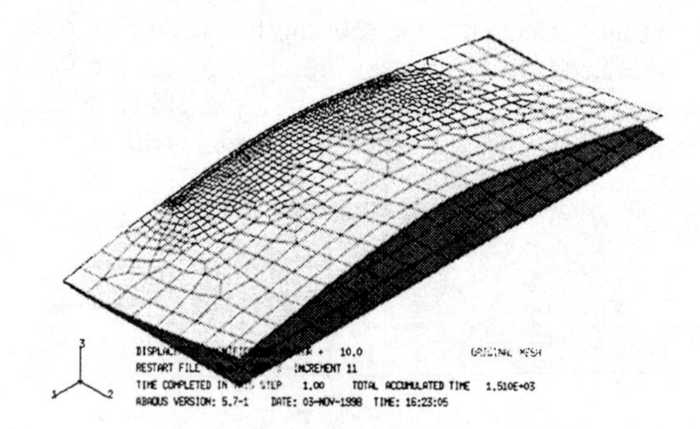

(a) Weld on a Flat Plate

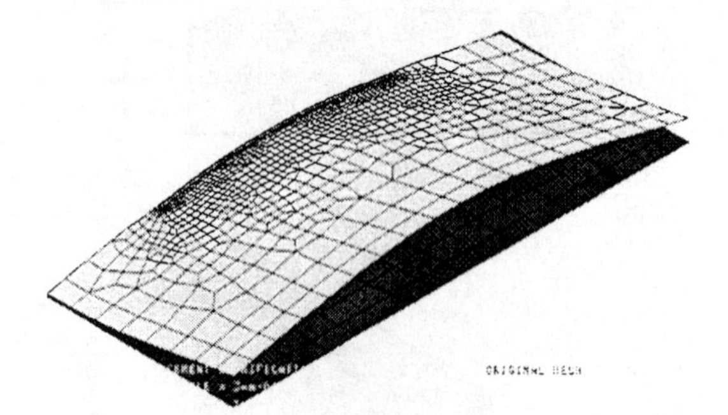

(a) Weld on a Convex Surface

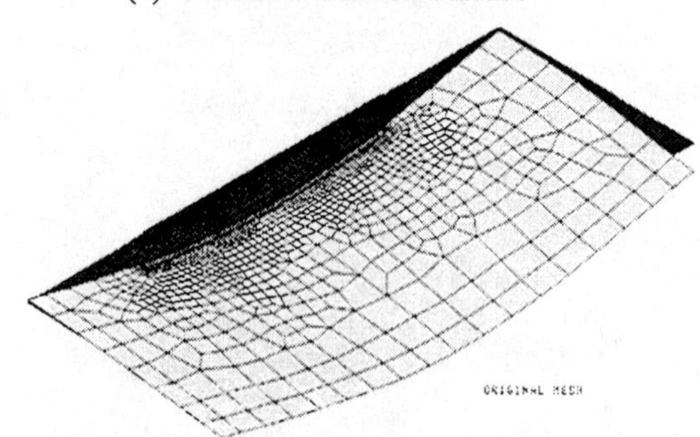

(c) Weld on a Concave Surface

Fig. 7 Effect of Plate Pre-Camber on Final Distortion

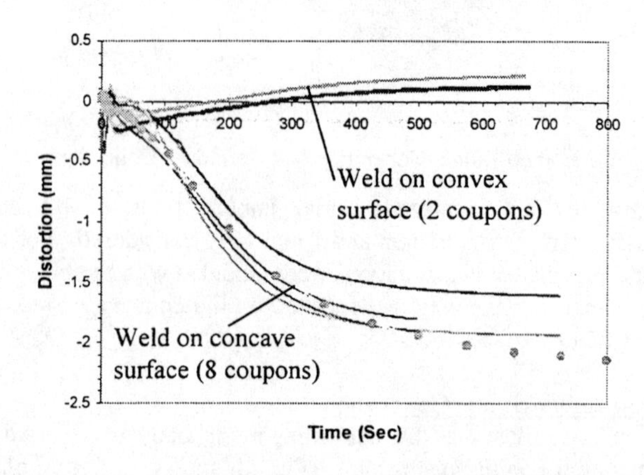

Fig. 8 Distortion Comparison between simulation and Experiment

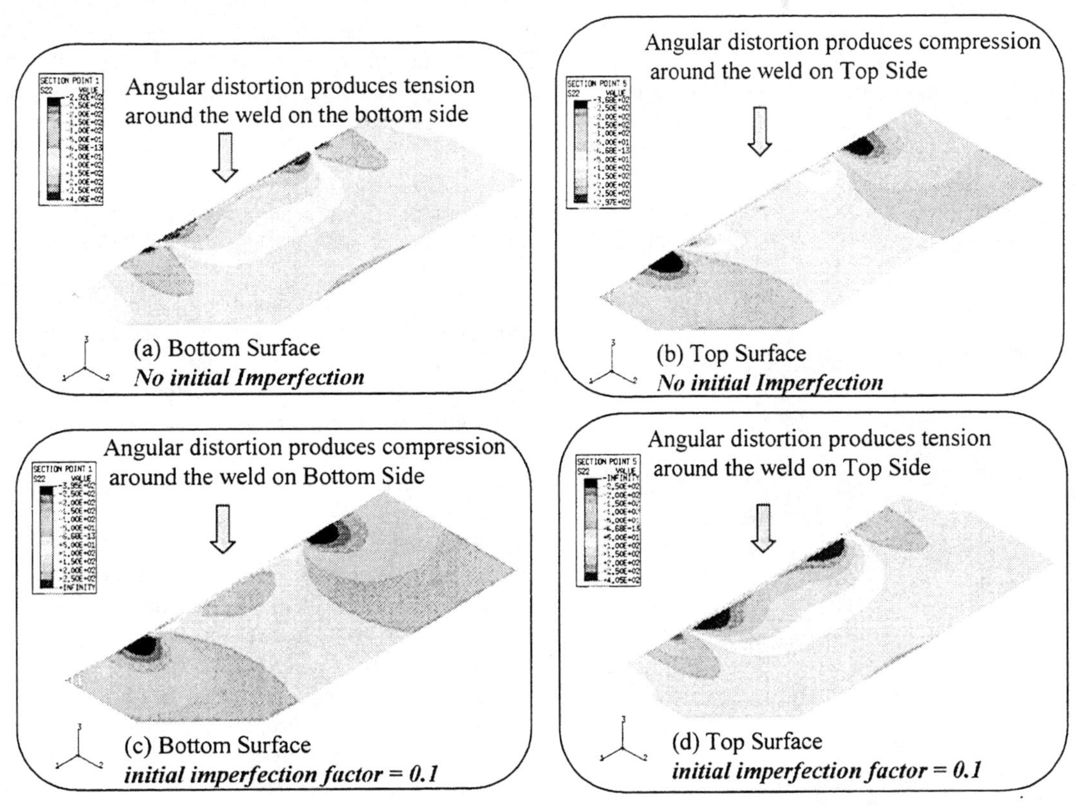

Fig. 9 Transverse Residual Stress Distribution on 3mm Plate

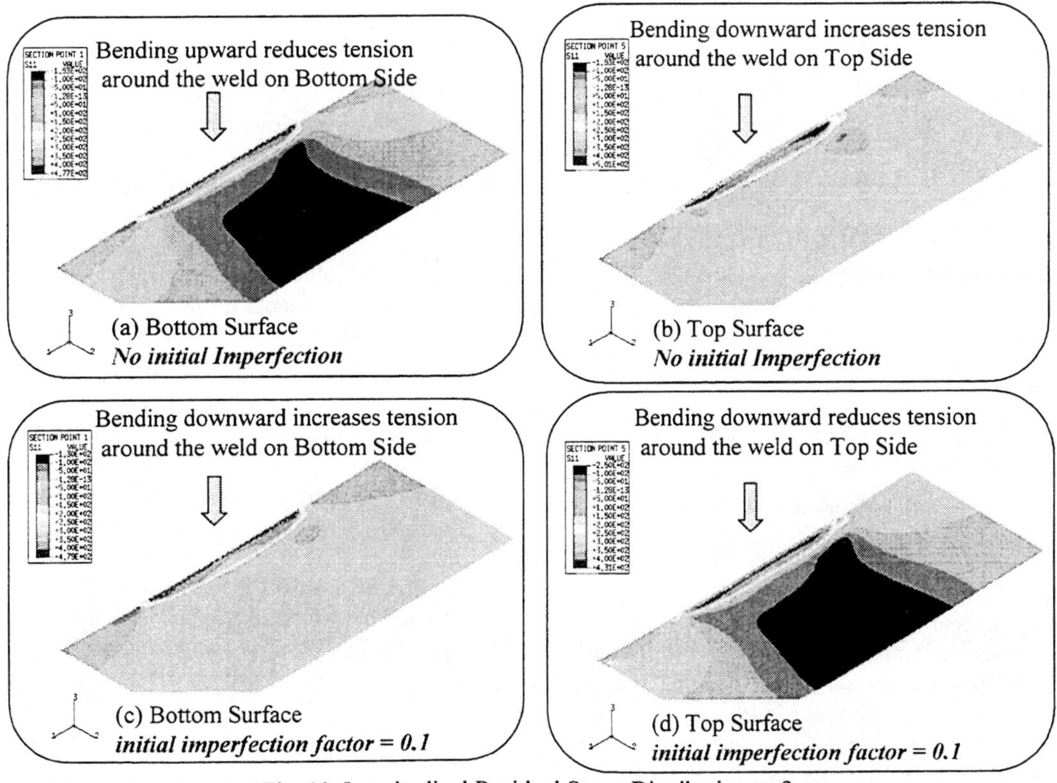

Fig. 10 Longitudinal Residual Stress Distribution on 3mm

also shows that the 3-mm plates had nearly complete fusion penetration.

Both Fig. 9 and Fig. 10 show that the residual stress is a combination of welding-induced residual stress and distortion-induced residual stress, especially for the transverse residual stress in the 3mm plate. The longitudinal residual stress of the weld area is uniform through the thickness in the 3-mm plate, which clearly contributes to the distortion instability of the 3-mm plate.

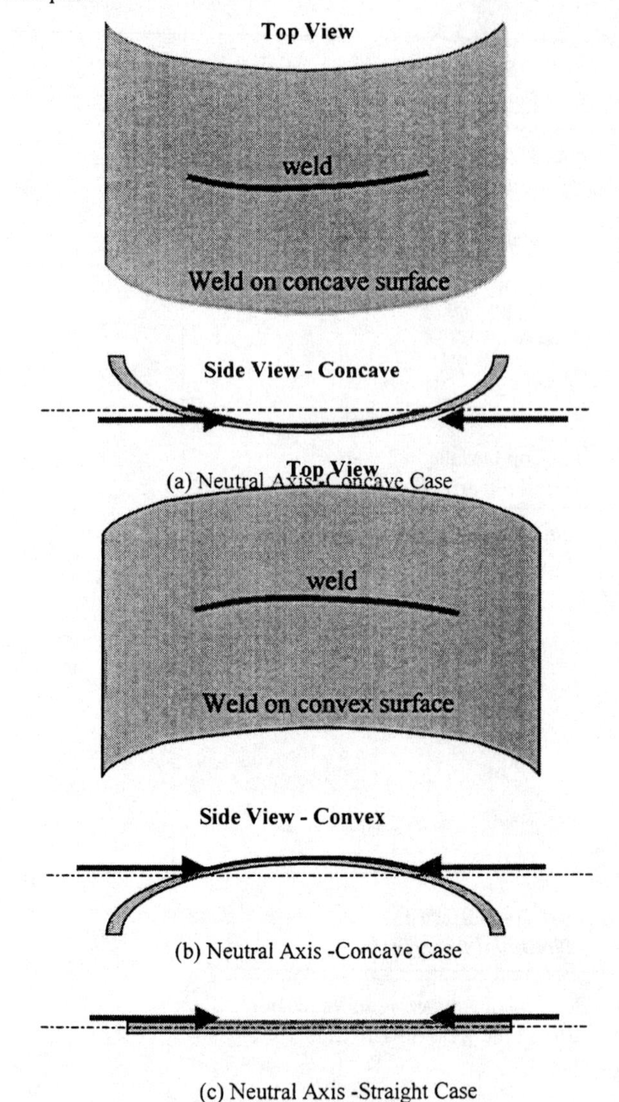

(a) Neutral Axis-Concave Case

(b) Neutral Axis -Concave Case

(c) Neutral Axis -Straight Case

Fig. 11 Effect of Pre-camber and Neutral Axis

Distortion Pattern Analysis

This scenario of distortion pattern inconstancy is illustrated in Fig. 11. Nearly full penetration as well as the initial downward or upward (pre-camber) bow of the plate raised or lowered the effective 'global' position of the neutral axis of the entire plate. As such, the plate edges went down or up. In the concave case, axial shrinkage is below neutral axis as shown in Fig. 11a, thus the plate should bow down. But in both convex and perfect flat plate case, neutral axis is above neutral axis as shown in Fig. 11b and Fig. 11c, thus the plate should bow up.

CONCLUSION

In summary, a modeling procedure was developed with consideration of bead shape and the initial imperfection of the plate for predicting buckling distortion in the welding of thin sheet. The factors contributing to the distortion behavior of the plate are the initial imperfection of the plate and the welding conditions. Welding residual stress is a combination of welding-induced residual stress and distortion-induced residual stress, especially for the transverse residual stress in thin plate welding. Using this procedure, welding-induced buckling distortion and residual stress can be predicted well.

REFERENCES

1. Brust, F. W. and Rybicki, E. F., "Computational Model of Backlay Welding for Controlling Residual Stresses in Welded Pipes," *Journal of Pressure Vessel Technology,* 1981, Vol. 103, pp. 294-299.
2. Brust, F.W., Dong, P., Zhang, J., Cao, Z, Yang, Y. P., and Hong, J.K., "Weld Process Modeling and It's Importance in a Manufacturing Environment", Proc. 1998 Earthmoving Industry Conference and Exposition, Peoria, Ill, April 8-9, 1998, SAE Technical Paper Series #981510.
3. Goldak, J, "Progress and Pacing Trends in Computational Weld Mechanics", *Proceedings of ICES'98*, October 7 - 9, Atlanta, 1998.
4. Dong, P., Hong, J. K., and Rogers, P., "Analysis of Residual Stresses in Al-Li Alloy Repair Welds and Mitigation Techniques," *Welding Journal,* 1998, Vol. 77 (11), pp. 439s-445s.
5. Masubuchi, K., "Analysis of Welded Structures," *Pergamon Press, Oxford,* 1980, pp. 235, 331.
6. Cao, Z., P. Dong, and Brust, F. W., "A Highly Efficient Heat-Flow Solution Procedure", *Proceedings of ICES'98*, October 7 - 9, Atlanta, 1998.
7. J. Zhang, P. Dong, F. W. Brust, *A 3-D Composite Shell Element Model for Residual Stress Analysis in Multi-Pass Welds*. Transactions of the 14th International Conference on Structural Mechanics in Reactor Technology (SMiRT 14), Lyon, France, Vol. 1, pp. 335-344 (1997).
8. ABAQUS, "User's Manual," *Hibbitt, Karlsson & Sorensen, Inc.,* 1997, Version 5.7, pp. 6.2.3-2

ACKNOWLEDGMENTS

The authors would like to acknowledge Tower Automotive for supporting this study. Adam Fisher, Robert Broman and Raj Thakkar participated the experimental work in this paper.

PIPING FRACTURE MECHANICS

Introduction

Gery Wilkowski
Engineering Mechanics Corporation of Columbus
Columbus, OH

Piping Fracture Mechanics sessions have been ongoing for many years as part of the PVP Conference. Various technical committees have sponsored these sessions, sometimes jointly, sometimes separately. This year there were two specific sessions from the Materials and Fabrication Committee on piping fracture mechanics that are included in this volume. The papers dealt with elastic-plastic fracture of straight pipes and elbows with circumferential surface flaws, axial surface cracks in elbows, creep assessments using time-dependant J analyses, a status report on a program involving cracks at bimetallic weld interfaces, fatigue behavior of workmanship flaws, use of neural network analyses to determine leak versus rupture of axially oriented corrosion flaws, accuracy assessments of crack-opening area estimation schemes for leak-before-break analyses, and general concept on structural integrity of nuclear piping. These papers do a fine job of increasing our understanding of piping fracture assessments, and are sure to be of archival value to our readers.

PVP-Vol. 410-1, Assessment Methodologies for Preventing Failure: Deterministic and Probabilistic Aspects and Weld Residual Stress
Volume 1
ASME 2000

Simplified *J*-Evaluation Method for Flaw Evaluation at High Temperature - Part I: Systematization of Evaluation Method

Naoki Miura
Materials Science Department
Central Research Institute of Electric Power Industry
Tokyo, Japan

Takashi Shimakawa
Research and Development Department
Kawasaki Heavy Industries
Tokyo, Japan

Yasunari Nakayama
Materials Science Department
Central Research Institute of Electric Power Industry
Tokyo, Japan

Yukio Takahashi
Materials Science Department
Central Research Institute of Electric Power Industry
Tokyo, Japan

ABSTRACT

J-integral is an effective inelastic fracture parameter for the flaw evaluation of cracked components at high temperature. The evaluation of *J*-integral for arbitrary crack configuration and arbitrary loading condition can be generally accomplished by detailed numerical analysis such as finite element analysis, however, it is time-consuming and requires a high degree of expertise for its implementation. Therefore, it is important to develop simplified *J*-estimation techniques from the viewpoint of industrial requirements. In this study, a simplified *J*-evaluation method is proposed to estimate the *J*-integral and its variant types of parameters; *J*-integral to describe ductile crack propagation, and fatigue *J*-integral range and creep *J*-integral to describe creep-fatigue crack propagation behavior. This paper presents the systematization of the simplified *J*-evaluation method incorporated with the reference stress method and the concept of elastic follow-up, then proposes a comprehensive evaluation procedure. The verification of the proposed method is presented in Part II (Shimakawa et al., 2000) of this paper.

INTRODUCTION

Structural components used at high temperature are to be properly designed to prevent crack initiation during a specified operation period due to creep-fatigue damage with appropriate safety margins. However, the existence of initial flaws such as weld defects can not be completely excluded, and initiated cracks do not lead to instantaneous failure of components. Therefore, evaluation of structural integrity under the presence of crack-like flaws is important to complement the "formal" design assessment. When we consider a rational maintenance of components, fine methodology to estimate crack propagation behavior is also essential to determine how a detected flaw is detrimental to continuous operation.

J-integral and its variant types of parameters (fatigue *J*-integral range and creep *J*-integral) are effective inelastic fracture parameters for the flaw evaluation of cracked components at high temperature. It was validated that ductile crack propagation at high temperature was controlled by *J*-integral (Miura et al., 1998) and fatigue crack propagation law and creep crack propagation law could be well described by fatigue *J*-integral range and creep *J*-integral, respectively (Nakayama et al., 1998). The evaluation of the *J*-integrals for arbitrary crack configuration and arbitrary loading condition can be generally accomplished by detailed numerical analysis such as finite element analysis, however, it is time-consuming and requires a high degree of expertise for its implementation. With this in mind, it is needed to develop simplified *J*-estimation technique especially from the viewpoint of industrial requirements.

In this study, a simplified *J*-evaluation method for flaw evaluation at high temperature is proposed to estimate *J*-integral, fatigue *J*-integral range, and creep *J*-integral. This paper focuses on the systematization of the simplified *J*-evaluation method, then proposes a comprehensive evaluation procedure. The verification of the proposed method is presented in Part II (Shimakawa et al., 2000) of this paper.

OVERVIEW OF SIMPLIFIED *J*-EVALUATION

Some simplified *J*-evaluation methods have been developed for the flaw evaluation of cracked components at high temperature (Roche, 1988; Iwasaki et al., 1991; Wakai et al., 1999; and Poussard et al., 1999). In these methods, fatigue *J*-integral range, ΔJ_f, is used to describe fatigue crack propagation behavior, and creep *J*-integral, J', or its equivalent parameter is used to describe creep crack propagation behavior. Simplified methods to evaluate ΔJ_f and J' are derived from simplified methods to evaluate *J*-integral, J, under monotonic loading. However, the existing methods are not enough generalized since they are applicable under limited condition regarding the combination of crack size, component configuration, and applied loading. The consideration of inelastic response of cracked components is one of the key concerns for flaw evaluation at high temperature. The existing methods, however, do not provide a consistent procedure to handle inelastic stress used for the evaluation. This inconsistency which may lead to the uncertainties of the evaluation methods.

The simplified *J*-evaluation method proposed in this study is based

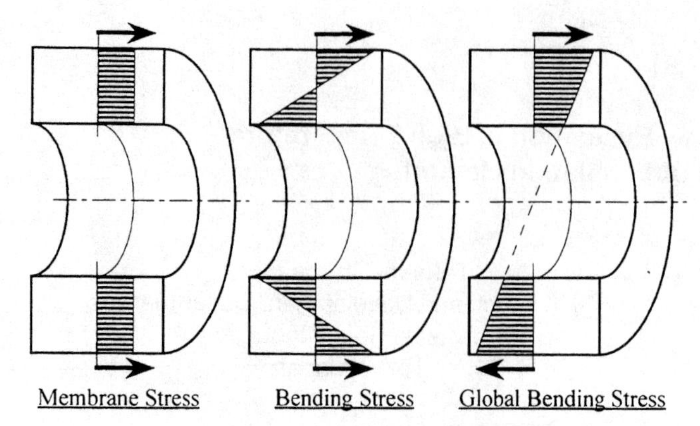

Figure 1 Schematic Stress Classification for Cylindrical Strucure

on the concept of the reference stress method. The reference stress method starts from the elastic J-integral obtained from the stress intensity factor. Inelastic response is then considered by multiplying the ratio of the elastic-plastic strain provided by the reference stress to the elastic strain. The idea of J-evaluation using the reference stress has been proposed by Milne et al. (1988). Later, Roche (1988) has introduced the consideration of elastic follow-up. However, this reference stress method gives rather a basic concept. It is not clear how the quantities needed for the evaluation can be determined. While the proposed simplified J-evaluation method states the way to determine the quantities clearly.

The simplified J-evaluation method proposed in this study is systematized including detailed provisions on the treatment of primary and secondary stresses and consistent evaluation procedure to realize reliable evaluation. The definite evaluation procedure will be presented in the following sections.

PRECONDITIONS OF EVALUATION METHOD

In advance of the systematization of the simplified J-evaluation method, following preconditions are assumed.

Elastic Stress

The elastic stress used for the evaluation of ΔJ_f and J' is defined as a half of the elastic stress range obtained from elastic finite element analysis (or an equivalent method) for uncracked model. The elastic stress used for the evaluation of J is directly obtained by the same analysis. The elastic stress is classified to primary stress and secondary stress by some appropriate method. Primary stress, such as pressure-induced stress, is load-controlled and is developed by imposed loading which is necessary to satisfy an equilibrium condition. Secondary stress, such as thermal expansion stress, is displacement-controlled and is developed by the constraint of adjacent material or by self-constraint of the model. Both of primary stress and secondary stress are further classified to membrane stress components and bending stress components. In addition, asymmetric global bending stress components are classified for cylindrical structures, since they are ones of the basic stress modes for symmetric structures. The schematic stress classification for a cylindrical structure is shown in Fig. 1. Elastic stress components used for the evaluation are listed in Table 1.

Stress Intensity Factor and Reference Stress

Solutions of stress intensity Factor, K_I, and reference stress, σ_R, used for the evaluation are given for a specific cracked component in advance. σ_R is the parameter provided by applied loading and the configuration of cracked component and is generally given as

$$\sigma_R = \frac{\text{(Applied Load)}}{\text{(Plastic Collapse Load)}} \times \text{(Yield Stress)}$$

K_I solutions and σ_R solutions listed in Table 2 have been prepared and compiled separately.

Monotonic Stress-Strain Relation

Monotonic stress-strain relation of the material of interest is approximated by the following Ramberg-Osgood equation:

$$\frac{\varepsilon}{\varepsilon_0} = \frac{\sigma}{\sigma_0} + \kappa\left(\frac{\sigma}{\sigma_0}\right)^n, \quad \varepsilon_0 = \frac{\sigma_0}{E} \tag{1}$$

where σ and ε are the stress and the strain, respectively, σ_0 and ε_0 are the base stress and the base strain, respectively, E is the Young's modulus, κ is the material constant, and n is the hardening exponent.

Cyclic Stress-Strain Relation

Cyclic stress-strain relation of the material of interest is

Table 1 Elastic Stress Components Used for Evaluation

Stress Classification		Definition for Evaluation of ΔJ_f and J'	Definition for Eavaluation of J
Primary Membrane Stress	$\sigma_{1,m}$	Half of Corresponding Stress Range Perpendicular to Crack	Maximum Corresponding Stress Perpendicular to Crack
Secondary Membtane Stress	$\sigma_{2,m}$		
Primary Bending Stress	$\sigma_{1,b}$		
Secondary Bending Stress	$\sigma_{2,b}$		
Primary Global Bending Stress	$\sigma_{1,bg}$		
Secondary Global Bending Stress	$\sigma_{2,bg}$		

Table 2 Type of Structures for which K_I Solutions and σ_R Solutions Are Provided

Structure	Crack Direction	Crack Type
Plate	-	Semi-Elliptical Surface Crack
		Single Edge Crack
		Through-wall Crack
Cylinder	Axial	Semi-Elliptical Internal/External Surface Crack
		Infinite Internal/External Surface Crack
		Through-Wall Crack
	Circumferential	Semi-Elliptical Internal/External Surface Crack
		Full-Circumferential Internal/External Surface Crack
		Through-Wall Crack
Elbow	Axial	Through-Wall Crack
	Circumferential	Through-Wall Crack

approximated by the following Ramberg-Osgood equation:

$$\frac{\Delta\varepsilon/2}{\varepsilon_0} = \frac{\Delta\sigma/2}{\sigma_0} + \kappa\left(\frac{\Delta\sigma/2}{\sigma_0}\right)^n, \quad \varepsilon_0 = \frac{\sigma_0}{E} \tag{2}$$

where $\Delta\sigma/2$ and $\Delta\varepsilon/2$ are a half of the stress range and a half of the strain range, respectively, σ_0 and ε_0 are the base stress and the base strain, respectively, E is the Young's modulus, κ is the material constant, and n is the hardening exponent. Note that the values of κ and n for the cyclic stress-strain relation in Eq. (2) are different from those for the monotonic stress-strain relation in Eq. (1).

Creep Strain Relation

Creep strain relation of the material of interest is approximated by the following Norton's Law:

$$\dot{\varepsilon}_c = B\,\sigma^m \tag{3}$$

where $\dot{\varepsilon}_c$ is the creep strain rate, σ is the stress, B is the material constant, and m is the creep exponent.

Elastic Follow-Up Factors

Elastic follow-up factor related to elastic-plastic deformation, q, is given by the following equation, which was derived from the analysis on the elastic follow-up behavior at the root of a cantilever of power law material (Iida et al., 1987).

$$q = \frac{n+2}{3} \tag{4}$$

Elastic follow-up factor related to creep deformation, q', is given by the following equation, which was derived from the analogy of the equation of q.

$$q' = \frac{m+2}{3} \tag{5}$$

EVALUATION OF J-INTEGRAL

J-integral, J, can be evaluated by the following equation.

$$J = F\,J_e \tag{6}$$

where F is the plastic correction factor and J_e is the elastic J-integral. κ and n used to evaluate J are for the monotonic stress-strain relation given in Eq. (1).

Elastic J-Integral

J_e is directly obtained from the stress intensity factor, K_I.

$$J_e = K_I^2 / E' \tag{7}$$

$$E' = \begin{cases} E & \text{, at surface point} \\ E/(1-\nu^2) & \text{, at internal point} \end{cases} \tag{8}$$

where E is the Young's modulus and ν is the Poisson's ratio. The solution of K_I is to be separately prepared.

Membrane stress, σ_m, bending stress, σ_b, and global bending stress, σ_{bg}, used for the evaluation of K_I are calculated by the following equation.

$$\sigma_m = \frac{\sigma_{1,m}+\sigma_{2,m}}{\sigma_{1,m}+\sigma_{2,m}+\sigma_{1,b}+\sigma_{2,b}+\sigma_{1,bg}+\sigma_{2,bg}}\,\sigma_3 \tag{9}$$

$$\sigma_b = \frac{\left(\frac{\sigma_b^*}{\sigma_0}\right)^3 + \frac{3(n+1)}{n+2}\kappa\left(\frac{\sigma_b^*}{\sigma_0}\right)^{n+2} + \frac{3n}{2n+1}\kappa^2\left(\frac{\sigma_b^*}{\sigma_0}\right)^{2n+1}}{\left[\frac{\sigma_b^*}{\sigma_0} + \kappa\left(\frac{\sigma_b^*}{\sigma_0}\right)^n\right]^2}\,\sigma_0 \tag{10}$$

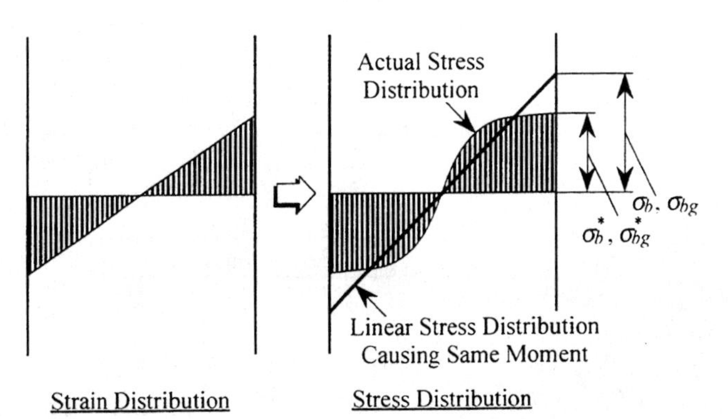

Figure 2 Derivation of Equivalent Linear Stress

$$\sigma_b^* = \frac{\sigma_{1,b}+\sigma_{2,b}}{\sigma_{1,m}+\sigma_{2,m}+\sigma_{1,b}+\sigma_{2,b}+\sigma_{1,bg}+\sigma_{2,bg}}\sigma_3 \tag{11}$$

$$\sigma_{bg} = \frac{\left(\frac{\sigma_{bg}^*}{\sigma_0}\right)^3 + \frac{3(n+1)}{n+2}\kappa\left(\frac{\sigma_{bg}^*}{\sigma_0}\right)^{n+2} + \frac{3n}{2n+1}\kappa^2\left(\frac{\sigma_{bg}^*}{\sigma_0}\right)^{2n+1}}{\left[\frac{\sigma_{bg}^*}{\sigma_0}+\kappa\left(\frac{\sigma_{bg}^*}{\sigma_0}\right)^n\right]^2}\sigma_0 \tag{12}$$

$$\sigma_{bg}^* = \frac{\sigma_{1,bg}+\sigma_{2,bg}}{\sigma_{1,m}+\sigma_{2,m}+\sigma_{1,b}+\sigma_{2,b}+\sigma_{1,bg}+\sigma_{2,bg}}\sigma_3 \tag{13}$$

σ_3, which is described later in this section, is the elastic-plastic applied stress in consideration of the elastic follow-up related to elastic-plastic deformation. σ_3 is proportionally divided into its membrane stress component, σ_m, bending stress component, σ^*_b, and global bending stress

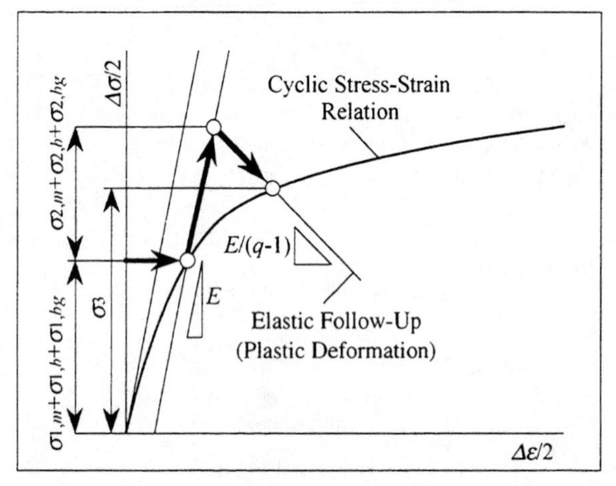

Figure 3 Evaluation Procedure of Elastic-Plastic Applied Stress, σ_3

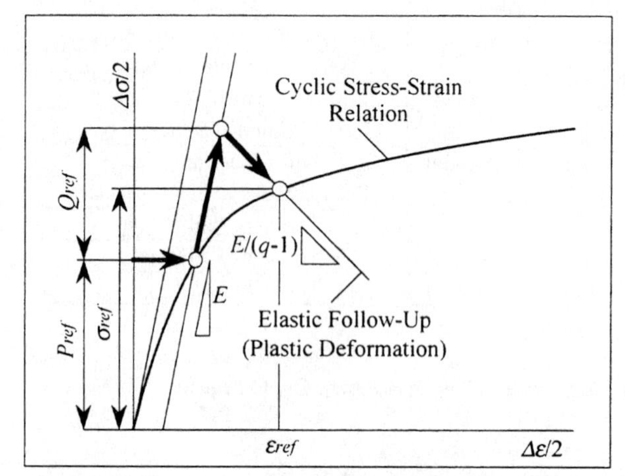

Figure 4 Evaluation Procedure of Elastic-Plastic Reference Stress, σ_{ref}, and Reference Strain, ε_{ref}

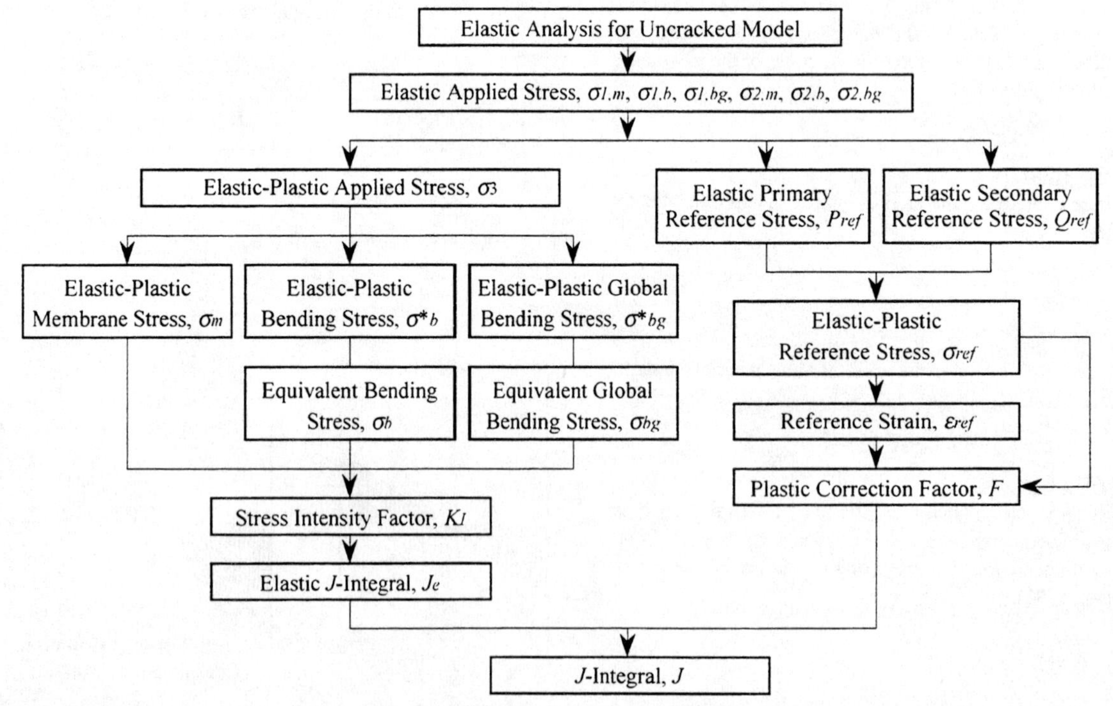

Figure 5 Evaluation Flow of J-Integral

component, σ^*_{bg}. σ_b and σ_{bg} are obtained from σ^*_b and σ^*_{bg}, respectively, by taking the nonlinear stress distribution at the cracked section into account as shown in Fig. 2. σ_3 is given as the solution of the following equation, which is schematically shown in Fig. 3.

$$q\left[\frac{\sigma_{1,m}+\sigma_{1,b}+\sigma_{1,bg}}{\sigma_0}+\frac{\sigma_{2,m}+\sigma_{2,b}+\sigma_{2,bg}}{\sigma_0}-\frac{\sigma_3}{\sigma_0}\right] = \kappa\left[\left(\frac{\sigma_3}{\sigma_0}\right)^n-\left(\frac{\sigma_{1,m}+\sigma_{1,b}+\sigma_{1,bg}}{\sigma_0}\right)^n\right] \tag{14}$$

Plastic Correction Factor

F is calculated by the following equation.

$$F = E\,\varepsilon_{ref}/\,\sigma_{ref} \tag{15}$$

σ_{ref} and ε_{ref} are the elastic-plastic reference stress and the reference strain, respectively. σ_{ref} and ε_{ref} are given as the solution of the following equations, which are schematically shown in Fig. 4.

$$q\left[\frac{P_{ref}+Q_{ref}}{\sigma_0}-\frac{\sigma_{ref}}{\sigma_0}\right]=\kappa\left[\left(\frac{\sigma_{ref}}{\sigma_0}\right)^n-\left(\frac{P_{ref}}{\sigma_0}\right)^n\right] \tag{16}$$

$$\varepsilon_{ref}=\frac{\sigma_{ref}}{E}+\kappa\frac{\sigma_0}{E}\left(\frac{\sigma_{ref}}{\sigma_0}\right)^n \tag{17}$$

P_{ref} is the primary reference stress and can be calculated using the solution of reference stress, σ_R, as a function of $\sigma_{1,m}$, $\sigma_{1,b}$, and $\sigma_{1,bg}$. The solution of σ_R is to be separately prepared. Q_{ref} is the secondary reference stress and can be similarly calculated as a function of $\sigma_{2,m}$, $\sigma_{2,b}$, and $\sigma_{2,bg}$. (Note that σ_{ref}, P_{ref}, and Q_{ref} denote actual values of reference stress. While σ_R means a function to identify an actual value of reference stress.)

Evaluation Flow

Figure 5 shows the evaluation flow of J. In the original reference stress method (Roche, 1988), Both K_I and F are calculated using the reference stress components. This procedure may duplicates the effect of the existence of crack. While in the proposed evaluation flow, the stress used for the calculation of K_I (elastic-plastic applied stress) and the stress used for the calculation of F (elastic-plastic reference stress) are exactly distinguished.

EVALUATION OF FATIGUE J-INTEGRAL RANGE

Fatigue J-integral, ΔJ_f, can be evaluated by the following equation.

$$\Delta J_f = 4\,U^2\,F\,J_e \tag{18}$$

where U is the crack opening ratio, F is the plastic correction factor, and J_e is the elastic J-integral. κ and n used to evaluate ΔJ_f are for the cyclic stress-strain relation given in Eq. (2).

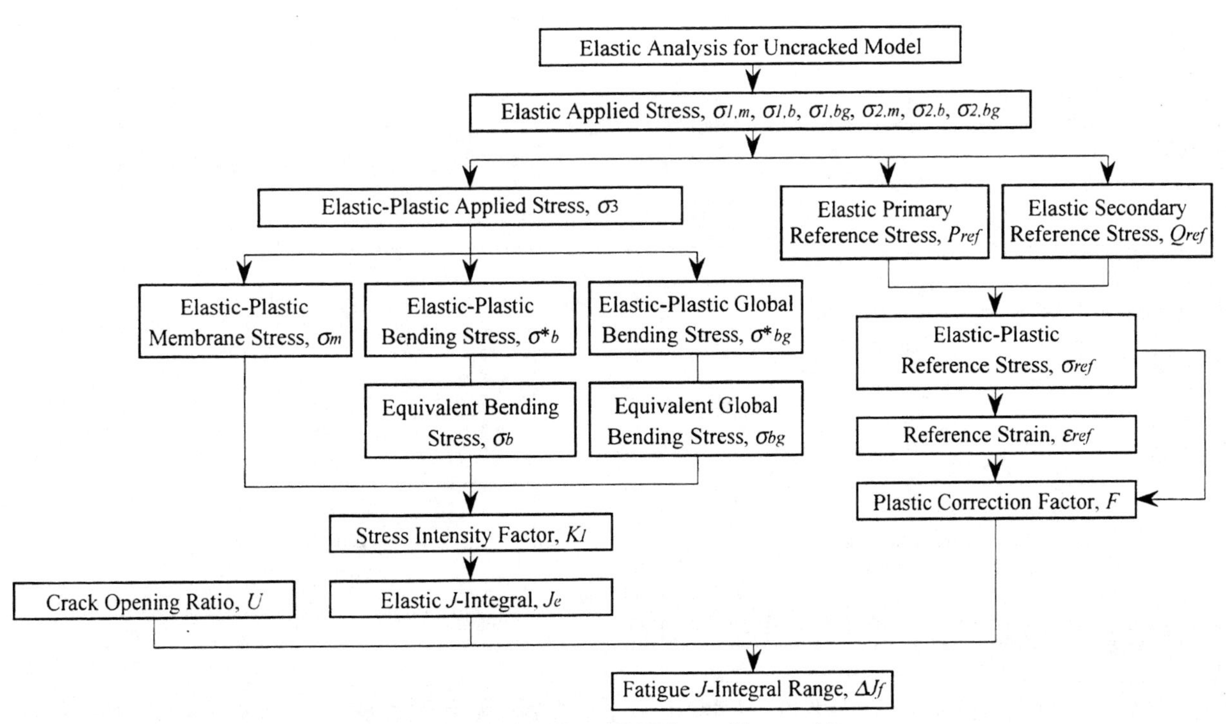

Figure 6 Evaluation Flow of Fatigue J-Integral Range

Elastic J-Integral

J_e is calculated by the same procedure as the case of the evaluation of J according to Eqs. (7) to (14). σ_3 used for the evaluation is schematically shown in Fig. 3. Note that the elastic stress for the evaluation of ΔJ_f is defined as a half of the corresponding elastic stress range.

Plastic Correction Factor

F is also calculated by the same procedure as the case of the evaluation of J according to Eqs. (15) to (17). σ_{ref} and ε_{ref} used for the evaluation are schematically shown in Fig. 4. The definition of the elastic stress is also the same as the case of the evaluation of J.

Crack Opening Ratio

In general, U ranges between 0 and 1, affected by many factors such as stress range, stress ratio, loading history, and so on. However, the effect of these factors on U is not enough comprehended at this time. Therefore, U is tentatively set as the following equation so that conservative evaluation can be assured.

$$U = 1.0 \tag{19}$$

Further efforts on U will be needed to accomplish more precise evaluation.

Evaluation Flow

Figure 6 shows the evaluation flow of ΔJ_f. The outline of the evaluation flow is the same as that for J except for the consideration of U.

EVALUATION OF CREEP J-INTEGRAL

Creep J-integral, J', can be evaluated by the following equation.

$$J' = F' J_e \tag{20}$$

where F' is the creep correction factor and J_e is the elastic J-integral. κ and n used to evaluate ΔJ_f are for the cyclic stress-strain relation given in Eq. (2).

Elastic J-Integral

J_e is obtained from K_I in consideration of stress relaxation during holding.

$$J_e = \left(\frac{\sigma'_3}{\sigma_3}\right)^2 K_I^2 / E' \tag{21}$$

σ_3 is the elastic-plastic applied stress at the beginning of holding in consideration of the elastic follow-up related to elastic-plastic deformation. σ_3 is given as the solution of Eq. (14), which is schematically shown in Fig. 7. σ'_3 is the elastic-plastic applied stress during holding in consideration of both elastic follow-up related to elastic-plastic deformation and elastic follow-up related to creep deformation. σ'_3 is calculated by the following equation including the holding time, t.

$$\sigma'_3 = \left[\frac{EB}{q'}(m-1)\,t + \sigma_3^{\,1-m}\right]^{1/(1-m)} \tag{22}$$

The lower limit of σ'_3 is bounded by $2\left[\sigma_{1,m}+\sigma_{1,b}+\sigma_{1,bg}\right]$ so that the relaxation is not below the sum of primary applied stress. The evaluation procedure of σ'_3 is also schematically shown in Fig. 7.

Creep Correction Factor

F' is calculated by the following equation.

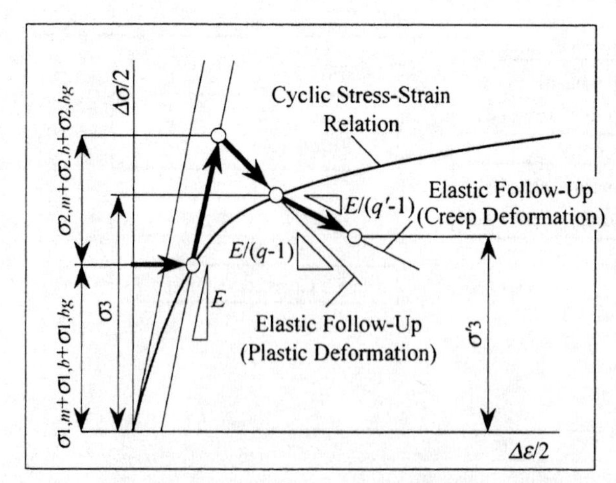

Figure 7 Evaluation Procedure of Elastic-Plastic Applied Stress, σ'_3, during Relaxation

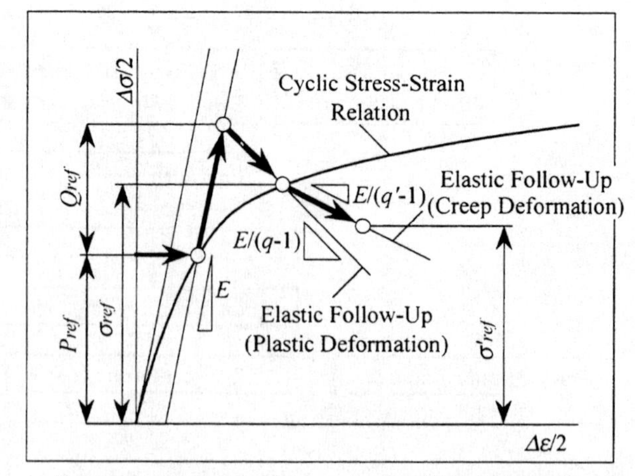

Figure 8 Evaluation Procedure of Elastic-Plastic Reference Stress, σ'_{ref}, during Relaxation

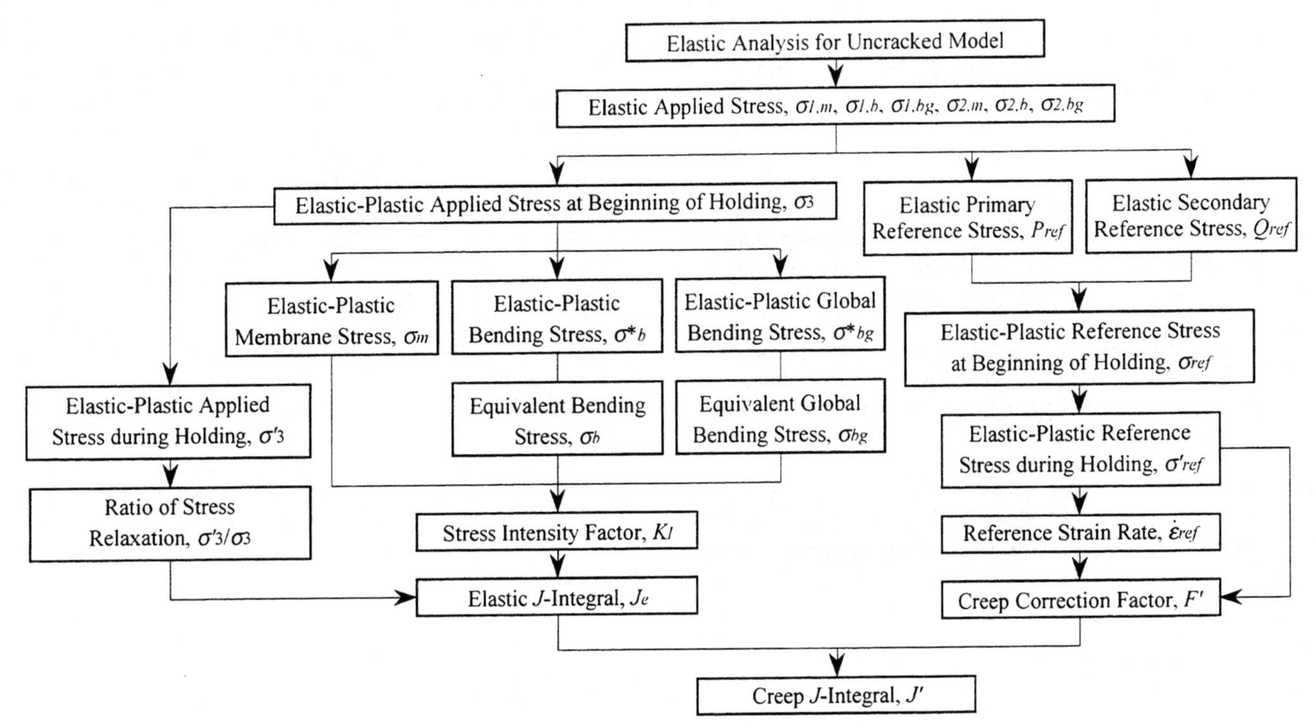

Figure 9 Evaluation Flow of Creep *J*-Integral

$$F' = E\,\dot{\varepsilon}_{ref} / \sigma'_{ref} \quad (23)$$

σ'_{ref} is the elastic-plastic reference stress during holding in consideration of both the elastic follow-up related to elastic-plastic deformation and the elastic follow-up related to creep deformation. σ'_{ref} is calculated by the following equation including t.

$$\sigma'_{ref} = \left[\frac{EB}{q'} (m-1)\,t + \sigma_{ref}^{\,1-m} \right]^{1/(1-m)} \quad (24)$$

σ_{ref} is the elastic-plastic reference stress at the beginning of holding in consideration of elastic follow-up related to elastic-plastic deformation. σ_{ref} is given as the solution of Eq. (16). The lower limit of σ'_{ref} is bounded by $2P_{ref}$ so that the relaxation is not below the primary reference stress. The evaluation procedure of σ_{ref} and σ'_{ref} is schematically shown in Fig. 8.

$\dot{\varepsilon}_{ref}$ in Eq. (23) is the reference strain rate. $\dot{\varepsilon}_{ref}$ can be evaluated according to the Norton's Law shown by Eq. (3).

Evaluation Flow

Figure 9 shows the evaluation flow of J'. The effect of the relaxation of both applied stress and reference stress is taken into account.

CONCLUSIONS

In this study, a simplified J-evaluation method for flaw evaluation at high temperature was investigated to estimate J-integral, fatigue J-integral range, and creep J-integral. The systematization of the simplified J-evaluation method incorporated with the reference stress method and the concept of elastic follow-up was presented, then the comprehensive evaluation procedure was proposed. This evaluation method has been incorporated in the draft of the guideline for structural integrity assessment of high-temperature components (Miura, et al., 1999) which presents a technical basis for the evaluation of the structural integrity of Fast Breeder Reactor main components using the latest fracture mechanics methodology.

ACKNOWLEDGMENT

This study was carried out as a part of the project of the Ministry of International Trade and Industry entitled "Verification Tests of Fast Breeder Reactor Technology," which has been conducted since 1987. The authors would like to express their sincere gratitude to all members of an advisory committee chaired by Professor Genki Yagawa of the University of Tokyo.

REFERENCES

Iida, K., Asada, Y., Okabayashi, K., and Nagata, T., 1987, "Simplified Analysis and Design for Elevated Temperature Components of Monju," *Nuclear Engineering and Design*, **Vol. 98**, pp. 305-317.

Iwasaki, R., Shimakawa, T., Nakamura, K., Takahashi, H., Uno, T., and Watashi, K., 1991, "Application of Simplified J-Estimation

Methods to Surface Cracked Structures under Creep-Fatigue Loadings," Proceedings, 11th International Conference on Structural Mechanics in Reactor Technology, **Vol. L, L10/2**, Tokyo, Japan, pp. 217-222.

Milne, I., Ainsworth, R. A., Dowling, A. R., and Stewart, A. T., 1988, "Assessment of the Integrity of Structures Containing Defects," *International Journal of Pressure Vessels and Piping*, **Vol. 32**, pp. 3-104.

Miura, N., and Shimakawa, T., 1998, "Ductile Fracture Behavior of Stainless Steel Cracked Pipes at High Temperature," Fatigue, Fracture, and High Temperature Design Methods in Pressure Vessel and Piping ASME PVP-**Vol. 365**, July, pp. 231-239.

Miura, N., Nakayama, Y., and Takahashi, Y., 1999, "Development of Flaw Evaluation Guideline for FBR Components," Proceedings, 15th International Conference on Structural Mechanics in Reactor Technology, **Vol. V, G04/3**, Seoul, Korea, Aug., pp. 137-144.

Nakayama, Y., Miura, N., Takahashi, Y., Shimakawa, T., Date, S., and Toya, Y., 1998, "Development of Fatigue and Creep Crack Propagation Law for 316FR Stainless Steel in Consideration of FBR Operating Condition," Fatigue, Fracture, and High Temperature Design Methods in Pressure Vessel and Piping ASME PVP-**Vol. 365**, July, pp. 191-198.

Poussard, C., Wakai, T., and Drubay, B., 1999, "A Comparison Between Japanese and French A16 Defect Assessment Procedures for Creep-Fatigue Crack Growth," Proceedings, 15th International Conference on Structural Mechanics in Reactor Technology, **Vol. V, G03/3**, Seoul, Korea, Aug., pp. 95-102.

Roche, R. L., 1988, "Modes of Failure - Primary and Secondary Stresses," *Transactions of the ASME, Journal of Pressure Vessel Technology*, **Vol. 110**, pp. 234-239.

Shimakawa, T., Ogawa, H., Miura, N., Nakayama, Y., and Takahashi, Y., 2000, "Simplified *J*-Evaluation Method for Flaw Evaluation at High Temperature Part II: Verification of Evaluation Method," to be presented at 2000 ASME Pressure Vessels and Piping Conference.

Wakai, T., Poussard, C., and Drubay, B., 1999, "A Comparison between Japanese and French A16 Defect Assessment Procedures for Fatigue Crack Growth," Proceedings, 15th International Conference on Structural Mechanics in Reactor Technology, **Vol. V, G03/1**, Seoul, Korea, Aug., pp. 79-86.

PVP-Vol. 410-1, Assessment Methodologies for Preventing Failure: Deterministic and Probabilistic Aspects and Weld Residual Stress
Volume 1
ASME 2000

Simplified *J*-Evaluation Method for Flaw Evaluation at High Temperature -Part II: Verification of Evaluation Method

Takashi Shimakawa, Hiroshi Ogawa
Research and Development Department
Kawasaki Heavy Industries, Ltd.
Tokyo, Japan

Yasunari Nakayama
Materials Science Department
Central Research Institute of Electric Power Industry
Tokyo, Japan

Naoki Miura
Materials Science Department
Central Research Institute of Electric Power Industry
Tokyo, Japan

Yukio Takahashi
Materials Science Department
Central Research Institute of Electric Power Industry
Tokyo, Japan

ABSTRACT

J-integral is the feasible parameter to represent fracture behavior in non-linear regime. However there are many difficulties to calculate accurate *J*-integral in actual components. CRIEPI has been developing a simplified *J*-evaluation method based on the reference stress approach (Miura, et al., 2000). This paper presents verifications of the proposed method through four representative problems. The first subject is the ductile crack growth problem for a straight pipe with through-wall crack. *J*-integral values calculated by the proposed method are compared with the results from finite element analysis and from some simplified methods. The second subject is the creep-fatigue crack propagation problem for test specimens such as CT and CCP. Calculated results are compared with those measured in experiments. Applicability to creep-fatigue problem is verified through this comparison. The third subject is the creep-fatigue crack growth problem for a plate with semi-elliptical surface crack. Estimated fatigue *J*-integral range and creep *J*-integral are compared with finite element analysis results. Applicability to the surface crack problem is confirmed. The fourth subject is the creep-fatigue crack growth problem for a cylinder under thermal stress. Applicability to thermal stress problem is also verified. These results show that *J*-integral values calculated by the proposed method coincide with the experimental and the finite element analysis results. It is confirmed that the proposed method is applicable for the flaw evaluation of cracked components at high temperature.

INTRODUCTION

Inelastic fracture mechanics, which uses J-integral, is effective as the structure integrity evaluation technique for cracks assumed in actual components at high temperature used under creep fatigue condition. Since the development of a simplified *J*-evaluation method was necessary to utilize the inelastic fracture mechanics as a practical design tool, Part I of this paper provided the outline of simplified evaluation method and the realistic calculation procedure that combined the reference stress method and the elastic follow-up concept. This paper presents the verification of the simplified evaluation method proposed in Part I through the following four representative problems.

(1) The ductile crack growth problem for a straight pipe with through-wall crack.
(2) The creep-fatigue crack propagation problem for test specimens such as CT and CCP.
(3) The creep-fatigue crack growth problem for a plate with semi-elliptical surface crack.
(4) The creep-fatigue crack growth problem for a cylinder under thermal stress.

VERIFICATION OF *J*-INTEGRAL FOR STRAIGHT PIPE WITH THROUGH-WALL CRACK

Analysis Problem

The first problem is a straight pipe with circumferentially through-wall crack under bending load. Figure 1 shows the subjected 4-point bending test loaded with the controlled displacement. Table 1 shows the material properties of 316FR base metal at 550°C.

Analytical parameters are pipe size and crack angle as summarized in Table 2.

Table 1 Material Properties of 316FR

Material	Heat Treatment	Temp.	0.2%Proof Stress $\sigma_{0.2}$ (MPa)	Ultimate Strength σ_u (MPa)	Flow Stress σ_f (MPa)	Design Flow Stress S_f (MPa)	Stress-Strain Relation** E (GPa)	Stress-Strain Relation** σ_0 (MPa)	Stress-Strain Relation** α	Stress-Strain Relation** n	*J-R* Curve*** C	*J-R* Curve*** m
316FR Base Metal	As Manufactured	550°C	172	412	292	244	154	172	3.163	4.698	0.524	0.475
	Thermal Aged*		143	399	271			117	5.441	3.386	0.878	0.800

* : Substituted by material properties from rolled plates
** : Stress-strain relation was formulated as $\varepsilon/\varepsilon_0 = \sigma/\sigma_0+\alpha(\sigma/\sigma_0)^n$, $\varepsilon_0=\sigma_0/E$
*** : *J-R* curve was formulated as $J=C(\Delta a)$, J in[MN/m], Δa in[mm]

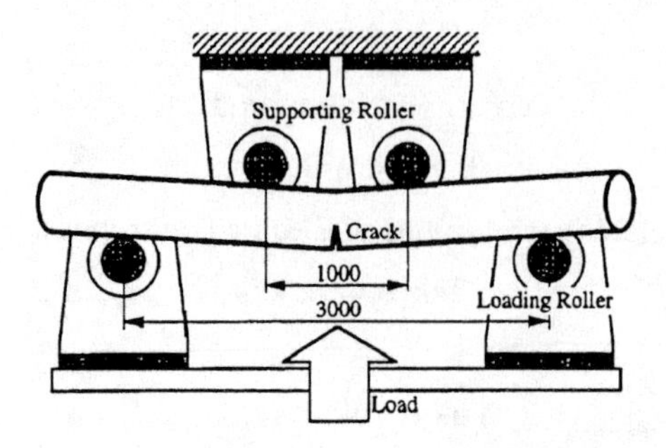

Figure 1 Circumferentially Through-Wall-Cracked Pipe Subjected to 4-Point Bending

Table 2 Summary of Object of Analysis

Case No.	Material	Temp.	Outer Diameter (mm)	Pipe Thickness (mm)	Total Crack Angle (degree)
1	316FR Base Metal	550°C	303.7	10	30
2					60
3					90
4			965.2	15.9	30
5					60
6					90

Application of Simplified *J*-Evaluation Method

The stress intensity factor, K_I and the reference stress, σ_{ref} were calculated by the following equations for through-wall-cracked pipe subjected to bending(Tada et al.,1985).

$$K_I=(\sigma_{bg}F_b)\,(\pi R\theta)^{1/2} \tag{1}$$

where,

$$F_b=1+6.8(\theta/\pi)^{3/2}-13.6(\theta/\pi)^{5/2}+20.0(\theta/\pi)^{7/2} \tag{2}$$

$$\sigma_{bg}=M/(\pi R^2 t) \tag{3}$$

$$\sigma_{ref}=\sigma_{bg}\frac{\pi}{4}\frac{1-\dfrac{t/R_i}{\left(1+\dfrac{t}{R_i}\right)\left(2+\dfrac{t}{R_i}\right)}}{\sin\left(\dfrac{\pi}{2}-\dfrac{c}{2R_i}\right)-\dfrac{1}{2}\sin\left(\dfrac{c}{R_i}\right)} \tag{4}$$

The reference stress σ_{ref} was obtained by assuming the elastic follow-up factor $q = 1.0$.

Verification of Simplified *J*-Evaluation Method

The finite element analysis was conducted using the computer code MARC with shell elements. Figure 2 shows an example of the finite element mesh. J was calculated along the five paths shown in Fig. 2. The path-independence of J was confirmed. The ratio of J for each path to the mean value ranged between 0.92 and 1.09. The finite element analyses as well as calculations by the J-evaluation method were performed for stationary cracks.

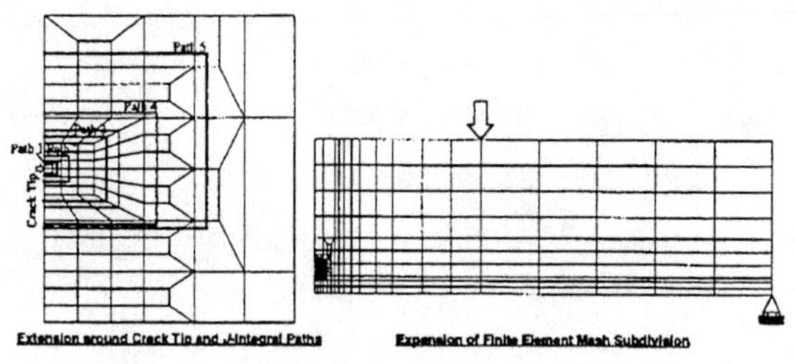

Figure 2 Example of Finite Element Mesh (case 2)

Elastic-plastic fracture analysis for a circumferential through-wall-cracked pipe subjected to bending can be commonly performed using simplified J-evaluation method. Four J-evaluation methods, LBB.NRC (Klecker et al., 1986), LBB.ENG2 (Brust, 1987), GE/EPRI (Kumar et al., 1984), and the proposed method (Miura et al., 2000) were also examined. The former three methods were well developed to predict the fracture performance of cracked pipes typical of Light Water Reactor plants. For Cases 4 to 6 in Table 2, it is noted that the ratio of the mean radius to the thickness, R_m/t is about 30, which is beyond the limitation of the compliance functions of the GE/EPRI method. These compliance functions were then calculated on the assumption that R_m/t takes its maximum value of 20 in this study. Similarly, some compliance functions for the value of the strain-hardening exponent, n beyond 5 were computed assuming that n was equal to 5.

Figure 3 shows the P (or Moment, M)-COD relations obtained from the LBB.NRC, LBB.ENG2, and GE/EPRI methods compared with those from the finite element analysis. The LBB.NRC and LBB.ENG2 methods give relatively higher P (M), while the predictions from the GE/EPRI method are in good agreement with those by the finite element analysis. The difference between P (M)-COD curves from different methods becomes smaller when the crack angle is larger.

Figure 4 shows the J-P (M) relations obtained from the LBB.NRC, LBB.ENG2, GE/EPRI, and the proposed methods compared with those from the finite element analyses. The GE/EPRI method gives the highest prediction in most cases; however, the choice of evaluation method does not significantly affect evaluated results. Thus the J-evaluation methods of interest are considered to be applicable to the cracked pipe fracture analysis.

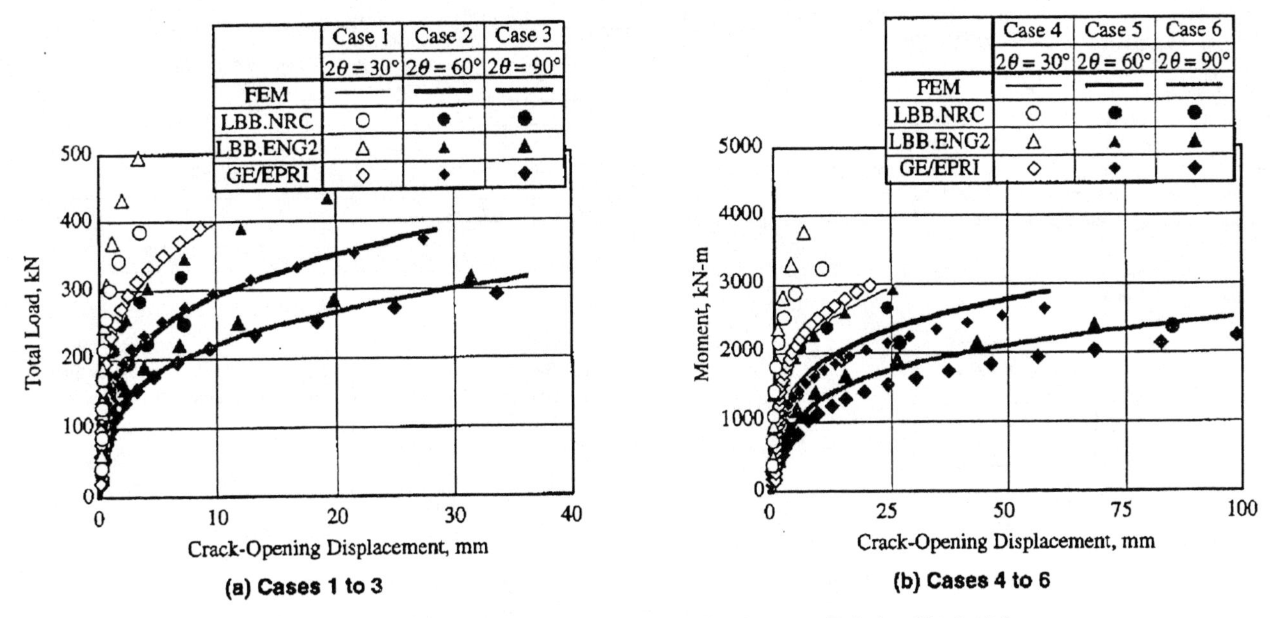

(a) Cases 1 to 3

(b) Cases 4 to 6

Figure 3 Load (Moment) and Crack-Opening-Displacement Relation Obtained from *J*-Evaluation method and Finite Element Analysis

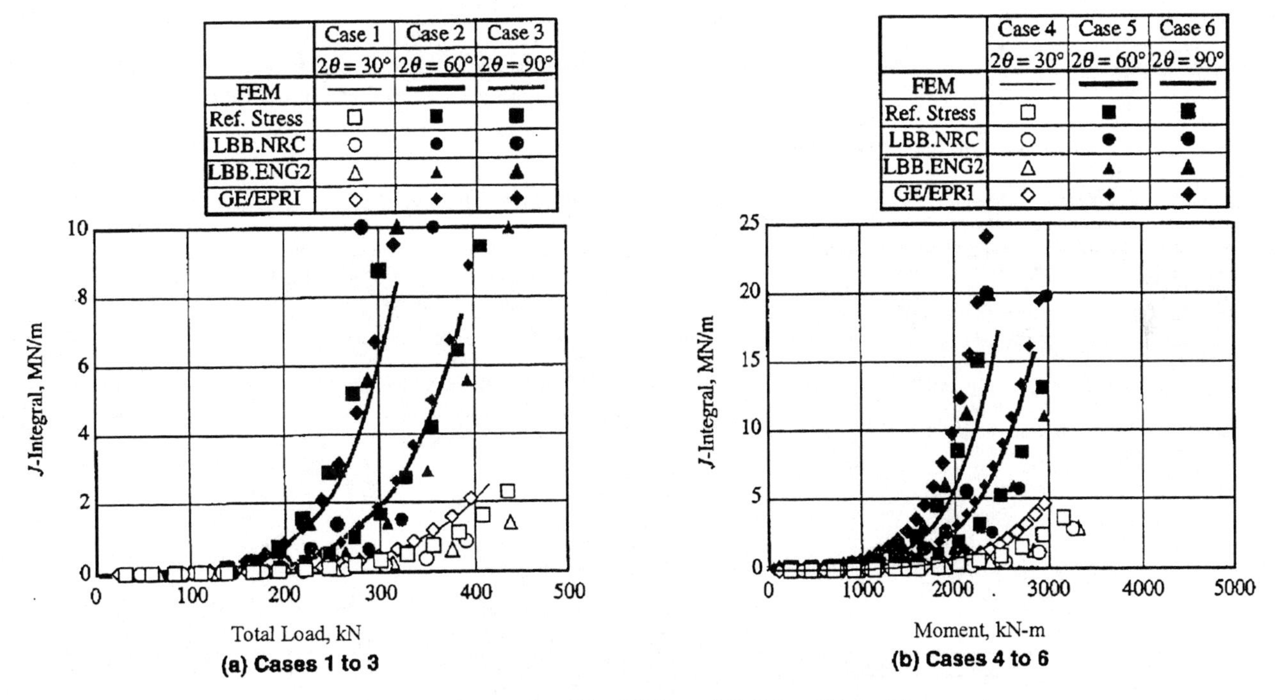

(a) Cases 1 to 3

(b) Cases 4 to 6

Figure 4 *J*-integral and Load (Moment) Relation Obtained from *J*-Evaluation method and Finite Element Analysis

VERIFICATION OF FATIGUE *J*–INTEGRAL RANGE AND CREEP *J*–INTEGRAL RANGE FOR 2D CRACK

Analysis Problem

Applicability of the proposed method to creep-fatigue crack propagation was verified by using CCP specimens and CT specimens. Figures 5 and 6 show the geometry and dimension of the test specimens, respectively. Four kinds of problems are selected to investigate the effects of specimen shape and the loading condition, load controlled or displacement controlled. Table 3 shows the loading conditions in the referred experiments, respectively. Tested material was Type 304 stainless steel as shown in Table 4.

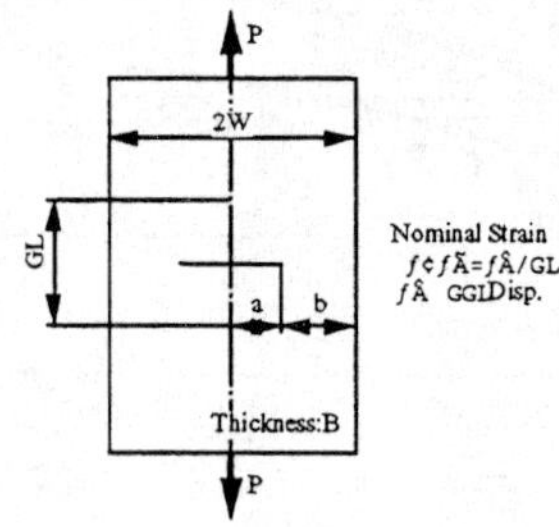

Figure 5 Center Cracked Plate Specimen (CCP)

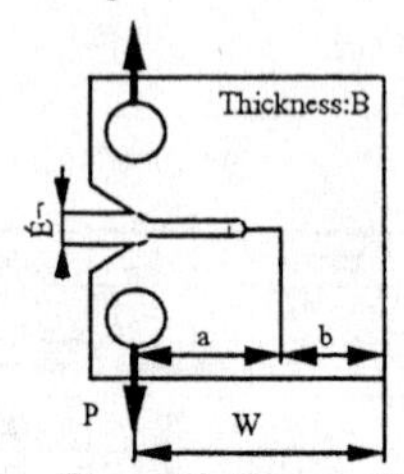

Figure 6 Compact Type Specimen (CT)

Application of Simplified *J*-Evaluation Method

Since the specimen thickness was relatively thin, the solution of the plane stress was used to quote the solution of two dimensions.
The stress intensity factor, K_I by (Murakami et al.):

1) CT:

$$K_I = \left(\frac{P}{2W}\right)\sqrt{\pi a}\cdot\sqrt{\sec\left(\frac{\pi\xi}{2}\right)} \tag{5}$$

2) CCP:

$$K_I = \frac{P}{\sqrt{W}}\cdot\frac{2+\xi}{(1-\xi)^{3/2}}\left(0.886+4.64\xi-13.32\xi^2+14.72\xi^3-5.6\xi^4\right) \tag{6}$$

where,

$$\xi = a/W \tag{7}$$

Table 4 Constitutive Equations of SUS304

Ramberg-Osgood type Stress-Strain Relation (Applied Strain Range $\Delta\varepsilon < 3.0\%$) $\Delta\varepsilon = (\Delta\sigma/E) + 2\varepsilon_0\kappa(\Delta\sigma/2\sigma_0)^n$					
T(°C)	E(N/mm²)	σ_0(N/mm²)	ε_0(mm/mm)	κ	n
650	143957	85.67	5.951×10^{-4}	0.0432	5.196

Norton Type Creep Strain Equation $\varepsilon = B\sigma^n$ (mm/mm/hr)			
T(°C)	tH (min)	B	n
650	10	7.125×10^{-16}	6.0504
650	60	1.250×10^{-15}	5.8650

The reference stress σ_{ref} was calculated by expressions by using P_e that was the load of input value or the apparent elastic load obtained from the displacement and using P_0 that was obtained from the GE/EPRI equation (Kumar et al., 1981).

$$\sigma_{ref} = \left(P_e/P_0\right)\sigma_0 \tag{8}$$

1) CT:

$$P_0 = 1.071\eta b\sigma_0 \tag{9}$$

$$\eta = \left[(2a/b)^2+2(2a/b)+2\right]^{1/2}-(2a/b+1) \tag{10}$$

2) CCP:

$$P_0 = 2b\sigma_0 \tag{11}$$

Since the elastic follow-up can be ignored for CCP specimens and CT specimens, the reference stress σ_{ref} was calculated by assuming the elastic follow-up factor $q = 1.0$.

Verification of Simplified *J*-Evaluation Method

The predicted result by the simplified *J*-evaluation method was compared with the experimental value ΔJ_f and ΔJ_c (Asada et al., 1992) in creep-fatigue crack propagation tests. ΔJ_c in the simplified evaluation was defined by the following expression:

$$\Delta J_c = \int J'dt \tag{12}$$

The simplified evaluation result for the tension type CCP specimen in case 1 and 2 is shown in Fig. 7. Values of ΔJ_f and ΔJ_c are almost the same in the load control (LC) in case 1.

Table 3 Analysis Condition of 2D Specimen

Case No.	Loading*	Specimen Type	W (mm)	B (mm)	GL (mm)	Temp. (°C)	Stress $\Delta\sigma$ (N/mm²)	Strain $\Delta\varepsilon$ (%)	Disp. δ (mm)	Hold Period (mm)	Crack Length a (mm)
1	LC	CCP	10	5	-	650	±127.5	-	-	60	0.5-5
2	DC	CCP	10	5	25		-	±0.25	-	60	0.5-8
3	LC	CT	50.8	12.7	-		-	±6.18	-	10	27-33
4	DC	CT	51	12.7	-		-	-	±0.7	10	25-38

*: LC: Load Controlled
DC: Displacement Controlled

The value of ΔJ_c with the Δ mark was slightly larger than that of ΔJ_f with the O mark. The same tendency was also obtained in the prediction by the simplified method. The experiment value notably increased where the half crack length reached about 2.5 mm. This reason was considered that the net section stress in the ligament section increased to be critical limit. Both predicted and experimental ΔJ_c were smaller than ΔJ_f with 1 order or more in the displacement control (DC) in case 2. In addition, calculated ΔJ_c by the simplified method was smaller than that by experiment. This reason was considered that the Norton-type creep strain approximation was used in the simplified evaluation although the primary creep might govern the stress relaxation process under the displacement control in the short 60 minutes retention time.

The simplified evaluation results for CT is shown in Fig. 8 for case 3 and 4. The holding period in case 3 under load control (LC) is shorter than that in case 1. Therefore, ΔJ_c is 1-order or more smaller than ΔJ_f. In addition, calculated ΔJ_c by the simplified evaluation was smaller than that of experimental value. This reason was also considered that the phenomenon governed by the primary creep was described by the Norton-type. On the contrary, the simplified evaluation value was in the safety side for case 4.

The above-mentioned verification using two dimensional specimens demonstrated that the simple evaluation method proposed by Part I could predict ΔJ_f and ΔJ_c of simple two-dimensional specimens in relatively high precision despite the error caused by modeling the stress-strain relationship and the creep strain rate expression.

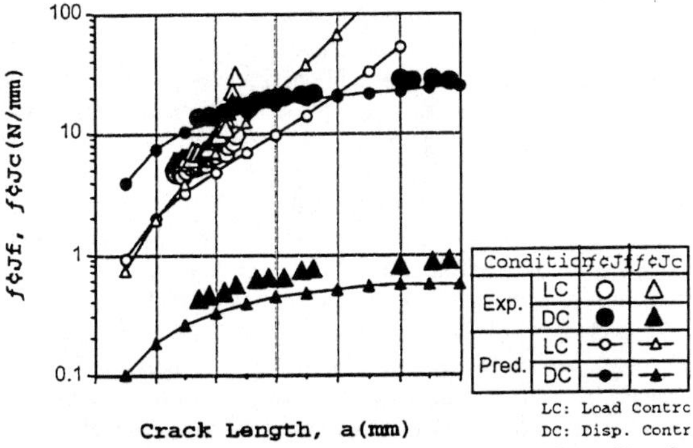

Figure 7 Comparison between experimental and predicted J-integral for CCP specimen

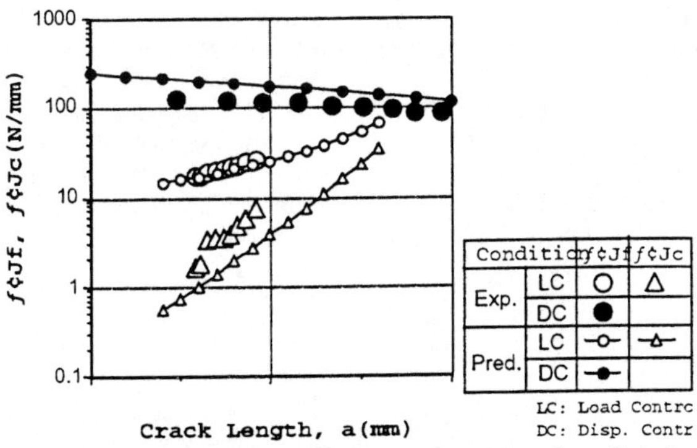

Figure 8 Comparison between experimental and predicted J-integral for CT specimen

VERIFICATION OF FATIGUE J-INTEGRAL RANGE AND CREEP J-INTEGRAL RANGE FOR 3D CRACK

Analysis Problem

A flat plate with a surface crack subject to bending load as shown in Fig. 9 was used for the three-dimensional problem. Parameters were the specimen size and holding period to set two kinds of problems as show in Table 5. Used material is SUS304 of which properties are shown in Table 4.

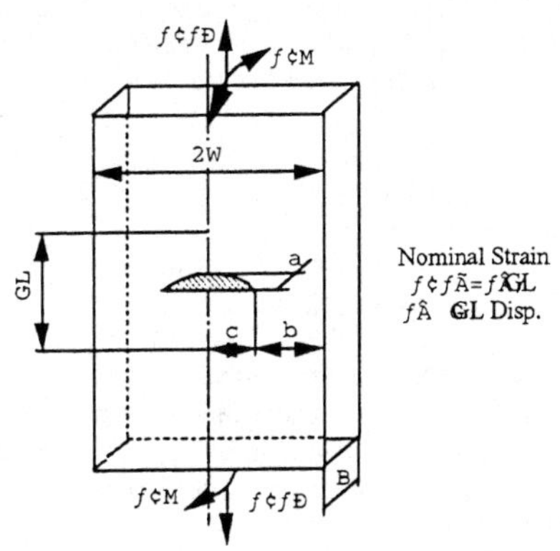

Figure 9 Surface Cracked Specimen

Table 5 Testing Condition of 3D Specimen

Case No.	Specimen (mm)			Temp. (°C)	Δε (%)	Hold Period (min.)	Crack Size a, c (mm)	Notes
	W	B	GL					
1	30	8	70	550	±0.4	300	2.0, 2.62	
							3.0, 4.91	
							3.5, 6.38	
							4.0, 8.18	
							4.5, 10.55	
2	250	50	10	550	±0.2	15	10, 40	Crack does not exist in gage length
							15, 60	
							20, 80	
							25, 100	
							26, 77	

Application of Simplified J-Evaluation Method

The reference stress is calculated by the following expression for flat plates (crack depth a, crack length, plate thickness, and plate width) that has a surface crack and receives the tensile stress and the bending stress (Willoughby et al., 1989):

$$\sigma_{ref} = \frac{\sigma_b/3+\sqrt{\sigma_b^2/9+(1-\zeta)^2\sigma_m^2}}{(1-\zeta)^2} \tag{13}$$

$$\zeta = \frac{ac}{t(c+t)} \tag{14}$$

The stress intensity factor used to calculate the elastic J-integral was obtained by the Raju & Newman equation (Raju et al., 1979). Since the elastic follow-up can be also ignored for the flat plate with a surface crack, the reference stress σ_{ref} was obtained by assuming the elastic follow-up factor $q = 1.0$.

Verification of Simplified J-Evaluation Method

Predicted J-integral value by simplified method was compared with the J-integral (Shimakawa et al., 1992) evaluated by the three-dimensional finite-element analysis.

The comparison between the finite-element analysis and the simplified evaluation result for case 1 is shown in Fig. 10. Predicted ΔJ_f by simplified method were slightly smaller in depth direction and larger in surface direction than those by the finite-element analysis. However, their agreements were satisfactory for an engineering use. The ΔJ_c predicted by the simplified evaluation method was about five times as large as the finite element analysis in both depth and surface directions. This reason was considered that ΔJ_c of the quoted finite element analysis result was obtained from (J' value after 300 hrs) (Holding period), while ΔJ_c of the simplified evaluation method was obtained by the cumulative integration of J' during stress relaxation.

The comparison between the finite element analysis result and the simple evaluation result for case 2 is shown in Fig. 11. Both ΔJ_f and ΔJ_c predicted by the simplified evaluation methods are slightly conservative than those of finite element analysis regardless of either direction of depth or surface. Since ΔJ_c of finite element analysis result quoted in case 2 is obtained by the cumulative integration of ΔJ_c in the stress relaxation as same as the simplified evaluation, the difference between the simplified evaluation method and the finite element method is smaller than that of case 1. The simplified evaluation method can represent tendencies of the applied displacement quantities difference (that of case 1 is about 2 times as large as that of case 2), the specimen size difference, the crack sizes difference (that of case 1 is about 1/10 as large as that of case 2), and the retention time (that of case 1 is about 5 times as large as that of case 2) between case 1 and 2.

It was confirmed that the prediction using the simplified evaluation method could provide almost same accuracy with the finite element analysis result.

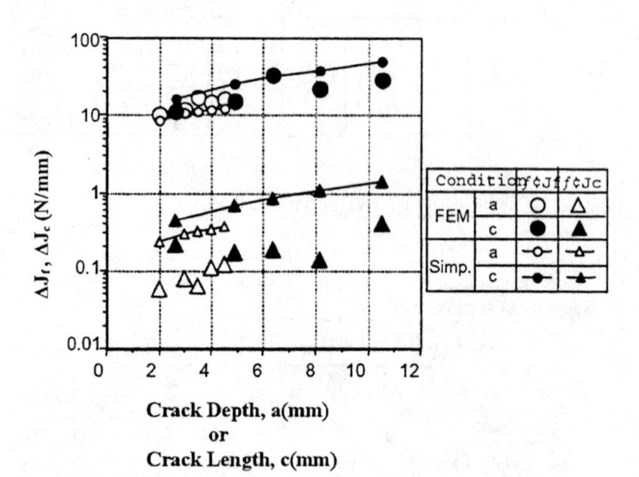

Figure 10 Comparison between measured and predicted J-integral for 3D-specimen (case 1)

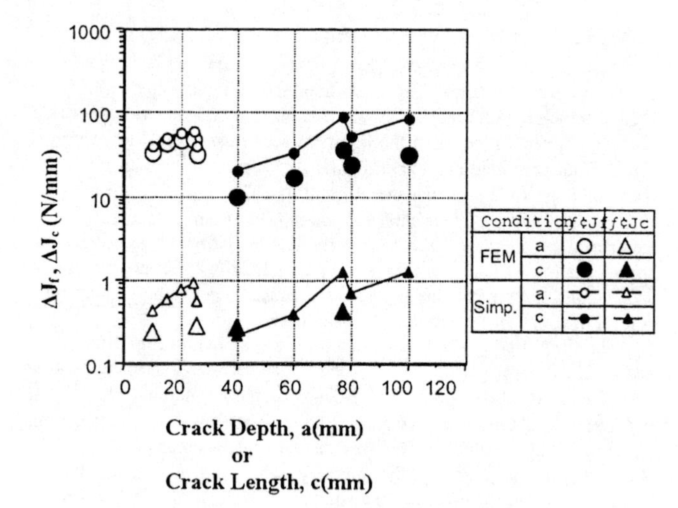

Figure 11 Comparison between measured and predicted J-integral for 3D-specimen (case 2)

VERIFICATION OF J-INTEGRAL UNDER THERMAL STRESS

Analysis Problem

The analysis specimen was a thin-wall cylinder of which the ratio of the inner radius to the wall thickness was 10 as shown in Fig. 12. A entire perimeter crack and a circumferentially perimeter crack of internal surface of the cylinder were assumed. The depth-to-length ratio of half oval crack was assumed to be 0.25. The analysis cases are shown in Table 6. 98MPa uniform load in the axial direction, maximum 600°C temperature gradient in the axial direction, and maximum 200°C temperature gradient in the wall thick direction were used in the three kinds of load condition of the uniform load in the axial direction, the temperature gradient in the axial direction, and the temperature gradient in the wall thick direction as shown in Fig. 13. The cyclic stress-strain relation of Type 304 stainless steel at 550°C was approximated by multiple lines as shown in Fig. 14.

The finite-element analysis was conducted using the computer code MARC (Version K5.4) and the calculation of virtual crack propagation method using the LORENZJ option (DeLorenzi, 1982) that had the path-independence when a specific strain was present as found in the thermal stress problem. An example of finite elements mesh to crack surfaces of semi-elliptical crack is shown in Fig. 15. The 20-node iso-parametric element was used. Both crack surface and the ligament section were assumed to have 7 layers and their divisions were so that the element width decreased by 1/2 decrement toward the crack end. 5 J-integral paths were defined in the sequence so that the area volume increased in the path from the crack end. Figure 16 shows the finite-element analysis result at the surface point of the semi-elliptical crack as an example J-integral path independency.

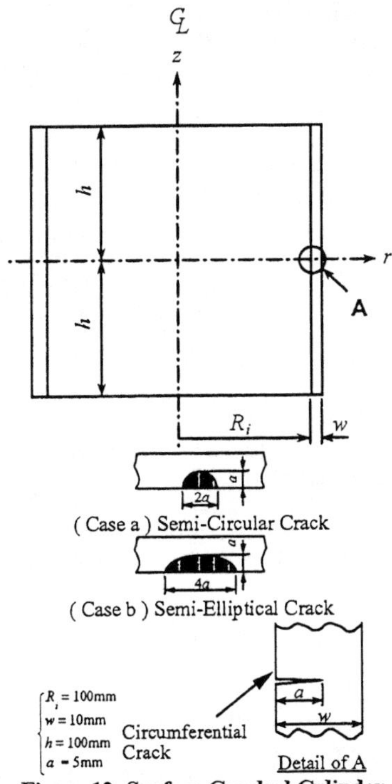

Figure 12 Surface Cracked Cylinder

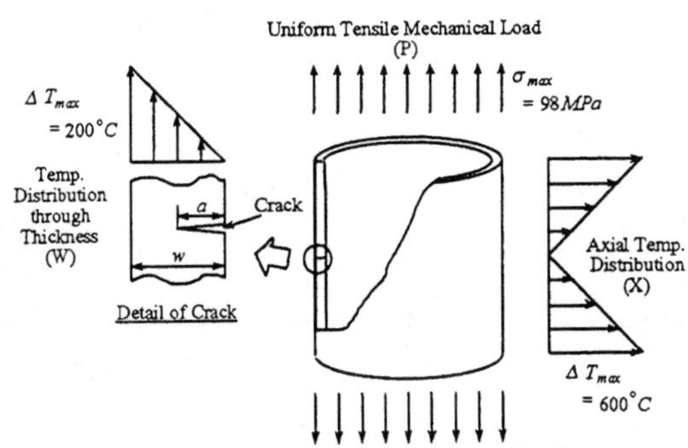

Figure 13 Loading condition of cylinder

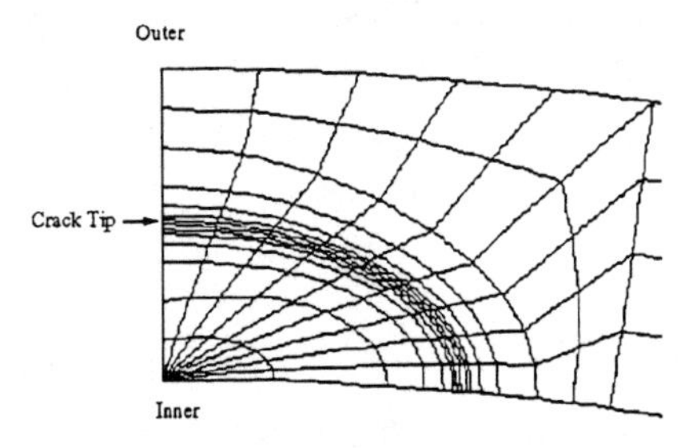

Figure 14 FEM mesh near crack tip

Table 6 Applied stress for cylinder

Stress / Applied Load	Primary Stress		Secondary Stress	
	Membrane P_m (MPa)	Bending P_b (MPa)	Membrane Q_m (MPa)	Bending Q_b (MPa)
Uniform Tensile Mechanical Load	98.1	0.0	0.0	0.0
Axial Linear Temperature Distribution	0.0	0.0	0.3	210.6
Linear Temperature Distribution Through Thickness	0.0	0.0	7.8	436.6

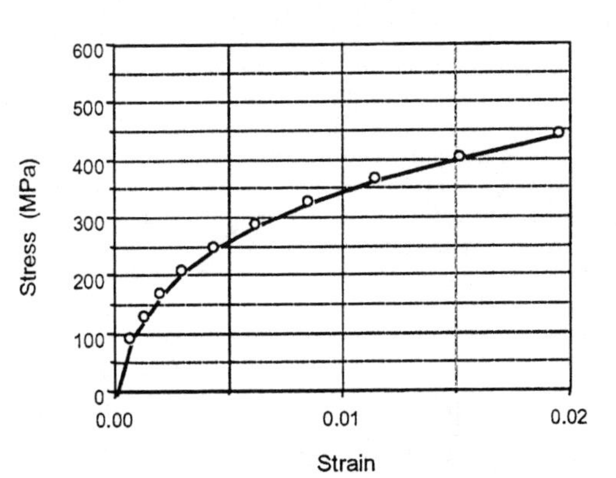

Figure 15 Stress-strain relation

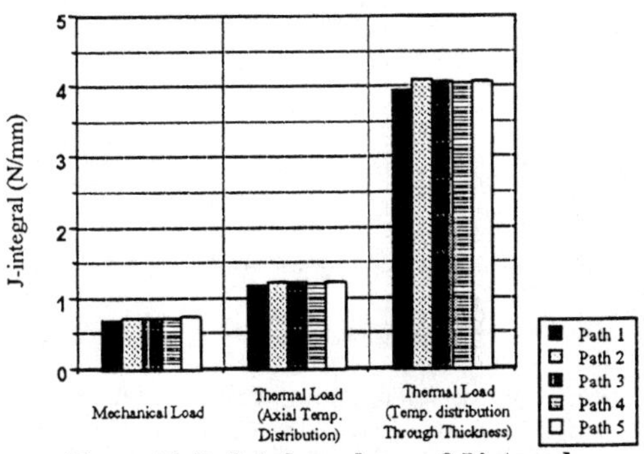

Figure 16 Path independency of *J*-integral at the surface point of elliptical crack

Application of Simplified *J*-Evaluation Method

The flat plate with surface crack was used as a model representing the cylinder with surface cracks in order to calculate the stress intensity factor. Total perimeter length of cylinder was used as the width of modeled flat plate. The modeling shown in Table 6 consists of the membrane stress range and the bending stress range calculated from the equivalent linear process using the elastic finite element analysis of the model with no crack.

The stress intensity factor was calculated by the Raju & Newman equation. In addition, the reference stress equation was also used by the flat plates equation (see Eqs. (13) and (14)) . The reference stress σ_{ref} was obtained by assuming the elastic follow-up factor $q = 1.0$.

Verification of Simplified *J*-Evaluation Method

Figure 17 shows the comparison between the *J*-integral obtained from the finite element analysis and the *J*-integral obtained from the proposed method. Sufficient agreement was recognized between the reference stress method prediction and the finite element analysis result for the mechanical load problem regardless of differences of crack shapes, surfaces, and insides. The simplified evaluation method well agrees with the finite element analysis of the deepest point in the thermal stress problem as regard to both cases of the axial-direction temperature gradient and the thickness-direction temperature gradient. However, the simplified evaluation method gives the prediction of surface point excessively larger than that given by the finite element analysis. The reason of this difference must be studied for both accuracy of the finite element analysis and the simplified evaluation method. However, the accuracy of finite element analysis can be considered to be reasonable since the simplified evaluation method matches the finite element analysis at least for the mechanical load at the surface point in addition to the favorable path independency obtained at the surface point. On the other hand, the elastic stress solution that was obtained from the model with no crack shown in Table 6 was used in the reference stress calculation and the surface strain release, due to the crack existence was not considered. Since the prediction of the reference stress method also tends to indicate to be in the safety side in accordance with the crack progress in the structure model test evaluation under thermal stress obtained so far, it seems necessary to study the thermal stress release according to the crack progress to further improve the accuracy in future.

Either of results is within the range of factor of 2. Therefore, it is considered that the simplified evaluation method can reasonably evaluate the J-integral under thermal stress.

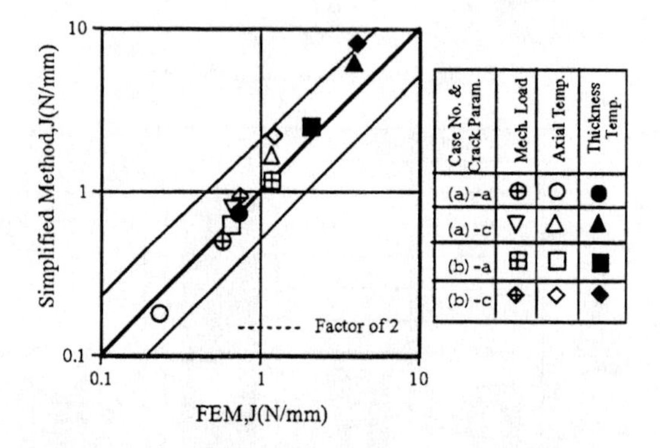

Figure 17 Comparison of J-integrals detailed from FEM and simplified analysis under thermal stress conditions

CONCLUSIONS

This paper presents verifications of the simplified J-evaluation method proposed in Part I through the following four representative problems.

It was confirmed that the simplified J-evaluation method proposed can provide satisfactory predicts as a result of the implemented simplified J-evaluation method based on the reference stress method in comparison with a finite element analysis result of a straight pipe with through-wall crack, a creep fatigue test result of a simple specimen, and a finite element analysis result of a structure under thermal stress. In addition, the validity of simplified J-evaluation method was confirmed using its predict in comparison with the detailed inelastic analysis performed to confirm the applicability to the thermal stress issue that was problematic in actual components.

ACKNOWLEDGEMENT

This study was carried out as a part of the project of the Ministry of International Trade and Industry entitled "Verification Tests of Fast Breeder Reactor Technology", which has been conducted since 1987. The authors would like to express their sincere gratitude to all members of an advisory committee chaired by Professor Genki Yagawa of the University of Tokyo.

REFERENCES

Asada, Y., et al., 1992, "Standardization of Procedures for The High Temperature Crack Grow Testing for FBR Materials in Japan," Nuclear Engineering and Design, 133, 465-473.

Brust, F. W., 1987, "Approximate Methods for Fracture Analysis for Through-Wall Cracked Pipes," NUREG/CR-4853.

DeLorenzi, H., G., 1982, "Energy Release Rate Calculations by the Finite Element Method," General Electric Company TIS Report 82 CRD205.

Klecker, R., et al., 1986, "NRC Leak-Before-Break (LBB.NRC) Analysis Method for Circumferential Through-Wall Cracked Pipes Under Axial Plus Bending Loads," NUREG/CR-4572.

Kumar, V., et al., 1981, "An Engineering Approach for Elastic-Plastic Fracture Analysis," EPRI NP-1931.

Kumar, V., et al., 1984, "Advances in Elastic-Plastic Analysis," EPRI Final Report NP-3607.

Miura, N., Shimakawa, T., Nakayama, Y., Takahashi, Y., 2000, "Simplified J-Evaluation Method for Flaw Evaluation at High Temperature Part I: Systematization of Evaluation Method," to be presented at 2000 ASME Pressure Vessels and Piping Conference.

Murakami, Y., et al., "Stress Intensity Factors Handbook," Committee on Fracture Mechanics, The Society of Materials Science, Japan.

Raju, I., S., and Newman, J., C., Jr, 1979, "Stress-Intensity Factors for a Wide Range of Semi-Elliptical Surface Cracks in Finite-Thickness Plates," Eng. Frac. Mech., 11, 817-829

Shimakawa, T., Iwasaki, R., Nakamura, K., Takahashi, H., Uno, T., Asada, Y., 1992, "Estimation of Surface Crack Growth under Creep-Fatigue Loading using Simplified Analysis," Journal of The Society of Materials Science Japan, Vol.41, No.468, pp.1433-1439.

Tada, H., Paris, A., and Irwin, G., 1985, "The Stress Analysis of Cracks Handbook, " DEL Research Corporation, St. Louis.

Willoughby, A., A., et al., 1989, "Plastic Collapse in Part-wall Flaws in Plates," ASTM STP 1020, ASTM, Philadelphia, 390-409.

PVP-Vol. 410-1, Assessment Methodologies for Preventing Failure:
Deterministic and Probabilistic Aspects and Weld Residual Stress
Volume 1
ASME 2000

BIMET : Structural Integrity of Bi-Metallic Components
Program status

Claude Faidy

Electricité de France -SEPTEN
12-14 Avenue Dutrievoz
69628 Villeurbanne Cedex - France
Tel: +33 (0)4 7282 7279, Fax: +33 (0)4 7282 7699
E-mail: claude.faidy@edf.fr

Abstract

The objective of this project is to contribute to the development and verification of analysis methods that describe the behaviour of an external circumferential defect at the interface of a bi-metallic weld (BMW). The complexity of the problem results from the prevailing mixed-mode loading conditions, the variation in material constitutive equations across the weld zone, and the presence of large residual stress field. Under these circumstances, classic fracture mechanics concepts are difficult to apply.

The project is formed around the performance of two benchmark 4-point bend pipe tests, conducted on a nominal 6 inch piping assembly, containing a ferritic to stainless steel (A508-308/309SS-304SS) BMW. The weld has been notched at the ferritic steel/buttering layer interface, to simulate plant cracking behaviour experience of such components. A range of analysis methodologies, including conventional flaw assessment methods, J approaches and Local Approach methods have been applied to predict the critical load for initiation of the defect, the extent of crack growth and the path followed by the crack through the weld up to the maximum load. Consideration will also be given to factors such as welding residual stresses, mixed-mode loading, and constraint effects.

The paper is focused on preliminary results in accordance with the program planning; the analytical and evaluation tasks are in progress.

Introduction

The objective of the project is to contribute to the development and verification of analysis methods that describe the behaviour of an external circumferential defect at the surface of a bi-metallic weld (BMW). The complexity of the problem results from the prevailing mixed-mode loading conditions, the variation in material constitutive equations across the weld zone, and the presence of large residual stress field. Under these circumstances, classic fracture mechanics concepts are difficult to apply.

The project is formed around the performance of two benchmark 4-point bend pipe tests, conducted on a nominal 6" piping assembly, containing a ferritic to stainless steel (A508-308/309SS-304SS) BMW. The weld has been notched at the ferritic steel/buttering layer interface, to simulate plant experiences of the cracking behaviour of such components. A range of analysis methodologies, including conventional flaw assessment methods, J approaches and Local Approach methods have been applied to predict the critical load for initiation of the defect, the extent of crack growth and the path followed by the crack through the weld up to the maximum load. Consideration will also be given to factors such as welding residual stresses, mixed-mode loading, and constraint effects.

The experimental program is now completed; the analytical program remains under evaluation, which explains the limited detailed results of the analytical work in this paper.

The test result has been of direct technological relevance to several of the project partners and the pressure equipment industry in general.

State of the Art

Bi-metallic welds are a necessity within Pressurised Water Reactor (PWR) and Boiling Water Reactor (BWR) designs where heavy section low alloy steel components are connected to stainless steel primary piping systems. For PWR's, the BMWs of particular interest are those attaching the piping system to the various nozzles of the Reactor Pressure Vessel (RPV), Steam Generators (SG) and Pressuriser.

Recent international surveys on BMW behaviour [1,2] have shown that there are several types of outer surface cracking problems encountered in both PWR's and BWR's which can be essentially grouped into the following two categories :

- those related to fabrication and
- those caused by surface corrosion.

At the same time, the occurrence in the literature of cases dealing with analytical and/or experimental aspects of such cracks in BMWs is very rare. The problem of integrity of cracked BMWs remains an important safety issue.

Weld Procurement

To guarantee the quality of the BMW's, a considerable experience and high degree of quality control has been required within the weld manufacturing process.

Five BMW specimens (Figure 1) have been manufactured of which two have been tested in 4-point bending, the remaining three will be utilised for the experimental determination of welding residual stresses and for the generation of material property data for input to the analysis procedures.

Weld production

The pipe dimension has been of 2.5 m length by 22 mm wall thickness by 168 mm outer diameter. The facility consists of supporting rollers that are compatible with a pipe of outer diameter of 168 mm. The inner span is fixed at 700 mm although the outer span is variable around 2 m. The maximum load, which can be applied by the facility, is 5 MN.

The welds have been produced according to a nuclear specification. During production, temperature and heat input history has been recorded for latter use in the modelling activities associated with the prediction of the residual stress fields.

Crack propagation has been measured during the experiment by the potential drop monitoring method. This has been calibrated against the post-test crack propagation measurement and also during the pre-test defect insertion operation. A single unique potential drop detection device has been used during the defect insertion and during the test.

Defect design and implantation

An initial design analysis has been undertaken to establish the size of defect required to guarantee propagation during the test and within the loading capacity of the proposed testing facility. These defects have been partial circumferential defect of depth of half the wall thickness for the first one and third of the thickness for the second one (Figure 3). The defect has been produced by means of electro-discharge machining. It is considered that the precise location of the defect within the weld has been a critical issue and great care has been needed to be applied during the procedure: in the buttering, as close as possible of the buttering to ferritic steel interface.

Defect measurement

The defect geometries were characterised accurately; X-Ray and UT examination has been performed to confirm the initial defect size.

Material characterisation

A key problem that has to be solved for the purposes of the present project is that there exists not only two mechanically different and distinct materials, but also, due to the heat affected zones, a gradient in material properties near to the fusion line [3,4]. For this reason, the analysis work requires extensive materials characterisation across the fusion line, in particular by miniature tensile tests. It is therefore an important result of this task to offer further user guidance with respect to the locations, across the BMW, for which material data have been obtained. The solution of this problem will provide a major step forward for the assessment of heterogeneous structures.

The primary objectives have been to characterise the strength; ductility and fracture behaviour of the weld heat affected zones and the dissimilar parent metals. Careful experimentation provided a basic understanding of the complex welded joint behaviour, essential input for the structural integrity assessments.

Chemical and metallographic analyses

The chemical composition of the weldment constituents has been determined.

The metallographic structure of the joint is complex and may vary from weld to weld and even within the same weld. These structures need to be characterised for the damage modelling and also for the mechanical test specimen preparation.

Conventional, miniature and notched tensile testing

Material properties, such as yield strength, elongation and stress-strain curves obtained from tensile tests, are required as inputs for analysis. The gradients of these properties near the fusion line have been measured in the three principal pipe orientations by means of miniature tensile specimens. Stresses and strains have been measured beyond necking by round tensile specimens extracted from the base metals and across the weld metal in a single direction.

Fracture mechanics testing

J_{IC}-values and fracture resistance (R) curves, characterising the crack initiation toughness and ductile crack growth, respectively, are required as inputs to the structural integrity analyses. The size of the welded joint and the need to precisely locate the crack tip, set special requirements for the specimen size and necessitates the use of mainly non-standard fracture mechanics specimens.

The materials characterisation task has required a considerable degree of innovation and the application of the most recent pre-normative testing technologies. Demonstration of a unified testing and analysis approach will offer guidance to the future analyses of similar components.

Residual Stress Characterisation

Bi-metallic welds between ferritic and austenitic steels exhibit a strong residual stress field, both as-welded and after PWHT (post welded heat treatment). Knowledge of the residual stress is an essential input to the structural integrity analysis of the joint, particularly where there is a locally embrittled region near the interface.

The objective has been to determine the residual stress field at the dissimilar weld in the test specimen. The neutron diffraction technique has been applied. It is intended to establish the capabilities and shortcomings of this technique as applied to the determination of residual stress fields in a medium-size BMW component.

Modelling

The modelling task involves the evaluation of the residual stress field using a simplified two-dimensional axisymetric model that determines the residual stress state of fabrication by simulating the cooling process from the final post-weld heat treatment (PWHT).

Neutron diffraction measurements

Neutron diffraction method has been used to measure the internal residual stresses in the specimen, with particular emphasis on the through-wall distribution adjacent to the ferritic-austenitic interface.

The comparison between measurements and evaluation confirms the adequacy of simple model simulation of residual stresses for our fracture purpose (Figure 5).

Benchmark Testing

Testing has consisted of submitting two cracked tubular BMW specimens to a 4-point bend loading with the objective of obtaining crack initiation and propagation events. The first test has been performed on an EDF facility (Figure 2) on October 1998, consisting of a servo-hydraulic actuator, combined with a load cell, and a 4-point bend bench. The test pieces have been EDM notched, without fatigue sharpened. The optimum crack geometry has been evaluated from pre-test design analysis. The tests has been performed at room temperature, to reduce the extent of the materials testing programme, to a single temperature and to realise a higher level residual stress field and mixed mode loading, at the weld interfaces.

The second test has been performed September 1999, similarly at room temperature with a smaller crack from 1/2 (#1) to 1/3 (#2) of the pipe thickness in order to assure larger ductile crack growth before maximum load (Figure 4).

Instrumentation

The specimen instrumentation and the data recording facility has been devised with the two main goals of:

- detecting and categorising the crack; and
- enabling the best comparison with analytical calculations.

The instrumentation has included the following:

- two clip gauges to measure the crack mouth opening and angle at mid defect position;

- several potential drop devices, to detect crack initiation and propagation at different points along the crack front;

- four inclinometers to measure the rotations of the specimens at specified points; and

- ten strain gauges located near the weldment on both sides of the BMW.

All these measurements have been recorded throughout the tests.

Testing programme

The programme will entail two tests on independent specimens:

- the first test was conducted during October 1998, on a 13.8-mm crack depth,

- the second test was conducted during September 1999, on a 9.28-mm crack depth.

The tests have been conducted under displacement control, to allow stable crack propagation beyond the maximum moment. In both cases the crack initiation occured before the maximum load, between 142 and 154 KN.m for #1 and between 140 and 150 KN.m for #2 (Figure 4).

Post-test examination

After testing, the specimens have been destructively examined to assess the size of the initial crack and the extent of propagation. A maximum crack growth of 2.2 mm at the deepest point and 1.2 to 1.4 mm at the surface points has been measured in the buttering of the first test. For the second test the maximum crack growth reached at the deepest point was 8.6 mm. Some fractographic analysis will also be undertaken. A full dimensional check has been made to assess the final shape of each of the specimens (ovality in some sections, curvature).

Analysis

The objectives has been to assess the capabilities of existing analysis techniques for the prediction of the behaviour of a BMW pipe component containing a crack, located parallel to, and at or close to, the fusion line. Three basic methodologies have been examined, conventional flaw assessment methods, more advanced J-methods and Local Approach based methods. The results of these analyses will be benchmarked against the results of the two pipe bend tests. Such an exercise is envisaged to increase the understanding of the role of interaction of two dissimilar (strength mismatched) materials in structural performance, and to provide input for the further development of models for assessing the structural integrity of bi-metallic welds.

Conventional and Codified Assessment methods

Engineering assessment methods are designed to allow evaluation of cracked structures without necessarily having to conduct detailed finite element calculations. A particularly versatile method is R6 [5] that is based upon the use of a failure assessment diagram (FAD). AEAT applied the R6 methodology with all these specificities. CEA and EDF will apply specific developments of this method done in France and included in RSE-M [6]. The first available results confirm a reasonable agreement between the experimental results and the prediction (Figure 6).

Engineering Treatment Model (ETM) developed in Germany

The ETM method developed by GKSS Research Centre serves for the estimation of the CTOD or the J-integral as driving-force parameters.

Due to the normalisation, the equation is a size and geometry independent formulation of the driving force. The ETM handbook is currently in print and will soon be available to the public.

The ETM has recently been extended to strength mismatched weld cases: ETM-MM.

The mechanical behaviour of the tested cracked dissimilar metal welds has been estimated by means of the ETM-MM method. The results both confirm the ETM-MM procedure and give rise to modifications to accommodate the special features of the dissimilar joints.

Three dimensional elasto-plastic finite element analysis

In such analyses, the crack driving-force is calculated accurately either in terms of the stress intensity factor based on elastic analysis, the J-integral or the crack-tip-opening displacement (CTOD) based on plastic analysis. The driving force is then compared with the measured fracture resistance K_{mat}, J_{mat}, or δ_{mat}.

From this analysis, the predictions have been made for the initiation of ductile crack growth; and the experimental results are in good agreement with the prediction (Figure 7). It has been checked whether predictions concerning the crack path during crack propagation are possible. The test expertise confirms the crack direction toward the lower yield strength material (buttering) and not toward the lower toughness material (ferritic steel).

Local Approach

Work conducted over the past two decades has demonstrated that assessment methods which reference micro-mechanical failure modes explicitly (often referred to as the "Local Approach") show considerable potential for the prediction of structural behaviour; where the physical mechanism of crack initiation and growth are modelled using constitutive equations derived from basic material properties.

Different models have been checked with experimental results: Damage Models, Cell Models, Beremin Model and Cohesive Zone Model.

There is a deep gradient in material properties near to the fusion line. For this reason, the analysis work requires extensive materials characterisation across the fusion line, in particular by miniature tensile tests. It will therefore be an important result to provide further guidance regarding the locations from where material data should be taken for a similar weld join.

The comparison of the results will provide a major step forward for the assessment of heterogeneous structures.

Conclusions

An "Evaluation" task is under progress to analyse together the results of the different tasks: residual stress determination, materials characterisation, benchmark testing and the findings of the analysis. Its primary role will be a comparison of the results of the analyses and of the benchmark experiment.

Conclusions from these comparisons and evaluations will lead to recommendations on the appropriateness of particular methodologies for the structural integrity assessment of cracked BMW components.

The BMW are needed to connect ferritic components with austenitic piping and are used in safety class 1 systems of all PWR and BWR plants. Their integrity whithout and with hypothetical cracks has to be justified in all condition for the life of the plant; they are also considered in the PWR severe accident studies where it's necessary to define the weakest point of the primary system during these conditions. Due to the complexity of the BMW for fabrication, inspection and analysis, specific programs as BIMET, has to be developed to confirm the quality of the present BMW design rules. This BIMET program is one important step considering tests and analysis of small diameter BMW at room temperature and can support the validation of different engineering methods, as R6 in U.K or RSE-M in France. A lot of innovative actions developed during this project can be used by different other industries, including non-nuclear industries.

Acknowledgement

List of participants to the project :
G. CHAS (EDF)-S. BHANDARI (FRAMATOME)-MP VALETA (CEA)-R. HURST – A. YOUTSOS (JRC)-
P. NEVASMAA (VTT)-W. BROCKS (GKSS)-
D. LIDBURY (AEAT)-C. WIESNER (TWI)

The author would like to thank European Community DGXII for their support to this work.

References

[1] H. Busboom, P.J. Ring, " Dissimilar-Weld Failure Analysis and Development - Comparative behavior of Similar and Dissimilar Welds", EPRI Report CS64666, July 1986

[2] P. Scott, R. Francini, S. Rahman, A. Rosenfield, G. Wilkoski, " Fracture Evaluations of Fusion Line Cracks in Nuclear Pipe Bimetallic Welds"; NUREG/CR-6297, April 1995

[3] " Mechanical Behaviour of Dissimilar Metal Welds ", Report EUR 13083 EN ,1990

[4] D Buckthorpe, C Escaravage, P Neri, P Pierantozzi, D Schmidt, " Study Contract on Bi-Metallic Weldments "; Report C9731/TR002, Issue V02, April 1997

[5] R/H/R6-Revision 3, "Assessment of the Integrity of Structures Containing Defects", Nuclear Electric Limited, July 1995.

[6] RSE-M ,"Surveillance and in-service inspection rules for mechanical components of PWR nuclear islands " edition 1997, AFCEN

[7] C. Faidy, « BIMET : Structural Integrity of Bi-Metallic Components - Program overview », SMIRT 15, paper G02-6, Seoul, KOREA August 1999.

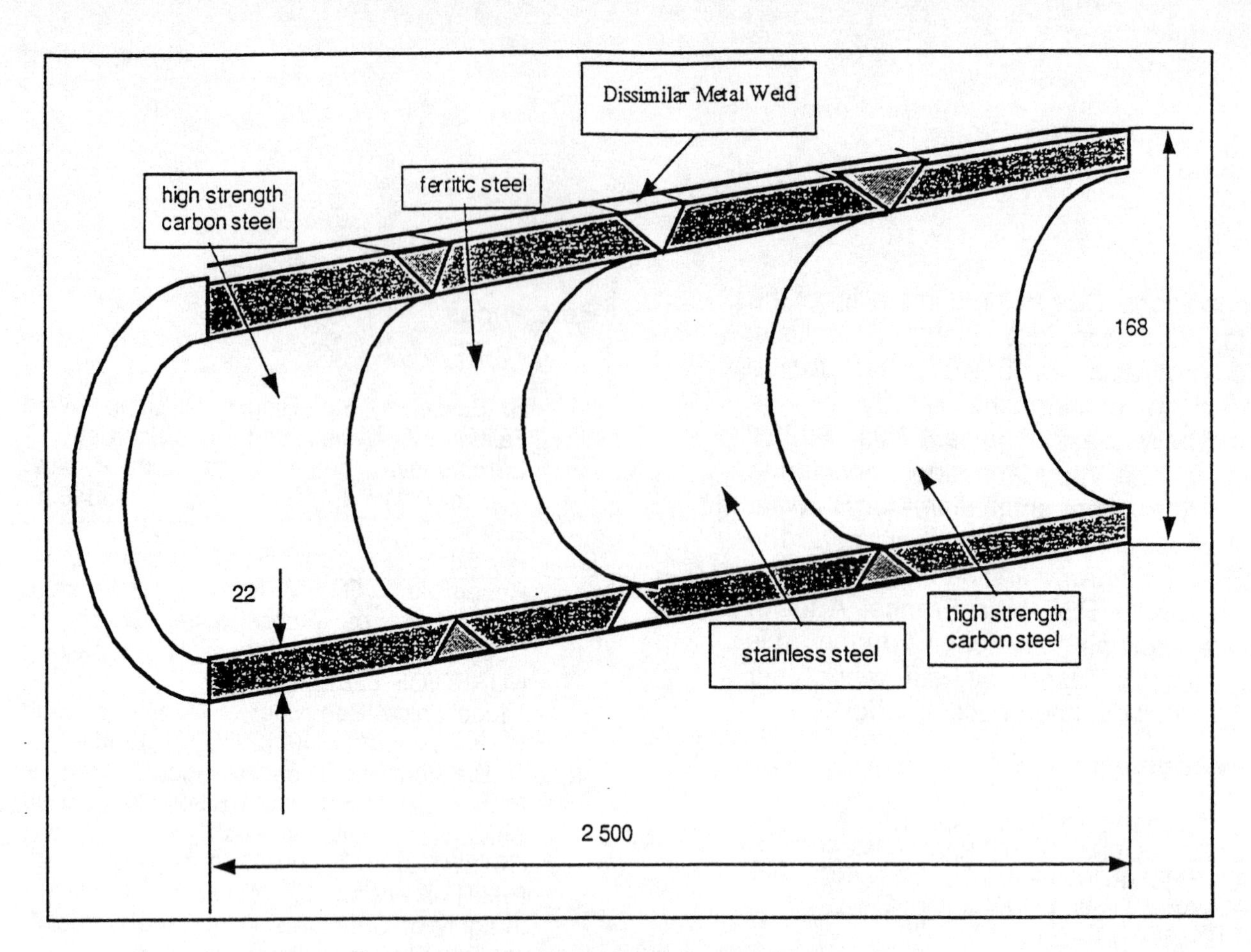

Figure 1 - Weld Assembly Design

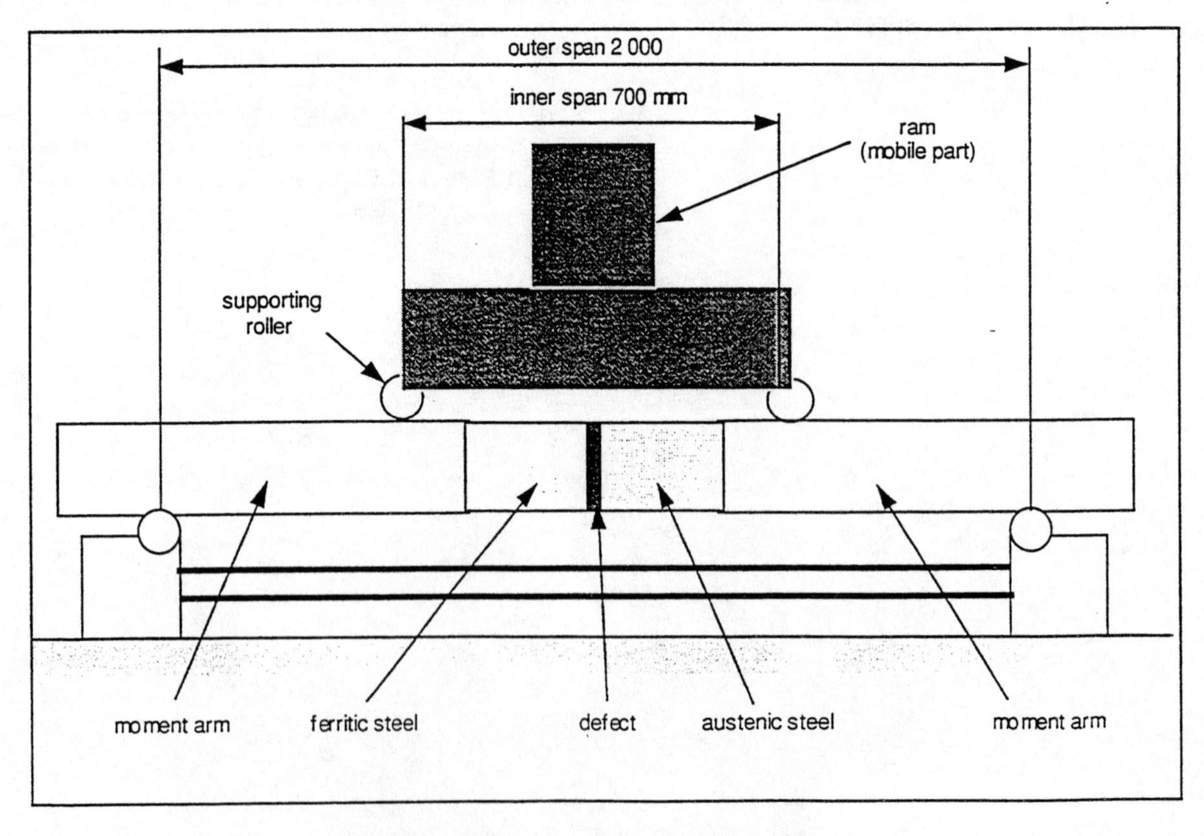

Figure 2 - Pipe Testing Facility

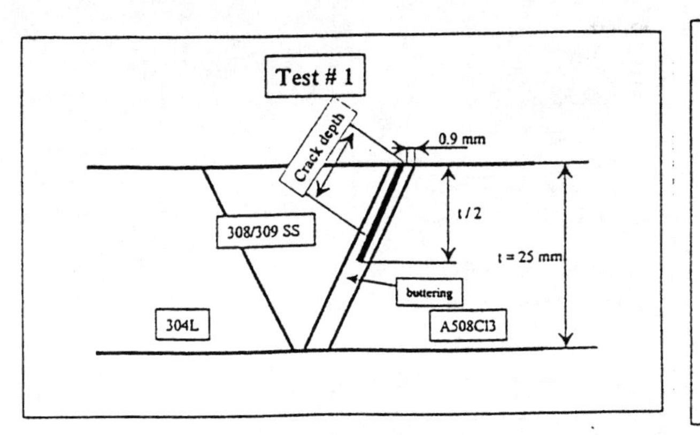

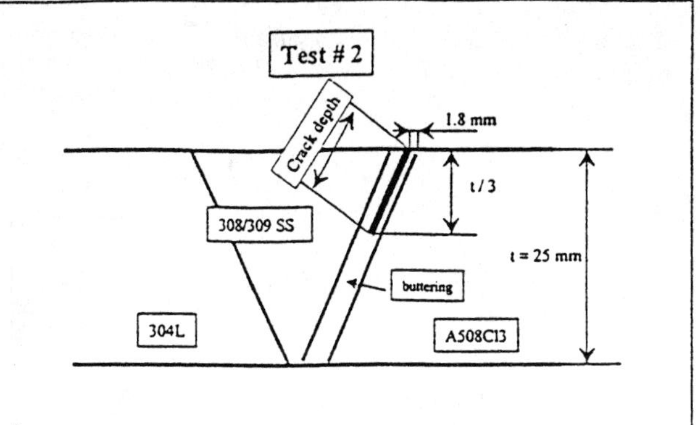

Figure 3 : Crack location and size

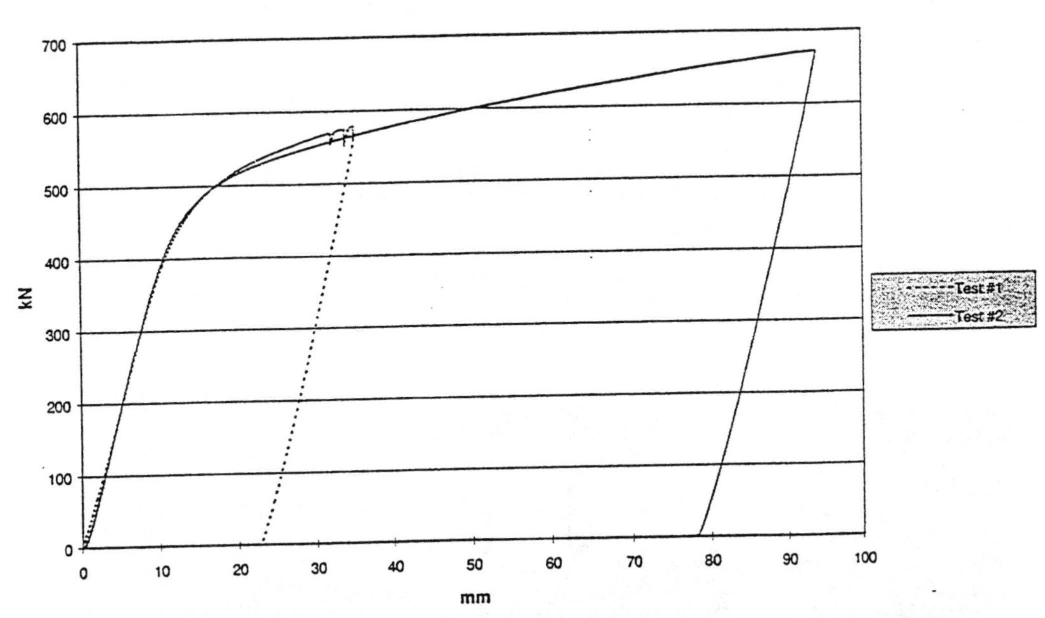

Figure 4 : Load versus ram displacement for test #1 and #2

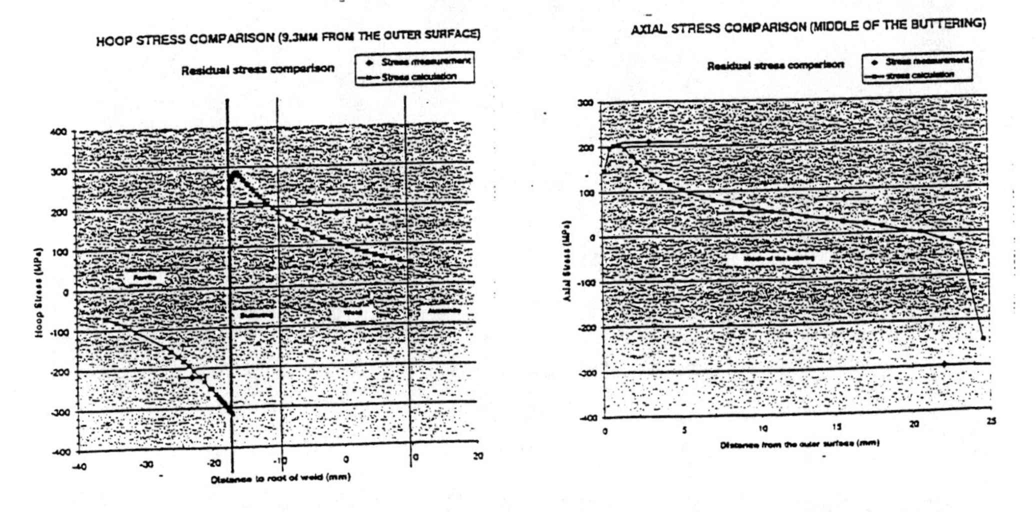

Figure 5 : Comparison of measured and computed residual stresses

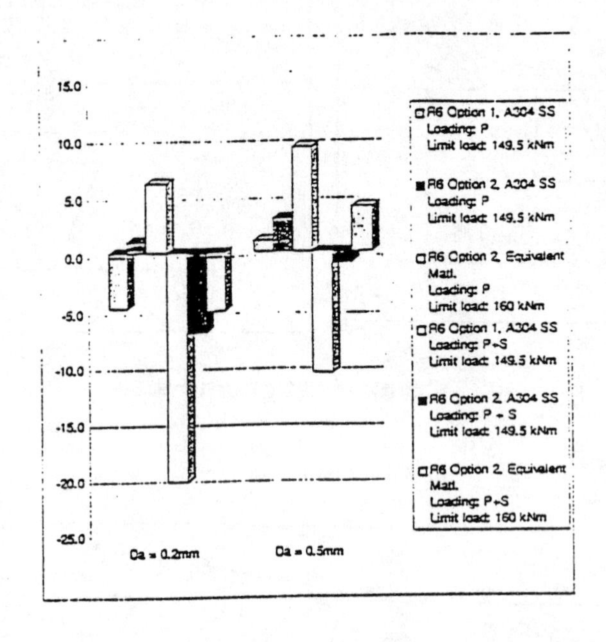

. Summary of results of R6 analyses to calculate initiation moment: percentage differences between observed (M_{INIT} = 154 kNm) and calculated values

Figure 6 : Evaluation of test # 1 through R6 rules

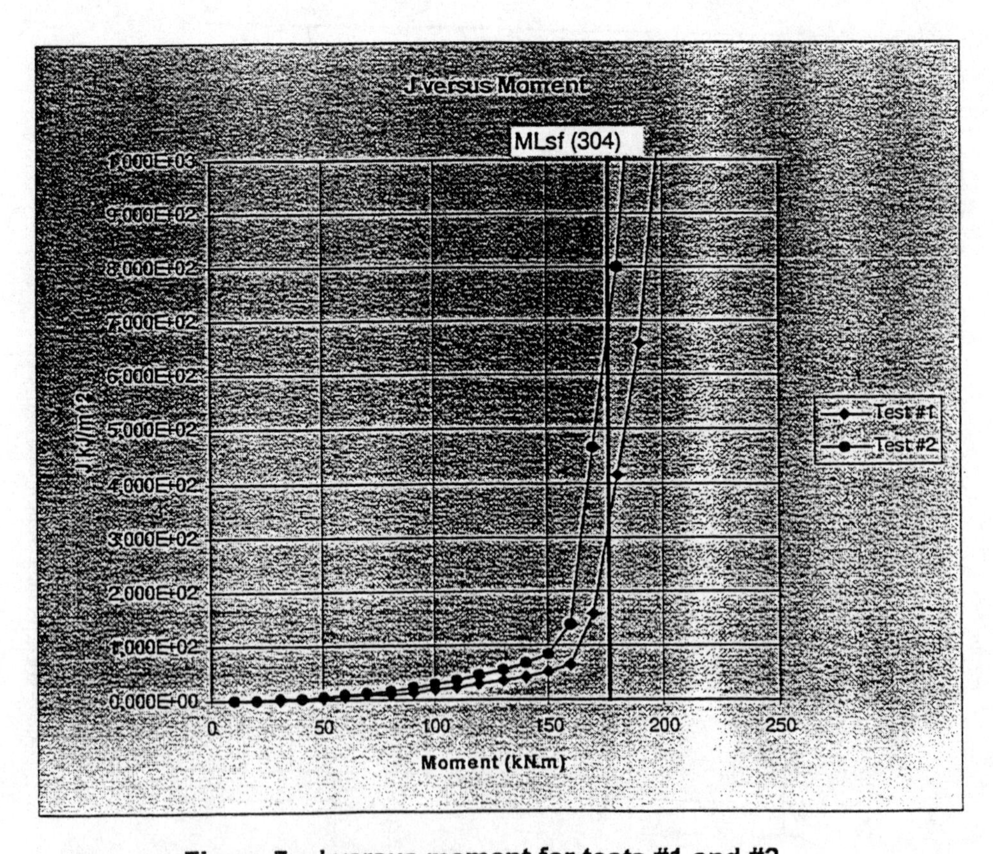

Figure 7 : J versus moment for tests #1 and #2

PVP-Vol. 410-1, Assessment Methodologies for Preventing Failure:
Deterministic and Probabilistic Aspects and Weld Residual Stress
Volume 1
ASME 2000

DUCTILE FRACTURE ASSESSMENT OF CRACKED ELBOWS USING ASME Z-FACTOR APPROACH

Suneel K. Gupta, V. Bhasin, K. K. Vaze and H. S. Kushwaha

Bhabha Atomic Research Centre
Reactor Safety Division
Trombay, Mumbai-400 085
India
Fax No. : (091)-(22)-5505151, 5519613
e-mail : kushwaha@magnum.barc.ernet.in

ABSTRACT

This paper is concerned with evaluation of Z-factor for 90° long-radius elbows subjected to bending loads. The Z-factor approach involves highly simplified procedure and provides quick ductile fracture assessment of power plant components. This approach yields conservative results as compared to usual J-Tearing approach. The use of Z-factor approach is suggested in ASME B&PV Code Sec. XI, for straight pipes. Based on similar philosophy, Z-factor for elbows were derived. Finally, a multiplying factor is suggested which, if applied to Z-factors of straight pipes, will give the Z-factors for elbows. Both circumferential and axial crack configurations have been considered. A circumferential crack was postulated at the extrados of elbow and axial crack at the crown of elbow.

INTRODUCTION

The Leak-Before-Break concept is a tool to provide early warning before major break in primary piping could develop. The LBB concept is required to restore some features of the original safety concept from the current point of view on maintaining primary circuit integrity. It is only feasible approach on providing reduction of the probability of primary breaks, which the nuclear power plant (NPP) design is not able to cope with. The NPP piping components are invariably made of ductile materials and their failure is governed either by ductile fracture or by plastic collapse. Hence, the resisting moment is taken as minimum of ductile fracture / unstable tearing moment (M_u) or plastic collapse moment (M_L), that is,

$$\text{Resisting moment} = \text{minimum} < M_u, M_L > \quad (1.1)$$

For ductile fracture assessment, different popular approaches have been used, however, most popular are J-Tearing approach, Z-factor approach etc. The Z-factor approach is simple to use and provides quick but conservative assessment as compared to J-Tearing approach. ASME B&PV Code Sec. XI Appendix H, [1], has suggested to evaluate the unstable ductile fracture load for straight pipes using Z-factor approach. Its usage involves determination of limit moment. The unstable tearing moment (or unstable ductile fracture moment) can be simply determined by dividing the limit moment (based on flow stress) by corresponding Z-factor.

In ASME B&PV Code Sec. XI, although Z-factors are intended for part-through-wall crack, however, they are evaluated assuming through-wall crack. Hence, the basic assumption is that for a given configuration the ratio of ductile tearing moment to limit moment is constant irrespective of whether the crack is a part-through-wall thickness or through-wall thickness. ASME Section XI has provided simplified equations to evaluate Z-factors for straight pipes.

The basic philosophy behind evaluation of Z-factors for straight pipes, with circumferential crack, has been discussed in detail, in Report No. EPRI NP-6045 for carbon/ferritic steels[2]. Based on similar philosophy, the Z-factor for elbows has been worked out. In EPRI NP-6045 report, GE/EPRI estimation scheme has been used to calculate the crack driving force or applied J-Integral (J_{app} or J) and tearing modulus (T_{app} or T) values. However, for the axially and circumferentially cracked elbows, this scheme cannot be used readily. Hence, the R-6 method was used to derive the Z-factors for elbows. Before applying the R-6 method to elbows, it was used for straight pipes with circumferential cracks. The aim was to assess the difference between Z-factors as determined using R-6 method and Z-factors as given in ASME Sec. XI, which are based on the GE/EPRI estimation scheme [2].

Finally, multiplying factors have been arrived, which when multiplied to pipe Z-factors will give the Z-factors for corresponding elbows.

PROCEDURE FOR EVALUATION OF Z-FACTOR

The basis of the ASME Z-factor for Straight pipes, with circumferential through wall crack, subjected to bending moment, is briefly discussed below, [2]:

- It is assumed that the pipe is having a circumferential through-wall crack, as shown in Fig. 1.
- The J-Tearing calculations are done, on a given pipe size of any material. From these calculations, the unstable tearing load/moment (UTL) 'M_u' is determined.

- These calculations are repeated for different crack sizes on same pipe and material.
- The ratio of 'M_U' to the limit moment 'M_L' (based on flow stress) is plotted versus through-wall circumferential crack size (θ/π).
- The reciprocal of minimum M_U/M_L is taken as equal to Z-factor. Only the maximum M_L/M_U versus (θ/π) ratio used a Z-value for that pipe size and material.
- The whole set of calculations can be repeated for different pipe sizes and different materials.
- In ASME Section XI pipe flaw analyses, the Z-factor versus nominal pipe size (NPS) relationship has been fitted in the form of simplified equations.
- ASME Section XI pipe flaw analyses have given the equations for two different broad range material sets.
- In the first set, the dependency of material properties has been simplified ignored, to a large extent. These are referred to as "material independent equations". In these only two broad categories have been considered, namely Category-I (Base metal) and Category-II (Weld metal).
- In second set, heat specific Z-factor equations have been provided. In these equations yield stress (σ_{ys}), flow stress (σ_f), design stress intensity (S_m) and initiation fracture toughness (J_I), parameters are also incorporated.

In deriving the material independent Z-factor equations for carbon steel, the ASME procedure has assumed certain properties, which are:

- Stress-strain curve can be represented as Ramberg-Osgood (R-O) curve fit.

$$\frac{\varepsilon}{\varepsilon_o} = \frac{\sigma}{\sigma_o} + \alpha\left(\frac{\sigma}{\sigma_o}\right)^n \tag{2.1}$$

The n and α are R-O curve fit parameter. For both the categories of material:
$n = 4.2$
$\alpha = 2.51$
σ_o = Reference Stress = taken as equal to σ_{ys}
ε_o = Reference Strain = σ_o / E
$\sigma_{ys} = 27.10$ ksi
$\sigma_f = 41.05$ ksi
$E = 26\times10^6$ ksi (Young's modulus of elasticity)

- Fracture properties of material such as initiation fracture toughness (J_I), J-Resistance (J-R) curve etc., are selected for reasonable lower-bound carbon steel base material (Category I) and weld material (Category II). The J-R curve (as J-Δa curve) is given in EPRI-NP- 6045 [2]. The resulting material J-Tearing curves (J-T curve) are also given.

The above procedure is schematically illustrated in Fig. 4. Based on this procedure, the resulting ASME Z-factor equations are reproduced below, [1] and [2]:

Material-Independent Z-factor Equations

Material Category I:

$$Z = 1.20\times[1 + 0.021(NPS - 4)A] \tag{2.2a}$$

Material Category II :

$$Z = 1.35\times[1 + 0.0184(NPS - 4)A] \tag{2.2b}$$

Where,

Z = the non-dimensional stress multiplier or Z-factor,
NPS = nominal pipe size, in inches,
$A = [0.125(R/t) - 0.25]^{0.25}$, for $5 \le (R/t) \le 10$, or
$A = [0.400(R/t) - 3.00]^{0.25}$, for $10 \le (R/t) \le 20$.

Heat Dependent Z-factor Equations

When $350 \le J_I < 600$ in-lb/in^2

$$Z = 2.566\ (\sigma_f/S_m)\ [1+0.0184\ (NPS - 4)A]/\sigma_y^{0.46} \tag{2.2c}$$

When $600 \le J_I < 1050$ in-lb/in^2

$$Z = 2.281\ (\sigma_f/S_m)\ [1+0.021\ (NPS - 4)A]/\sigma_y^{0.46} \tag{2.2d}$$

When $J_I \ge 1050$ in-lb/in^2

$$Z = 1.958\ (\sigma_f/S_m)\ [1+0.0152\ (NPS - 4)A]/\sigma_y^{0.46} \tag{2.2e}$$

Where,

σ_y = 0.2% yield strength (in ksi)

Application Of Z-Factor Procedure To Circumferentially Cracked Pipes

As mentioned before, the ASME Z-factor data and simplified equations were derived using GE/EPRI fracture estimation scheme and certain assumed properties of typical carbon/ferritic steels. In GE/EPRI scheme the J_{app} (or J) is estimated as follows:

$$J = J_e + J_p \tag{2.3}$$

$$J_e = \frac{K^2}{E} \tag{2.4a}$$

$$J_p = \frac{\alpha\sigma_o^2}{E} R\theta\left(1-\frac{\theta}{\pi}\right) h\left(\frac{\theta}{\pi}, n, \frac{R}{t}\right)\left[\frac{M}{M_o}\right]^{n+1} \quad (2.4b)$$

Where,

J_e , J_p	= Elastic and plastic part of J_{app}
α, n	= R-O curve fit parameter
R, t	= Mean radius and thickness of pipe
K	= Stress intensity factor (SIF)
M	= Applied moment
M_o	= Reference limit moment based on σ_o or σ_{ys} [given by Eq. (2.4c)]
$h(\theta/\pi$, n, R/t)	= Dimensionless factor, which depends upon crack size (θ/π), pipe geometry (R/t) and pipe material (n).

$$M_o = 4\sigma_o R^2 t\left[\cos\left(\frac{\theta}{2}\right) - \frac{1}{2}\sin(\theta)\right] \quad (2.4c)$$

Knowing, J_e and J_p, the T_{app} can be evaluated as:

$$T_{app} = \left(\frac{dJ_{app}}{da}\right)\frac{E}{\sigma_f^2} \quad (2.5)$$

Equation (2.5) will yield applied J-T curve. Finally, the Z-factors can be evaluated using the procedure as discussed earlier. One important aspect is briefly discussed here. The GE/EPRI scheme is based on finite element (FE) analysis results. The J_p is expressed as in Eq. (2.4b) and the difference in value with respect to FE results is adjusted using $h(\theta/\pi,n,R/t)$ factor. Hence, if the material truly follows the R-O law, then the GE/EPRI scheme will lead to an accurate estimation of J_p. Kumar, German, et. al., [3] and [4], have tabulated the values of $h(\theta/\pi$, n, R/t) factors. Zahoor has also presented them in Ductile Fracture Handbook, [5].

For power plant piping components such as elbows, nozzles, tee-junctions etc., h-factors are not available widely. Hence, the GE/EPRI scheme for determination of Z-factors is limited, since, these factors are required to compute J_p and its derivative. The accurate evaluation of derivative requires h-factors to be available for a significant number of crack sizes and other parameters.

Therefore, to calculate Z-factors for elbows, it was decided to use the R-6 method [6]. The R-6 method gives three options. For the present study, option –2 was applied [7], which takes into account the actual stress-strain curve. Ainsworth, [7], has shown the derivation of option-2 R-6 from standard GE/EPRI scheme. Ignoring the crack tip plasticity correction factor, the J can be expressed as follows:

$$J = \left[1+\phi\alpha\left(\frac{M}{M_o}\right)^{n-1}\right] J_e \quad (2.6a)$$

The ϕ factor in the above equation is proportional to ratio of the h-function value (for any 'n') to h-function value (for 'n=1'). In standard R-6 method this factor is taken as equal to 1.0. Therefore, the J equation becomes:

$$J = \left[1+\alpha\left(\frac{M}{M_o}\right)^{n-1}\right] J_e \quad (2.6b)$$

However, in general, ϕ is always less than 1.0. Therefore, the option-2 R-6 method will, in general, lead to an overestimation of the applied J and hence an underestimation of ultimate tearing loads (M_u).

Before application of the R-6 method to elbows, it was first applied to, straight pipes with circumferential through-wall cracks. This was done in order to assess the difference between Z-factors using R-6 method and Z-factors given in ASME Sec. XI equations. Later the method was extended to 90° long-radius elbows with either a circumferential crack at extrados or an axial crack at crown.

For straight pipes, applied J solution using the R-6 method is given by Eq. (2.6b). However, for more accurate solution, one can use Eq. (2.6a) along with the ϕ factor. The approximate value of ϕ factor may be obtained using following equation:

$$\phi = \left[1.4\left(\frac{\theta}{\pi}\right)^{0.14}\right]^{1-n} - 0.2(1-n)\left(\frac{\theta}{\pi}\right)^2 + 0.001\left(\frac{R}{t}\right)^2 \quad (2.7)$$

The details derivation of ϕ factor are discussed in reference, [8], for straight pipes. The usage of the ϕ factor reduces the conservatism in standard the R-6 equation [Eq. (2.6b)]

The application of the method to the pipe is discussed in following steps. The material properties used are the same as those assumed in the ASME Sec. XI, [1] and EPRI NP-6045[2].

Step 1: *Evaluation of Applied J-integral versus Load Curve Equation*

The J_e is determined using following equations [5]:

$$J_e = F_2\left(\frac{M}{M_o}\right)^2 \quad (2.8)$$

Where, F_2 and M_o, are defined as follows:

$$F_2 = f_b^2\left(\frac{\theta}{\pi}\right)\left(\frac{M_o^2}{ER^3t^2}\right) \quad (2.9)$$

$$f_b = 1 + A\left[4.5967\left(\frac{\theta}{\pi}\right)^{1.5} + 2.6422\left(\frac{\theta}{\pi}\right)^{4.24}\right] \quad (2.10)$$

$$A = \left(0.125\frac{R}{t} - 0.25\right)^{0.25} \quad \text{for } 5 \le \frac{R}{t} < 10$$

$$A = \left(0.400\frac{R}{t} - 3.00\right)^{0.25} \quad \text{for } 10 \le \frac{R}{t} \le 20 \quad (2.11)$$

$$M_o = 4\sigma_o R^2 t[\cos(0.5\theta) - 0.5\sin(\theta)] \quad (2.12)$$

Step 2: *Evaluation of Applied J-integral versus Tearing Modulus Curve Equation*

It is assumed that applied J versus applied T is a straight line. The equation for this line is given as, [8]:

$$T_{app} = F_1 J_{app} \quad (2.13)$$

Where,

$$T_{app} = \left(\frac{dJ_{app}}{da}\right)\frac{E}{\sigma_f^2} \quad (2.13a)$$

$$F_1 = F\left[S_{fe} + (n-1)\frac{\theta}{2}\left(1 + 4\frac{\theta}{\pi}\right) - 0.1n\right] \quad (2.14)$$

$$S_{fe} = \left(1 + 5\frac{\theta}{\pi}\right)\left(0.85 + 0.01\frac{R}{t}\right) \quad (2.14a)$$

and

$$F = \frac{E}{\sigma_f^2 R\theta} \quad (2.14b)$$

The details regarding derivation of Eq. (2.14), are discussed in reference, [8].

Step 3: *Evaluation of J-material versus T-material curve equation*

The material J-T curve which is given in EPRI NP-6045 [2], was fitted in the form of straight line as given below:

$$J_{mat} = mT_{mat} + C \quad (2.15)$$

The straight line fit is strictly true only for low T_{mat} or high J_{mat} values. However, in the zone of interest, the value of the T_{mat} would be low or the value of J_{mat} would be high.

For Category I Material: $m = -9.375$, $C = 2006.25$ lb-in/in^2
For Category II Material: $m = -14.0625$, $C = 1500.00$ lb-in/in^2

Step 4: *Evaluation of J-critical*

The intersection point of J_{app} versus T_{app} and J_{mat} versus T_{mat} curves yields J-critical. The $J_{critical}$ can also be evaluated by solving Eq. (2.13) and Eq. (2.15) and the value is given below:

$$J_{critical} = \frac{C}{1 - mF_1} \quad (2.16)$$

Step 5: *Evaluation of Unstable Tearing Moment (M_u)*

Unstable tearing moment (M_u) is equal to the applied moment when J_{app} become equal to J-critical, that is,

$$M_u = M, \text{ when } J_{app} = J_{critical}$$

Solving Eq. (2.16) and Eq. (2.6a), we get:

$$J_{critical} = \left[\left(\frac{M_u}{M_o}\right)^2 + \phi\alpha\left(\frac{M_u}{M_o}\right)^{n+1}\right]F_2 \quad (2.17)$$

Let thus define

$$F_3 = \frac{1}{\phi\alpha}, \quad F_4 = \frac{J_{critical}F_3}{F_2} \quad \text{and} \quad X = \frac{M_u}{M_o}$$

In factor F_4, the $J_{critical}$ is taken from Eq. (2.16). In the term X, the M_u gets normalized with respect to M_o. Then Eq. (2.17) can be re-written as:

$$X^{n+1} + F_3X^2 = F_4 \quad (2.17a)$$

The solution of this equation will yield X, which is the ratio (M_u/M_o).

Step 6: *Evaluation of Z-factor*

By using the above methodology M_u/M_o for different crack sizes are evaluated for the given pipe. Then M_u/M_o is plotted against crack size (θ/π) and the minimum value of M_u/M_o is designated as $(M_u/M_o)_{cr}$. The Z-factor is then evaluated as:

$$Z = \frac{\sigma_f/\sigma_o}{(M_u/M_o)_{cr}}$$

The calculations were performed for 15 different pipe sizes, belonging to ferritic material Category I (base) and Category II (weld). For both the cases, the Z-factors was evaluated for two conditions:

Condition A : Z-factor using standard R-6 method, i.e., $\phi = 1.0$
Condition B : Z-factor using modified R-6 method, i.e., $\phi \neq 1.0$ and the value given by Eq. (2.7).

The results are shown in Table 1 for Category I material and in Table 2 for Category II material. The table also shows the comparison with corresponding ASME Z-factors, which were evaluated using Eq. (2.2a) and Eq. (2.2b), respectively[1].

The results, given in Table 1 and Table 2, show that Z-factors evaluated using modified R-6 method [Eq. (2.7a)], are quite close to the ASME values. The average difference is ~ 5%. The Z-factors calculated using standard R-6 method (ϕ = 1.0) are also reasonably close to the ASME values. The average difference is ~10%. Most of the values are on higher side, since standard R-6 method generally over predicts the applied J values. Moreover, the proposed methodology is simple and quick since it requires very little computational effort.

Determination Of Z-Factor For 90° Long-Radius (Lr) Elbows With Cracks

In view of the reasonably close comparison, it was decided to use the standard R-6 method [Eq. (2.6b)] for evaluation of Z-factors of elbows. For elbows, two crack orientations are considered which are, either axial crack at the crown or circumferential crack at the extrados. It was assumed that:

(1) Internal Pressure is zero. The only load is bending moment.
(2) Material for elbow is same as that of pipe material. Hence, same material properties are used.

In order to determine Z-factors for elbows, the calculations were done for seamless elbows only because they are the ones, which are mostly used in nuclear power plant piping. However, the procedure developed is not limited to the seamless elbows only. Hence, Z-factors for only Category-I material were evaluated.

Elbows with Through Wall Axial Crack at Crown- In Plane Bending Moment

The crack configuration is shown in Fig. 2. The steps involved in evaluation of Z-factor are given below:

Step 1: *Applied J-integral versus Load Curve Equation*

The SIF for axial through wall crack at crown of 90° LR Elbow is given as, [5]:

$$K = \left(\frac{4M}{\pi D_o^2 t}\right)(\pi D_o)^{0.5} F_b \qquad (2.18)$$

Where, F_b is expressed as:

$$F_b = \left[0.05607 + 0.2489\left(\frac{a}{D_o}\right)\right]\left(\frac{D_o}{t}\right)^{0.59} \qquad (2.19)$$

The above equation if for range, $0.08 \le \left(\frac{a}{D_o}\right) \le 0.5$

The M_o, based on σ_{ys}, for elbow with axial cracks at crown is given as, [5]:

$$M_o = 0.935\sigma_{ys} D^2 t \lambda^{\frac{2}{3}}\left[1 - 0.0824\left(\frac{a}{d}\right) - 1.3755\left(\frac{a}{d}\right)^2 + 0.9327\left(\frac{a}{d}\right)^3\right]$$

(2.20)

Where,

$$\lambda = \text{Elbow parameter} = \frac{rt}{R^2}$$

r = bend radius = 3R, for a LR Elbow

Therefore, $\lambda = \frac{3t}{R}$

$$J_e = \frac{K^2}{E} = \frac{2M^2 F_b^2}{\pi E R_o^3 t^2} \qquad (2.21)$$

Let us define, F_2 as:

$$F_2 = \frac{2M_o^2 F_b^2}{\pi E R_o^3 t^2} \qquad (2.21a)$$

Hence, J_e can be expressed as:

$$J_e = F_2\left(\frac{M}{M_o}\right)^2 \qquad (2.22)$$

then the applied J can be evaluated using standard R-6 method.

$$J = \left[1 + \alpha\left(\frac{M}{M_o}\right)^{n-1}\right] J_e \quad (2.23)$$

***Step 2:** Applied J-integral versus Tearing Modulus Curve Equation*

Tearing modulus is defined as:

$$T = \frac{E}{\sigma_f^2}\frac{dJ}{da}$$

From Eq. (2.23), we get:

$$\frac{dJ}{da} = \frac{d}{da}\left[\left\{1 + \alpha(L_r)^{n-1}\right\}J_e\right]$$

or

$$\frac{dJ}{da} = \left[1 + \alpha(L_r)^{n-1}\right]\frac{dJ_e}{da} + (n-1)J_e\alpha(L_r)^{n-2}\frac{dL_r}{da}$$

In above expressions the 'L_r' is the ratio of applied moment to limit moment i.e., (M/M_o). After simplification, the expression for tearing modulus can be reduced to:

$$T = F_1 J \quad (2.24)$$

Where,

$$F_1 = F[S_{fe} + (n-1)g(a)] \quad (2.25)$$

and F is defined as:

$$F = \frac{E}{\sigma_f^2 a} \quad (2.26)$$

S_{fe} can be expressed as:

$$S_{fe} = \frac{a}{J_e}\frac{dJ_e}{da} = \frac{2a}{F_b}\frac{dF_b}{da}$$

or

$$S_{fe} = \frac{2a}{F_b}\frac{dF_b}{da} = \frac{0.4978a}{0.05607D_o + 0.2489a} \quad (2.27)$$

and, g(a) can be expressed as:

$$g(a) = \frac{0.0824\left(\frac{a}{D}\right) + 2\times 1.3755\left(\frac{a}{D}\right)^2 - 2.7981\left(\frac{a}{D}\right)^3}{1 - 0.0824\left(\frac{a}{D}\right) - 1.3755\left(\frac{a}{D}\right)^2 + 0.9327\left(\frac{a}{D}\right)^3} \quad (2.28)$$

Step 3, 4, 5 and 6 are similar to the corresponding steps followed for the case of straight pipes. The only difference is in F_1, F_2, F_3 and F_4 factors. Here, F_1 and F_2 are given by Eq. (2.25) and Eq. (2.21) respectively. Let us define

$$F_3 = \frac{1}{\alpha}, \text{ and } \quad F_4 = \frac{J_{critical}F_3}{F_2}$$

Finally, as in the case of pipes, we get following equation:

$$X^{n+1} + F_3X^2 = F_4$$

The solution to the above equation and usage of Step 6, as discussed earlier, gives the elbow Z-factors. The results are given in Table 3. It is observed, that in general the elbow Z-factor values are less than those of corresponding straight pipes.

Elbows With Through Wall Circumferential Crack At Extrados - In Plane Bending Moment

The crack configuration is shown in Fig. 3. The steps involved in evaluation of Z-factor are given below:

***Step 1:** Applied J-integral versus Load Curve Equation*

$$J = \left[1 + \alpha\left(\frac{M}{M_o}\right)^{n-1}\right] J_e \quad (2.29)$$

Where,

$$J_e = F_2\left(\frac{M}{M_o}\right)^2$$

$$F_2 = f_b^2\left(\frac{\theta}{\pi}\right)\left(\frac{M_o^2}{ER^3t^2}\right) \quad (2.30)$$

The expressions for f_b and A are same as those for straight pipe and are given by Eq. (2.10) and Eq. (2.11), respectively. The limit moment based on σ_{ys} is given below [5]:

$$M_o = 0.93\,\sigma_{ys} D^2 t\lambda^{\frac{2}{3}}\left[1 - 0.2137\left(\frac{a}{D}\right) - 0.0485\left(\frac{a}{D}\right)^2 - 1.0559\left(\frac{a}{D}\right)^3\right]$$

Since, a = Rθ, hence M_o can further be expressed as follows:

$$M_o = 0.935\,\sigma_{ys} D^2 t\lambda^{\frac{2}{3}}\left[1 - 0.10685\theta - 0.012125\theta^2 - 0.1319875\theta^3\right] \quad (2.31)$$

Step 2: Applied J-integral versus Tearing Modulus Curve Equation

Now T is defined as:

$$T = \frac{E}{\sigma_f^2 R}\frac{dJ}{d\theta}$$

From Eq. (2.29), we get

$$\frac{dJ}{d\theta} = \frac{d}{d\theta}\left[\left\{1 + \alpha(L_r)^{n-1}\right\}J_e\right]$$

or

$$\frac{dJ}{d\theta} = \left[1 + \alpha(L_r)^{n-1}\right]\frac{dJ_e}{d\theta} + (n-1)J_e\alpha(L_r)^{n-2}\frac{dL_r}{d\theta}$$

In above expressions the 'L_r' is the ratio of applied moment to limit moment i.e., (M/M_o). After simplification, the expression for tearing modulus can be reduced to:

$$T_{app} = F_1 J_{app} \quad (2.32)$$

Where,

$$F_1 = F\left[S_{fe} + (n-1)g(\theta)\right] \quad (2.33)$$

and F is defined as:

$$F = \frac{E}{\sigma_f^2 R\theta}$$

S_{fe} is same as that for straight pipe and is given by Eq. (2.14a). The $g(\theta)$ is expressed as follows :

$$g(\theta) = (n-1)\left[\frac{0.10685\theta + 0.024225\theta^2 + 0.3959625\theta^3}{1 - 0.10685\theta - 0.012125\theta^2 - 0.1319875\theta^3}\right] \quad (2.34)$$

Step 3, 4, 5 and 6 are similar to the corresponding steps followed for the case of straight pipes. The only difference is in F_1, F_2, F_3 and F_4 factors. Here F_1 and F_2 are given by Eq. (2.33) and Eq. (2.30), respectively. Let us define,

$$F_3 = \frac{1}{\alpha}, \text{ and } \quad F_4 = \frac{J_{critical}F_3}{F_2}$$

Finally, as in the case of pipes, we get following equation:

$$X^{n+1} + F_3X^2 = F_4$$

The solution of above equation and usage of Step 6, as discussed for straight pipes, gives the elbow Z-factors. The results are given in Table 3. It is observed, that in general the elbow Z-factor values are less than those of corresponding straight pipes.

DETERMINATION FOR MULTIPLYING FACTOR FOR 90° LONG-RADIUS (LR) ELBOWS

Based on above-mentioned derivations, the multiplying factors for 90° LR Elbows have been derived. These factors are such that if multiplied by Pipe Z-factors, they will yield Z-factor for the corresponding elbows.

It is observed that, for all the cases considered, the form of the equation of unstable tearing moment (M_u) is similar (for standard R-6 method, i.e., $\phi = 1$) and is given by Eq. (2.17). Using Eq. (2.16) and Eq. (2.17), we get:

$$\frac{C}{1 - mF_1} = \left[\left(\frac{M_u}{M_o}\right)^2 + \alpha\left(\frac{M_u}{M_o}\right)^{n+1}\right]F_2 \quad (3.1)$$

It should be noted that, at critical condition $M = M_u$. Taking the ratio of the left-hand side (LHS) and right hand side (RHS), of above equation, for pipe and corresponding elbow, we get:

$$\frac{\left\{\frac{C}{1-mF_1}\right\}_P}{\left\{\frac{C}{1-mF_1}\right\}_E} = \frac{\left\{\left[\left(\frac{M_u}{M_o}\right)^2 + \alpha\left(\frac{M_u}{M_o}\right)^{n+1}\right]F_2\right\}_P}{\left\{\left[\left(\frac{M_u}{M_o}\right)^2 + \alpha\left(\frac{M_u}{M_o}\right)^{n+1}\right]F_2\right\}_E} \quad (3.2)$$

Here the subscripts 'P' and 'E' refer to pipe and elbows receptively. From the numerical values it has been observed that the LHS of Eq. (3.2) is approximately equal to 1.0.

$$\frac{\left\{\left[\left(\frac{M_u}{M_o}\right)^2 + \alpha\left(\frac{M_u}{M_o}\right)^{n+1}\right]F_2\right\}_P}{\left\{\left[\left(\frac{M_u}{M_o}\right)^2 + \alpha\left(\frac{M_u}{M_o}\right)^{n+1}\right]F_2\right\}_E} \approx 1 \quad (3.2a)$$

Assuming $\left(\frac{M_u}{M_o}\right)^2 < \alpha\left(\frac{M_u}{M_o}\right)^{n+1}$ Eq. (3.2a) reduces to:

$$\frac{\left\{\left(\frac{M_u}{M_o}\right)^{n+1} F_2\right\}_P}{\left\{\left(\frac{M_u}{M_o}\right)^{n+1} F_2\right\}_E} \approx 1$$

or

$$\frac{\left\{\left(\frac{M_u}{M_o}\right)\right\}_P}{\left\{\left(\frac{M_u}{M_o}\right)\right\}_E} = \left[\frac{\left(F_2\right)_E}{\left(F_2\right)_P}\right]^{\frac{1}{n+1}} \quad (3.3)$$

The Z factor of pipe and elbows is defined as:

$$\left[\frac{\sigma_o M_u}{\sigma_f M_o}\right]_{minimum} = \frac{1}{Z}$$

Then performing the minima operation on Eq. (3.3), the multiplying factor for 90° long-radius elbows with circumferential crack orientation is given as:

$$MFc = [2\lambda]^{4/3\,(n+1)} \quad (3.4)$$

The multiplying factor for 90° long-radius elbows with axial crack at crown is given as:

$$MFa = [\lambda/8]^{4/3(n+1)}\,[R_o/t]^{1.18/\,(n+1)} \quad (3.5)$$

Where,

λ = Elbow parameter
R_o = Outer radius
t = Nominal pipe thickness.

Therefore, the Z-factor for elbows with circumferential crack at extrados is equal to 'MFc × Corresponding Pipe Z-factor'. Z-factor for elbows with axial crack at crown is equals to 'MFa × Corresponding Pipe Z-factor'

The approximations involved in deriving these multiplying factors are realistic and are not likely to cause large errors. In Fig. 5 the Z-factors for elbows as given in Table 3, are plotted against Z-factors as evaluated by using multiplying factors with pipe Z-factors [Table 1, with $\phi = 1$]. From these figures, it is observed that almost all the points lie on 45° line.

It may be noted that these factors have been derived using standard R-6 method where ϕ factor is taken as 1.0 and hence J_{app} values are overestimated. But multiplying factors have been derived from ratios, hence effect of considering the actual value of ϕ may not be significant.

Therefore, it can be concluded that, Z-factors for elbows, with above crack orientations, can be simply obtained using multiplying factors with the corresponding ASME pipe Z-factor equations.

DISCUSSION AND CONCLUSION

The aim of this study was to evaluate the Z-factors for 90° long-radius elbows with either circumferential crack at the extrados or axial crack at the crown. The usage of Z-factors leads to quick but conservative ductile fracture assessment of power plant piping components. The procedure was similar to that of straight pipes. The straight pipe procedure is discussed in EPRI NP-6045 report [2]. ASME B&PV Code, Section XI has provided a set of formulae for Z-factors of straight pipe. EPRI NP-6045 report [6] indicates that these Z-factors were evaluated using J-Tearing method, where GE/EPRI scheme was utilized for estimating applied J or T parameter.

In case of elbows, the GE/EPRI scheme cannot be readily applied due to unavailability of h-factor values. The h-factor is one of the parameters required to estimate the plastic part of applied J-integral and is obtained using finite element analysis. Hence, for first order assessment, the estimation can be based on option-2 R-6 method. Initially this method is applied to straight pipes, in order to ascertain the difference with respect to ASME Z-factor equations. Since, standard option-2 R-6 method, in general leads to overestimation of applied J hence, the Z-factors values were higher than those given by ASME Z-factor equations. On an average the values were higher by ~ 10%. Modified R-6 method, which includes the factor ϕ, was shown to yield closer results. Here the Z-factor were higher by ~5% (average) as compared to GE/EPRI method.

Hence, it was decided to use the R-6 method for elbows also. The analysis was done on 16 different Elbow sizes. The material properties are the same as those assumed in the ASME Z-factor equations of pipes.

The results indicate that in general Z-factor for elbows are less than those of corresponding straight pipes. From the derived formulae, used to calculate the Z-factors, the ratio with respect to corresponding pipe, were also derived for both crack orientations discussed earlier. The ratio is termed as multiplying factor (MF). Hence, if Z-factor of straight pipe is multiplied by MF then one can get Z-factor for corresponding elbows.

The derived MF, for the two crack configurations, are as follows:

- Multiplying factor for 90° long-radius elbows with circumferential crack at the extrados.

$$MFc = [2\lambda]^{4/3(n+1)}$$

- Multiplying factor for 90° long-radius elbows with axial crack at the crown

$$MFa = [\lambda/8]^{4/3(n+1)} [R_o/t]^{1.18/(n+1)}$$

The ASME Z-factor equations are based on n = 4.2. Hence, one can easily evaluate Z-factor for elbows for given elbow parameter λ. It is also observed that MFa weakly depends on λ and R_o/t since, they nullify each other. For n = 4.2 the MFa can be conservatively assumed to around 0.75 to 0.8, (n = 4.2 is the value assumed by ASME for pipe Z-factor equation).

REFERENCES

1. "ASME Boiler & Pressure Vessel Code, Section XI, Appendix H", American Society of Mechanical Engineers.
2. "Evaluation of Flaws in Ferritic Piping" Report No. EPRI NP-6045, Electric Power Research Institute, Palo Alto, CA, 1988.
3. Kumar, V., German, M.D. and Shih, C.F., "An Engineering Approach for Elastic-Plastic Fracture Analysis", Report No. EPRI NP-1931, Electric Power Research Institute, Palo Alto, CA, 1981.
4. Kumar, V., German, M.D., Wilkening, W.W., Andrews, W.R., de lorenzi, H.G., and Mowbray, D.F., "Advances in Elastic-Plastic Fracture Analysis", Report No. EPRI NP-3607, Electric Power Research Institute, Palo Alto, CA, 1984.
5. Zahoor, A., "Ductile Fracture Handbook, Volume I : Circumferential Through Wall Cracks", Report No. EPRI NP-6301-D, Electric Power Research Institute, Palo Alto, CA, 1989.
6. Milne, I.S., Ainsworth, R. A,Dowling, A. P. and Stewart, A. T., "Assessment of the Integrity of Structures Containing Defects", Report No. CEGB/R/H/R-6-Rev.3.
7. Ainsworth, R. A., "Assessment of Defect in Structures of Strain Hardening Material", Engineering Fracture Mechanics, vol. 19, No. 4, pp. 633-642,1984.
8. Gupta, Suneel K., Bhasin, V., Vaze, K.K., and Kushwaha H.S., "Development of An Engineering Approach for Ductile Fracture Assessment of Straight Pipes". Submitted to ASME PVP-2000 Conference.

Table 1: Z-factors for straight pipes of Category I material (ferritic base material) and comparison with corresponding ASME Z-factors

Case No.	Pipe Size	ASME Z-factor [Eq. (2.2a)]	Z-Factor using Modified R-6 method [ϕ given by Eq. (2.7)]		Z-Factor using Standard R-6 method [ϕ=1]	
			Z-factor	% Difference w.r.t. ASME [Z-factor]	Z-factor	% Difference w.r.t. ASME [Z- factor]
1	6" Sch 40	1.256	1.375	9.5	1.442	14.8
2	6" Sch 80	1.245	1.280	2.8	1.367	9.8
3	8" Sch 40	1.322	1.491	12.8	1.550	17.2
4	8" Sch 80	1.294	1.364	5.4	1.450	12.1
5	10" Sch 60	1.385	1.566	13.1	1.625	17.3
6	12" Sch 60	1.417	1.550	9.3	1.625	14.7
7	16" Sch 80	1.492	1.573	5.4	1.660	11.3
8	16" Sch 100	1.472	1.532	4.1	1.627	10.5
9	16" Sch 120	1.455	1.502	3.2	1.602	10.1
10	20" Sch 80	1.593	1.662	4.3	1.749	9.8
11	20" Sch 100	1.564	1.616	3.3	1.713	9.5
12	20" Sch 120	1.543	1.584	2.7	1.685	9.2
13	24" Sch 80	1.693	1.742	2.9	1.829	8.0
14	24" Sch 100	1.656	1.690	2.1	1.788	8.0
15	24" Sch 120	1.626	1.653	1.7	1.756	8.0

Table 2: Z-factors for straight pipes of Category II material (ferritic weld material) and comparison with corresponding ASME Z-factors

Case No.	Pipe Size	ASME Z-factor [Eq. (2.2a)]	Z-Factor using Modified R-6 method [ϕ given by Eq. (2.7)]		Z-Factor using Standard R-6 method [ϕ=1]	
			Z-factor	% Difference w.r.t. ASME [Z-factor]	Z-factor	% Difference w.r.t. ASME [Z-factor]
1	6" Sch 40	1.405	1.509	7.4	1.580	12.5
2	6" Sch 80	1.395	1.402	0.5	1.494	7.1
3	8" Sch 40	1.470	1.635	11.2	1.696	15.4
4	8" Sch 80	1.443	1.491	3.3	1.581	9.6
5	10" Sch 60	1.532	1.716	12.0	1.777	16.0
6	12" Sch 60	1.564	1.695	8.4	1.772	13.3
7	16" Sch 80	1.638	1.717	4.8	1.807	10.3
8	16" Sch 100	1.618	1.671	3.3	1.770	9.4
9	16" Sch 120	1.602	1.638	2.2	1.741	8.7
10	20" Sch 80	1.737	1.816	4.5	1.906	9.7
11	20" Sch 100	1.709	1.764	3.2	1.864	9.1
12	20" Sch 120	1.688	1.729	2.4	1.833	8.6
13	24" Sch 80	1.836	1.906	3.8	1.995	8.7
14	24" Sch 100	1.799	1.847	2.7	1.947	8.2
15	24" Sch 120	1.770	1.805	2.0	1.910	7.9

Table 3: Z-factors for Category I material (ferritic base material) elbows and comparison with corresponding straight pipe Z-factors [based on Standard R-6 Method]

Elbow No.	90° L.R. Elbow Size	Category I St. pipe Z-Factor using Standard R-6 method [ϕ=1]	Elbow Z-Factor using Standard R-6 method [ϕ=1]	
			Axial crack at crown	Circumferential crack at extrados
1	6" Sch 40	1.442	1.0316	1.2113
2	6" Sch 80	1.367	1.0252	1.3069
3	8" Sch 40	1.550	1.0835	1.2490
4	8" Sch 80	1.450	1.0791	1.3342
5	10" Sch 60	1.625	1.1300	1.2969
6	12" Sch 60	1.625	1.1677	1.3676
7	16" Sch 80	1.660	1.2196	1.4713
8	16" Sch 100	1.627	1.2145	1.5405
9	16" Sch 120	1.602	1.2079	1.6006
10	20" Sch 80	1.749	1.2772	1.5339
11	20" Sch 100	1.713	1.2715	1.6137
12	20" Sch 120	1.685	1.2650	1.6747
13	24" Sch 80	1.829	1.3274	1.5901
14	24" Sch 100	1.788	1.3213	1.6791
15	24" Sch 120	1.756	1.3139	1.7487

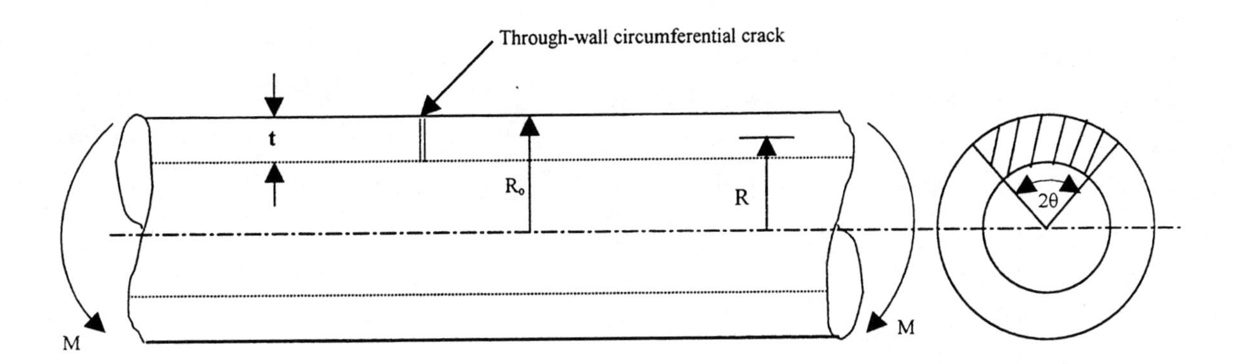

Figure 1: Straight pipe with through-wall circumferential crack

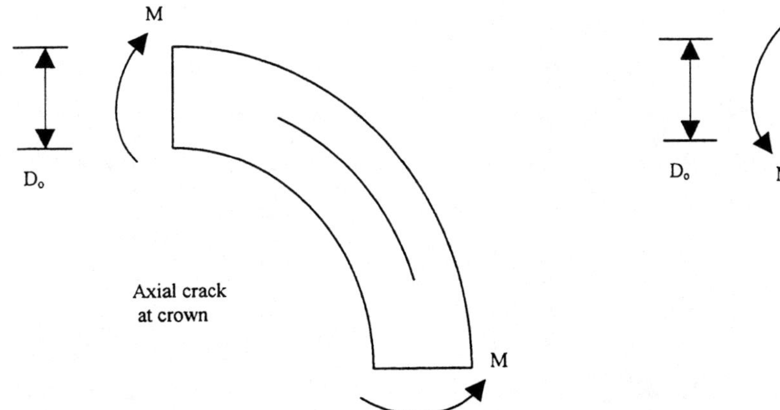

Figure 2: 90° Long-radius through-wall axially cracked elbows

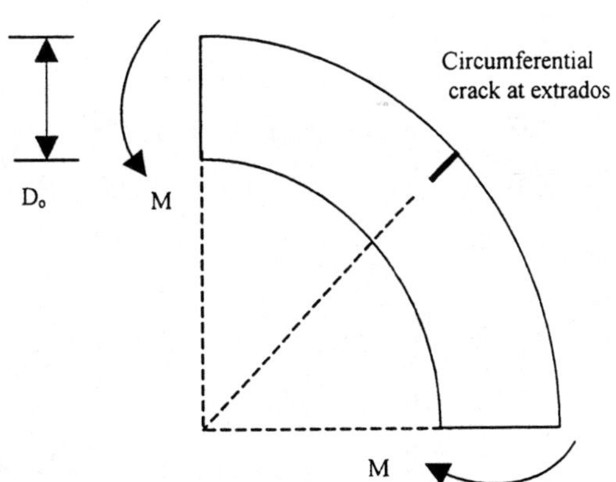

Figure 3: 90° Long-radius through-wall circumferentially cracked elbows

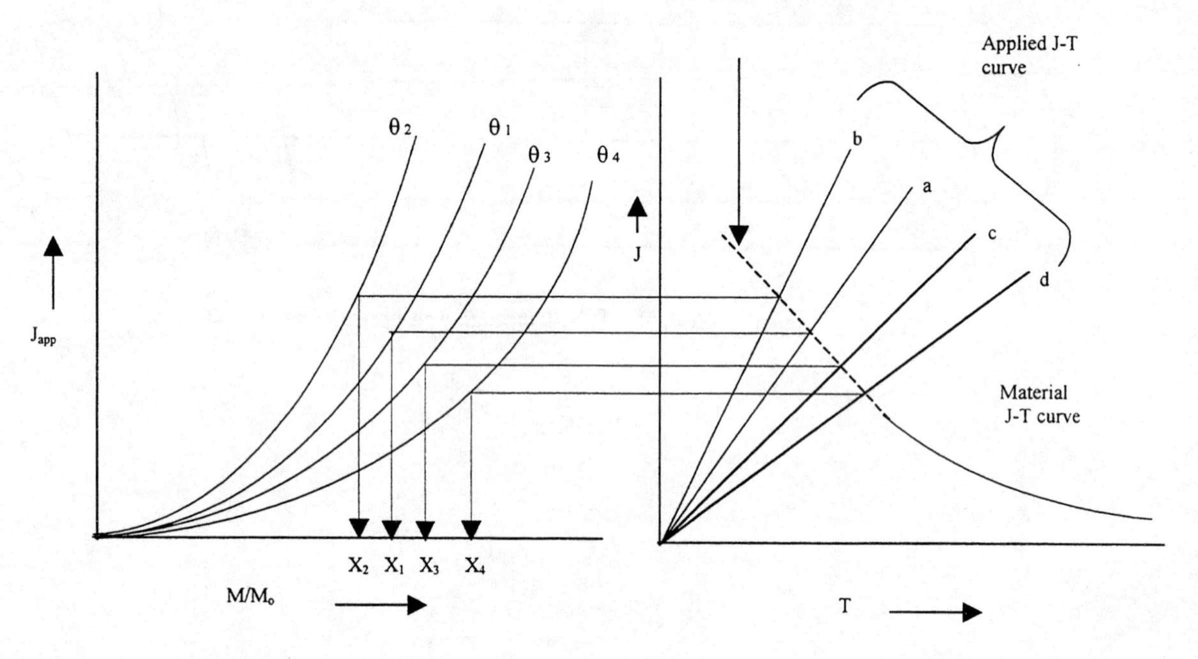

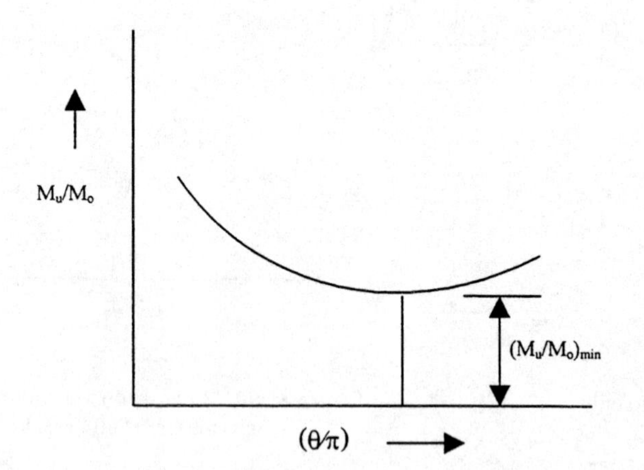

M : Applied bending moment
M_L : Plastic collapse moment
$\theta_1, \theta_2, \theta_3, \theta_4, \ldots$: Different crack sizes. $[\theta_1 > \theta_2 > \theta_3 > \theta_4]$
T : Tearing modulus
M_u : Unstable tearing moment
a,b,c,d : Applied J-T curve, corresponding to crack sizes, $\theta_1, \theta_2, \theta_3, \theta_4$ (a typical trend)
X_1, X_2, X_3, X_4 : UTL, corresponding to crack sizes, $\theta_1, \theta_2, \theta_3, \theta_4$

Figure 4: Schematic diagram for determination of Z-factors

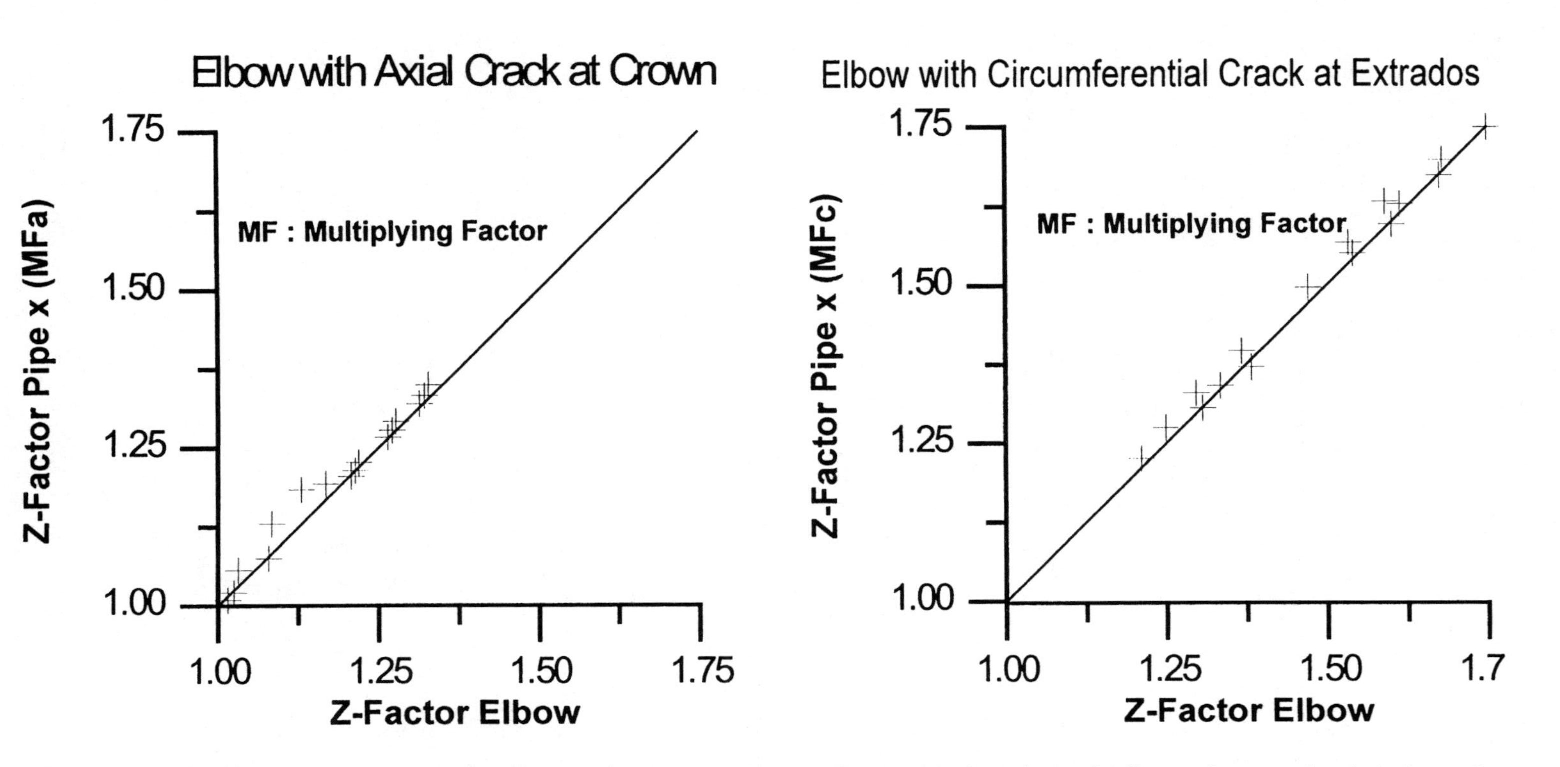

Figure 5 : Comparison of Z-factor for elbows obtained directly and obtained using multiplying factor with corresponding pipe Z-factor [based on standard R-6 method]

PVP-Vol. 410-1, Assessment Methodologies for Preventing Failure:
Deterministic and Probabilistic Aspects and Weld Residual Stress
Volume 1
ASME 2000

EXPERIMENTAL STUDIES IN INFLUENCE OF FLAW WITH NEAR ALLOWABLE SIZE ON PIPE FRACTURE FOR NUCLEAR POWER PLANTS

Katsumi Hosaka
Japan Power Engineering and Inspection Corporation

Masaaki Kikuchi
TOSHIBA Corporation

Hiroshi Ueda
Ishikawajima-Harima Heavy Industries Co., Ltd.

Yasuhide Asada
Professor Emeritus of the University of Tokyo

ABSTRACT

A lot of experimental data of fracture tests in flawed pipes have been published, but there is few data about pipes with small size flaw such as allowable size by ASME Code Section XI[1]. Pipe Fracture tests were performed to clarify the influence of small size flaw on the pipe failure.

Large diameter carbon steel pipes (JIS STS410) whose nominal diameter was 318.5mm(12B) and nominal wall thickness was 19.3mm (Sch100) were used for specimens. Several different size of semi-elliptical cracks and 360 degree shallow cracks in circumferential direction were introduced in inner surface at the center of each specimens. The size of cracks were selected as the allowable flaw size of ASME Code Sec. XI. All tests were carried out in four point bending mode in air at 288℃. The load, cross head displacement and diameters in vertical and horizontal direction at the center of pipe were measured.

The maximum loads and pipe shape change measured in cracked pipe were almost equal to those of the non-cracked pipe. The maximum stress of cracked pipes was satisfied with the safety factor considered in a design of non-cracked pipe.

INTRODUCTION

Light water reactor (LWR) plants have been in operation in Japan for over 25 years and their utility rate has reached one of the highest levels in the world. They have aged steadily and thus further improvement in the integrity and reliability of the plants and related equipment is essential. To this end, a national project, the Structural Assessment of Flawed Equipment (SAF), was initiated in 1991. The purpose of the SAF is to assess the integrity of the aged plants during operation, assuming that the equipment and piping of the aged plants have small cracks. The research was carried out by the Japan Power Engineering and Inspection Corporation (JAPEIC) which was entrusted by the Ministry of International Trade and Industry (MITI). Three tests: 1) a 12 inch nominal diameter piping model verification test, 2) a 1/2.5-scale RPV vessel model verification test, and 3) study on flaw evaluation criteria using a small flat plate test are to be carried out.

A lot of experimental data of fracture tests in flawed pipes have been published, and many evaluation methods of pipe fracture, such as net section criteria, R6 Option 3[2] and DPFAD[1] were developed based on those data.

But there is few data about pipes with small size flaw which are allowed by ASME Code Section XI and the influence of such small size flaw on pipe fracture is not clarified. This paper provides the experimental data of full-scale pipe fracture test with small size flaw.

EXPERIMENTAL CONDITIONS

Material and Specimen

Large diameter carbon steel pipe (JIS STS410) whose nominal diameter was 318.5mm (12inch nominal diameter) was used. Table 1 lists chemical compositions and mechanical properties of the steels of STS410. The pipes with the lower bound energy impact property and the mechanical properties were selected based on the survey of pipes used in services.

Table 1 Chemical composition and mechanical properties of test materials

(a) Chemical composition

	C	Si	Mn	P	S
Base Metal (STS410)	0.19	0.23	0.93	0.019	0.010

(b) Mechanical properties

	Base Metal (STS410)	
Test Temp., (℃)	RT	288
Yield Stress, σ_y (MPa)	337	199
Tensile Strength, σ_u (MPa)	489	457
Elongation, δ (%)	35	30

Figure 1 shows the pipe model specimen, 318.5mm in outer diameter and Sch 100 (19.3mm thickness). Thicker pipes (Sch 160) were welded at the both sides of the specimen to reinforce the loading point. Three pipe model specimen were made. One is the non-cracked pipe (FR-1) to compare with other specimens. The others had semi-elliptical fatigue pre-crack (FR-2, 3.3 mm depth and 13.2 mm length) and 360 degree shallow cracks in circumferential direction (FR-3, 2 mm depth) in inner surface at the center of each specimens. The size of cracks were selected as the allowable flaw size of ASME Code Sec. XI. The comparison between crack sizes and the allowable flaw size of ASME Code Sec. XI were shown in Figure 2.

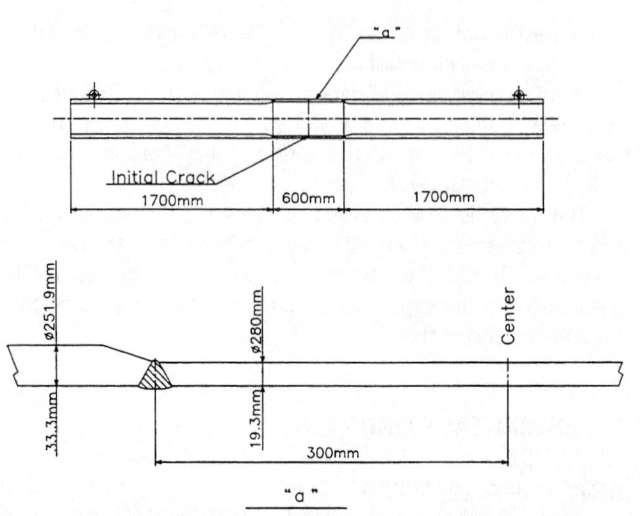

Figure 1 Configuration of pipe model specimen

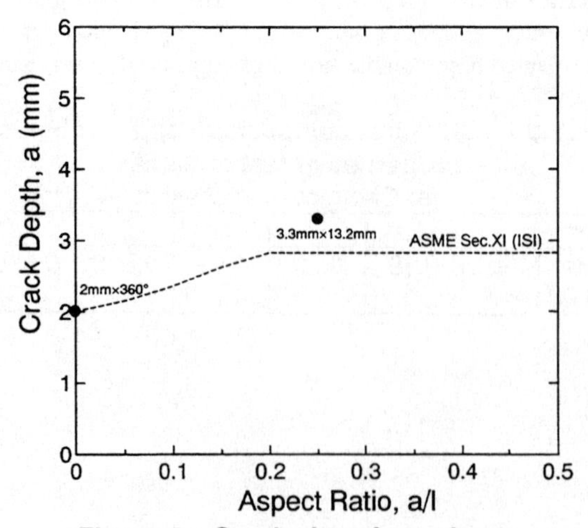

Figure 2 Crack size of specimen

Test Condition

Loading conditions and environmental features are shown in Table 2. The specimen was heated to 288C by a electric heater placed on its outer surface. After the temperature of each specimen became steady-state, each tests were carried out. The test setup is shown in Figure 3.

Load , displacement of actuator and temperature of specimens were measured throughout the test.

Table 2 Test conditions

(a)Loading condition

Loading	4 point bending
Control mode	Stroke
Loading speed	0.1mm/sec

(b)Environmental condition

Environment	in Air
Temperature	288 ℃
Pressure	—

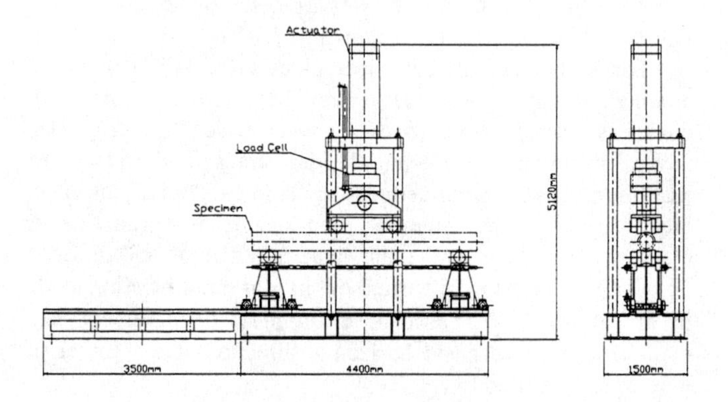

Figure 3 Test equipment

TEST RESULTS

The specimens were loaded until the load-displacement curve showed maximum value and then unloaded. On completion of the test, each crack surface broken at subzero temperature was examined.

Figure 4 shows the load-displacement curves and figure 5 shows the deformation of each specimen after the test. There were no differences with each specimen on the load-displacement curves and deformation of specimen. Table 3 shows the maximum stress of each specimen. Table 3 also shows the prediction of collapse stress, buckling moment[3] and fracture stress by R6 Option 3. All specimens were not buckled until load exceeded maximum value and those results are agree with the prediction of collapse and buckling stress. The prediction of collapse and R6 Option 3 are almost equal because specimens fractured with collapse. The example of fracture assessment diagram is shown in Figure 6. The maximum applied stresses were about 1.5 times of collapse stress based on the flow stress concept. The reason of this difference is the margin of the flow stress (σ_f=(σ_y + σ_u)/2).

Figure 7 shows macro-graphs of crack surfaces. According to the photographs, the ductile crack growth at deepest point were about 0.5mm.

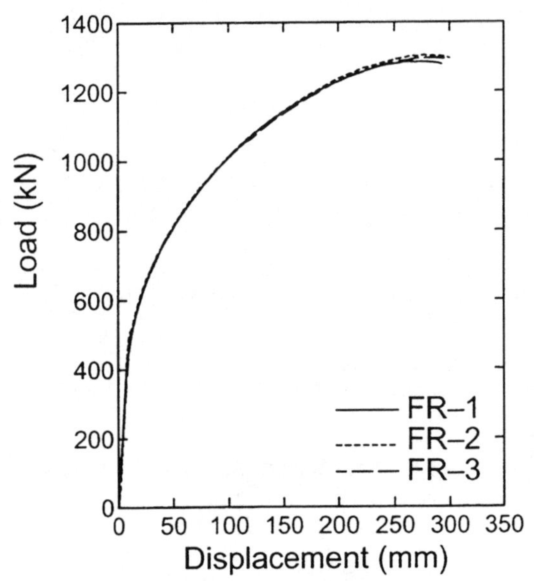

Figure 4 Load-displacement curve

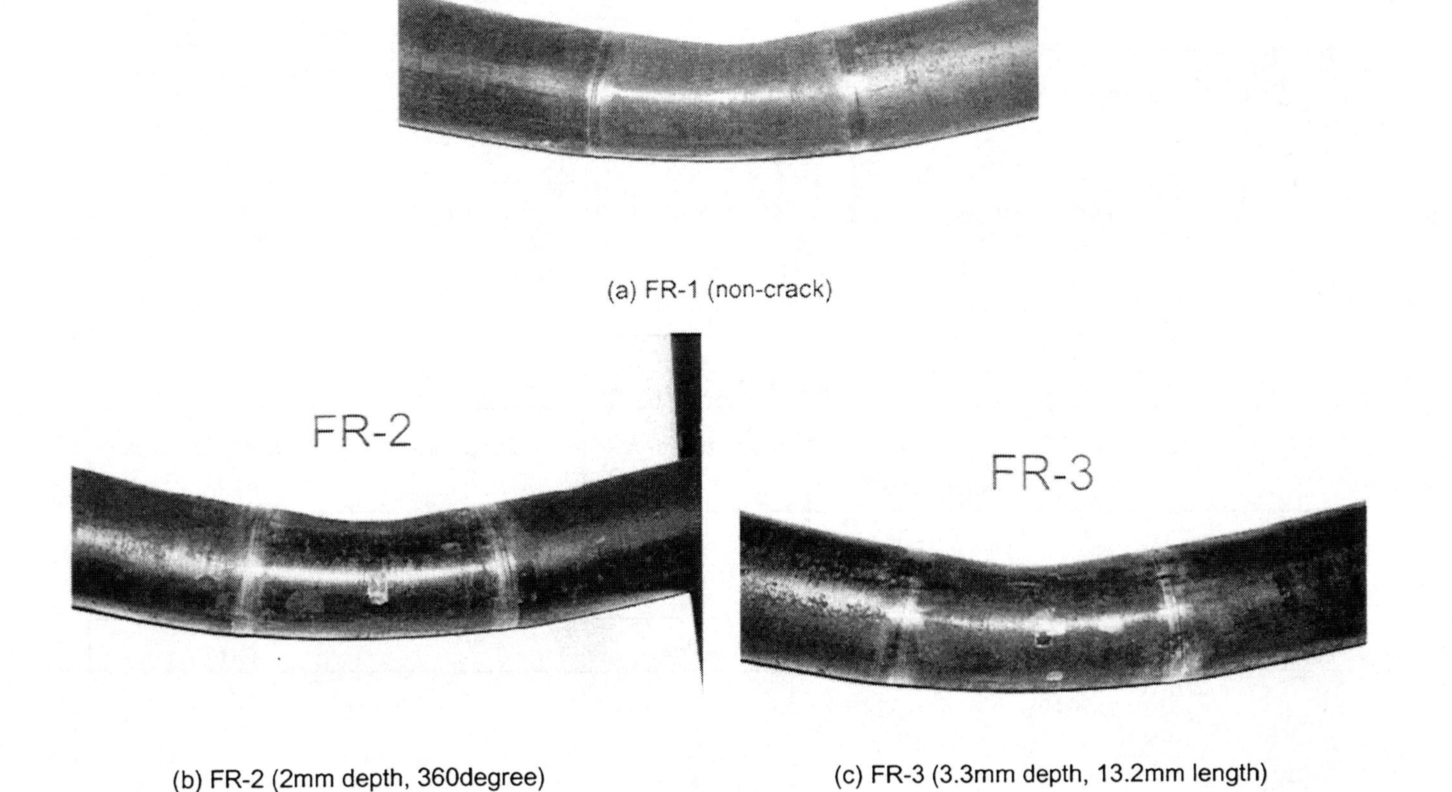

(b) FR-2 (2mm depth, 360degree)

(c) FR-3 (3.3mm depth, 13.2mm length)

Figure 5 Deformation of specimen

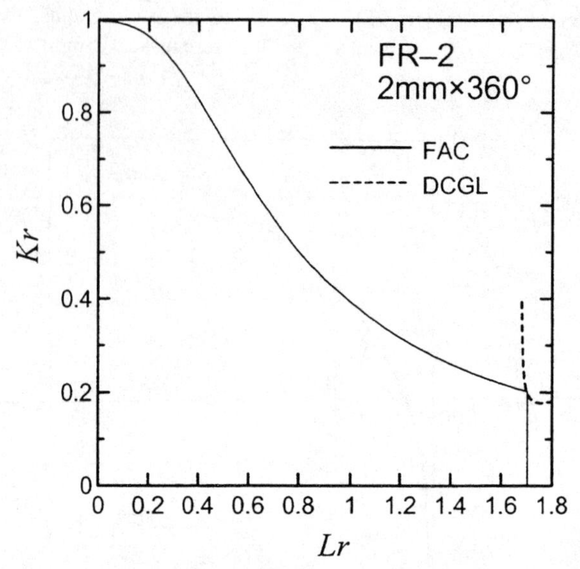

Figure 6 Fracture assessment diagram

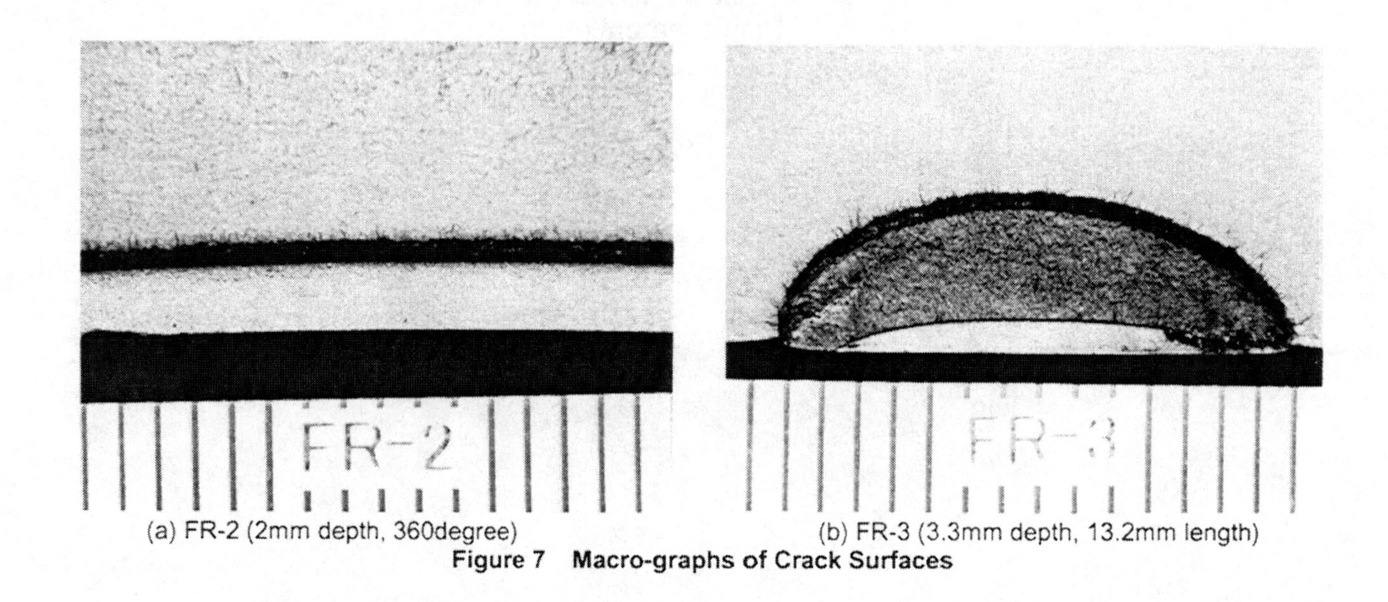

(a) FR-2 (2mm depth, 360degree) (b) FR-3 (3.3mm depth, 13.2mm length)

Figure 7 Macro-graphs of Crack Surfaces

Table 3 Fracture stress

	Crack size	Test results σ_{exp} (N/mm^2)	Collapse stress* σ_{NSC} (N/mm^2)	Buckling stress σ_{Mesloh} (N/mm^2)	R6 Option 3 σ_{R6} (N/mm^2)
FR-1	non-crack	588	412	429	–
FR-2	2 mm, 360 degree	609	389	–	384
FR-3	3.3 mm, 13.2 mm	600	411	–	410

* σ_{NSC} : Nominal stress related to the flow stress $((\sigma_y+\sigma_u)/2)$ at cracked cross section.

CONCLUSIONS

Pipe fracture tests were performed to clarify the influence of small size flaw on the pipe failure. The following results were obtained:

(1) There was no influence of flaws with near allowable size on pipe fracture.

(2) The maximum stresses with pipe fracture tests were about 1.5 times of collapse stress from net section criteria.

ACKNOWLEDGMENTS

This project is being conducted under the guidance of independent learned scientists and those with cooperation of Japanese Utilities (Tokyo Electric Power Company and Kansai Electric Power Company) and related companies. We wish to thank the Agency of Natural Resources and Energy for sponsorship and support, and technical officials in charge of SAF project for valuable suggestions and the critical reading of this manuscript.

REFERENCES

1. ASME Boiler & Pressure Vessel Code Section XI, "Rules for In-service Inspection of Nuclear Power Plant Components 1995 EDITION", 1995.
2. I. Milne et al., R/H/R6-Rev.3, CEGB Rep., 1986.
3. P. Krishnaswamy et al., "Fracture Behavior of Short Circumferentially Surface-Cracked Pipe", NUREG/CR-6298, 1995.

PVP-Vol. 410-1, Assessment Methodologies for Preventing Failure:
Deterministic and Probabilistic Aspects and Weld Residual Stress
Volume 1
ASME 2000

COMPARISON OF ESTIMATION SCHEMES AND FEM ANALYSIS PREDICTIONS OF CRACK-OPENING DISPLACEMENT FOR LBB APPLICATIONS

David L. Rudland, Yong-Yi Wang and Gery Wilkowski

Engineering Mechanics Corporation of Columbus
3518 Riverside Dr. – Suite 202
Columbus, OH 43221

ABSTRACT

This work was conducted as part of the process of recommending procedures for a new LBB Regulatory Guide for nuclear piping applications. This regulatory Guide may replace the current NRC Draft Standard Review Plan 3.6.3. This paper reviews the efforts conducted in a number of past NRC funded programs on the accuracy of COD predictions for leak-rate analyses. Additionally a series of nonlinear FEM analyses were conducted to assess the accuracy of crack-opening-estimation schemes for both austenitic and ferritic piping. Pipe diameters of 102, 305, and 914 mm (4, 12, and 36 inches) were assessed with crack lengths of 8, 16, and 32 percent of the circumference. Recommendations of which estimation COD scheme should be used are presented.

INTRODUCTION

In Leak-Before-Break (LBB) evaluations for nuclear piping applications, there are typically two analysis steps[1]. The first is to determine what crack size could exist for a given leak rate at normal operating conditions. The next step is to determine if that leaking crack would be stable at a transient load such as seismic loading. Safety factors on the leak rates and the transient loads are applied. From probabilistic LBB analyses conducted in Reference 2, it was shown that the failure probabilities were more sensitive to the leak rate analyses than the fracture analysis. Hence for the ongoing effort for the USNRC to help establish a Regulatory Guide for LBB analyses, it was deemed important to establish the accuracy of frequently used crack-opening area methodologies that are part of the leak-rate analyses.

BACKGROUND

A significant amount of work has been conducted in the past to establish and verify analysis procedures to predict the crack-opening area of a circumferential through-wall crack in a pipe under bending or combined pressure (axial tension) and bending. The most commonly used methods in past LBB applications in the U.S. are the GE/EPRI estimation scheme[3, 4], the Tada/Paris method[5], and finite element analyses.

Efforts were undertaken in Reference 6 to compare the accuracy of these methods to experimental results. In that report, the experimental/predicted COD values were compared for a variety of pipe experiments at a particular load. (An experimental/predicted ratio of greater than 1.0 is conservative for LBB analyses.) The load level chosen in Reference 6 was arbitrarily set at 40-percent of the predicted load from the Net-Section-Collapse analysis. It was felt that the COD at 40-percent of limit-load comparison might be representative for LBB application cases, although that value was not justified.

Table 1 shows that there was not much difference between the Original GE/EPRI and the Battelle-modified GE/EPRI COD h_2-functions developed in Reference 6. The Tada/Paris analysis underpredicted the experimental COD values by a factor of three at the load corresponding to 40-percent of the Net-Section-Collapse load. Of additional concern was the very high coefficient of variance (COV), where one standard deviation was about equal to the mean. (The COV equals the standard deviation/mean.) One thought about the high COV was that the accuracy of the experimental data might be the problem. The experimental COD values were not taken in the older pipe tests with high-precision instrumentation. Furthermore, the COD values were measured some distance off the surface of the pipe, whereas the analysis COD values are for the mid-thickness location. This concern with the experimental data was cited in Reference 6, but corrections to the experimental values were beyond the scope of that project.

Table 1 Comparison of experimental-to-predicted COD values from 15 TWC pipe experiments[6]

Fracture Analysis Method	Experimental/Predicted COD	
	Mean	Coefficient of Variance (COV), percent
Original GE/EPRI	1.01	72.8
Battelle-modified GE/EPRI	1.02	86.5
Tada/Paris	2.96	146

To make an assessment of the accuracy of the GE/EPRI and Tada/Paris COD analyses, a sensitivity study was conducted. This sensitivity study involves determining the accuracy of these estimation schemes relative to FEM COD values as a function of load. Hence, the accuracy of the COD analyses was determined throughout all load levels that might be practically used in LBB analyses. This analysis was limited to pipe with R/t = 10 in all cases. The analysis matrix is given in Table 2.

This analysis matrix gives nine FEM meshes times two materials for 18 runs. One additional FEM run was made using the Ramberg-Osgood fit of the stainless-steel stress-strain curve rather than the actual stress-strain curve. Figure 1 shows a comparison the actual stress stress-strain curve and the Ramberg-Osgood curves for the stainless steel material. ABAQUS was used for the FE analysis with 8-noded shell elements with seven integration points through the thickness. The crack tip was modeled using collapsed quad-elements to allow for the crack tip blunting behavior to occur. This mesh has been used in many previous programs and has been though extensive mesh refinement studies. The loading consisted of internal pressure (hoop stress loading) of $S_m/2$ (S_m is material design stress[1]), longitudinal membrane stress corresponding to pressure loading (crack face pressures were ignored), and bending load up to 80-percent of Net-Section-Collapse predicted load. (In many of the past membrane and bending analyses in the literature, the hoop stress was ignored.) In the FEM analysis, the ends of the pipe are free to rotate and allowed to ovalize as needed, which is the same boundary condition as in the estimation schemes. These boundary conditions were achieved using a customized multi-point constraint equation that also allowed for application of the far-field bending moment. Figure 2 shows the deformed state of the FE mesh for an intermediate crack length case. (Deformations were magnified by factor of 7 in this figure.)

The original GE/EPRI COD h-functions[3,4] are used in the PICEP[7] and NRCPIPE[8] computer codes. Only the NRCPIPE computer code had the Tada/Paris solution.

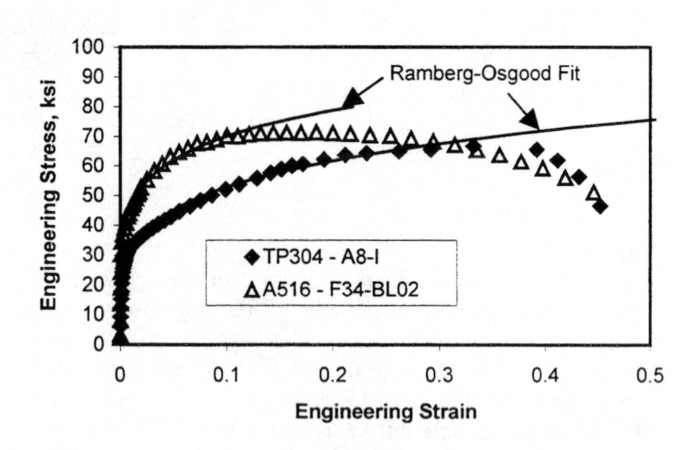

Figure 1 Comparison of actual and Ramberg-Osgood curves for the TP304 stainless steel

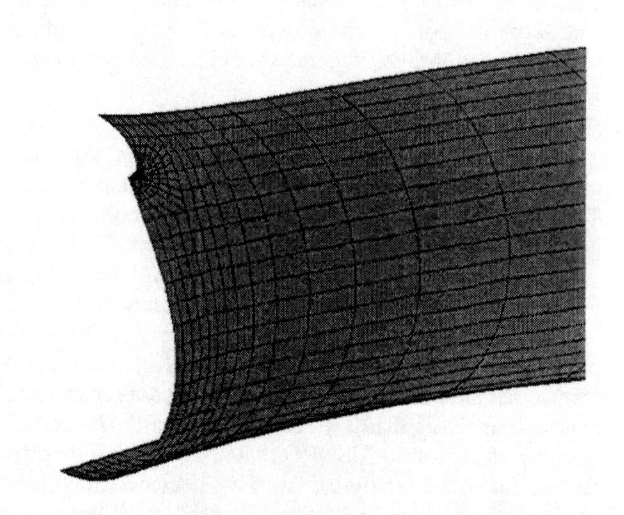

Figure 2 Deformed mesh of 16% TWC in 914-mm (36-inch) diameter stainless steel pipe (At load step of 80% of NSC moment.)

---

1 The S_m for TP304 stainless steel at 288C (550F) = 125 MPa (18.1 ksi), and for A516 Gr 70 ferritic steel at 288C (550F) S_m = 135 MPa (19.6 ksi).

TABLE 2 VARIABLES IN COD ANALYSIS

Outside diameter, mm (inch)	114.3, 323.85, and 914.4 (4.5, 12.75, and 36) – R/t = 10 in all cases
Crack lengths, % of circumference	8, 16, and 32
Materials	TP304 stainless steel (Pipe A8-I stress-strain curve at 288C [550F]), and A516 Gr70 (Pipe F34-BL02 stress-strain curve at 550F).

The Battelle-modified GE/EPRI solutions for combined pressure and bending were also used NRCPIPE (Version 3.0) for the calculation of the original and Battelle-modified GE/EPRI method as well as the Tada/Paris predictions. Comparisons were also made between the Battelle-NRCPIPE code and EPRI's PICEP Code for the GE/EPRI method. The PICEP to NRCPIPE comparison was deemed necessary, since practical applications are for pipes under combined load. The GE solutions were for either pure bending or pure tension[3], with very few combined bending and pressure solutions[4]. Neither of these solutions included the hoop stress effects that might stiffen the shells.

An additional uncertainty in the two different combined load GE/EPRI COD codes is how they developed the combined load solution. In the NRCPIPE code, there is no explicit documentation of the technical approach, but it was determined that the following equations were used.

$$M_o^B = 4\sigma_o t R^2[\cos(\theta/2) - 0.5\mathrm{Sin}\theta] \quad (1)$$

$$M_o^{B+T} = 4\sigma_o t R^2\{\cos[\theta/2+\pi\sigma_T/(2\sigma_o)] - 5\mathrm{Sin}\theta\} \quad (2)$$

$$h_2^{B+T} = h_2^B + (\sigma^T/\sigma^{T+B})\, h_2^T \quad (3)$$

$$\sigma_o^{B+T} = M_o^{B+T}(R/I) \quad (4)$$

$$COD = \alpha\varepsilon_o a h_2^{B+T}(M/M_o^{B+T})^n \quad (5)$$

Where,

M = applied moment
M_o^B = limit-load moment under pure bending
M_o^{B+T} = limit-load moment under combined axial tension and bending stresses
θ = half crack angle
σ_o = reference stress (typically yield)
σ_T = axial tension stress
σ^{T+B} = combined axial tension and bending stress
ε_o = reference strain
α = Ramberg-Osgood parameter
a = half crack length
COD = crack-opening displacement
I = moment of inertia
R = mean pipe radius
t = pipe thickness
h_2^B = parameter relating COD to stress under pure bending
h_2^{B+T} = parameter relating COD to stress under pure bending
n = strain-hardening exponent

In the PICEP code, there are two pressure and bending options. One is pure elastic superposition where

$$COD_{Total} = COD_{elastic\ axial} + COD_{elastic\ bending} + COD_{plastic\ tension} + COD_{plastic\ bending} \quad (6)$$

This would give conservative results if plasticity occurs.

The second option in PICEP is a combined option case, which is the default case. There was no description of the combined option analysis procedure in the PIFRAC code itself. However, Reference 9 describes the superposition method as well as two different combined-option procedures. One of the procedures, referred to as the Kashida method, is given in the following equations.

$$h_2^{B+T} = h_1^{B+T}(1\text{-}a/b)(K_T+\lambda K_B)/(1+\lambda) \quad (7)$$

$$K_T = h_2^T/[h_1^T(1\text{-}a/b)] \quad (8a)$$

$$K_B = h_2^B/[h_1^B(1\text{-}a/b)] \quad (8b)$$

$$\lambda = P_b/(2P_m) \quad (9)$$

Where,

b = πR
P_b = bending stress
P_m = membrane stress
h_1^{B+T} = parameter relating J to stress under pure bending

However, in Reference 9 they say that the h_1 function should be for combined tension and bending, but do not supply those values.

The alternate combined-load solution in Reference 9 was a method by Zahoor where the equations used are given below.

$$h_2^{T+B} = h_1(1\text{-}a/b)\lambda_1(K_T + \lambda K_B)/(1+\lambda) \quad (10)$$

Where,

$$\lambda_1 = 0.5[1 + (1+\lambda)^2)^{0.5}/(1+\lambda)] + \{0.05\lambda/(1+\lambda)^2\}(n\text{-}1)(1\text{-}2a/b)] \quad (11)$$

From inspection of Equation 11, it can be seen that to make this equation unambiguous, several additional parenthesis are needed. Furthermore the combined-load h_1 functions were still not defined.

Because of the uncertainities in both the NRCPIPE and PICEP combined load solutions, comparisons were made to FEA results.

RESULTS

Some typical FEA and estimation scheme results are shown in Figures 3 thru 5 for the 114-mm (4.5-inch) diameter TP304 pipe. In these figures, the bending moment is normalized by the Net-Section-Collapse predicted bending moment versus the center-crack-opening displacement (COD). The flow stress used was average of yield and ultimate strength of the actual material properties, which agrees well with experimental data[10].

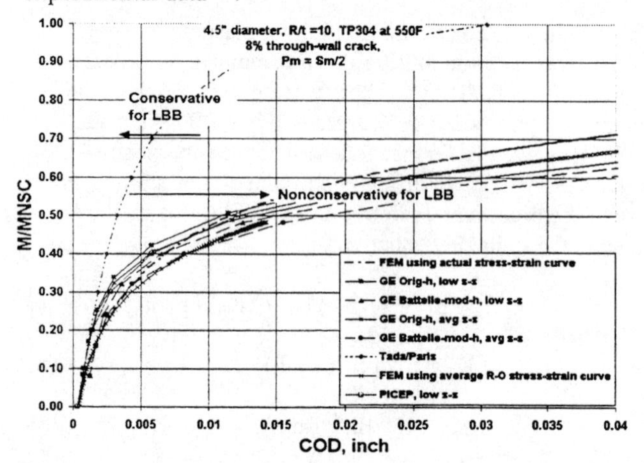

Figure 3 Results for 114-mm (4.5-inch) diameter TP304 pipe with TWC 8-percent of circumference

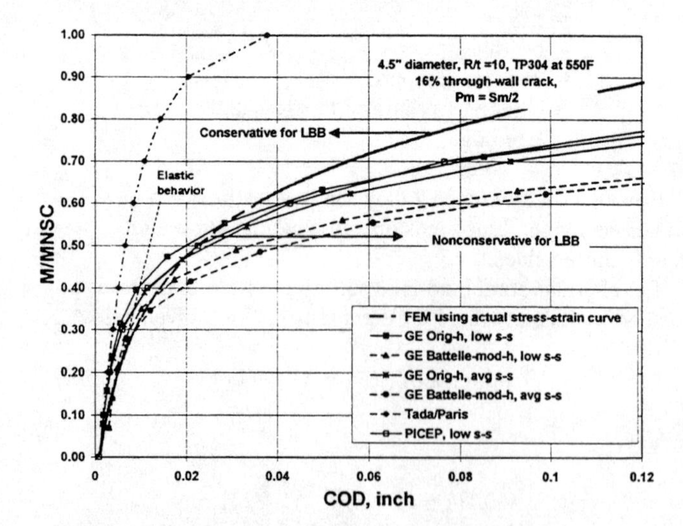

Figure 4 Results for 114-mm (4.5-inch) diameter TP304 pipe with TWC 16-percent of circumference

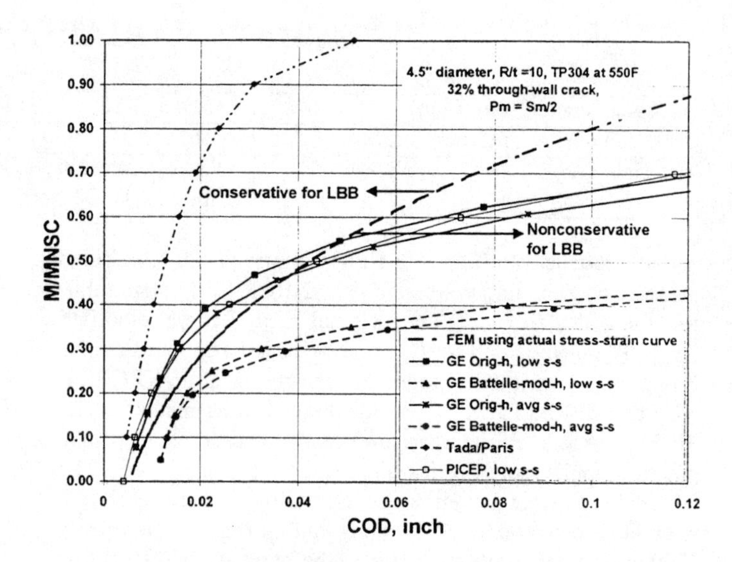

Figure 5 Results for 114-mm (4.5-inch) diameter TP304 pipe with TWC 32-percent of circumference

The following key observations can be seen in the comparisons in these figures.

- In Figure 3, we used both the actual stress-strain curve and the average Ramberg-Osgood fit (from 1-percent strain to the strain at 80 percent of the ultimate strength) in the FEM analyses. This difference showed that using the Ramberg-Osgood fit gives a slightly large COD than the actual behavior, i.e., the actual behavior would be slightly conservative for LBB analysis when compared to using the Ramberg-Osgood mean fit of the stress-strain curve in the FEM analyses.
- The COD at zero load is the COD due to the pressure. The Tada/Paris and original GE/EPRI COD values due to pressure agreed well with the FEM results, but the Battelle-modified GE/EPRI values were higher, probably due to the Battelle-modified functions being only developed for a pressure level of 17.24 MPa (2,500 psig). The Battelle-modified GE/EPRI functions give nonconservative over predictions for LBB applications in these cases.
- In the purely elastic bending range, there is little difference between the original GE/EPRI and Tada/Paris COD values as the load increases above the elastic range, the Tada/Paris method gives conservative predictions relative to the GE/EPRI and FEM analyses. The degree of conservatism depends on the load level, and in the worst case could be overly conservative by a factor of 10 to 15.
- As the crack length increased, the most accurate analysis changed from the original GE/EPRI to the Battelle-modified GE/EPRI analysis.
- There was not much difference in using the low-strain or average range Ramberg-Osgood fit of the actual stress-strain curve for the TP304 stainless steel in the GE/EPRI analyses.

The results from these three figures and the rest of the results are shown in fully normalized figures where (estimation scheme moment)/(Net-Section-Collapse moment) versus (estimation scheme COD)/(FEM COD) curves are shown in Figures 6 to 8 for the three different pipe sizes, three different analyses, and two different materials.

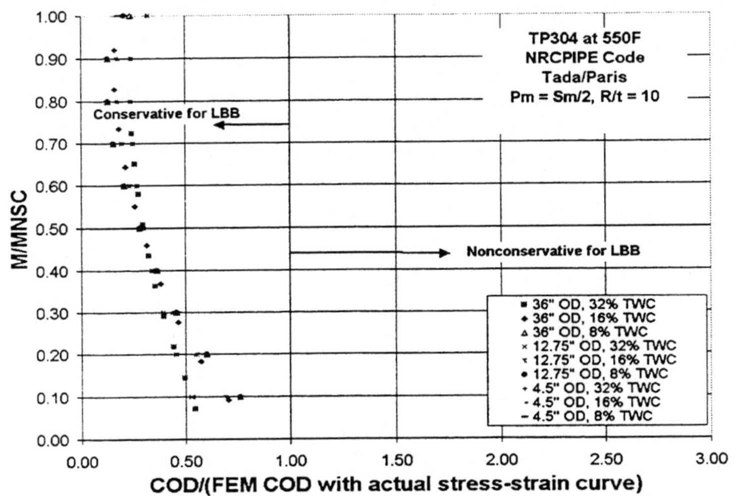

Figure 6a Tada/Paris solutions from the NRCPIPE code for TP304 pipes at 288 C

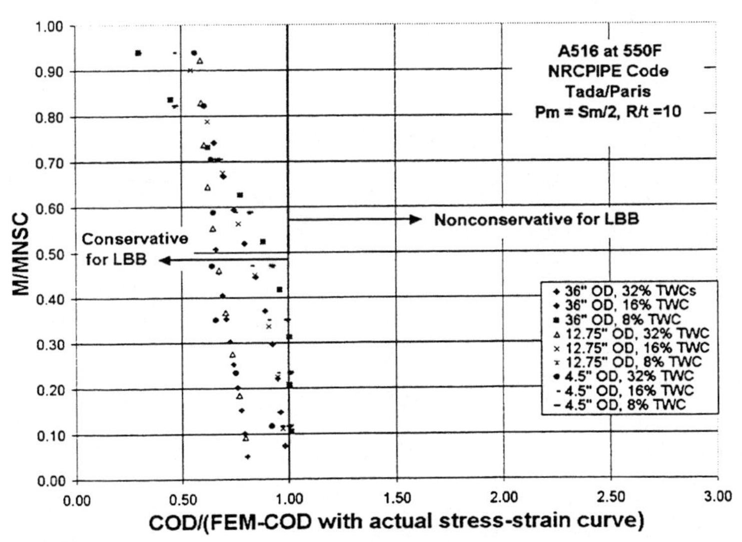

Figure 6b Tada/Paris solutions from the NRCPIPE code for A516 Gr. 70 pipes at 288C

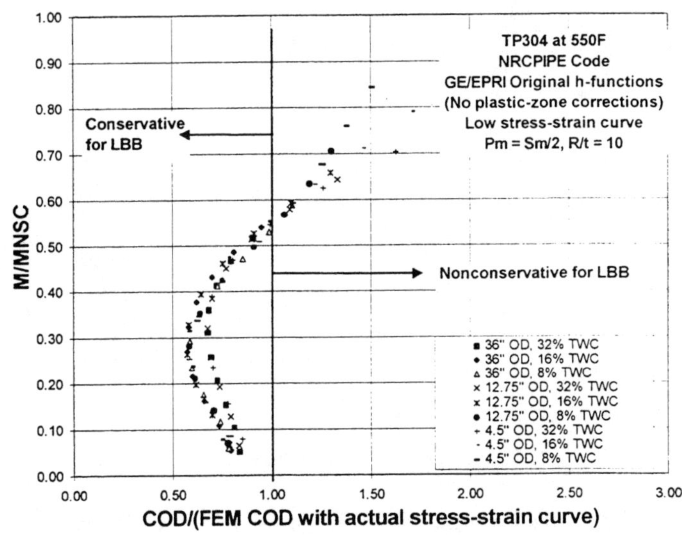

Figure 7a Original GE/EPRI solutions from the NRCPIPE code for TP304 pipes at 288 C

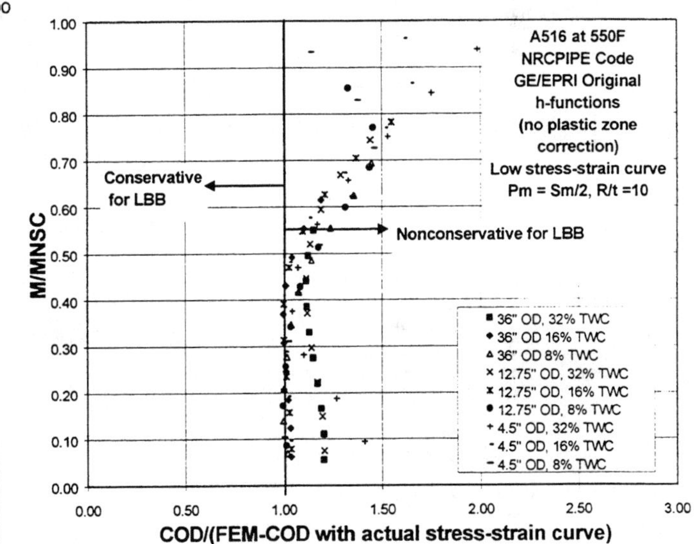

Figure 7b Original GE/EPRI solutions from the NRCPIPE code for A516 Gr. 70 pipes at 288C

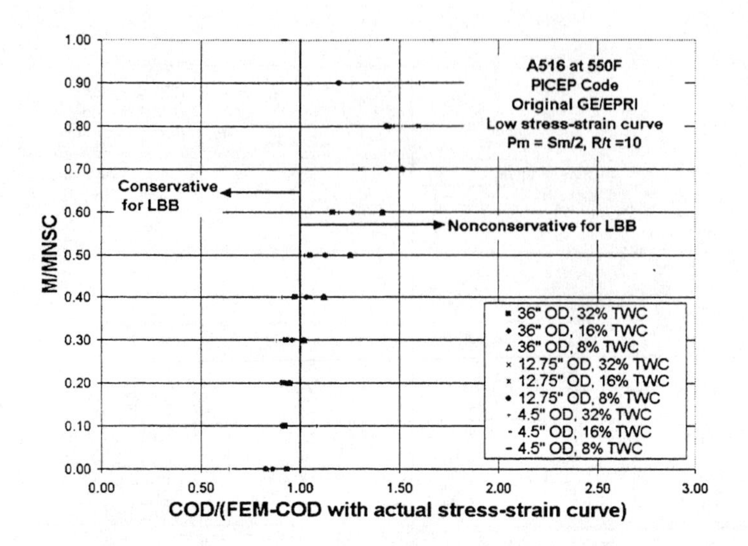

Figure 8b Original GE/EPRI solutions from the PICEP code for A516 Gr. 70 pipes at 288C

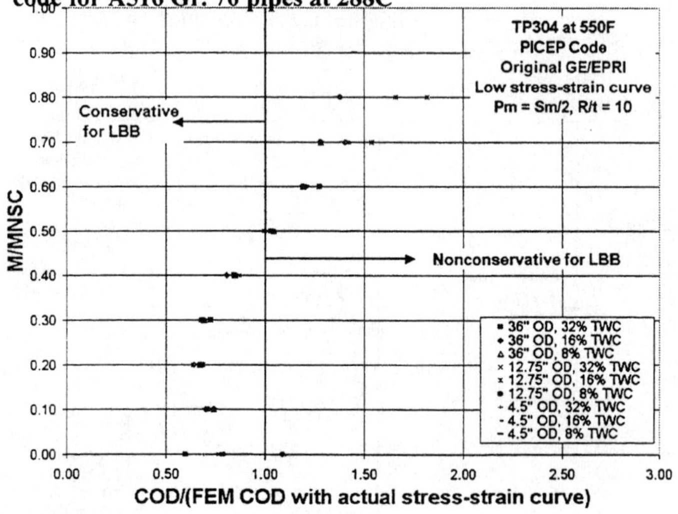

Figure 8a Original GE/EPRI solutions from the PICEP code for TP304 pipes at 288 C

SUMMARY AND CONCLUSIONS

After all the uncertainities of the actual formulations used in both the NRCPIPE and PICEP codes for determining the combined-load COD values, it can be seen by comparing Figures (a) and (b) in Figures 7 and 8 that there is little difference between the two methods. Both codes have the same trends, although the trends are different for the austenitic pipe than the ferritic pipe. Both methods can be conservatively used in LBB analyses if the bending load is less than half of the combined-load Net-Section-Collapse analysis moment. For loads greater than 50-percent of the Net-Section-Collapse limit moment, a safety factor of 1.5 to 2 would cover the inaccuracies of the COD predictions for these analyses. This factor of 1.5 to 2 does not account for inaccuracies from other factors such as restraint of pressure-induced bending[6], weld residual stresses[6], crack morphology variables[2], or inaccuracies in the leak-detection equipment.

REFERENCES

(1) Solicitation for Public Comment on Standard Review Plan 3.6.3 "Leak-Before-Break Evaluation Procedures," *Federal Register*, Vol. 52, No. 167, pp 32633, Friday, August 28, 1987.

(2) Rahman, S., and others, "Probabilistic Pipe Fracture Evaluations for Leak-Rate-Detection Applications," NUREG/CR-6004, April 1995.

(3) Kumar, V., and others, "An Engineering Approach for Elastic-Plastic Fracture Analysis," EPRI Report NP-1931, July 1981.

(4) Kumar, V. and German, M., "Elastic-Plastic Fracture Analysis of Through-Wall and Surface Flaws in Cylinders," EPRI Report NP-5596, January 1986.

(5) Paris, P. C., and Tada, H., "Application of Fracture Proof Methods Using Tearing Instability Theory to Nuclear Piping Postulating Circumferential Through Wall Cracks," NUREG/CR-3464, September 1983.

(6) Rahman, S., and others, "Refinement and Evaluation of Crack-Opening-Area Analyses for Circumferential Through-Wall Cracks in Pipes," NUREG/CR-6300, March 1995.

(7) Norris, D., and others, "PICEP: Pipe Crack Evaluation Program," EPRI Report NP-3596-SR, 1984.

(8) "NRCPIPE User's Guide," Version 3.0 for Windows, April 30, 1996. Battelle document for the Second International Integrity Research Group Program (IPIRG-2).

(9) Zahoor, A., and Kashida, K., "Crack-Opening Area Calculation for Circumferential Through-wall Pipe Cracks," EPRI report NP-5959-SR, August 1988.

(10) Wilkowski, G. M., Olson, R. J., and Scott, P. M., "State-of-the-Art Report on Piping Fracture Mechanics," U.S. Nuclear Regulatory Commission report NUREG/CR-6540, BMI-2196, February 1998.

PVP-Vol. 410-1, Assessment Methodologies for Preventing Failure:
Deterministic and Probabilistic Aspects and Weld Residual Stress
Volume 1
ASME 2000

A GENERAL CONCEPT TO ENSURE THE INTEGRITY OF COMPONENTS

E. Roos, K.-H. Herter, F. Otremba
Staatliche Materialprüfungsanstalt (MPA) University of Stuttgart
Pfaffenwaldring 32, 70569 Stuttgart, Germany

ABSTRACT

The failure behaviour of pipes and piping components (e.g. straight pipe, pipe bend, T-joint) with and without cracks under different loading conditions has been investigated in numerous experimental and analytical/numerical research projects. The results of these projects were used to adjust and to verify different methodologies and procedures to calculate the failure loads, the respective critical crack sizes as well as the leak area and the leak rates.

On the basis of the actual material characteristics, the actual as-built configurations and design of the piping systems, the knowledge of possible failure mechanism, a concept for the assessment of the integrity of the systems could be developed for the different actual as well as for postulated loading conditions. An important part of this integrity analysis is the leak before break assessment and the break preclusion concept.

In the present paper based on the essential research results, the developed procedures, and methodologies for the assessment of the critical crack sizes as well as the critical loading conditions will be reported and discussed. With these results and methodologies the proof of the integrity of piping systems, especially the leak before break and break preclusion concept used in Germany will be demonstrated.

NOMENCLATURE

- A pipe cross section area
- M pipe bending moment
- S safety factor
- a crack depth
- c crack length
- $\dot{m}$ leakage flow rate
- s wall thickness
- Θ pipe bending angle
- α circumferential crack angle
- σ stress

INTRODUCTION

The long term experience in the field of conventional and chemical pressure vessel and piping technology made an essential contribution to establish the Basis Safety Concept (BSC) in Germany into nuclear practice, Kussmaul [1, 2]. With the BSC it is essential to assess and quantify the integrity of components in a mechanistic/deterministic way. Through stringent measures for all safety related components, in the choice of optimised materials and processing, design, stress analysis, manufacture, operation, testing and inspection, it is possible to create the necessary prerequisites for redundancies which make catastrophic failure incredible (exclusion of catastrophic failure or break preclusion concept).

In the last 25 years world wide, Wilkowski [3], as well as within the German reactor safety research programs tremendous efforts were made to develop and to verify methods to describe the load bearing capacity as well as the fracture behaviour of primary circuit piping of nuclear power plants, Kussmaul et al. [4], Bartholomé et al. [5] and Schulz [6].

Parallel to the research activities in the late 1970's in Germany the BSC, also designated as break preclusion concept, was developed and adopted in principle by the German Reactor

Safety Commission (RSK guidelines 1982) [7]. The background of the technical thinking and basis was published in original papers, Kussmaul et al. [1, 2, 8]; Bartholomé et al. [9]; Schulz [6]. In Germany thus the break preclusion concept became a legal requirement. For practical reasons an upper limit of the leakage area of 0.1A (A corresponds to the cross section area of the pipe) was chosen. In general the break preclusion concept is applied to the large diameter primary piping, *Fig. 1*, and to the branch connections down to a nominal diameter of 200 mm. The concept was in principle also applied to pipes down to a diameter of 50 mm, *Fig. 2 and 3*. ($M_{b,3Sm,act}$ and $M_{b,3Sm,KTA}$ - bending moment calculated on a elastic basis for a stress level according to the actual 3Sm value of the material respectively to the material data given in the KTA code; $M_{b,Oper.+Upset}$ - bending moment developed by the pipe system analysis for operating plus upset conditions).

The aim of the following discourse is to state and describe in detail a general concept based on the Basis Safety Concept to ensure the integrity of components. Although the procedure may differ in a plant and component specific way, the concept can be used generally.

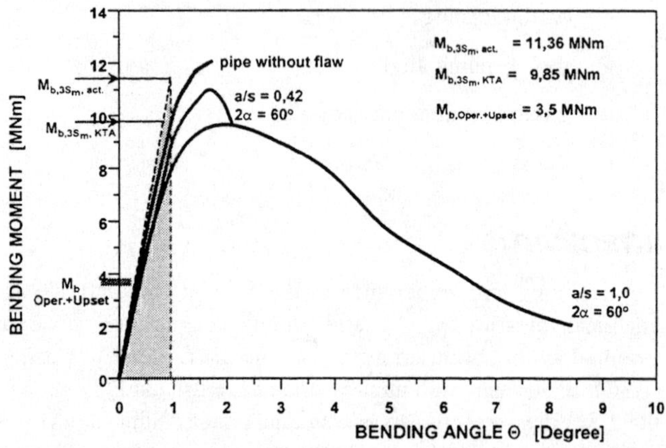

Fig. 1 Load bearing behaviour of ferritic pipes with nominal diameter DN800 (outer diameter 800 mm, wall thickness 47 mm) and internal pressure 15 MPa at room temperature.

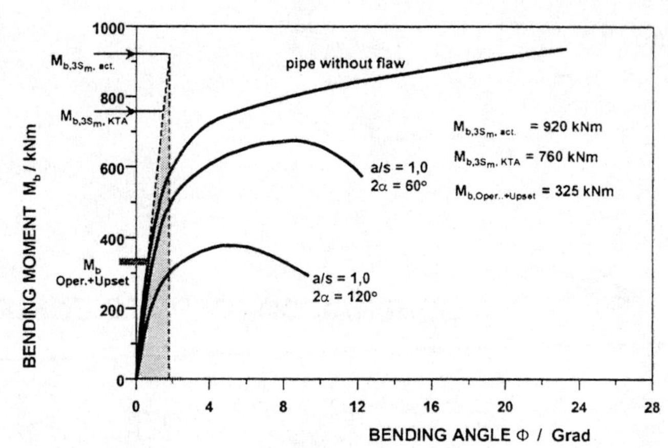

Fig. 2 Load bearing behaviour of austenitic pipes with nominal diameter DN300 (outer diameter 331 mm, wall thickness 32 mm) and internal pressure 16 MPa at room temperature

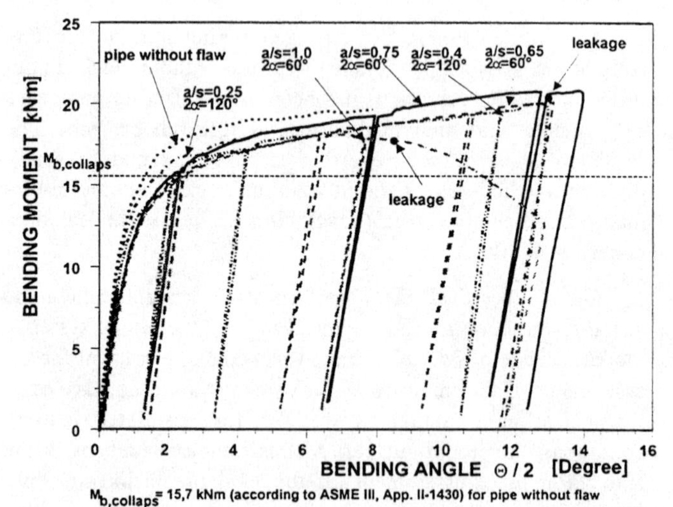

Fig. 3 Load bearing behaviour of austenitic pipes with nominal diameter DN80 (outer diameter 88.9 mm, wall thickness 8.8 mm) and internal pressure 16 MPa at room temperature

BASIS OF VALUATION

Principles for the safety-related requirements taken as a basis for the design of nuclear power plants (NPP), especially with respect to the state-of-the-art, are detailed in the "Safety Criteria for Nuclear Power Plants" of the BMI [10]. Criterion 1.1 "Principles of Safety Precautions" of the BMI safety criteria requires, besides others, a comprehensive quality assurance for fabrication, erection and operation. Criterion 2.1 "Quality Assurance" requires, besides others, the application, preparation, and observation of design rules, material specifications, construction rules, testing and inspection as well as operating instructions and the documentation of quality assurance. Criterion 4.1 "Reactor Coolant Pressure Boundary" principally requires, besides others, the exclusion of dangerous leakage, rapidly extending cracks and brittle fractures with respect to the state-of-the-art.

Moreover the German Reactor Safety Commission (RSK) prepared guidelines [7] as a compilation of the safety-related requirements that, in the Commission's opinion, have to be compiled with the design, construction and operation of a NPP with pressurised water reactor (PWR). In relation with these safety-related requirements postulated leaks and breaks for the main coolant pipes are lay down as well as for the main steam and feed water piping.

PREMISE TO ENSURE THE INTEGRITY OF COMPONENTS

During the development of the BSC safety-related requirements were established which make catastrophic failure incredible by assessing and quantifying the integrity of components in a mechanistic/deterministic way. The requirements will provide the components with a "Basis Safety" (BS) that will preclude any disastrous failure of the component as a result of manufacturing defects (the "quality through production" principle), Kussmaul [2]. A comparable statement for components already in operation implies that the assumptions taken during design, especially for the loading conditions (mechanical, thermal, corrosive), are monitored and compared to the loading conditions included in the design specifications. For the practical application of the BSC (break preclusion concept) under operational conditions, the existence of the independent redundancies is demanded, but their application has to be balanced in a way case by case. The third redundancy ("continuous in-service monitoring and documentation" principle) in the BSC consists of continuous in-service verification that design conditions are not exceeded during plant operation (continuous plant monitoring, repeated testing) which is of important significance. This makes evident that "Basis Safety" (the "quality through production" principle) exclusively can not ensure the integrity of components for all the life time and therefore the independent redundancies are demanded in a balanced way.

Within the general analysis of the mechanical behaviour of the components it has to be demonstrated, that stresses (mechanical and thermal loading) as well as the usage factor (fatigue) will not exceed the limits given by the KTA safety standards [11]. Furthermore already during design the supporting of piping system has to be optimised using fracture mechanics analysis taking into consideration the minimum detectable flaw sizes, the examination intervals, the specified loading conditions, the material characteristics and the in-service monitoring. Taking care of these aspects it is possible to demonstrate leak-before-break (LBB) behaviour for the succeeding operation. By this it can be avoided that critical crack sizes become too small due to high loads resulting from non optimised supporting of components or of the piping system and that leak flow rates are too small for detection (a critical crack size will be reached before the leak flow rates can be detected).

For components and piping systems already in operation or not fulfilling the requirements of the basis safety (BS) in a complete extent the integrity of components can be demonstrated using the independent redundancies of the BSC ("multiple party testing" principle, "worst case" principle, "continuous in-service monitoring and documentation" principle, "validation" principle) in a balanced way associated with the operational experience of the NPP's and using the results of ongoing research programs world wide. In this way a "Quasi Basis Safety" can be demonstrated.

As a premise for a systematically approach to ensure the integrity of components it is indispensable to show that the as-built status of quality (design, construction, loading) is according to the requirements given in the guidelines and standards, to show that sufficient knowledge of possible failure mechanism (e.g. no inadmissible dynamic loading, no corrosion[1]) is available and to show that the as-built status of quality can be guaranteed for the succeeding operation.

In detail the following aspects have to be treated: (a) evaluation of the as-built status of quality (design, construction, material, fabrication; results of recurrent non destructive examinations up to now, operational experience, match the requirements of the basis safety); (b) determination of the relevant loading conditions by means of in-service monitoring (monitoring of the mode of operation, the water chemistry, the mechanical, thermal and corrosive stresses, the dynamic loading); (c) evaluation of the as-built status of quality with respect to the relevant loading conditions (stress analysis - limitation of the stresses; fatigue analysis - determine the usage factor; fracture mechanics analysis - determination of crack growth and critical crack and loading conditions); (d) evaluation and extent of the in-service monitoring to guarantee the succeeding operation (recurrent non destructive examination - minimum detectable flaw sizes, examination area, examination intervals; leak detection system - leak area and flow rate); (e) proof of the closed general concept with graduated measures (independent redundancies) by a summarising evaluation of the integrity.

METHOD OF PROOF TO ENSURE THE INTEGRITY OF COMPONENTS

The method of proof to ensure the integrity of components differentiates between (1) the proof for components under construction (new design and new fabrication), *Fig. 4*, and (2) the proof for components already under operation (differentiates between cases where the real loading conditions are well known by in-service monitoring and cases where no in-service monitoring is available), *Fig. 5*.

[1] It has to be demonstrated, that there is no safety relevant corrosion cracking / crack growth.

(1) Method of proof for components under construction (new design and new fabrication)

Evaluation of the as-built status of quality

Concerning the as-built status of quality documents must be checked with respect to design, material and fabrication (it is a prerequisite that the requirements of the RSK-guidelines, of the Basis Safety and of the KTA-standards are fulfilled) and possible failure mechanism must be well known. Based on the operational experience and the state-of-the-art the failure mechanism must be identified and monitored.

Relevant loading conditions

The knowledge of the relevant loading for components under construction is of decisive importance. This forms the relevant input for stress analysis, fatigue analysis and fracture mechanics analysis. The relevant loadings (normal and upset conditions, reaction forces resulting from a 0.1A leak) are included in the load specifications of the design.

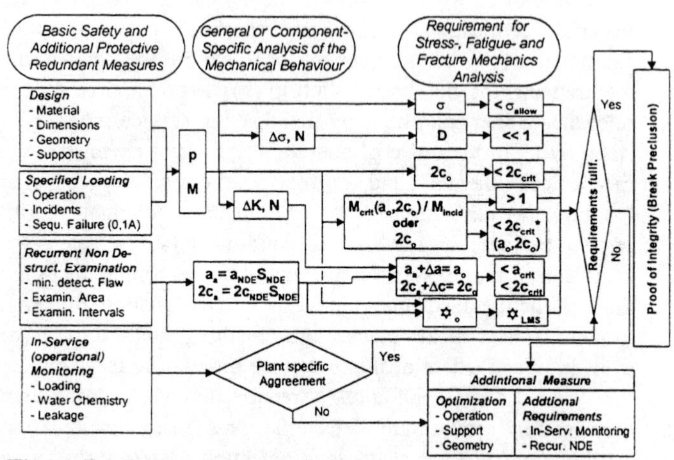

Fig. 4 Procedure (schematically) to proof the integrity of components during design

Determination of stresses and limitation

Fracture mechanics analysis must be performed for a crack size and shape detectable by non-destructive examination (NDE) methods, Wellein [12]. Stimulated by discussion with experts and the licensing authorities as well as the results from research programs in the following the different steps for the fracture mechanics procedures to be used for the proof of integrity (break preclusion) are explained and shown in Fig. 6.

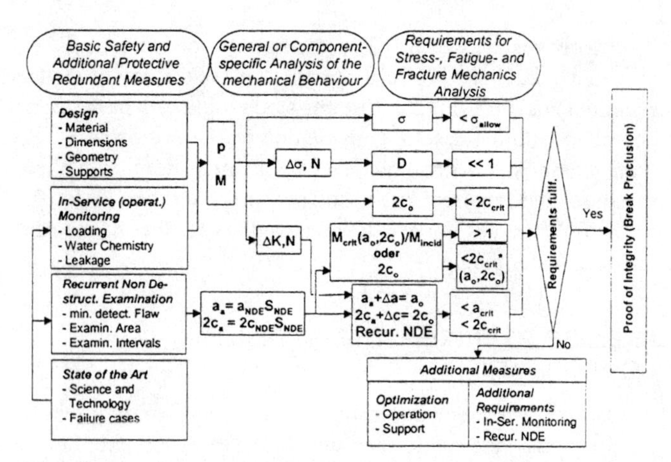

Fig. 5 Procedure (schematically) to proof the integrity of components for components or systems in operation

Step1: Calculation of the critical crack length $2c_{crit}$

The critical crack length of a through wall crack $2c_{crit}$ has to be calculated for the maximum loads specified for plant lifetime (plant and system specific maximum load combination for normal operating and emergency/faulted conditions). This can be achieved by limit load calculations or fracture mechanics approaches verified by component testing, Fig. 7 to 10.

Step 2 Definition of the initial crack size (crack depth a_a and crack length $2c_a$)

Component specific definition of the initial crack size well detectable by non-destructive examination (NDE) methods (the safety factor S_{NDE} has to be determined in a component and examination specific way) is

crack depth $a_a = a_{NDE} \bullet S_{NDE}$

crack length $2c_a = 2c_{NDE} \bullet S_{NDE}$

Step 3:Crack growth calculations (Δa and $2\Delta c$)

Crack growth calculation (Δa, $2\Delta c$) for the initial crack shape (depth a_a and length $2c_a$) with loads specified for normal operating conditions or for loads determined by in-service monitoring considering the appropriate load cycles. The final crack shape for the period to be considered will be

crack depth $a_o = a_a + \Delta a$

crack length $2c_o = 2c_a + 2\Delta c$

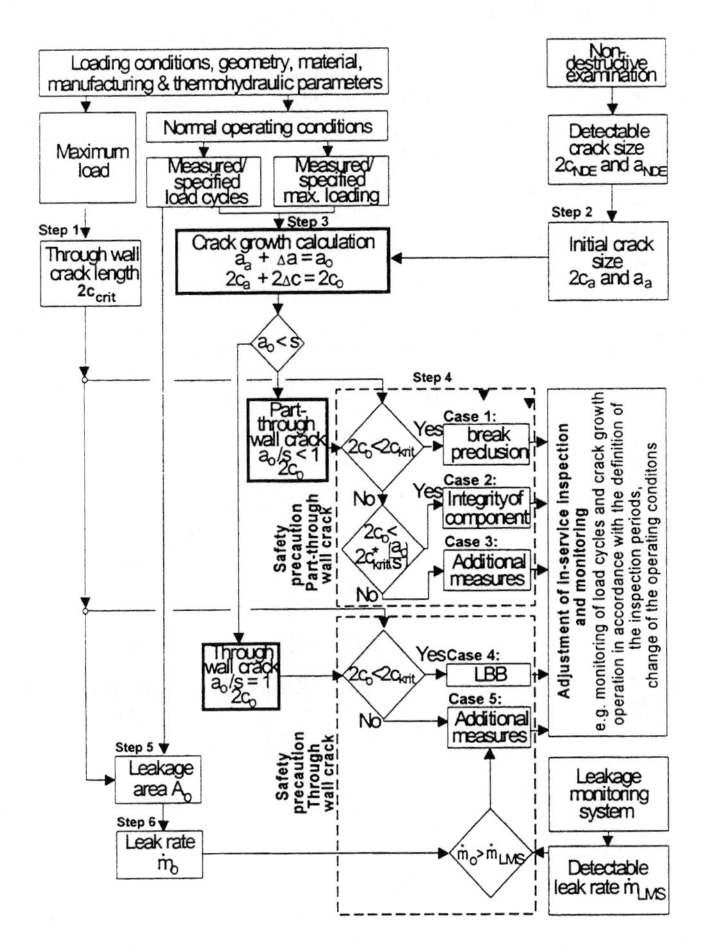

Fig. 6 Fracture mechanics procedure (schematically) to proof the leak-before-break behaviour of components

Crack growth shall be calculated for a period safely covering the intervals of the recurrent inspections. The concept for recurrent inspections and in-service monitoring has to be adjusted to the results of the crack growth calculation.

Step 4: Leak before break (LBB) behaviour

a) Part through crack ($a_o/s < 1$)

The final crack length $2c_o$ must be less than the critical through wall crack length $2c_{crit}$, that means $2c_o < 2c_{crit}$. For the case $2c_o \geq 2c_{crit}$ it has to be demonstrated that $2c_o$ is less than the critical crack length $2c_{krit}$* of a part through crack with crack depth a_o, that means $2c_o < 2c_{crit}*(a_o/s)$ or it has to be demonstrated that the critical moment for the initial crack shape (a_a and $2c_a$) is higher than the moment corresponding to the critical through wall crack length $2c_{crit}$, that means M_{crit} (a_a, $2c_a$)/ M_{crit} (a/s=1) > 1.

b) Through wall crack ($a_o/s = 1$)

The final crack length $2c_o$ must be less than the critical through wall crack length $2c_{crit}$, that means $2c_o < 2c_{crit}$.

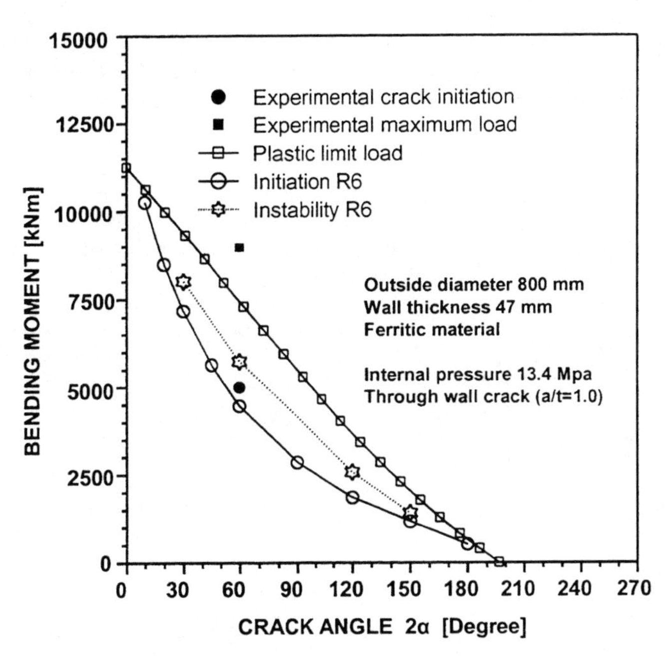

Fig. 7 Critical crack sizes R6-method and limit load calculations (σ_{fl}=[$R_{p0,2}$ +R_m]/2) for pipes with nominal diameter 800 mm

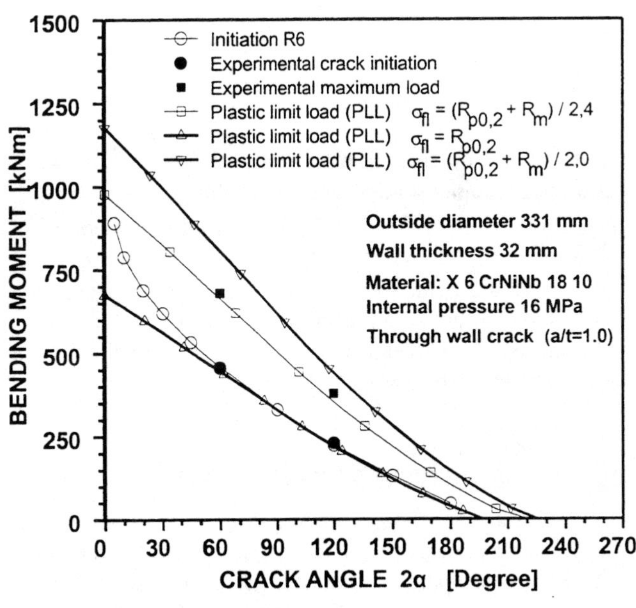

Fig. 8 Critical crack sizes R6-method and limit load calculations for pipes with nominal diameter 300 mm

Step 5: Calculation of leak area A_O

The leak area A_o has to be calculated for the critical through wall crack length $2c_{crit}$ (Step 1) and the crack opening (COD) for loads under normal operating conditions.

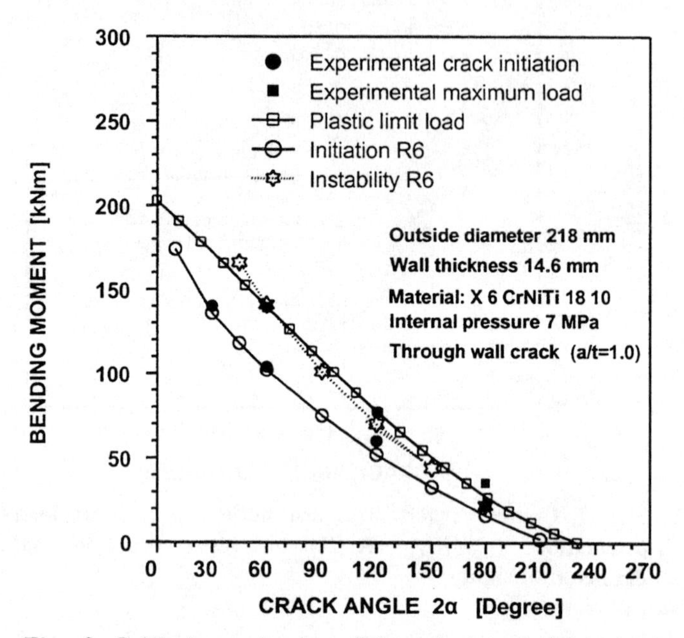

Fig. 9 Critical crack sizes R6-method and limit load calculations (σ_{fl}=[$R_{p0,2}$ +R_m]/2.4) for pipes with nominal diameter 200 mm

Step 6: Calculation of leak rate $\dot{m}_o$

It has to be demonstrated that for loads under normal operating conditions the leak rate becomes $\dot{m}_o /S > \dot{m}_{LÜS}$ (detectable leak rate by leak rate monitoring system) with a safety factor S. The safety factor S has to be determined in a component and plant specific way. The concept for recurrent inspections has to be adjusted to the results of the leak rate calculations. The leak rate shall be calculated based on a model verified by experimental data.

Recurrent inspections and in-service monitoring

By recurrent inspections and in-service monitoring it has to guarantee that assumptions of design, especially the loading conditions (mechanical, thermal, corrosive) do not change during operation. The requirements of KTA safety standard 3201.4 must be kept. In particular the in-service monitoring must detect the relevant local and global loading. The status of quality after fabrication ("Basis Safety") must be guaranteed for the succeeding operation. Therefore in-service monitoring is of decisive importance.

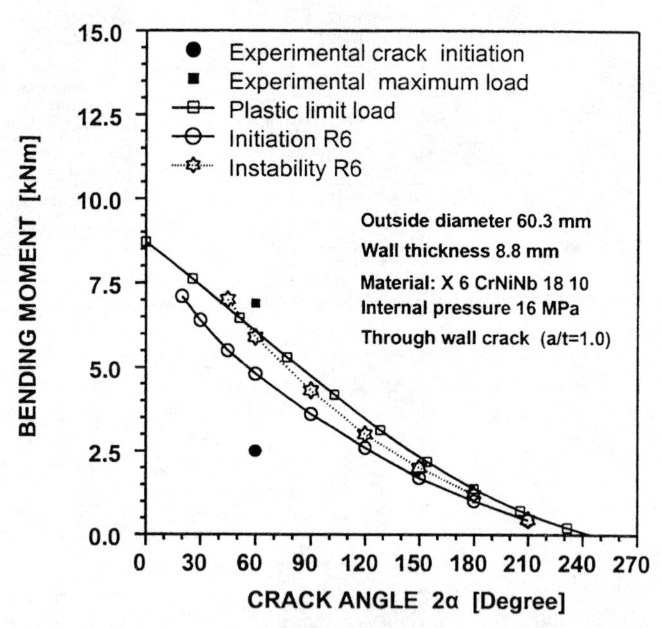

Fig. 10 Critical crack sizes R6-method and limit load calculations (σ_{fl}=[$R_{p0,2}$ +R_m]/2.4) for pipes with nominal diameter 50 mm

Evaluation of the integrity

The evaluation of the integrity is based on the results of the procedure shown and thereby the following aspects are of importance: (a) monitoring of the variables responsible by sufficient detection of the causes of possible operational failure mechanisms (protective provisions, avoidance or control of the causes); (b) well-timed detection of operational failure mechanisms by appropriate examination methods, examination areas and examination intervals; (c) use of the state-of-the-art and the operational experience.

(2) Method of proof for components under operation

Evaluation of the as-built status of quality

If the requirements of the "Basis Safety" are fulfilled, the procedure is following the method of proof for components under construction. If the requirements are not fulfilled documents must be checked with respect to design, material and fabrication (geometry, weldment, supporting, snubbers, function of active components) and the possible failure mechanisms (material, design, processing boundary conditions, mode of operation, results of recurrent non destructive examinations, operational experience).

As a result of the evaluation of the as-built status of quality areas for additional measures concerning the loading conditions, the in-service monitoring and the recurrent non-destructive examinations can be determined.

Relevant loading conditions

It differentiates between two cases: (a) In the previous operation the relevant loading is known by in-service monitoring, the upset conditions are lay down in the specifications of the design; (b) Relevant loading conditions are only available in the specifications of the design.

Determination of stresses and limitation

The procedure is following the one described in chapter "method of proof for components under construction" (step 1 up to step 6).

Recurrent inspections and in-service monitoring

Going beyond the procedure described in chapter "method of proof for components under construction", additional measures for recurrent inspections and in-service monitoring are necessary, especially for the case where in the previous operation no in-service monitoring was installed.

Evaluation of the integrity

Going beyond the procedure described in chapter "method of proof for components under construction" for the succeeding operation additional measures are indispensable. It must be guaranteed by recurrent non-destructive examinations and in-service monitoring that during operation no failure mechanism will occur.

CONCLUSIONS

Based on the German Basis Safety Concept a general concept to ensure the integrity of pressurised components is developed. The concept can be applied to components under construction as well as to components already in operation. The calculation methods and fracture mechanics approaches are verified by numerous experimental data. The main points are on the one hand to demonstrate the actual as-built status of quality and on the other hand in-service monitoring and recurrent non-destructive examinations to guarantee the ongoing operation of the plants.

REFERENCES

1. K. Kussmaul and D. Blind; "Basis Safety – A Challenge to Nuclear Technology," IAEA Spec. Meeting, Madrid, March 5.-8. 1979; Ed. in *Trends in Reactor Pressure Vessel and Circuit Development* by R.W. Nichols 1979, Applied Science Publishers LTD, Barking Essex, England

2. K. Kussmaul; "German Basis Safety Concept Rules Out Possibility of Catastrophic Failure," *Nuclear Engineering International* 12 (1984), 41-46

3. G.M. Wilkowski, R. J. Olson and P. M. Scott; "State-of-the-Art Report on Piping Fracture Mechanics," NUREG/CR-6540, BMI-2196, January 1998

4. K. Kussmaul, D. Blind, E. Roos and D. Sturm; "The leak before break behaviour of pipes," VGB Kraftwerks-technik 70, No. 7, July 1990, 465-477

5. G. Bartholomé, E. Bazant, R. Wellein, W. Stadtmüller and D. Sturm; "German experimental programs and results," *Spec. Meeting on LBB*, Lyon, France October 1995

6. H. Schulz; "Current position and actual licensing, decisions on LBB in the Federal Republic of Germany," CSNI Report No. 82, NUREG/CP-0050, Monterey, September 1983

7. Guidelines of the German Reactor Safety Commission (RSK-Guidelines) for Pressurised Water Reactors; 3rd Edition, October 14, 1981; Revised Section 21.1 of 03/1984 and Section 21.2 of 12/1982; associated general specification "Basis Safety of Pressurised Components; Editor: Gesellschaft für Reaktorsicherheit (GRS)

8. K. Kussmaul et. al.; "Exclusion of fracture in piping of pressure boundary, Part 1: Experimental investigations and their interpretation"; *Int. Symposium on reliability of reactor pressure components*, IAEA-SM-269/7, Stuttgart, March 1983

9. G. Bartholomé et. al.; "Exclusion of fracture in piping of pressure boundary, Part 2: Application to the primary coolant piping"; *Int. Symposium on reliability of reactor pressure components*, IAEA-SM-269/7, Stuttgart, March 1983

10. "Safety Criteria for Nuclear Power Plants" by the Minister of the Interior (BMI); notification of October 21, 1977

11. "Safety Standards" of the Nuclear Safety Standards Commission (KTA); KTA 3201.2 Components of the Reactor Coolant Pressure Boundary of Light Water Reactors, Part 2: Design and Analysis, 06/1996

12. R. Wellein and G. Preusser; "State of Engineering of the Leak-Before-Break Procedure in Germany"; *7th German-Japanese Joint Seminar*, September 29-30, 1997, MPA Stuttgart

PVP-Vol. 410-1, Assessment Methodologies for Preventing Failure:
Deterministic and Probabilistic Aspects and Weld Residual Stress
Volume 1
ASME 2000

DEVELOPMENT OF AN ENGINEERING APPROACH FOR DUCTILE FRACTURE ASSESSMENT OF STRAIGHT PIPES

Suneel K. Gupta, V. Bhasin, K. K. Vaze and H. S. Kushwaha
Bhabha Atomic Research Centre
Reactor Safety Division
Trombay, Mumbai-400 085
India
Fax No.: (091)-(22)-5505151, 5519613
e-mail: kushwaha@magnum.barc.ernet.in

ABSTRACT

This paper is concerned with the development of an approximate formula for calculation of unstable tearing load for straight pipes with through-wall circumferential crack under bending moment. The derivation is based on GE/EPRI estimation scheme and R-6 method. The final formula depends upon geometry of pipe, crack size, material property, load magnitude and crack resistance curve in the form of power law. This formula will be helpful in quick assessment of ductile fracture in straight pipes, especially at design stage. The formula has been extensively validated against standard GE/EPRI scheme. The parameters studied, covered the wide range of values expected for power plant piping systems. Several cases have also been checked against the Finite Element analysis results.

INTRODUCTION

Nuclear power plant (NPP) piping is normally made of ductile materials. Their rigorous fracture assessment is essential for demonstration of leak-before-break (LBB) capability and safety margin against fracture failure, under safe shutdown earthquake (SSE) loading. The dominant failure mode of high energy NPP piping, is either plastic collapse or ductile fracture. Hence, the resisting moment is taken as minimum of ductile fracture / unstable tearing moment (M_u) or plastic collapse moment (M_L), that is,

$$\text{Resisting moment} = \text{minimum} < M_u, M_L > \qquad (1.1)$$

In order to assess the possibility of ductile fracture, one has to perform detailed fracture calculations of pipes with postulated flaws/cracks. The calculation involves evaluation of crack-driving force or applied J (or J_{app}), its derivative with respect to crack size and knowledge of relevant material data. The material properties, required for assessment, are stress-strain curve, yield strength (σ_{ys}), ultimate tensile strength (σ_{uts}), material J-Resistance (J-R or J-Δa) curve etc. Seamless pipes have circumferential girth welds and will experience severe stresses during SSE event, therefore only circumferential cracks are postulated.

The leak-before-break (LBB) methodology is now accepted design procedure for eliminating the postulation of double-ended guillotine breaks (DEGB) in high energy piping systems. For assurances of LBB capability, the through-wall circumferential (TWC) cracks are postulated, which are of size equal to leakage size crack (LSC). The required margin against ductile fracture shall be assured against the severe design basis accident (DBA) loads. Ductile fracture assessment essentially requires evaluation of unstable tearing loads.

The necessary parameters required for ductile fracture assessment such as J_{app}, its derivative etc., can be evaluated using finite element method (FEM). However, the computational cost and time required is prohibitive to its usage at design stage, when the basic pipe parameters are at evolving stage. As an alternative, simplified estimation schemes can be used. A number of such schemes are available, to perform the required calculations. These are GE/EPRI scheme, [1] and [2], R-6 method, [3] and [4], A-16 method, [5] etc. Although usage of these schemes leads to quick assessment however, they do require a set of ductile tearing calculations before the unstable tearing load/moment (UTL) can be obtained.

The aim of this study was to develop a formula, which can yield the value of UTL straight away and involves least amount of calculations. The proposed formula has been obtained by using GE/EPRI scheme and R-6 method.

In this paper, initially the procedure for J-tearing calculation is discussed using GE/EPRI scheme. Subsequently the derivation of approximate formula is discussed.

J-TEARING EVALUATION USING GE/EPRI SCHEME

The J-tearing evaluation procedure is schematically illustrated in Fig. 2, for straight pipes with through wall circumferential cracks. The cracked pipe is subjected to bending moment (M), Fig. 1. The crack length is 2a (a = Rθ), where R is the mean radius and 2θ is the angle subtended, by crack, at the center.

First the applied J (or J_{app}) versus applied moment (M) is determined for the given leakage size crack (LSC), as shown in Fig. 2. Then, applied J versus applied tearing modulus (T_{app}) and material-J

(J_{mat}) versus material tearing modulus (T_{mat}) curves are also plotted, as shown in Fig. 2. The material J-T curve is generally obtained from C-T specimen test. Since, in small C-T specimens, the tearing is limited to 5-6mm hence, the information is available up to a certain J_{mat} value, called J_{max}. Beyond this point, the material J-T curve is often extrapolated as straight line with slope equal to the slope of J-T curve at $J = J_{max}$. This extrapolation is known to be conservative.

The equations, necessary for determination of critical load based on J-Tearing scheme, will be discussed now. In the GE/EPRI scheme applied J is represented as:

$$J = J_e + J_p \tag{2.1}$$

Where,

J_e = Elastic part of applied J
J_p = Plastic part of applied J

If the crack tip plasticity correction is neglected then, J_e can be expressed as, [6]:

$$J_e = \frac{K^2}{E} = f_b^2\left(\frac{\theta}{\pi}\right)\left(\frac{M^2}{ER^3t^2}\right) \tag{2.2}$$

Where,

K = Stress intensity factor (SIF).
θ = Half the angle subtended by TWC Crack, Fig. 1.
M = Applied bending moment.
E = Young's modulus of elasticity.
R = Mean radius.
t = Nominal pipe thickness.
f_b = Crack geometry factor.

The geometry factor, f_b is given as, [6]:

$$f_b = 1 + A\left[4.5967\left(\frac{\theta}{\pi}\right)^{1.5} + 2.6422\left(\frac{\theta}{\pi}\right)^{4.24}\right] \tag{2.3}$$

and A is given as, [6]:

$$A = \left(0.125\frac{R}{t} - 0.25\right)^{0.25} \quad \text{for} \quad 5 \le \frac{R}{t} < 10$$

$$A = \left(0.400\frac{R}{t} - 3.00\right)^{0.25} \quad \text{for} \quad 10 \le \frac{R}{t} \le 20 \tag{2.4}$$

The plastic part of the J, i.e. , J_p, is expressed as, [8]:

$$J_p = \frac{\alpha\sigma_o^2}{E}R\theta\left(1 - \frac{\theta}{\pi}\right)h\left(\frac{\theta}{\pi}, n, \frac{R}{t}\right)\left[\frac{M}{M_o}\right]^{n+1} \tag{2.5}$$

Here it is assumed that stress-strain curve can be expressed as Ramberg-Osgood (R-O) curve fit, as given below:

$$\frac{\varepsilon}{\varepsilon_o} = \frac{\sigma}{\sigma_o} + \alpha\left(\frac{\sigma}{\sigma_o}\right)^n \tag{2.6}$$

Where,

α and n = R-O curve fit parameters.
σ_o = Reference stress = σ_{ys} = Yield strength.
ε_o = Reference strain = σ_o/E
σ = Stress corresponding to strain ε
M_o = Reference Limit Moment based on σ_o

Here M_o is expressed as, [6]:

$$M_o = 4\sigma_o R^2 t[\cos(0.5\theta) - 0.5\sin(\theta)] \tag{2.7}$$

The h(θ/π, n, R/t) parameter, in Eq. (2.5), is a non-dimensional influence parameter which in turn is a function of crack size (θ/π), material strain hardening coefficient (n) and pipe geometry (R/t). Based on finite element (FE) analysis results, Kumar, German et. al., [1] and [2], have obtained the value of h as the tuning parameter. The h-factor values for different values of R/t, n, θ/π have been tabulated in EPRI-NP-1931, [1] and EPRI-NP-3607, [2], reports. These values have also been presented by Zahoor, [6], in Ductile Fracture Handbook. Rahman, [8], has also given h-factor but in the form of polynomial equation. This equation was obtained by fitting the FE analysis results of Brust et al., [7], and is reproduced below:

$$h\left(\frac{\theta}{\pi}, n, \frac{R}{t}\right) = \left[1 \quad \left(\frac{\theta}{\pi}\right) \quad \left(\frac{\theta}{\pi}\right)^2 \quad \left(\frac{\theta}{\pi}\right)^3\right][C_{ij}]_{4\times4}\left[1 \quad n \quad n^2 \quad n^3\right]^T \tag{2.8}$$

The coefficient matrix, $[C_{ij}]_{4x4}$ in above equation, is a function of R/t and its values for R/t equal to 5, 10 and 20 have been tabulated by Rahman, [8]. The h-factor as given by Eq. (2.8) is approximate but convenient to handle analytically.

Using Eq. (2.1) through Eq. (2.8), the applied J versus M curve can be completely evaluated, for the given leakage size crack (LSC). The next step is to obtain the applied J versus T curve, which can be derived as follows:

$$T = \frac{E}{\sigma_f^2}\frac{dJ}{da} = \frac{E}{\sigma_f^2 R}\frac{dJ}{d\theta} \tag{2.9}$$

Where,

$$\sigma_f = 0.5(\sigma_{ys} + \sigma_{uts}) = \text{Flow stress}$$

Using Eq. (2.1), applied T can be further expressed as:

$$T = \frac{E}{\sigma_f^2 R}\frac{d}{d\theta}\left[J_e + J_p\right] = \frac{E}{\sigma_f^2 R}\left[\frac{dJ_e}{d\theta} + \frac{dJ_p}{d\theta}\right] \tag{2.10}$$

The above equation can be re-written as:

$$T = F\left[S_{fe}J_e + S_{fp}J_p\right] \tag{2.11}$$

The terms F, S_{fe} and S_{fp}, in Eq. (2.11), can be expressed as follows:

$$F = \frac{E}{\sigma_f^2 R\theta} \tag{2.12}$$

$$S_{fe} = \frac{\theta}{J_e}\frac{dJ_e}{d\theta} = \frac{1 + A\left[18.387\left(\frac{\theta}{\pi}\right)^{1.5} + 25.048\left(\frac{\theta}{\pi}\right)^{4.24}\right]}{1 + A\left[4.5967\left(\frac{\theta}{\pi}\right)^{1.5} + 2.6422\left(\frac{\theta}{\pi}\right)^{4.24}\right]} \tag{2.13}$$

and

$$S_{fp} = \frac{\theta}{J_p}\frac{dJ_p}{d\theta} = \frac{\pi - 2\theta}{\pi - \theta} + \frac{\theta}{h}\frac{dh}{d\theta} - (n+1)\frac{\theta}{M_o}\frac{dM_o}{d\theta} \tag{2.14}$$

From the above expressions it is clear that S_{fe} is a dimensionless factor and function of only the crack size (θ/π) and pipe geometry (R/t).

Using Eq. (2.7) and Eq. (2.8), $(\theta/h)(dh/d\theta)$ and $(n+1)(dM_o/d\theta)$ can be expressed as:

$$\frac{\theta}{h}\frac{dh}{d\theta} = \frac{\left\lfloor 0, \left(\frac{\theta}{\pi}\right), 2\left(\frac{\theta}{\pi}\right)^2, 3\left(\frac{\theta}{\pi}\right)^3 \right\rfloor \left[C_{ij}\right]_{4\times 4} \left\lfloor 1, n, n^2, n^3 \right\rfloor^T}{\left\lfloor 1, \left(\frac{\theta}{\pi}\right), \left(\frac{\theta}{\pi}\right)^2, \left(\frac{\theta}{\pi}\right)^3 \right\rfloor \left[C_{ij}\right]_{4\times 4} \left\lfloor 1, n, n^2, n^3 \right\rfloor^T} \tag{2.15}$$

$$(n+1)\frac{\theta}{M_o}\frac{dM_o}{d\theta} = -(n+1)g(\theta) \tag{2.16}$$

Where,

$$g(\theta) = \theta\left[\tan(0.5\theta) + 0.5\sec(0.5\theta)\right] \tag{2.17}$$

From Eq. (2.14) through Eq. (2.17) it can be observed that S_{fp} is independent of applied moment. Secondly, the value of T, as given by Eq. (2.11) depends upon S_{fe} and S_{fp}. Since, both S_{fe} and S_{fp} are independent of applied moment, therefore, their relative importance will be dictated by the value of J_e and J_p. At higher M/M_o values J_p dominates over J_e, therefore, only S_{fp} is important and vice versa for very low values of M/M_o. An important point to be observed is that accuracy of T depends upon accuracy of $dh/d\theta$ values. If h-factor data or its polynomial fit is based on scanty results, then large errors are likely to occur in $dh/d\theta$ values.

J-T curve can be evaluated using Eq. (2.11) through Eq. (2.17). The J_{mat} versus T_{mat} curve is obtained from material J-Δa (or J-R) curve. The typical form of material J-Δa curve is given below:

$$J_{mat} = C_1(\Delta a)^{C_2} \tag{2.18}$$

T_{mat} is defined as:

$$T_{mat} = \frac{E}{\sigma_f^2}\frac{dJ_{mat}}{d(\Delta a)} = \frac{E}{\sigma_f^2}\frac{dJ_{mat}}{da} \tag{2.19}$$

The material J-T curve can be expressed as:

$$T_{mat} = D_1(J_{mat})^{D_2} \quad \text{for } J_{mat} < J_{max} \tag{2.20a}$$

(from direct measurement)

$$J_{mat} = mT_{mat} + C \quad \text{for } J_{mat} \geq J_{max} \tag{2.20b}$$

(from extrapolation, see Fig. 2.)

The complete J-Tearing calculation involves evaluation of Eq. (2.1) through Eq. (2.20b) and then determining the point of intersection of applied and material J-T curve, as shown in Fig. 2. The J value, corresponding to this point of intersection, is designated as $J_{critical}$. The bending moment corresponding to $J_{critical}$ is ultimate tearing load/moment (UTL).

Now, based on these equations and the R-6 method, the derivation of approximate equations, for determination of UTL, would be discussed.

DEVELOPMENT OF APPROXIMATE APPLIED J VERSUS MOMENT EQUATION

The applied J can also be estimated using R-6 method, [3]. R-6 method offers three different options. Here, option-2 was selected. Ainsworth, [4], has shown the derivation of option-2 R-6, from GE/EPRI scheme. After neglecting the crack tip plasticity correction, the applied J can be expressed as:

$$J = \left[1 + \phi \alpha \left(\frac{M}{M_o} \right)^{n-1} \right] J_e \quad (3.1)$$

If ϕ factor is equal to 1.0, then:

$$J = \left[1 + \alpha \left(\frac{M}{M_o} \right)^{n-1} \right] J_e \quad (3.1a)$$

In option-2, R-6 the value of ϕ factor, of Eq. (3.1), is taken as unity as shown in Eq. (3.1a). The actual value of ϕ factor is proportional to ratio of h-function value [for given n] to h-function value [for n=1]. In general, the value of $\phi < 1.0$. Therefore, the standard R-6 method (i.e., $\phi = 1$), in general, leads to overestimation of applied J. On the other hand, the applied J as estimated by GE/EPRI scheme is more accurate, provided the material truly follows Ramberg-Osgood curve fit. Therefore, if actual value of ϕ factor were included in Eq. (3.1), then the resulting applied J values would be realistic. The Eq. (2.1), of GE/EPRI scheme, can be re-written as:

$$J = J_e + J_p = \left(1 + \frac{J_p}{J_e} \right) J_e \quad (3.2)$$

Using Eq. (2.1) through Eq. (2.7), the Eq. (3.2) can be expressed as:

$$J = \left[1 + \phi_g \alpha \left(\frac{M}{M_o} \right)^{n-1} \right] J_e \quad (3.3)$$

The ϕ_g is given by following equation:

$$\phi_g = \frac{(\pi - \theta)h}{\left[4 f_b (\cos(0.5\theta) - 0.5\sin(\theta)) \right]^2} \quad (3.4)$$

Now Eq. (3.1) and Eq. (3.3) are quite similar. In order to remove the dependency of ϕ_g on h-function, extensive numerical analysis was done and finally an approximate expression for ϕ is suggested. The resulting expression is simple but slightly conservative and is given in Eq. (3.5). The value of ϕ factor required in Eq. (3.1), can be evaluated using Eq. (3.5).

$$\phi_{approx} = \left[1.4 \left(\frac{\theta}{\pi} \right)^{0.14} \right]^{1-n} - 0.2(1-n) \left(\frac{\theta}{\pi} \right)^2 + 0.001 \left(\frac{R}{t} \right)^2 \quad (3.5)$$

This approximate expression for ϕ factor is based on the wide range of parameters. The parameters studied are as follows:

R/t: from 5 to 20; θ/π : from 0.1 to 0.4 and n : from 2 to 8

These parameters cover all the possible, pipe sizes and material likely to be encountered in design. If this ϕ_{approx} factor is used in standard R-6 method, then applied J can be more realistically evaluated as:

$$J = \left[1 + \phi_{approx} \alpha \left(\frac{M}{M_o} \right)^{n-1} \right] J_e \quad (3.6)$$

The Eq. (3.6) has been validated rigorously against applied J as estimated using GE/EPRI scheme. About 8000 cases were studied which covered parameter values R/t from 5 to 20, n from 2 to 8, α from 2 to 10, θ/π from 0.1 to 0.4 and M/M_o from 0.1 to 1.5. The results are shown in Fig. 3(a) through Fig. 3(d), corresponding to θ/π = 0.1, 0.2, 0.3 and 0.4, respectively. In these figures, the applied J as evaluated from Eq. (3.6) has been plotted against applied J evaluated using GE/EPRI scheme, that is, Eq. (2.1). It can be seen that out of all the 8000 cases most of the points lie on or around 45° line. However, in few cases, there is a slight scatter, but generally on conservative side.

The Eq. (3.6) has also been validated against several finite element results, given in reference [9] and [10]. This reference gives finite element results of straight pipe of primary heat transport (PHT) system, belonging to Darlington Nuclear Generation System (NGS 'A'), Canada and Tarapur Atomic Power Project (TAPP) 500 MWe. The results are shown in Fig. 4, for steam generator (SG) pipe, pump suction (PS) pipe and pump discharge (PD) pipe. The internal pressure has been considered to be zero in all the above problem. For each of the pipe, the results are compared for two crack sizes, which are equal to 1xLSC and 2xLSC. The Darlinton NGS pipes are made of SA-106 Gr. B material and TAPP piping is made of SA-333 Gr.6 material. Their sizes and material properties have been taken from the reference, [9] and [10]. It is observed that comparison with finite element results is also quite close.

DEVELOPMENT OF APPROXIMATE APPLIED J-T EQUATION

The applied T, as per GE/EPRI scheme, is given by Eq. (2.11):

$$T = F\left[S_{fe} J_e + S_{fp} J_p \right] \quad (2.11)$$

From the basic definition of applied T and standard form of R-6, we get:

$$T = \frac{E}{\sigma_f^2 R} \frac{dJ}{d\theta}$$

In R-6 method $\phi = 1$, then differentiating Eq. (3.1) with respect to θ we get:

$$\frac{dJ}{d\theta} = \frac{d}{d\theta}\left[\left\{1 + \alpha\left(L_r\right)^{n-1}\right\}J_e\right]$$

or

$$\frac{dJ}{d\theta} = \left[1 + \alpha\left(L_r\right)^{n-1}\right]\frac{dJ_e}{d\theta} + (n-1)J_e\alpha\left(L_r\right)^{n-2}\frac{dL_r}{d\theta} \quad (4.1)$$

In above expressions the 'Lr' is the ratio of applied moment to limit moment i.e., (M/M_o). After simplification, the expression for tearing modulus, as per R-6 method, can be expressed as:

$$T = F_1 J = F\left[S_{fe} + (n-1)g(\theta)\right]J \quad (4.2)$$

In the above equation expressions for Sf_e and $g(\theta)$ is given by Eq. (2.13) and Eq. (2.17).

Comparing the Eq. (2.11) and Eq. (4.2) and numerically approximating the functions S_{fe} and $g(\theta)$, a simplified equation is proposed for applied J-T line slope calculation. The equation is given as:

$$T_{approx} = F_1 J = F\left[S_{fe\text{-}approx} + (n-1)\frac{\theta}{2}\left(1 + 4\frac{\theta}{\pi}\right) - 0.1n\right]J \quad (4.3)$$

Where,

$$S_{fe\text{-}approx} = \left(1 + 5\frac{\theta}{\pi}\right)\left(0.85 + 0.01\frac{R}{t}\right) \quad (4.4)$$

and

$$F = \frac{E}{\sigma_f^2 R\theta} \quad (4.5)$$

The Eq. (4.3) can also be expressed as:

$$T = [E/(\sigma_f^2 R\theta)] \times [\text{Correction factor}] \times J \quad (4.6)$$

Where, the correction factor is function of θ/π, n and R/t.

The Eq. (4.3) and Eq. (4.4) have been derived after extensive numerical evaluation on the wide range of parameters. The values of which was varied as,: R/t from 5 to 20, n from 2 to 8, θ/π from 0.1 to 0.4. Hence, it covers all the possible pipes sizes, materials and crack sizes, likely to be encountered in practice.

The applied J-T slope values, as evaluated from Eq. (4.3), have been compared against those obtained using GE/EPRI scheme. About 110 cases were studied, which covered the above mentioned parameter ranges. While evaluating the applied J-T slopes using GE/EPRI scheme, it was assumed that J versus T curve is a straight line, the justification of which was discussed above. The results are shown in Fig. 5(a) through Fig. 5(d), for R/t = 5, 10, 15 and 20, respectively. It is observed that out of all the 110 points, most of them lie on or around 45° line.

DEVELOPMENT OF AN EQUATION FOR UTL CALCULATION

Based on the derived expressions of Applied J-M curve and Applied J-T curves, an equation has been derived to directly evaluate the unstable tearing load/moments. From Eq. (3.5) and Eq. (3.6)

$$J = \left[1 + \phi_{approx}\,\alpha\left(\frac{M}{M_o}\right)^{n-1}\right]J_e \quad (3.6)$$

Where,

$$\phi_{approx} = \left[1.4\left(\frac{\theta}{\pi}\right)^{0.14}\right]^{1-n} - 0.2(1-n)\left(\frac{\theta}{\pi}\right)^2 + 0.001\left(\frac{R}{t}\right)^2 \quad (3.5)$$

J_e can be expressed as:

$$J_e = F_2\left(\frac{M}{M_o}\right)^2 \quad (5.1)$$

Where, F_2 is given as:

$$F_2 = f_b^2\left(\frac{\theta}{\pi}\right)\left(\frac{M_o^2}{ER^3t^2}\right) \quad (5.2)$$

At the same time, from Eq. (4.3) and Eq. (4.4), the T_{app} versus J_{app} can be expressed as:

$$T_{app} = F_1 . J_{app} \quad (5.3)$$

Where,

$$F_1 = F\left[S_{fe\text{-}approx} + (n-1)\frac{\theta}{2}\left(1 + 4\frac{\theta}{\pi}\right) - 0.1n\right] \quad (5.4)$$

In the above equation the expressions for $S_{fe\text{-}approx}$ and F is given by Eq. (4.4) and Eq. (4.5) receptively. The material J-T curve is described by Eq. (2.20). At the point of intersection of applied J-T and material J-T curve,

$$J_{app} = J_{mat} = J_{critical} = \text{Critical value of J}$$

And

$$M = M_u = \text{Unstable tearing load/moment}$$

Hence, from Eq. (2.20), Eq. (5.3) and Eq. (5.4), $J_{critical}$ can be expressed as:

$$J_{critical} = \left[\left(\frac{F_1}{D_1}\right)^{\frac{1}{D_2-1}}\right] \quad ; \text{ if } J_{critical} < J_{max} \qquad (5.5a)$$

or

$$J_{critical} = \left[\frac{C}{1-mF_1}\right] \quad ; \quad \text{if } J_{critical} > J_{max} \qquad (5.5b)$$

Also from Eq. (3.6) and Eq. (5.1), we get:

$$J_{critical} = \left[1 + \phi_{approx}\,\alpha\left(\frac{M_u}{M_o}\right)^{n-1}\right] F_2 \left(\frac{M_u}{M_o}\right)^2 \qquad (5.6)$$

Now, let us define:

$$F_3 = \frac{1}{\phi_{approx}\alpha}, \quad F_4 = \frac{J_{critical}F_3}{F_2} \quad \text{and} \quad X = \frac{M_u}{M_o}$$

In factor F_4, the $J_{critical}$ is taken from Eq. (5.5a) or Eq. (5.5b). Then Eq. (5.6) can be re-written as:

$$X^{n+1} + F_3X^2 = F_4 \qquad (5.7)$$

The solution of above equation directly gives the value of X, which is nothing but M_u/M_o. Since, M_o is known hence M_u, the unstable tearing moment can be evaluated straight away. The Eq. (5.7) is of polynomial form, with fractional powers; hence its general solution is impossible. Therefore, a recursive technique may be adopted for its solution. One of the suggested schemes is as follows:

To start with assume X = 1.0, in square term, then:

$$X_1 = [F_4 - F_3]^{\frac{1}{n+1}} \qquad (5.8)$$

In the next iteration put X = X1, in square term, then:

$$X_2 = [F_4 - F_3X_1^2]^{\frac{1}{n+1}} \qquad (5.9)$$

Repeating once again:

$$X_3 = [F_4 - F_3X_2^2]^{\frac{1}{n+1}} \qquad (5.10)$$

Above recursion may be further repeated, till insignificant change occurs in the values of X. Generally 2 to 3 iterations are sufficient. Hence,

$$M_u = X_3M_o \qquad (5.11)$$

The M_u, as evaluated from Eq. (5.7) has been rigorously compared against the values as evaluated using GE/EPRI scheme (article 3.0). The comparison is shown in Fig. 6(a) through Fig. 6(d), in which M_u from Eq. (5.7) is plotted against M_u from GE/EPRI scheme. Fig. 6(a) through Fig. 6(d) correspond to R/t = 5, 10, 15 and 20, respectively. About 1100 cases were studied covering the wide range of parameters. The values of parameters investigated are – R/t from 5 to 20, n from 2 to 8, α from 2 to 10 and θ/π from 0.1 to 0.4. It is observed that results compare reasonably well. The error analysis of data indicated that maximum error in M_u, as evaluated using Eq. (5.7), lies between +5% to –10%. Hence, if the final value of M_u is multiplied by 0.95, then it will always be on conservative side.

The value of M_u, as evaluated using approximate equation, was also compared against the finite element results of Darlington NGS [9] and TAPP 500 MWe [10] PHT pipes. The comparison is done for the case of zero internal pressure, and is shown in Table 1.

CONCLUSIONS AND DISCUSSION

In this paper, an approximate formula for determination of unstable tearing moment, has been derived. This depends on pipe geometry, material property, crack size, loading and crack resistance curve. This formula has been derived for straight pipes with Through Wall Circumferential cracks subjected to bending moment. Initially, derivations of approximate expressions for applied J and applied T were discussed. Using them an equation for unstable tearing moment was derived. Its solution directly gives the unstable tearing moment value. All these derived expressions were thoroughly validated against GE/EPRI scheme results. The validation covered wide range of parameters, whose values are:

Pipe geometry	R/t	: from 5 to 20
R-O Curve fit Parameter	n	: from 2 to 8
R-O Curve fit Parameter	α	: from 2 to 10
Crack size	θ/π	: from 0.1 to 0.4
Normalized Moment	M/M_o	: from 0.1 to 1.5

For validation of applied J about 8000 cases, for applied J-T curve slope about 110 cases and for unstable tearing moment about 1100 cases, were studied. The results have been presented in terms of graphs. From these results, its was observed comparison is satisfactory. The maximum error, in prediction of unstable tearing moment, lies between +5% and –10%. If the obtained value is multiplied by a factor 0.95, then it will always be on conservative side.

Some cases were also compared against published finite element results. The comparison of results was satisfactory. Hence, it is concluded that the derived approximate equation can be used for Ductile Fracture assessment without much loss of accuracy. It would be specifically helpful at design stage, where quick results are required.

REFERENCES

1. Kumar, V., German, M.D., and Shih, C.F., "An Engineering approach for Elastic-Plastic Fracture Analysis", Report No. EPRI NP-1931, Electric Power Research Institute, Palo Alto, CA, 1981.
2. Kumar, V., German, M.D., Wilkening, W.W., Andrews, W.R, de Lorenzi, H.G., and Mowbray, D.F., "Advances in Elastic-Plastic Fracture Analysis", Report No. EPRI NP-3607, Electric Power Research Institute, Palo Alto, CA, 1984.
3. Milne, I.S., Ainsworth, R.A., Dowling, A.P, and Stewart, A.T., "Assessment of the Integrity of Structures Containing Defects", Report No. : CEGB/E/H/R6-Rev.3.
4. Ainsworth, R.A., "Assessment of Defects in Structures of Strain Hardening Material", Engineering Fracture Mechanics, Vol. 19, No. 4, pp. 633-642, 1984.
5. "A15 : Guide for Defect Assessment and Leak Before Break", CEA report, 1995.
6. Zahoor, A., "Ductile Fracture Handbook, Volume I, Circumferential Through Wall Cracks", Report No. EPRI NP-6301-D, Electric Power Research Institute, Palo Alto, CA, 1989.
7. Brust, F.W., Scott, P., Rahman, S., Ghadiali, N., Kilinski, T., Francini, B., Marschall, C.W., Mura, N., Krishnaswamy, P. and Wilkowski, G.M., "Assessment of Short Through-Wall Circumferential Cracks in Pipes-Experiments and Analysis" Report No.: NUREG/CR-6235, U.S. Nuclear Regulatory Commission, Washington, D.C., 1995.
8. Rahman, S, "Probabilistic Fracture Analysis of Cracked Pipe with Circumferential Flaws", Int. J. of Pressure Vessel and Piping, Vol. 70, pp. 223-236, 1997.
9. Lin, T.C., Kozluk, M.J., Misra, A.S., Gilbert, K.R., Vanderglas, M.L., et al., "Assessment of Through-Wall Crack Tolerance of Darlington NGS "A" Large Diameter Heat Transport Piping" Report No. 89511, Rev. 0, Design and Development Division-Generation, Ontario Hydro, 1990. [Internal report]
10. Chattopadhyay, J., Dutta, B.K., Kushwaha, H.S., "Leak-Before-Break-Qualification of 500 MWe PHWR Pipes By J-integral-Tearing Modulus and Limit load method," BARC, external report, BARC/1997/E/017, 1997. [Internal report]

Table 1 : Comparison with Results From Eq. (5.7) with FE results.

Pipe Name and Crack Size	Unstable tearing moment (M_u) in kN-m			
	TAPP, Tarapur, India		Darlinton NGS 'A', Canada	
	FE Values, [10]	Eq. (5.7) Values	FE Values, [9]	Eq. (5.7) Values
SG Pipe – 1xLsC	2094	2133	1647	1735
SG Pipe – 2xLSC	1610	1551	1149	1207
PD Pipe – 1xLSC	1498	1552	2014	2034
PD Pipe – 2xLSC	1159	1135	1451	1452
PS Pipe – 1xLSC	3870	3780	2293	2401
PS Pipe – 2xLSC	3020	2799	1683	1765

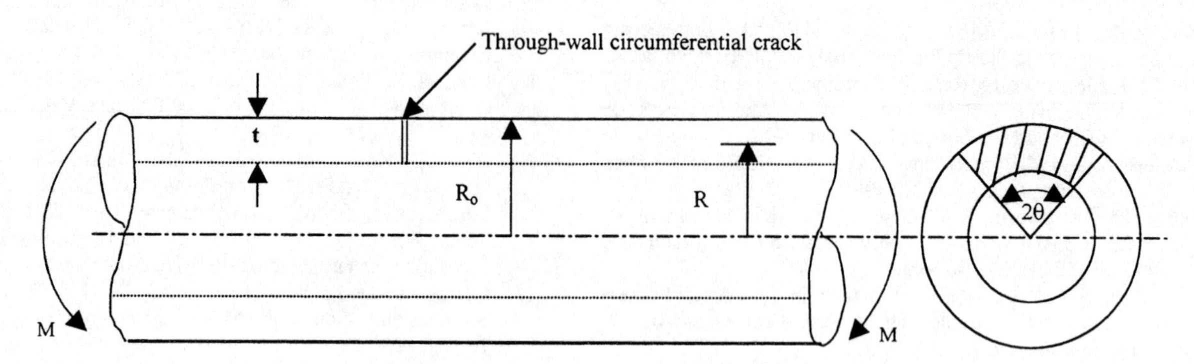

Figure 1: Straight pipe with through-wall circumferential crack

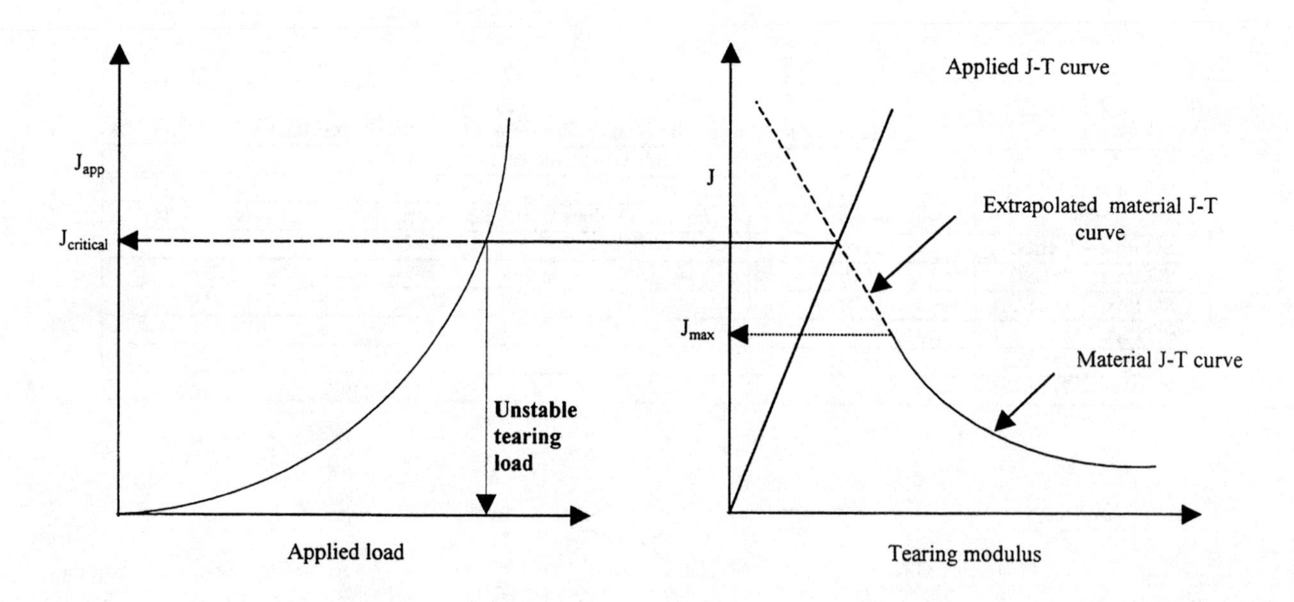

Figure 2: Schematic for determination of unstable tearing load

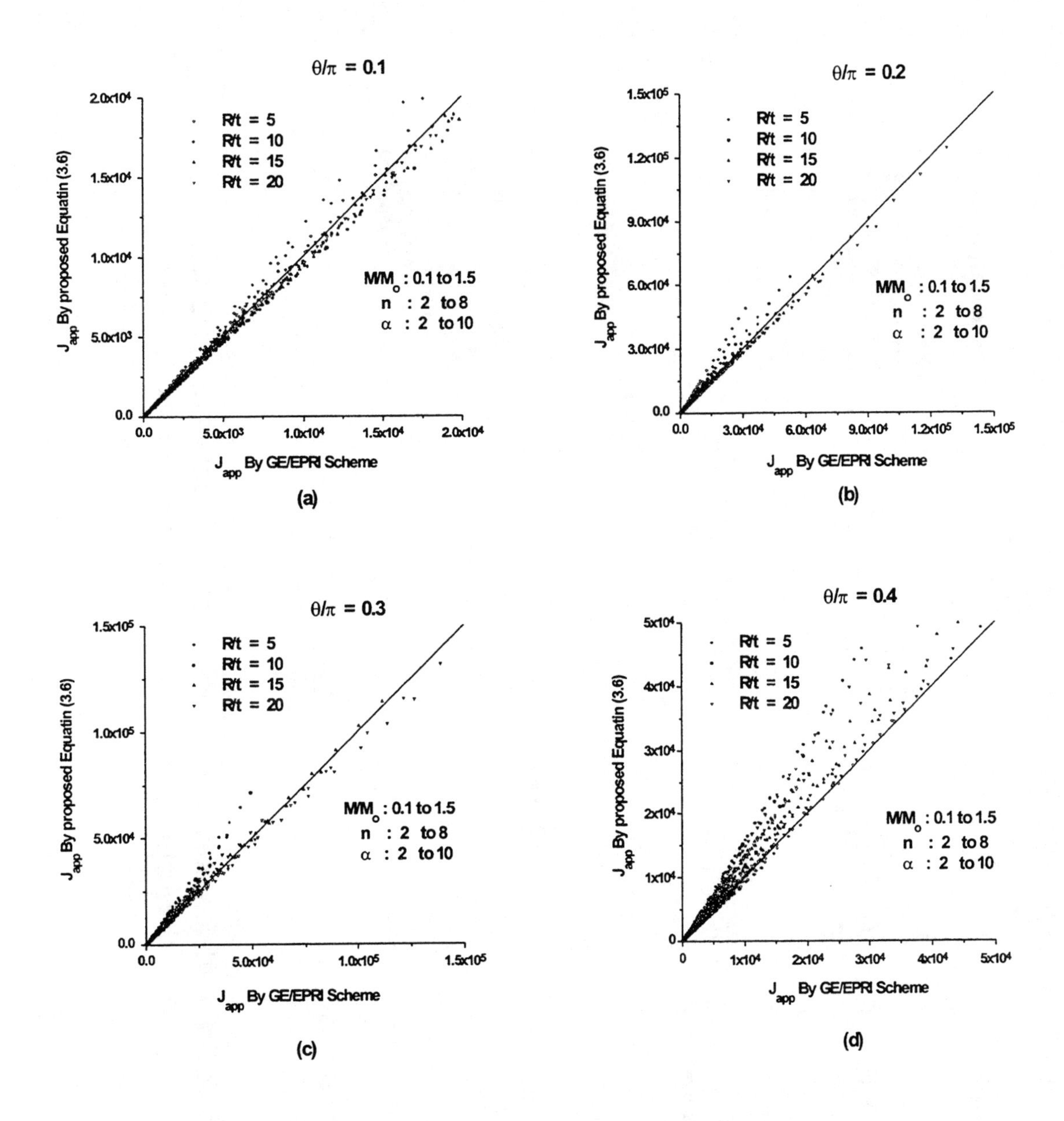

Figure 3: Comparison of applied J_{app} as evaluated using GE/EPRI scheme and proposed equation (3.6)

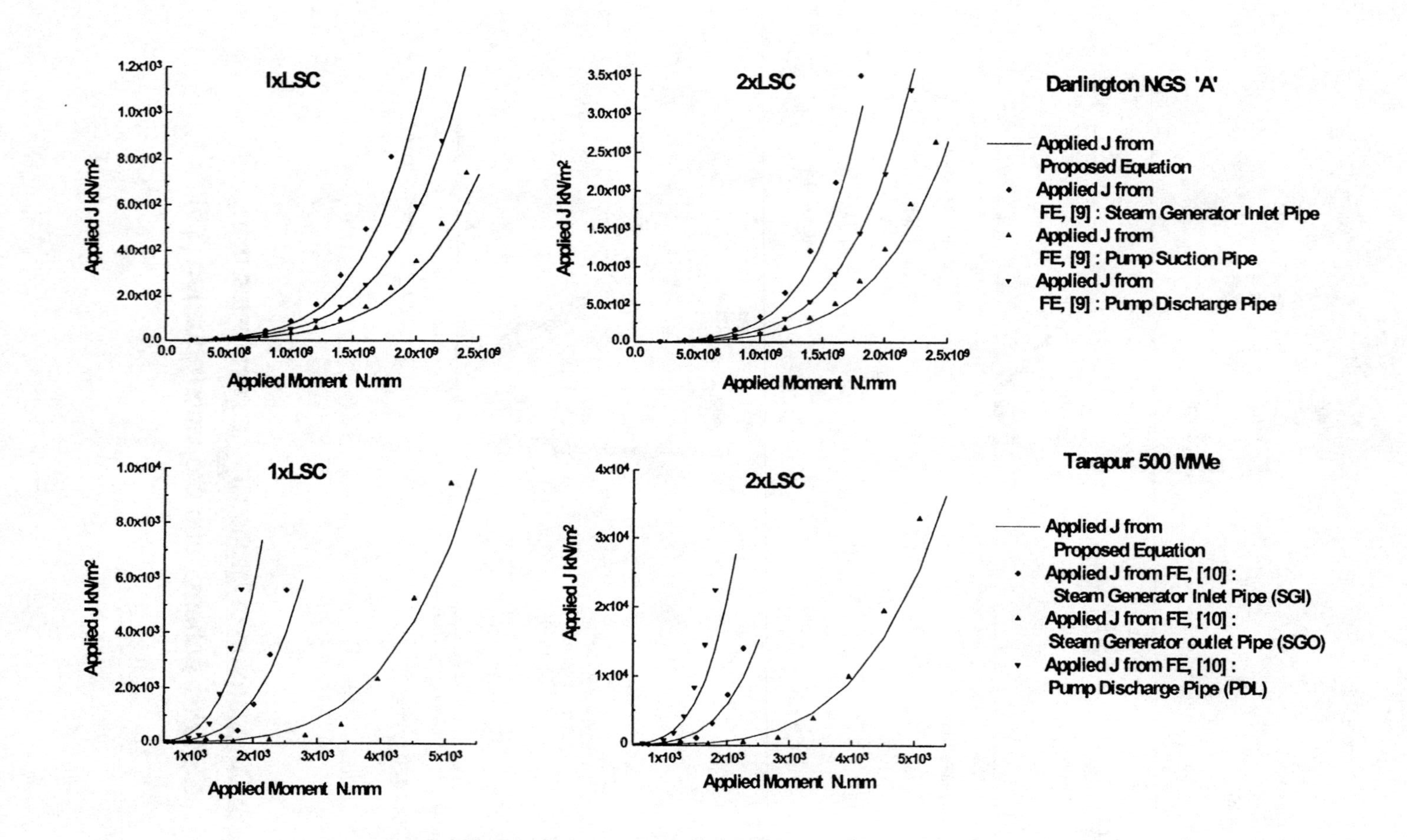

Figure 4: Comparision of applied J versus applied moment as evaluated by proposed equation (3.6) and finite element results, [9] and [10]

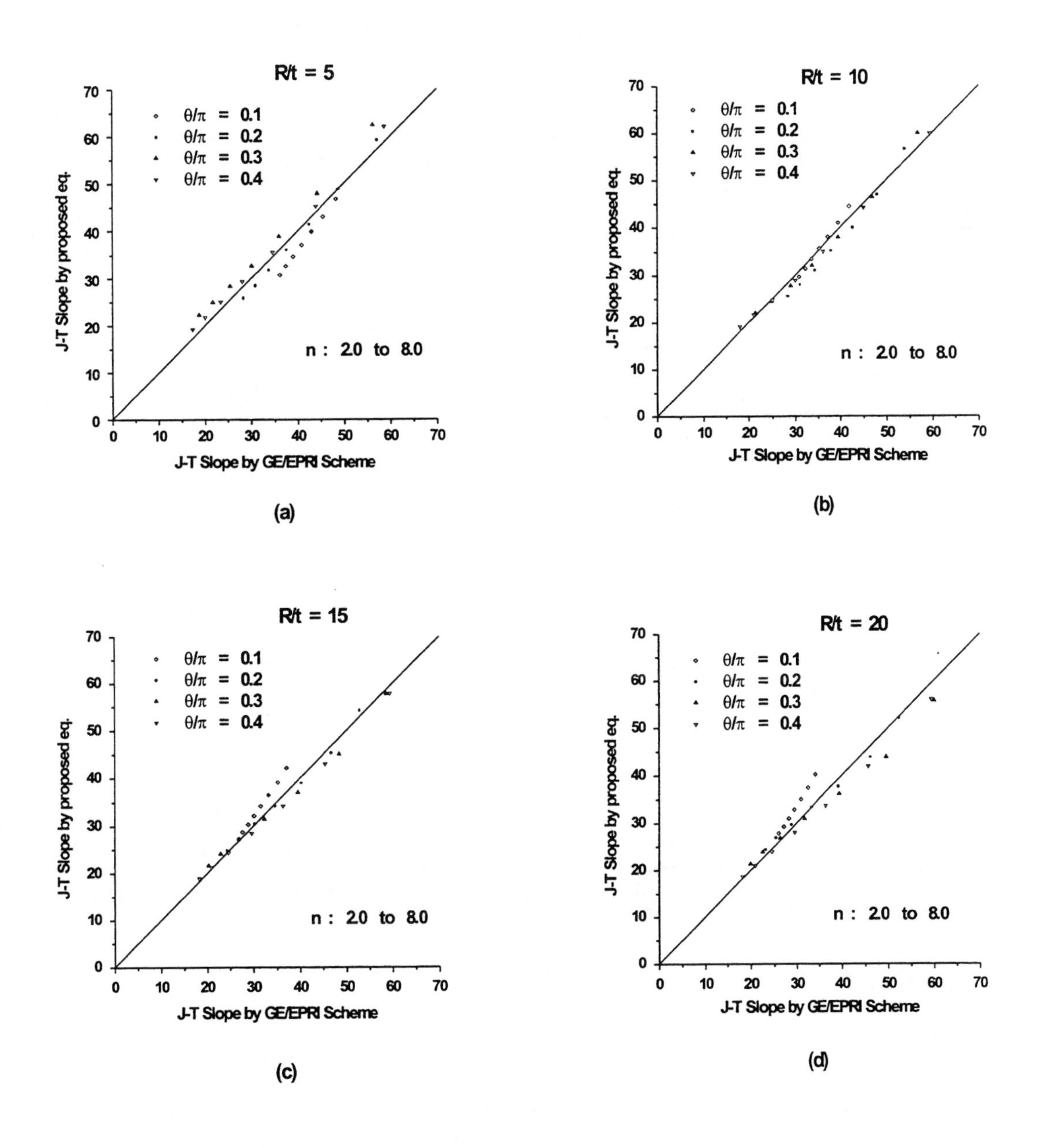

Figure 5: Comparison of Applied J-T Slope as evaluated using GE/EPRI Scheme and Proposed Equation (4.3) for slope

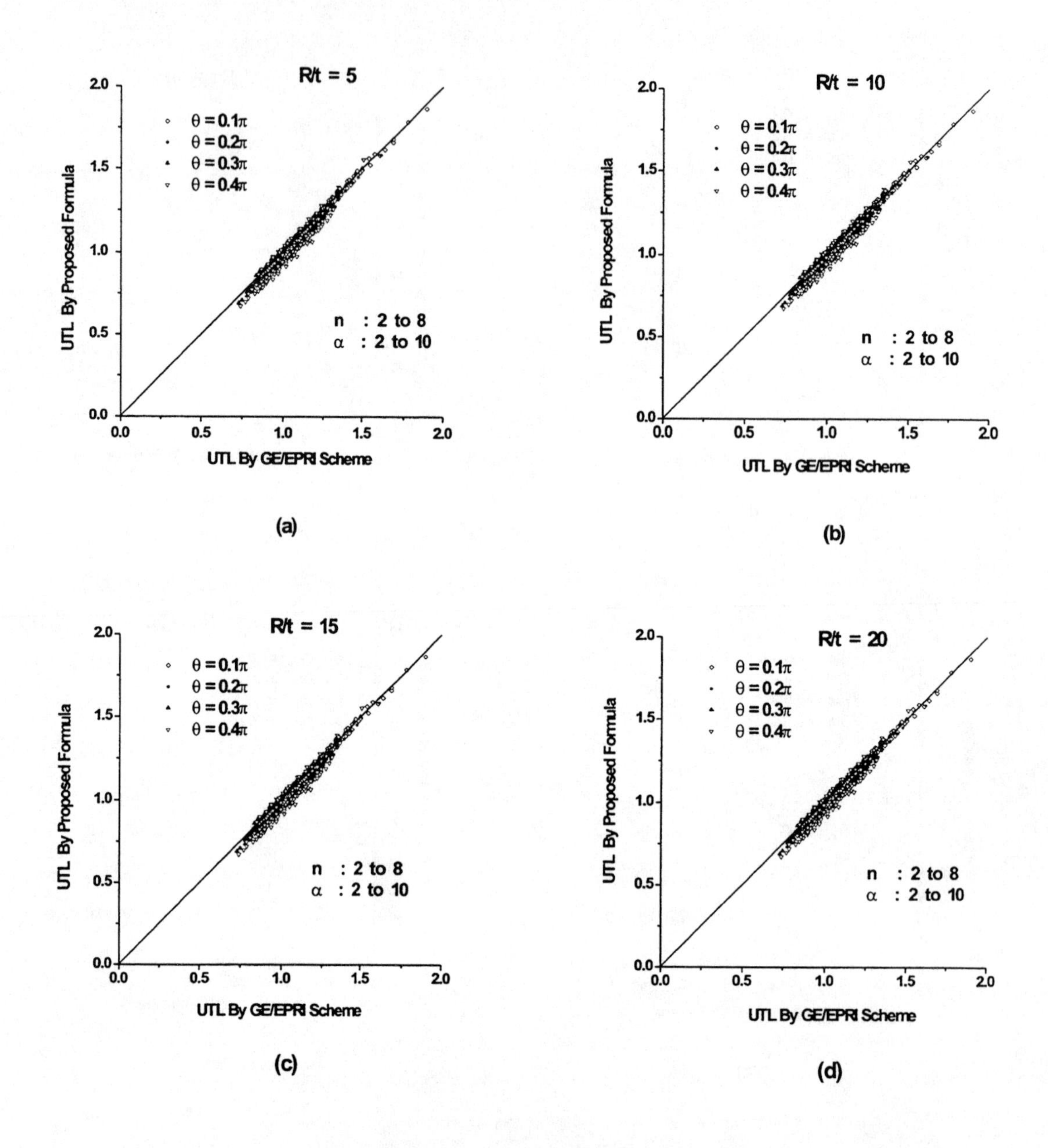

Figure 6: Comparison of unstable tearing load (UTL) as evaluated using GE/EPRI scheme and proposed equation (5.7)

PVP-Vol. 410-1, Assessment Methodologies for Preventing Failure: Deterministic and Probabilistic Aspects and Weld Residual Stress
Volume 1
ASME 2000

Neural Network Application in Failure Mode Recognition of the Flawed Pipelines

Lianghao Han Yu Zhou
CAD, Institute of Nuclear Energy Technology, Tsinghua University, Beijing 100084, P.R.C.

Liangjie Han, Cengdian Liu
East-China University of Sci.&Techn. Shanghai 200237, P.R.C.

ABSTRACT: Artificial neural networks (ANN) are currently enjoying tremendous popularity across many disciplines. ANNs are massively parallel, densely interconnect systems of simple processing elements (neurons) that are based on known and postulated properties of biological information processing systems. In the past few years, neural network technique has been very proficient in pattern recognition. The failure modes of corroded pipes can also be regarded as a problem of pattern recognition. The two kinds of failure modes of pressurized pipes containing defect are leak and rupture. In order to prevent catastrophic accidents, it is very important to predict failure modes of structures timely and correctly in engineering. In this paper, an artificial neural network (ANN) is used for representing the complex nonlinear mapping relationship between defect dimensions and failure modes of pressurized cylinders containing defects. The neural network is trained based on pressure tests of pressurized cylinders, then the trained neural network is applied to predict failure modes of flawed cylinders. At the same time, the influence of the hidden layer unit on the accuracy of prediction is discussed. The results show that ANN based pattern recognition method can be used to identify failure modes of flawed structures.

1. INTRODUCTION

Artificial neural networks (ANN) are currently enjoying tremendous popularity across many disciplines. ANNs are massively parallel, densely interconnect systems of simple processing elements (neurons) that are based on known and postulated properties of biological information processing systems. Pattern classifiers based on neural network models have been proposed and demonstrated [1,2]. In the past few years, neural network technique has been very proficient in pattern recognition. The advantages most often stated in favor of a neural network approach to pattern recognition are that: (1) it requires less input of knowledge about problem than other approaches, (2) it is capable of implementing more complex partition of feature space, and (3) it is amenable to high-performance parallel processing implementations. The failure modes of corroded pipes are also regarded as the problem of pattern recognition.

Two kinds of failure modes of pressurized pipes containing defect are leak and rupture. In order to prevent catastrophic accidents, predicting the failure mode in a timely and correct manner is very important. In this paper, the artificial neural network is introduced to identify the failure modes of pressurized pipes with defects.

2. NEURAL NETWORK BASED PATTERN RECOGNITION

Recent developments in the field of ANN have provided potential alternatives to the traditional techniques of pattern recognition, ANNs are capable of many functions, among them optimization, clustering, mapping and classification. ANNs have been applied largely to solve pattern recognition problem. Their popularity has been enhanced by the fact that they are relatively easy to use in principle, and that they have been rather successful.

The neural network classifier used in this study is a multi-layer feed forward ANN, which is typically called a "multi-layer perceptron" [3,4]. The network consists of three types of layers, each of which is composed of various number of units, units in adjacent layers are connected through links whose associated weights determine the contribution of units on one end to the overall activation of units on the other end. Units in the input layer bear much resemblance to the sensory units in a classical perceptron. Each of them is connected to a component in the input vector. The output units, analogous to the response units in a perceptron, represent different classes of patterns. Arbitrarily many hidden layers may be depending on the desired complexity. Each unit in the hidden layer is connected to every unit in the layer immediately above and below.

2.1 NETWORK ARCHITECTURE

Figure 1 displays a three-layer feed forward neural network, which has been used to identify the failure mode of pressurized cylinder. It consists of an input layer to receive the feature parameters, one hidden layer to internalize a representation of relationships, and an output layer to produce the failure mode. All connections were weighed, and each neuron received a net signal that was the linear weighted sum of all its inputs. We trained the above network with the back-propagation algorithm (BP) discussed by Han[5]. If the network is "over-trained", then its generalization ability becomes poor whereas if it is "under-trained", its recognition accuracy falls. Therefore the network was required to be trained to an "optimal accuracy range".

2.2 B-P ALGORITHM

The error back-propagation algorithm (B-P algorithm), a well-known learning algorithm, is selected as the learning algorithm for the neural network in this paper. When the neural network undergoes training, the errors between the results of the output neurons and the corresponding target values are propagated backward through the neural network. The backward propagation of error signals is used to update the connection weights in which the "information content" of the neural network is embodied. To the G-th sample, its error function is as follows:

$$E_G = \frac{1}{2}\sum_{k=0}^{P-1}(O_k - V_k)^2 \qquad (1)$$

Where E_G is the difference between the actual output V_k and the target output O_k of the G-th sample. Then we get error function of all samples:

$$E = \sum_G E_G \qquad (2)$$

In which E is the total error of the whole samples.

The B-P algorithm is described as follows:

Step 1. Initialize all θ_j, w_{ji}, θ_k and w_{kj} to random values.

Step 2. Standardize and input learning samples - the learning set (it includes neural network input data and target output data).

Step 3. Calculate the output of every neuron of the neural network by turns from input layer to output layer according to Eqs. (3) and (4).

$$U_j(n) = f\left[\sum_{i=0}^{N-1} W_{ji}(n)X_i(n) - \theta_j(n)\right], \quad j=0,1,...,Q-1 \qquad (3)$$

$$V_k(n) = f\left[\sum_{j=0}^{Q-1} W_{kj}(n)U_j(n) - \theta_k(n)\right], \quad k=0,1,...,P-1 \qquad (4)$$

Where $x_i(n)$, $U_j(n)$ and $v_k(n)$ is the output information of the i-th neuron of input layer, the j-th neuron of hidden layer and the k-th neuron of output layer respectively; w_{ji} is the linking strength between i-th neuron of input layer and j-th neuron of hidden layer; w_{kj} is the linking strength between j-th neuron of hidden layer and k-th neuron of output layer; θ_j and θ_k is the threshold of j-th neuron of hidden layer and k-th neuron of output layer respectively.

Step 4. Calculate Δw_{kj}, $\Delta\theta_k$, Δw_{ji} and $\Delta\theta_j$ and update the w_{kj}, θ_k, w_{ji} and θ_j from the output layer to hidden layer according to Eqs. (5), (6) and (7):

$$\left.\begin{aligned} W_{kj}(n+1) &= W_{kj}(n) + \alpha\,\delta_k(n)U_j(n) + \eta(n)\Delta W_{kj}(n) \\ W_{ji}(n+1) &= W_{ji}(n) + \alpha\,\delta_j(n)X_i(n) + \eta(n)\Delta W_{ji}(n) \\ \theta_k(n+1) &= \theta_k(n) + \beta\,\delta_k(n) + \eta(n)\Delta\theta_k(n) \\ \theta_j(n+1) &= \theta_j(n) + \beta\,\delta_j(n) + \eta(n)\Delta\theta_j(n) \end{aligned}\right\} \begin{matrix} j=0,1,...,Q-1 \\ i=0,1,...,N-1 \\ k=0,1,...,P-1 \end{matrix} \qquad (5)$$

Where

$$\delta_k(n) = V_k(n)[1 - V_k(n)][O_k(n) - V_k(n)] \quad k=0,1,...,P-1 \qquad (6)$$

$$\delta_j(n) = U_j(n)[1 - U_j(n)]\sum_{k=0}^{P-1}\delta_k(n)W_{kj}(n) \quad j=0,1,...,Q-1 \qquad (7)$$

α and β is the learning rate respectively,

$\eta(n)$ is the momentum coefficient.

Step 5. Calculate the error function according to Eqs. (1) and (2). If E is less than or equal to the desired value of error, go to Step 6. If not, update learning time *n* and go back to Step 3.

Step 6. Output and Stop.

2.3 FEATURE EXTRACTION AND CLASSIFICATION

In pattern classification, only a small number of carefully selected features are used. We select the most sensitive factors to act as characteristic input, these factors include cylinder diameter, wall thickness, notch geometry and material property. According to failure modes in pressure tests, failure mode is classified two classes, characteristic output is 0.01 and 0.8.

2.4 THE STANDARDIZATION OF CHARACTERISTIC INPUT

Because the span of input parameters is different, some parameters have greater values than others. The characteristic input data must be standardized in order to avoid exaggerating the effect of some parameters, the method is shown as follow:

Suppose the total number of samples is G: $(x_0(P), x_1(P),..., x_{N-1}(P))$, $P = 1,...,G)$

$$\text{Let } x_i = \sum_{p=1}^{G} x_i(P)/G \qquad i = 0,1,...,N-1 \tag{8}$$

$$S_i = \sqrt{\sum_{P=1}^{G} \left(\left[(x_i(P) - x_i \right]^2 / G \right)} \qquad i = 0,1,...,N-1 \tag{9}$$

x_i and S_i are the mean value and mean square deviation of ith parameter.

$$x_i(P) = \frac{(x_i(P) - x_i)}{S_i} \qquad P = 1,...,G; i = 1,..., N-1 \tag{10}$$

After standardization of characteristic inputs, the mean value of input data will equal to zero and all input data are non-dimension.

3. RESULTS

3.1 EXPERIMENTAL DATA

All training data came from Kiefner [6], these experimental data were obtained from pressure tests of full-scale end-capped pipes with surface flaws. The surface flaws were made by machining grooves into pipes with cutter wheels as shown in Figure 2.

The number of experimental data is 46, only 20 of them are presented in Table 1 and Table 2. Table 1 and Table 2 contain essential data including pipe geometry, material properties, flaw parameters, test results and predication results.

3.2 THE TRAINING SET

The 46 experiment data are used as training set. The input data are pipe diameter, wall thickness, notch length, notch depth, and flow stress of material. The output data are failure modes of cylinders. The feature values of failure modes in network are determined by the training sample distribution. If the number of some samples belonging to one class in training set is larger than the number of other samples belonging to other classes, the effect of this class will be exaggerated, the output results after training will be close to this class, the accuracy of the prediction will fall. In order to avoid such cases, the feature values of failure modes will be decreased at the same time. In this paper, the samples are divided into two classes, leak and rupture. The number of leaks is 16, the sample number of second class is 30. So the feature values of these two class samples are not used with 0 and 1 but 0.01 and 0.8, respectively. The neural network is trained until the mean square error converged to a pre-specified low value. The network is then tested both on the training and test set of patterns.

3.3 TEST SET

Ten samples are selected arbitrarily from 46 samples to act as "unknown" samples, these samples are input in trained network and the output results (predicted failure mode) will be obtained. The predicated failure modes are displayed in Table 2. According to the approximate degree to the desired output value, failure modes will be determined for flawed cylinders.

4. DISCUSSION AND CONCLUSION

4.1 ACCURACY OF PREDICATION WITH NEURAL NETWORK

The accuracy rate of predication is 90%(see Table 2). After learning, the neural network is supposed to have learned the knowledge about the failure modes of pipes with surface defects, and the information is memorized in network. When new input is sent, the result will be given.

In order to evaluate the property of network, the Cross-Validation method is used. This method can be described as three processes: (1) some samples are arbitrarily extracted from sample set, and the remaining samples are used to build network, then the samples extracted are input into trained network to predicate results, (2) Repeat Step (1) until all samples are extracted and only once. In this paper, the total number of samples is 46, considering the distribution of samples, 7 networks are built with same network parameters. The accuracy rate of predication is shown in Table 3.

The failure mode recognition is very complex nonlinear problem. However, because of the high nonlinear and parallel of neuron network, the satisfactory predication results still can be obtained.

4.2 THE EFFECT OF HIDDEN LAYER UNITS

The effect of the number of hidden layer unit has great influence on neuron network performance. But there still has no mature theory to guide the selection of unit number of hidden layer. Generally, increasing the unit number of hidden layer can improve accuracy rate, but the training time will increase. In this paper, the effect of hidden layer unit number is studied, and results are shown in Table 4.

When unit number of hidden layer is up to 12, the information of sample set has been mastered, and the increase of unit number in hidden layer will have no effect to improve predication accuracy.

4.3 CONCLUSION

The failure mode recognition of pressurized cylinders is a very complex nonlinear problem; the failure mode is affected by many factors such as diameter, wall thickness, material property, and defect geometry. Pattern recognition based on neural network can attain the model identification well suited for training data because of its non-linearity and learning function.

In this paper, the pattern recognition method based on neural network is applied to predicate the failure mode of cylinders with defects. After training, the network has higher discrimination. In order to evaluate the property of network, cross-validation method is adopted. The testing results show that neural network built has a good behavior and property.

REFERENCES

1. Lippman R.P. "Pattern Classification Using Neural Network", *IEEE Communications Magazine*, pp47-64, November 1989.
2. Rumelhart D.E., Hinton G.Z., and Williams R.J., "Learning Internal Representation by Error Propagation", *Parallel Distributed Process, Vol 1.* MIT Press, Cambridge, MA. 1986.
3. Lippmann R.P., An Introduction to Computing with Neural Nets, IEEE ASSP, April 1994
4. Pao, Y-H, Adaptive Pattern Recognition and Neural Networks. Addison-Wesley Publishing Co., 1989.
5. Han Lianghao, " Neural Network Applied to the Prediction of the Failure Stress for Pressurized Cylinder Containing Defects". *International Journal of Pressure Vessel and Pipe, Vol. 76 No.4*, pp 215-219,1999.
6. Kiefner, J.F. et al.; Failure Stress Levels of Flaws in Pressurized Cylinders, *ASTM STP536*, 1973, pp 461-481.

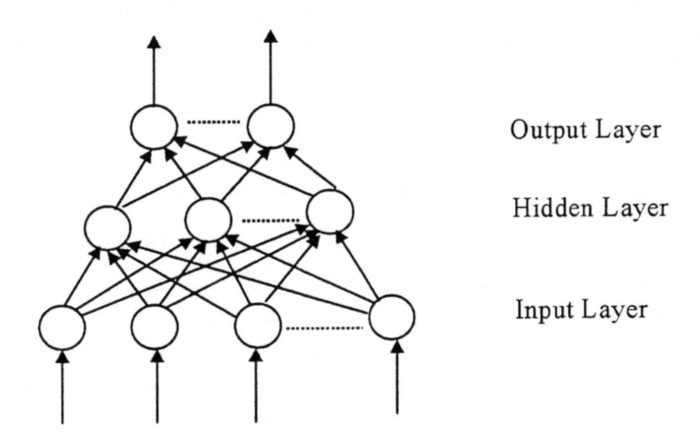

Figure 1. Three-layered neural network

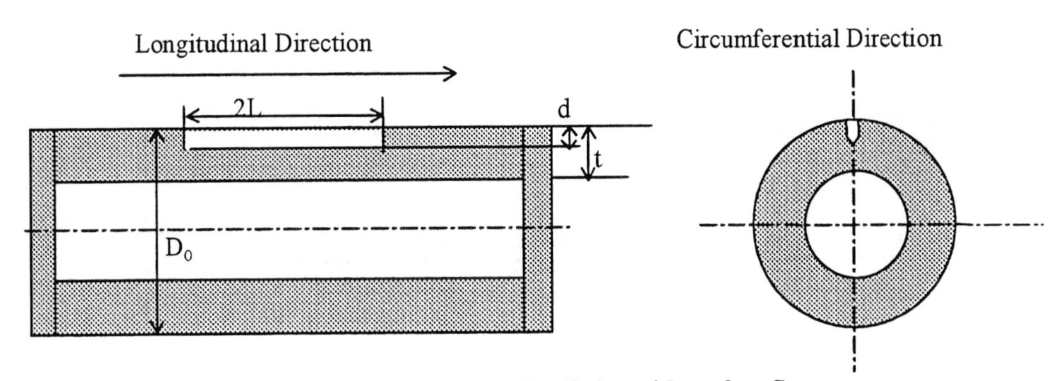

Figure 2. The pressurized cylinder with surface flaw

Table 1 Training Set

Test No	Grade	Outside Diameter D(mm)	Wall thickness t(mm)	Notch length 2c(mm)	Notch depth d(mm)	Flow stress •$_0$(MPa)	Diameter Of Cutter Wheel(mm)	Actual failure mode	Predicted Failure mode	The Output value
1	X52	762	9.73	369.82	5.92	489.19	69.85	Rupture	Rupture	0.7973
2	X52	762	9.96	224.02	9.17	489.19	69.85	Leak	Leak	0.0021
3	X52	762	9.63	83.82	3.91	509.17	69.85	Rupture	Rupture	0.8032
4	X52	406.4	6.35	185.42	3.18	451.25	69.85	Rupture	Rupture	0.8002
5	X60V	914.4	11.15	139.70	6.83	447.85	101.6	Leak	Leak	0.0022
6	X65	863.6	12.80	406.40	3.23	533.98	101.6	Rupture	Rupture	0.8011
7	X65	863.6	12.93	609.60	11.23	533.98	101.6	Leak	Leak	0.0013
8	X60	1066.8	12.04	292.1	5.97	547.76	101.6	Rupture	Rupture	0.8039
9	X60	863.6	11.13	127	7.57	466.45	101.6	Leak	Leak	0.0087
10	X65	863.6	9.91	406.4	4.95	553.27	101.6	Rupture	Rupture	0.8007

Table 2 Test set

Test No	Grade	Outside Diameter D (mm)	Wall thickness t (mm)	Notch length 2c(mm)	Notch depth d (mm)	Diameter of Cutter Wheel (mm)	Flow stress •₀ (MPa)	Actual failure mode	Predicated failure mode	The output value
1	X60V	914.4	11.3	139.7	8.66	101.6	447.85	Leak	Leak	0.001
2	X52	762	9.58	224.0	7.747	69.85	486.43	Leak	Leak	0.0023
3	X60V	914.4	9.98	121.9	7.54	101.6	518.13	Leak	Leak	0.006
4	X65	914.4	9.83	152.4	6.86	101.6	553.27	Leak	Leak	0.0812
5	X60	914.4	11.12	127	7.57	101.6	466.45	Leak	Leak	0.007
6	X52	762	9.53	312.4	4.78	69.85	498.84	Rupture	Rupture	0.8002
7	X52	762	9.80	219.2	5.99	69.85	489.19	Rupture	Rupture	0.8643
8	X60C	762	9.47	152.4	4.52	101.6	516.75	Rupture	Rupture	0.7850
9	X60	1066.8	10.26	165.1	6.50	101.6	503.66	Rupture	Rupture	0.7154
10	X65	762	9.88	370.8	8.10	69.85	508.48	Rupture	Leak	0.0236

Table 3 The property testing of network with Cross-Validation method

The sample number of training set	The sample number of test set	The number of Failure	Accuracy rate
39	7	1	87.5%
39	7	1	87.5%
39	7	0	100%
39	7	1	87.5%
39	7	1	87.5%
42	4	0	100%

Table 4 The effect of hidden layer unit number to neural network

Unit number of hidden layer	Number of test sample	Number of failure	Accuracy rate
6	7	1	87.5%
10	7	1	87.5%
20	7	0	100%
24	7	0	100%
30	7	0	100%

PROBABILISTIC FATIGUE, FRACTURE, AND RELIABILITY

Introduction

Mohammed Khaleel
Pacific Northwest National Lab.
Richmond, WA

Sharif Rahman
University of Iowa
Iowa City, IA

Robert W. Warke
Southwest Research Institute
San Antonio, TX

Probabilistic aspects of fatigue and fracture are known to be of significance in the development of appropriate predictive methodologies for life assessment and structural reliability. For over five years, probabilistic fatigue and fracture has been topic of interest to the pressure vessel and piping community. Two sessions on this subject were organized for the 2000 ASME-PVP Conference in Seattle, WA. Seven of the papers presented in these sessions are included in this volume.

The paper by Ardillon et al. presents the results of the EDF research findings in support of the safety acceptance criteria of the RSE-M code. The paper also details a general procedure and the scientific aspects of code development for safe operation of nuclear components. Chen et al. present an efficient method to conduct shape sensitivity analysis of the elastic energy release rate of cracked bodies. One of the advantages of this technique is that the results are independent of mesh discretization. Some examples are included.

The next paper by Cioclov and Kröning presents a methodology that attempts to integrate nondestructive inspection parameters into a probabilistic failure assessment diagram. The authors demonstrate the practical significance of this integrated procedure. Malyukova and Timashev present methods for deriving deterministic and probabilistic life assessment of a cracked pipeline subjected to cyclic loading and local corrosion.Using the maximum entropy method in conjunction with Monte Carlo simulation techniques, they calculate the least predetermined PDF of remaining life of the cracked pipeline.

Simonen et al. discuss the procedures for the probabilistic analysis of fatigue crack initiation in pressure boundary components in commercial light water reactors. They further combine this analysis with a probabilistic fracture mechanics analysis to predict catastrophic failure. The results presented in this paper are of use in risk assessment of components with high cycle usage. In the next paper, Cioclov and Wiedemann present a probabilistic analysis of failure in a high temperature water pipe in a fossil power plant. The outline failure analysis offers an example of the increasing insight that can be achieved when fracture mechanics principles combined with probabilistic methods are used in risk evaluation of pressure vessel and piping systems. The last paper in this section deal with the reliability analysis of structures with defects. The framework and procedure established in the paper would be very useful to the reliability analysis of structures containing defects and easy to use by practicing engineers

The editors hope that the ASME-PVP community will benefit from the valuable research presented in these papers.

PVP-Vol. 410-1, Assessment Methodologies for Preventing Failure: Deterministic and Probabilistic Aspects and Weld Residual Stress
Volume 1
ASME 2000

PROBABILISTIC APPROACHES AS METHODOLOGICAL SUPPORT TO INDUSTRIAL CODIFICATION IN NUCLEAR ENGINEERING

Emmanuel ARDILLON
ELECTRICITE DE FRANCE-
Research and Development Division-
6, quai Watier - 78401 Chatou Cedex (FRANCE).
☎ (+33) 1 30 87 85 89 fax (+33) 1 30 87 82 13
🖳 emmanuel.ardillon@edf.fr

Bruno BARTHELET
ELECTRICITE DE FRANCE-
Nuclear Generation Division-
Cap Pleyel 93200 Saint-Denis (FRANCE)
☎ (+33) 1 43 69 34 33
🖳 bruno.barthelet@edf.fr

Eric MEISTER
ELECTRICITE DE FRANCE
Basic Design Department
12-14, Avenue Dutriévoz, 69628 Villeurbanne Cedex, (FRANCE)
☎(+33) 4 72 82 75 97
🖳 eric.meister@edf.fr

ABSTRACT

Margin assessments carried out on the components of operating installations are an important step in lifetime evaluations. They are based on codified rules relying on deterministic criteria supposed to ensure the safe operating of the industrial installations, e.g. of the nuclear power plants for Electricité de France (EDF).

This paper is based on a work carried out at EDF in the framework of an internal research project aimed at supporting the safety acceptance criteria of the RSE-M code. It proposes a general methodology to be used as a support to industrial codification of structural integrity assessment criteria involving safety coefficients. This approach is derived from the practice that has been applied at Electricité de France in the field of nuclear engineering. It also has a close relationship to the methodologies currently applied in other industrial sectors. It should be noted that this methodology is only the scientific background of the codification process, which involves many other aspects, as for instance organisational procedures that are specific to each branch.

After reminding the stakes of the codification of safety margins, the paper presents a general code development procedure (as for its scientific aspects). This procedure is based on a methodology that nowadays resorts to a probabilistic approach. This approach is presented and illustrated by some basic and simple examples.

1 LIFETIME EVALUATION OF INDUSTRIAL STRUCTURES AND SAFETY MARGINS

The lifetime evaluations of industrial facilities involve many aspects. One of them is the safe operating of the industrial facility : for many industrial branches the safe operating of the installations is a prerequisite for their lifetime to be consistent with the objective of industrial competitiveness imposed by the economic environment. This is due to two main reasons :

- the safe operating on an appropriate period guarantees that no structural breakdown happens and that the facilities can work as expected and provide the expected benefits ;
- the safe operating on an appropriate period is necessary to ensure the social acceptance of the corresponding industrial facilities ; and the cost of lack of social acceptance can be extremely high.

The latter point is important especially for the nuclear industry, which is submitted to strict controls and high safety demands. That is why EDF, the French electricity utility that basically requires to nuclear energy (80%), is particularly concerned with the safe operating of its facilities.

2 CONTEXT AND STAKES OF THE USE OF SAFETY MARGINS

For many centuries the method used to build structures has been relying on the ratio of the maximum stress the material can tolerate to the maximum stress actually experienced by the material over the anticipated life of the structure. This ratio must be sufficient for the structure to be considered safe. This ratio is called the **safety factor** or the **(safety) margin**. It is the basis of deterministic safety assessments.

Safety factors are therefore applied in all industrial branches:

- Civil engineering and building works (bridges, tunnels, etc.), offshore engineering and metal constructions;
- Transport: rail, air, road and sea;
- Heavy industry: mining, steelmaking, petrochemicals, nuclear, etc.
- High-tech industries: space, armaments, etc.

More generally, the current codes supposed to ensure the safe operating of industrial facilities (e.g. nuclear power plants) mostly resort to deterministic criteria relying on safety margins, the form of which is close to the aforementioned ratio.

So, the issue at stake in estimating safety margins is to guarantee the safety of the constructions concerned, or more specifically the various types of safety that can be envisaged, depending on the installations concerned:

- safety of the personnel working in the installation;
- safety of the surrounding populations;
- safety of future populations;
- non-failure of the equipment during its planned service life (availability and satisfactory performance),
- preservation of the environment.

A *rational* estimation of safety margins is aimed at guaranteeing the *desired level* of safety, without building in excessive margins, in order to minimize the associated costs and ensure that the industrial branch concerned remains competitive. This is the case of the French nuclear power plants in operation owned by EDF, the French electricity utility.

3 HOW TO PROVIDE A RATIONAL ESTIMATION OF MARGINS?

This is the methodological question underlying the problem mentioned above.

This paper is based on a work carried out at Electricité de France (EDF) in the framework of an internal project aimed at supporting the flaw acceptance criteria of the RSE-M code (Barthelet B., Le Delliou P., Héliot J., Faidy C., Drubay B. [3]). Note that the RSE-M Code is the code used by EDF to ensure the safe operating of its nuclear pressurized water reactors ; it provides rules and requirements for maintenance operations and in service inspections.

This paper presents this approach to the rational estimation of safety margins, also called the **codification support approach or pro-codification approach.** It is essentially based on the approach conducted in support of the RSE-M code. However, it should be noted that this methodology is a generic one, applied in other industrial branches. The examples given are not specific to the fracture margins problem in the RSE-M.

The various aspects of the codification support approach, i.e. the approach in support of the flaw acceptance criteria codified in the RSE-M code, are presented in order to trace the approach step-by-step and make it understandable to engineers who are not specialized in margin-related problems, and to enable other teams not concerned by the RSE-M code to use it to estimate margins. Emphasis is placed on the reliability-based methodology.

The following steps of the method have been adopted :

1. clarification of the notions of codification ;
2. description of the general pro-codification approach and of its steps ;
3. description of the reliability-based method employed in certain steps of the general approach.

4 METHOD (STEP 1): CLARIFICATION OF CODIFICATION VOCABULARY

The clarification of certain terms commonly employed in the pro-codification approach is in response to the need felt by practitioners from nuclear industry and also by those from other industrial branches. They highlight a number of terminological ambiguities to be avoided, for example the term "Margin", for which 4 different meanings at least have been identified. They are also intended to render the field more accessible to other teams.

The definitions proposed herebelow try to represent the meaning currently accepted by the members of the internal EDF research Project. These definitions can change, and maybe this paper will be an opportunity to have a discussion about them. Moreover, only a few basic terms are proposed in this section. Note also that the definition work has been carried out in French basically and that here some terms are translated from French to English. It should be checked if this translation is appropriate.

Term	Definition
Characteristic value	Same as reference value, but includes a probabilistic meaning (definition as a distribution fractile for example)
Explicit margin	same as safety coefficient
Implicit margin	same as reference value
Margin	same as safety margin
Overall safety factor	Ratio between a parameter characterizing the material strength and a parameter characterizing the effect of actions (loading)
Reference value	Value of an analysis parameter that is used in the margin equation before application of safety coefficient. It is generally a pessimistic value representing the parameter in the analysis.
Reserve factor	Difference between the threshold value of the criterion and the value derived from the calculation carried out with the problem data including their safety coefficients
Safety coefficient	❶(Sense 1) same as Overall safety factor ❷(Sense 2) = partial safety coefficient = minimum coefficient to be applied to the characteristic values
Safety factor	same as safety coefficient
Safety margin	❶(Sense 1) same as overall safety factor ❷(Sense 2) Equation arising from the physical model, with the same form as ❶ but that can be applied to whatever variable value. In this sense, the margin can be a random variable ❸(Sense 3) same as partial safety coefficient ❹(Sense 4) same as reserve factor

It should be noted that the term « margin » has plenty of senses. Sometimes this can cause some communication problems.

5 METHOD (STEP 2) : THE GENERAL PRO-CODIFICATION APPROACH

5.1 Description of the general pro-codification approach

This section has a close relationship to (Nowak A.S. and Lind N.C. [7], Madsen H.O., Krenk K. and Lind N.C. [5]). These authors had proposed a classical description of the approach for design codes. For the RSE-M code, which is relevant to operating structures and not to structures to be designed, the same general description can be used.

The general code development procedure relies on 5 steps :

❶definition of the code scope and of the data space,

❷definition of the code objectives,

❸determination of the frequency of occurrence of the situations,

❹selection of the code format,

❺evaluation of the code performances (comparison between objectives and supposed realisations).

They are presented in figure 1.

5.2 Some good reasons to use the reliability-based approach

❶ Presence of uncertainties or contingencies

Reliability-based approaches are ideally suited to problems involving uncertainties or contingencies. They are in fact a methodological tool for the quantification and rational processing of uncertainties. Several different sources of uncertainty are involved in the analysis of a mechanical structure :

- the intrinsic variability of the values characterising the state of the structure and its environment (geometry, material, loading);
- statistical uncertainties which appear when estimating the input value distribution parameters; they depend on the number of data used for the estimation;
- measurement or implementation errors (or errors in description between what was planned and what was achieved);
- modelling errors (simplifications, working assumptions, etc.).

The first type of uncertainty plays a particular role. This is the type of uncertainty covered by the term contingency. It concerns the uncertainties inherent in the system studied, given the scale at which it is considered, or the physical limits set for the problem. We consider that it is impossible to modify them. The origin of these uncertainties can be spatial (properties of a material) or temporal (variability of a load over time).

This type of uncertainty is present in the various industrial branches, albeit to varying degrees: for example, the properties of the materials used in civil engineering or mining structures have a far greater dispersion than the steel used in industrial engineering structures.

However, the other three types of uncertainty depend on the level of knowledge of the structure and its environment (information, modelling, control).

❷ Objectives:

- To understand: the reliability-based approach aims to explain and make predictions on the basis of the information contained in the models and in the distribution profiles of the input values (the "causes");
- To decide:

in the case of EDF: the issues are related both to controlling the safety of electricity production installations and controlling per-kWh production costs, with the feared failure events generally having a low probability of occurrence. A reliability-based approach provides the means of evaluating the failure risks liable to impair safety and making decisions by comparing these risks with predetermined thresholds.

It should be noted that in the case of codification, the two reasons mentioned above exist side by side:

- presence of uncertainties (all types mentioned);
- codes are decision-making tools (concerning mechanical strength), with important underlying safety and economic implications.

Those reasons are probably responsible for the fact that probabilistic approaches have been used more and more extensively for the last decade in various fields of nuclear engineering. They have therefore been integrated as support to codification and applied to the French code RSE-M

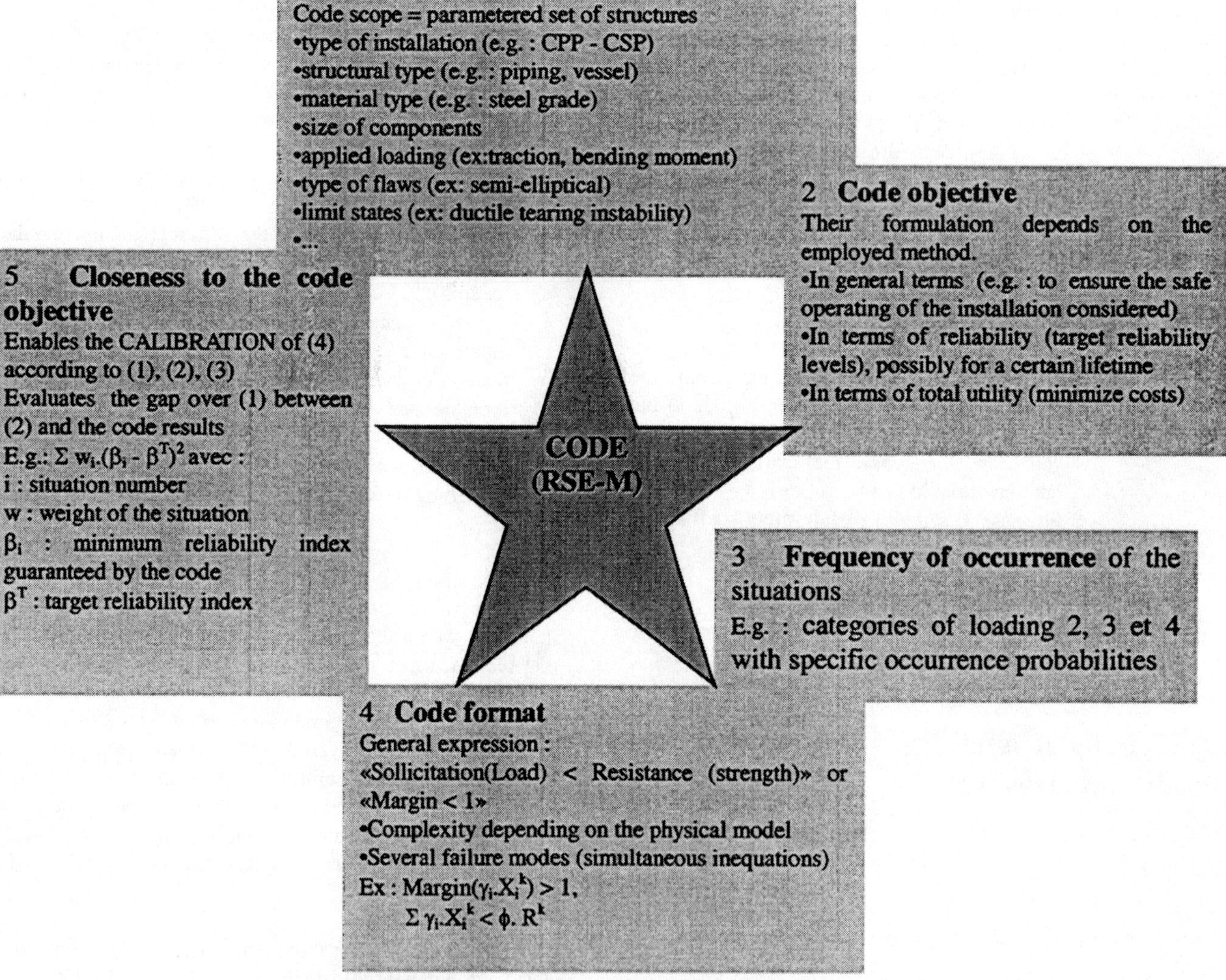

Figure 1 – The general pro-codification approach

5.3 Steps of the general approach potentially resorting to the reliability-based approach

Almost every step can be concerned by the reliability-based approach.

Step 1 : it is the description of the code scope that will change (using reliability notions), not the code scope itself.

Step 2 : the formulation of the objectives depends on the selected approach ;

Step 3 : the formulation of the frequency of the situations clearly resorts to probabilistic notions, but can remain implicit in a deterministic code ;

Step 4 : the most important application field of the probabilistic approach, if this approach is selected ;

Step 5 : its formulation strongly depends on the selected method.

6 METHOD (STEP 3) : THE RELIABILITY-BASED APPROACH

As the most important role for the reliability-based approach concerns step 4, the presentation focuses on step 4.

6.1 Reliability-based code calibration : various methods

The calibration of the code format involves assigning specific values to the general parameters in the code format, such as partial safety factors and reference values.

It is the **degree of sophistication** given to the code which will determine the type of calibration method used.

From a **reliability-based** viewpoint, there are **four levels of design code** (Nowak A.S. and Lind N.C. [7], Madsen H.O., Krenk K. and Lind N.C. [5]). This distinction is also valid for codes designed for the operation of existing installations, such as the RSE-M code.

Level i is a particular case of level i+1. The four code levels are as follows:

- Level 1:
  - ⇒ traditional deterministic approach based on characteristic values and safety factors (partial or otherwise);
  - ⇒ determination is based on expert judgement, including an initial analysis of feedback;
- Level 2: introduction of the notion of reliability index, but the variables are assumed to be Gaussian, which means that we only need to know their first two moments, and possibly the covariances between correlated variables;
- Level 3:
  - ⇒ variables are random and of any type, which allows distribution queues to be taken into account more satisfactorily, which is useful when the target reliability levels are high;
  - ⇒ introduction of the notion of generalised reliability index;
  - ⇒ introduction of the most probable failure point, which provides the basis for determining partial safety factors using reliability-based methods;
  - ⇒ calculation of failure probability using FORM/SORM reliability-based methods or simulation methods (Monte Carlo and variance reduction techniques);
  - ⇒ the objective is to get as close as possible to the target reliability level within the code scope.
- Level 4:
  - ⇒ overall optimisation process based on maximising a utility function (or by minimising a cost function);
  - ⇒ the target reliability is an output of the calibration process and not an input as is the case in level 3: in this case the target reliability corresponds to the minimum overall cost of the installation over its forecast service life;
  - ⇒ the problem lies in evaluating an overall cost for which all the necessary data are not available, and which can be liable to inaccuracy.

At present, many codes use the format for level 1, but call on levels 2 and 3 to calibrate safety factors and the corresponding characteristic values. This is notably the case with the RSE-M Version 2 (Ardillon E., Pitner P., Barthelet B., Rémond A. [1], Barthelet B., Le Delliou P., Héliot J., Faidy C., Drubay B. [3]). This type of approach is referred to as a **mixed approach**.

The degree of sophistication does not only concern the choice of reliability-based methods. To varying degrees, it also concerns the following progress factors:

- inclusion of new materials or new technologies;
- better knowledge of the input data, through the incorporation of feedback. This improvement is useful from both the deterministic and reliability-based viewpoints;
- the choice of deterministic calculation methods: simplified methods or finite element-based methods;
- the inclusion of new structural failure modes.

6.2 Code calibration through a classical example : the (R/S) case

6.2.1 General characteristics of the case

The (R, S) case is the most simplified representation of a mechanical structure characterized by its Resistance R and submitted to the Stress S from its environment. It has been widely studied (Nowak A.S. and Lind N.C. [7], Ardillon E., Pitner P., Barthelet B., Rémond A. [1]). It has the following characteristics:

* **simplicity**: it is the most basic case of a mechanical model for a structure subject to a failure hazard;
* **accessibility**: the results obtained in this case require no special probability tools and can be found by any engineer;
* **comprehensiveness**: the case can be used to address all the main notions involved in reliability-based approaches in support of codification: overall safety factor, partial safety factors, partial safety factor optimisation, choice of characteristic values, failure probability, reliability index, etc.
* **generic nature**: the methods used here to calibrate the safety factor(s) used in the level 1 semi-probabilistic formats will be used on the more sophisticated mechanical models.

Variable or parameter	Signification	Value
R	Resistance (strength)	Gaussian random variable : $R \sim N(\mu_R, \sigma_R)$
S	Stress	Gaussian random variable : $S \sim N(\mu_S, \sigma_S)$
μ_R, μ_S	Mean (of the resistance or of the stress)	variable parameter, superior to 0.
σ_R, σ_S	Standard deviation (of the resistance or of the stress)	variable parameter, superior to 0.
V_R, V_S	Coefficient of variation(of the resistance or of the stress)	$V_R = \sigma_R/\mu_R$ $V_S = \sigma_S/\mu_S$
r^k , s^k	Characteristic values (of the resistance or of the stress)	$r^k = \mu_R - k_R \cdot \sigma_R$, $s^k = \mu_S + k_S \cdot \sigma_S$ Those values are located in the pessimistic part of the distribution
β	reliability index	$\beta = \dfrac{\mu_r - \mu_s}{\sqrt{(\sigma_r^2 + \sigma_s^2)}}$ (1) (Madsen H.O., Krenk K. and Lind N.C. [5]) $\beta = \dfrac{\Theta - 1}{\sqrt{(\Theta^2 . V_R^2 + V_S^2)}}$ (2) (Madsen H.O., Krenk K. and Lind N.C. [5])
Θ	Overall safety factor (see §4) relying on the mean values of the distribution Central safety factor [MAD 86]	$\Theta = \dfrac{\mu_R}{\mu_S}$ $\Theta = \dfrac{1 + \beta . \sqrt{(V_R^2 + V_S^2 - \beta^2 . V_R^2 . V_S^2)}}{1 - \beta^2 . V_R^2}$ (3) (Madsen H.O., Krenk K. and Lind N.C. [5])
Θ^k	Overall safety factor (see §4) safety factor relying on characteristic values	$\Theta^k = r^k / s^k = \Theta . \dfrac{1 - k_R \cdot V_R}{1 + k_S \cdot V_S}$
P_f	Failure probability	$P_f = \Phi(-\beta)$
Φ	Cumulative density function of the normal standardized distribution	Between 0 and 1

Table 1 : Mathematical definition of the (R/S) case

In this simple example, R and S are 2 Gaussian random variables. It is important to notice that in this case there is an **explicit bijection** between the **reliability index** β (synonymous with failure probability) and the **overall safety factors Θ or Θ^k**.

This link is a well-known result (Madsen H.O., Krenk K. and Lind N.C. [5], Ardillon E., Pitner P., Barthelet B., Rémond A. [1]). It is the basis of code calibration depending on reliability target values.

6.2.3 One important result : the role played by the choice of the characteristic values

As aforementioned, there is a bijection between the reliability index β and the overall safety factors Θ or Θ^k. However, as shown by relations (2 - 3) of the table of last section 6.2.2, this link depends on the variation coefficients V_R and V_S. This link is therefore a bijection only for one particular situation ; but in general, a code scope is a set of many different situations. Moreover, many code formats rely on the codification of values of overall safety factors (traditional format). These factors can be calibrated by a reliability-based approach, depending on reliability target levels. It is therefore interesting to see if a particular type of overall safety factor, that is one overall safety factor relying on specific characteristic values, is recommended to ensure a reliability level as homogeneous as possible over the whole code scope.

Note that this issue is not always debated, as in many cases multiple (partial) safety factor formats (Madsen H.O., Krenk K. and Lind N.C. [5], Turner R. C. [8]) are considered (as in the RSE-M code for example), and the main question is to calibrate the partial safety factors given predetermined characteristic values. In the simplified approach presented we consider only one overall safety factor, and the question is to select the optimized type of overall safety factor. We assume that the code scope is a set of situations distributed around one « median » situation.

The exercise presented below shows that the characteristic values of the overall safety factor should be chosen as close as possible to the most probable failure point (also called design point). This notion of design point is supposed to be known. It is described in (Madsen H.O., Krenk K. and Lind N.C. [5], Ardillon E., Barthelet B., Sorensen J.D [2], Barthelet B., Ardillon E. [4]). Let us remind that the most probable failure point represents the way the structure will fail most likely.

Assumptions used in the exercise:

⇒ Case (R, S) - normal variables

⇒ 9 situations numbered in the following table (the central situation is the situation n°5):

	V_R=0.05	V_R=0.1	V_R=0.15
V_S=0.15	1	2	3
V_S=0.2	4	5	6
V_S=0.25	7	8	9

Table 2 : Definition of the 9 situations

⇒ the symbols (V_R, V_S, k_R, k_S...) are defined in the table of section 6.2.2 ;

⇒ Θ_k factors for different characteristic values, chosen so as β=3.72 (i.e. a failure probability $P_f = 10^{-4}$) for the "central" situation (V_R = 0.1, V_S=0.2).

Objective:

To compare the homogeneity of the guaranteed reliability level provided with various Θ_k factors for the nine situations.

Results: (the guaranteed β values are given in the following tables)

($k_R = k_S = 0$)	**V_R=0.05**	**V_R=0.1**	**V_R=0.15**
V_S=0.15	5.876	4.188	3.105
V_S=0.2	4.755	3.720	2.899
V_S=0.25	3.958	3.299	2.686

Table 3 - Case 1: average values ($k_R = k_S = 0$) - Θ = 2.071

($k_R = 1$; $k_S = 1$)	**V_R=0.05**	**V_R=0.1**	**V_R=0.15**
V_S=0.15	4.973	3.958	3.155
V_S=0.2	4.319	3.720	3.099
V_S=0.25	3.865	3.505	3.028

Table 4 : Case 2: values with 1 standard deviation ($k_R = 1$; $k_S = 1$) - Θ_κ = 1.55

($k_R = 2$; $k_S = 2$)	**V_R=0.05**	**V_R=0.1**	**V_R=0.15**
V_S=0.15	4.109	3.785	3.307
V_S=0.2	3.819	3.720	3.354
V_S=0.25	3.618	3.647	3.374

Table 5 : Case 3: values with 2 standard deviations ($k_R = 2$; $k_S = 2$) - Θ_κ = 1.18

($k_R = 2.676$; $k_S = 2.584$)	**V_R=0.05**	**V_R=0.1**	**V_R=0.15**
V_S=0.15	3.540	3.702	3.481
V_S=0.2	3.440	3.720	3.571
V_S=0.25	3.366	3.711	3.628

Table 6 : Case 4: values at most probable failure point (k_R = 2.676 ; k_S = 2.584) - Θ_κ = 1

($k_R = 1$; $k_S = 4.314$)	**V_R=0.05**	**V_R=0.1**	**V_R=0.15**
V_S=0.15	4.241	3.511	2.869
V_S=0.2	4.319	3.720	3.099
V_S=0.25	4.357	3.850	3.258

Table 7 : Case 5: values at a point on the failure surface (k_R = 1 ; k_S = 4.314) - Θ_κ = 1

⇒ Remarks:

* after constructing the Θ_K factors, the index value found for central situation n°5 is exactly equal to the target value (3.72);
* for the last two cases, the Θ_K factor must be equal to 1 as the reference point is on the failure surface;
* the purpose of case 5 is to check whether choosing the most probable failure point is preferable to choosing a random point on the failure surface;
* the fractiles corresponding to the characteristic values of these five cases are as follows:

	Nb_sigma_R	Nb_sigma_S	Fractile R	Fractile S
Case 1	0	0	50.00%	50.00%
Case 2	-1	1	15.87%	84.13%
Case 3	-2	2	2.28%	97.72%
Case 4	-2.676	2.584	0.37%	99.51%
Case 5	-1	4.314	15.87%	99.9992%

Table 8 : Fractiles of the characteristic values used in the Θ^k coefficients

The graph below summarises the numerical results for the five choices of factor Θ_k and contained in the five tables above. The situations are numbered as in the table in which they are defined.

We can see that:

1. As for the 9 situations concerned (see the situation definition in the first table of section 6.2.3) :
   - the **situations** for which the guaranteed reliability level is least uniform are n°s 1, 4 and 7: they correspond to a low V_R factor;
   - for certain situations the guaranteed reliability level is systematically lower than the target value: the cases concerned are n°s 3, 6 and 9, which correspond to a high V_R factor;
2. As for the selection of the best safety factor definition (i.e. the best case) :
   - case 4 (**most probable failure point**) emerges as the **best choice**;
   - the closer the chosen characteristic values to the most probable failure point, the better the guaranteed uniformity: case 1 < case 2 < case 3 < case 4;
   - the choice of **a random point on the failure surface is irrelevant,** as shown in case 5.

⇒ *An ideal choice, but hard to apply*

However, the choice of characteristic values taken exactly at the most probable failure point is not always feasible. The codifiable values must be **observable fractiles** of the variables considered, and as such located not too far down the distribution line. The relevance of this remark naturally depends on the number of data available. In the example, the most probable failure point corresponds to the 0.37% fractile for R and 99.52% for S: these values are not codifiable.

Consequently, when using a code whose format is based on a single factor, one should attempt to codify characteristic values which are as close as possible to the most probable failure point, while being realistically observable.

6.2.4 Summary of the (R/S) case

The results presented here rely on a more developed study carried out in the framework of the EDF internal research project. Only the issue relative to the selection of characteristic values could be presented in this paper (see §6.2.3).

Main results

The main results are set and used as the basis for calibration of the safety factors on more sophisticated models:

- there is an **explicit bijection** between the **reliability index** β (synonymous with failure probability) and the **overall safety factors Θ and Θ^k**;
- it is preferable to use the **reliability index** rather than the overall safety factor as a basis, as the latter does not offer a sufficiently uniform reliability level,
- with a code whose format is based on an **overall safety factor Θ^k**, one should seek to codify a factor covering characteristic values as close as possible to the **most probable failure point** but which are realistically **observable**;
- **partial safety factors** must satisfy the general equation $\Theta = \theta_R{}^{\mu}.\theta_S{}^{\mu}$ which leaves them a degree of freedom and relates them to the target level of reliability;
- it is possible, for a given situation (i.e. a set choice of β, V_R, and V_S), to calculate explicitly the partial safety factors obtained at the most probable failure point;
- when **several situations** have to be covered by the same calibrated code format, it is possible to optimise the Θ^k factor to be codified – in this case with no need for sophisticated tools.

Summary of the various factor calibration methods for the case (R, S) for a target reliability level

Method 1: overall safety factor Θ^k

Approach: $\Theta^k \leftrightarrow \Theta \leftrightarrow \beta$

For a given situation: $\Theta =$

$$\frac{1+\beta.\sqrt{\left(V_R^2+V_S^2-\beta^2.V_R^2.V_S^2\right)}}{1-\beta^2.V_R^2} \; ; \; \Theta^k = \frac{1-k_R \cdot V_R}{1+k_S \cdot V_S}.\Theta$$

Method 2: partial safety factors

For a given situation:

* Method 2.1: Relationship between the partial safety factors: $\Theta = \theta_R{}^k.\theta_S{}^k$
* Method 2.2: Most probable failure point: analytical calculation of factors $\theta_R{}^k$ and $\theta_S{}^k$
* *Method 3:* For multiple situations: optimisation of the overall safety factor Θ^k

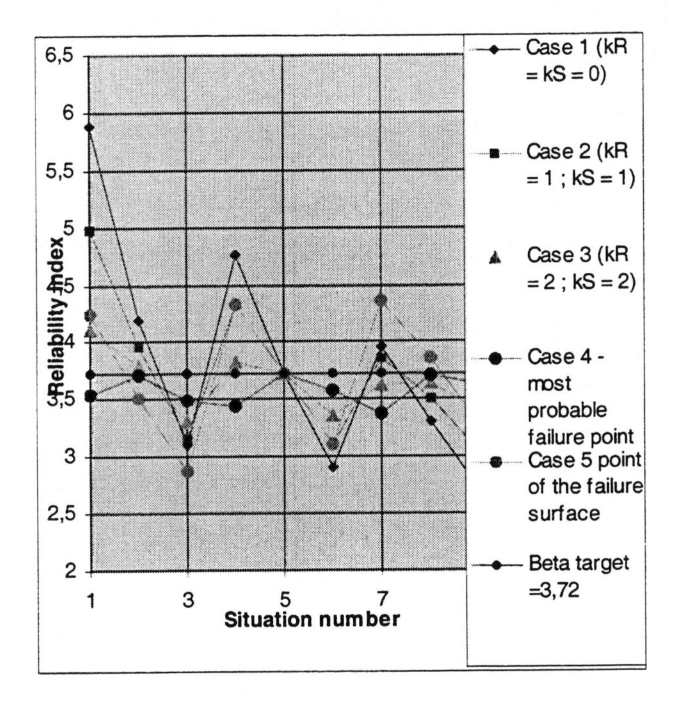

Figure 2 : Various types of Θ^k factors – Comparison of the homogeneity of the reliability level guaranteed over a set of situations around a central situation

7 CONCLUSION

This paper has proposed an overview of the code calibration process. The problem of code calibration arises from the need to ensure the safety of industrial facilities (e.g. the nuclear power plants of EDF) with optimized costs.

The general pro-codification approach is the scientific background of the solution to this problem. It is presented in this paper. First of all, some terms and notions need to be clarified. For example, at least 4 different senses have been identified for the term « margin ». Secondly, this general approach relies on 5 steps : definition of the code scope and of the data space, definition of the code objectives, determination of the frequency of occurrence of the situations, selection of the code format, evaluation of the code performances (comparison between objectives and supposed realisations).

For some steps of this approach the reliability-based methods have been used more and more extensively, especially for the problem of code calibration (step 4). The common practice is to calibrate a deterministic format code by higher level methods based on structural reliability. This is the approach used at EDF for the RSE-M code's flaw acceptance criteria. It is illustrated on the basic Resistance / Stress (R/S) case, for which some generic considerations can be formulated although it is basic.

One of them is developed : the selection of the type of overall safety factor to be codified should lead to codify factors relying on characteristic values as close as possible to the most probable failure point. However, this selection is not always feasible. The characteristic values must be realistically observable in the distribution tails.

The main conclusions of the whole R/S case study are given, although they could not be presented in the paper itself. They have been proposed in the framework of an internal research project launched to support the RSE-M code's flaw acceptance criteria.

8 REFERENCES

/1/ Ardillon E., Pitner P., Barthelet B., Rémond A. - *Probabilistic calibration of safety coefficients for flawed components in nuclear engineering* - SMIRT 13, August 1995,Porto Alegre, Brazil.

/2/ Ardillon E., Barthelet B., Sorensen J.D. - *Probabilistic calibration of safety factors for nuclear operating installations* - Paper presented at ASME - PVP 1998, San Diego, CA, U.S.A. (july 1998).

/3/ Barthelet B., Le Delliou P., Héliot J., Faidy C., Drubay B. - *RSE-M Code progress in the field of examination evaluation and flaw acceptance criteria* - SMIRT 13, August 1995, Porto Alegre, Brazil.

/4/ Barthelet B., Ardillon E. - *Simplified probabilistic approach to determine safety factors in deterministic flaw acceptance criteria* - SMIRT 14, Workshop common to Divisions G/M, Lyon, August 1997.

/5/ Madsen H.O., Krenk K., Lind N.C. - Methods of Structural Safety - 1986, Prentice-Hall, Inc., Englewood Cliffs, New Jersey 07632, Etats-Unis.

/6/ Milne, I., R.A. Ainsworth, A.R. Dowling & A.T. Stewart 1988. *Assessment of the integrity of structures containing defects.* International Journal Pressure Vessels & Piping n°32.

/7/ Nowak A. S., Lind N. C. - *Probability -based design codes* - Probabilistic structural mechanics handbook , edited by C. Sundararajan, 1995.

/8/ Turner R. C. - Partial Safety factor Calibration for the North Sea Adaptation of API RP2A-LRFD , 1st ISOPE Conference, Edinburgh, United Kingdom, 1992.

PVP-Vol. 410-1, Assessment Methodologies for Preventing Failure:
Deterministic and Probabilistic Aspects and Weld Residual Stress
Volume 1
ASME 2000

SHAPE SENSITIVITY ANALYSIS OF LINEAR-ELASTIC CRACKED STRUCTURES

Guofeng Chen, Sharif Rahman, and Young Ho Park
Center for Computer-Aided Design
The University of Iowa
Iowa City, IA 52242

ABSTRACT

An efficient method was developed for shape sensitivity analysis of elastic energy release rate (ERR) for a crack in a homogeneous, isotropic, and linear-elastic body subject to model-I loading condition. The material derivative of continuum mechanics is utilized to conduct continuum-based shape sensitivity analysis. For the sensitivity equation, the domain integral form of J-integral is invoked to represent ERR and then the material derivative of the variational form of equilibrium equation is obtained using the design velocity field concept to describe shape design changes. The direct differentiation method is employed to develop sensitivity coefficients. Unlike the virtual crack extension techniques, no mesh perturbation is needed in the proposed method. Since the governing variational equation is differentiated before the process of discretization, the resulting sensitivity equations are independent of any approximate numerical techniques, such as the finite element method, boundary element method, meshless method, and others. Two numerical examples on two-dimensional plane stress problem are presented to calculate the sensitivity of J-integral using the proposed method. The results agree very well with the reference solutions from the finite difference method. The maximum error in calculating the sensitivity of J-integral is less than two percent.

INTRODUCTION

For structures containing cracks, the stress-intensity factor (SIF) and energy release rate (ERR) are the well-known fracture parameters in linear-elastic fracture mechanics (LEFM). If an asymptotic crack-tip solution with a free amplitude parameter exists, the strength of the stress and strain fields near the crack tip can be expressed in terms of these parameters. Hence, these fracture parameters provide a mechanistic relationship between the residual strength of a structural component to the size and location of a crack – either real or postulated – in that component. However, in some applications of fracture mechanics, derivatives of SIF or ERR with respect to crack size are needed for predicting the stability and arrest of crack propagation. Having differential forms of these parameters are useful in calculating fatigue crack growth in structures under cyclic loads. Another major use of the derivatives of SIF or ERR is in the reliability analysis of cracked structures. For example, the first- and second-order reliability methods [1,2], typically used in probabilistic fracture-mechanics analysis [3-5], require the gradient and Hessian of the performance function with respect to the crack length. In LEFM, this performance function is built on SIF or ERR. Hence, both first- and second-order sensitivities of SIF or ERR are needed for performing reliability analysis. Therefore, an important requirement of some fracture-mechanics analysis is to evaluate the rates of SIF and ERR accurately and efficiently.

For predicting the sensitivities of SIF or ERR, some methods have already appeared. In 1988, Lin and Abel [6] introduced a direct-integration method for virtual crack extension technique that employs the variational formulation and finite element analysis to calculate the first derivative of SIF for a structure containing a single crack. This method maintains all of the advantages of the similar virtual extension techniques introduced by deLorenzi [7,8], Haber and Koh [9], and Barbero and Reddy [10], but adds the capability of calculating the derivatives of SIF. Subsequently, Hwang *et al.* [11] generalized this method to calculate both first- and second-order derivatives for structures involving multiple crack systems, axisymmetric stress state, and crack-face and thermal loading. A salient feature of this method is that the SIF and their derivatives can be evaluated in a single analysis. However, this method requires mesh perturbation – a fundamental requirement of all virtual crack extension techniques. For second-order derivatives, the number of elements surrounding the crack tip that are affected by mesh perturbation has a significant effect on the solution accuracy [11]. Recently, Feijóo *et al.* [12] applied the concepts of shape sensitivity analysis [13] to calculate the first-order derivatives of the potential energy. Since ERR is the first derivative of potential energy, the ERR can be calculated by this approach without any mesh perturbation. Later on, Taroco [14] extended this approach to formulate second-order sensitivities of potential energy to predict derivative of ERR. This is, however, a formidable task since it involves calculation of second-order sensitivities of stress and strain. It is worth mentioning that no numerical results of the sensitivity of ERR were reported by Taroco [14].

This paper presents a new method for predicting the first-order sensitivity of energy release rate and stress-intensity factor for a crack in a homogeneous, isotropic, and linear-elastic structure subject to mode-I loading condition. The method involves variational formulation of continuum mechanics, domain integral representation of ERR, and shape sensitivity analysis by direct differentiation. Two numerical examples are presented to calculate the first-order derivatives of *J*-integral using the proposed method. The results from this method are compared with the reference solutions based on the finite difference method.

SHAPE DESIGN SENSITIVITY ANALYSIS OF LINEAR-ELASTIC SOLID

The method of shape design sensitivity analysis is used to study the behavior of functional when the shape of a body is changed. This method is widely discussed by Haug *et al.* [13] and Cèa [15].

Design Velocity Field

Consider a general three-dimensional body defined as a of uncountable infinity of points, called material points, that can be mapped homeomorphically into the closure of open, connected subsets of Euclidean vector space, $\mathcal{E}^3$. Each such homeomorphism defines a configuration of the body. Consider one particular configuration called reference configuration with domain $\Omega \subset \mathcal{E}^3$ and identify material points of the body with their position vectors, $\boldsymbol{x} \in \Omega$. As shown in Figure 1, consider a motion of the body from a configuration with domain Ω and boundary Γ into another configuration with domain Ω_τ and boundary Γ_τ. This dynamic process of deformation can be represented by

$$\boldsymbol{T} : \boldsymbol{x} \rightarrow \boldsymbol{x}_\tau, \quad \boldsymbol{x} \in \Omega \tag{1}$$

where $\boldsymbol{x}$ and $\boldsymbol{x}_\tau$ are the position vectors of a material point in the reference and deformed configurations, respectively, $\boldsymbol{T}$ is a transformation mapping, τ plays the role of time, with

$$\begin{aligned} \boldsymbol{x}_\tau &= \boldsymbol{T}(\boldsymbol{x},\tau) \\ \Omega_\tau &= \boldsymbol{T}(\Omega,\tau). \\ \Gamma_\tau &= \boldsymbol{T}(\Gamma,\tau) \end{aligned} \tag{2}$$

A design velocity field, $\boldsymbol{V}$, can then be defined as

$$\boldsymbol{V}(\boldsymbol{x}_\tau,\tau) = \frac{d\boldsymbol{x}_\tau}{d\tau} = \frac{d\boldsymbol{T}(\boldsymbol{x},\tau)}{d\tau} = \frac{\partial \boldsymbol{T}(\boldsymbol{x},\tau)}{\partial \tau} \tag{3}$$

In the neighborhood of initial time $\tau = 0$, assuming a regularity hypothesis and ignoring high-order terms, $\boldsymbol{T}$ can be approximated by

$$\boldsymbol{T}(\boldsymbol{x},\tau) = \boldsymbol{T}(\boldsymbol{x},0) + \tau \frac{\partial \boldsymbol{T}(\boldsymbol{x},0)}{\partial \tau} + O(\tau^2) \cong \boldsymbol{x} + \tau \boldsymbol{V}(\boldsymbol{x},0) \tag{4}$$

where $\boldsymbol{x} = \boldsymbol{T}(\boldsymbol{x},0)$ and $\boldsymbol{V}(\boldsymbol{x}) = \boldsymbol{V}(\boldsymbol{x},0)$.

Design Sensitivity Analysis

The variational governing equation for a structural component with the domain Ω can be written as [13]

$$a_\Omega(z, \bar{z}) = \ell_\Omega(\bar{z}), \quad \text{for all } \bar{z} \in \boldsymbol{Z} \tag{5}$$

where z and $\bar{z}$ are the actual displacement and virtual displacement fields of the structure, respectively, $\boldsymbol{Z}$ is the space of kinematically admissible virtual displacements, $a_\Omega(z, \bar{z})$ and $\ell_\Omega(\bar{z})$ are energy bilinear and load linear forms, respectively. The subscript Ω in Equation 5 is used to indicate the dependency of the governing equation on the shape of the structural domain.

The pointwise material derivative at $\boldsymbol{x} \in \Omega$ is defined as [13]

$$\dot{z} = \lim_{\tau \to 0} \left[\frac{z_\tau(\boldsymbol{x} + \tau \boldsymbol{V}(\boldsymbol{x})) - z(\boldsymbol{x})}{\tau} \right] \tag{6}$$

If z_τ has a regular extension to a neighborhood of Ω_τ, then

$$\dot{z}(\boldsymbol{x}) = z'(\boldsymbol{x}) + \nabla z^T \boldsymbol{V}(\boldsymbol{x}) \tag{7}$$

where

$$z' = \lim_{\tau \to 0} \left[\frac{z_\tau(\boldsymbol{x}) - z(\boldsymbol{x})}{\tau} \right] \tag{8}$$

is the partial derivative of z and $\nabla = \{\partial/\partial x_1, \partial/\partial x_2, \partial/\partial x_3\}^T$ is the vector of gradient operators. From Equation 7,

$$z'(x) = \dot{z}(x) - \nabla z^T V(x). \tag{9}$$

One attractive feature of the partial derivative is that, with smoothness assumption, it commutes with the derivatives with respect to x_i , i =1, 2, and 3, because they are derivatives with respect to independent variables, *i.e.*,

$$\left(\frac{\partial z}{\partial x_i}\right)' = \frac{\partial}{\partial x_i}(z'),\ i = 1,2, \text{and } 3. \tag{10}$$

Let ψ_1 be a domain functional, defined as an integral over Ω_τ,

$$\psi_1 = \int_{\Omega_\tau} f_\tau(x_\tau) d\Omega_\tau \tag{11}$$

where f_τ is a regular function defined on Ω_τ. If Ω is C^k regular, then the material derivative of ψ_1 at Ω is [13]

$$\dot{\psi}_1 = \int_\Omega [f'(x) + \text{div} f(x) V(x)] d\Omega . \tag{12}$$

For a functional form of

$$\psi_2 = \int_{\Omega_\tau} g(z_\tau, \nabla z_\tau) d\Omega_\tau , \tag{13}$$

the material derivative of ψ_2 at Ω using Equations 10 and 12 is

$$\dot{\psi}_2 = \int_\Omega [g_z z' + g_{\nabla z} \nabla z' + \text{div}(gV)] d\Omega , \tag{14}$$

where $g_z = \{\partial g/\partial z_1, \partial g/\partial z_2, \partial g/\partial z_3\}^T$, and

$$g_{\nabla z} = \begin{bmatrix} \dfrac{\partial g}{\partial(\partial z_1/\partial x_1)} & \dfrac{\partial g}{\partial(\partial z_1/\partial x_2)} & \dfrac{\partial g}{\partial(\partial z_1/\partial x_3)} \\ \dfrac{\partial g}{\partial(\partial z_2/\partial x_1)} & \dfrac{\partial g}{\partial(\partial z_2/\partial x_2)} & \dfrac{\partial g}{\partial(\partial z_2/\partial x_3)} \\ \dfrac{\partial g}{\partial(\partial z_3/\partial x_1)} & \dfrac{\partial g}{\partial(\partial z_3/\partial x_2)} & \dfrac{\partial g}{\partial(\partial z_3/\partial x_3)} \end{bmatrix}. \tag{15}$$

Using Equation 9, Equation 14 can be rewritten as

$$\dot{\psi}_2 = \int_\Omega \left[g_z \dot{z} - g_z(\nabla z^T V) + g_{\nabla z} \nabla \dot{z} - g_{\nabla z} \nabla(\nabla z^T V) + \text{div}(gV)\right] d\Omega \tag{16}$$

In Equation 16, the material derivative $\dot{z}$ is the solution of the sensitivity equation obtained by taking the material derivative of Equation 5.

If no body force is involved, the variational equation (Equation 5) can be written as

$$a_\Omega(z, \bar{z}) \equiv \int_\Omega \sigma_{ij}(z) \varepsilon_{ij}(\bar{z}) d\Omega = \ell_\Omega(\bar{z}) \equiv \int_\Gamma T_i \bar{z}_i d\Gamma \tag{17}$$

where $\sigma_{ij}(z)$ and $\varepsilon_{ij}(\bar{z})$ are the stress and strain tensors of the displacement z and virtual displacement $\bar{z}$, respectively, T_i is the ith component of the surface traction, and $\bar{z}_i$ is the ith component of $\bar{z}$.

Taking the material derivative of both sides of Equation 17, using Equation 9, and noting that the partial derivatives with respect to τ and x_i commute with each other,

$$a_\Omega(\dot{z}, \bar{z}) = \ell'_V(\bar{z}) - a'_V(z, \bar{z}),\ \forall \bar{z} \in Z \tag{18}$$

where the subscript V is used to indicate the dependency of the terms on the design velocity field. The terms $\ell'_V(\bar{z})$ and $a'_V(z, \bar{z})$ can be further derived as

$$\ell'_V(\bar{z}) = \int_\Gamma \left\{-T_i(\bar{z}_{i,j} V_j) + \left[(T_i \bar{z}_i)_{,j} n_j + H(T_i \bar{z}_i)\right](V_i n_i)\right\} d\Gamma \tag{19}$$

and

$$a'_V(z, \bar{z}) = \int_\Omega [\sigma_{ij}(z)(\bar{z}_{i,k} V_{k,j}) + \sigma_{ij}(\bar{z})(z_{i,k} V_{k,j})] d\Omega - \int_\Omega [\sigma_{ij}(z) \varepsilon_{ij}(\bar{z}) \text{div} V] d\Omega \tag{20}$$

where V_i is the ith component of V, n_i is the ith component of unit normal vector, n, and H is the curvature of the boundary, and $z_{i,j} = \partial z_i/\partial x_j$, $\bar{z}_{i,j} = \partial \bar{z}_i/\partial x_j$, $V_{i,j} = \partial V_i/\partial x_j$.

To evaluate the design sensitivity expression of Equation 16, a numerical method is needed to solve Equation 17. In this study, a standard finite element method (FEM) was used. If the solution z of Equation 17 is obtained using a FEM code, the same code can be used to solve Equation 18 for $\dot{z}$. This solution of $\dot{z}$ can be obtained efficiently since it requires only the evaluation of the same set of FEM matrix equations with a different fictitious load, which is the right hand side of Equation 18. In this study, the ABAQUS [16] finite element code was used in all numerical calculations to be presented in a forthcoming section.

THE *J*-INTEGRAL AND ITS SENSITIVITY

The *J*-integral

Consider a body with a crack of length, a, which is subject to mode-I loading. Using an arbitrary counter-clockwise path, Γ around the crack tip, as shown in Figure 2(a), a formal definition of J under mode-I condition is [17]

$$J \stackrel{\text{def}}{=} \int_{\Gamma} \left(W n_1 - T_i z_{i,1} \right) d\Gamma \tag{21}$$

where, $W = \int \sigma_{ij} d\varepsilon_{ij}$ is the strain energy density with σ_{ij} and ε_{ij} representing components of stress and strain tensors, respectively, z_i and $T_i = \sigma_{ij} n_j$ are the ith component of displacement and traction vectors, n_j is the jth component of unit outward normal to integration path, $d\Gamma$ is the differential length along contour Γ, and $z_{i,1} = \partial z_i / \partial x_1$ is the differentiation of displacement with respect to x_1. Here, the summation convention is adopted for repeated indices.

Domain Integral Form of *J*

For numerical calculation of J, the energy domain integral methodology [18-21] can be used in the finite element analysis. This methodology is versatile, as it can be applied to both quasi-static and dynamic fracture problems with elastic, plastic, or viscoplastic material response, as well as thermal loading. In addition, the domain integral formulation is relatively simple to implement numerically in a finite element code.

Using the divergence theorem, the contour integral defined in Equation 21 can be expanded into an area integral in two dimensions, and volume integral in three-dimensions, over a finite domain surrounding the crack tip or crack front. For a linear or nonlinear elastic material under quasi-static condition, in the absence of body forces, thermal strains, and crack-face tractions, Equation 21 for two-dimensional problem reduces to [18-21]

$$J = \int_A \left[\sigma_{ij} \frac{\partial z_j}{\partial x_1} - W \delta_{1i} \right] \frac{\partial q}{\partial x_i} dA \tag{22}$$

where δ_{1i} is the Kronecker delta, q is an arbitrary but smooth weighting function that is equal to *unity* on Γ_0 and *zero* on Γ_1, and A is the annular area enclosed by the inner contour Γ_0 and outer contour Γ_1 as shown in Figure 2(b). In this study, the inner contour Γ_0 coincides with the crack tip. Hence, A becomes the area inside the outer contour Γ_1. On further expansion,

$$\begin{aligned} J = & \int_A \left(\sigma_{11} \frac{\partial z_1}{\partial x_1} + \sigma_{12} \frac{\partial z_2}{\partial x_1} \right) \frac{\partial q}{\partial x_1} dA \\ & + \int_A \left[\left(\sigma_{21} \frac{\partial z_1}{\partial x_1} + \sigma_{22} \frac{\partial z_2}{\partial x_1} \right) \frac{\partial q}{\partial x_2} \right] dA \, . \\ & - \int_A W \frac{\partial q}{\partial x_1} dA \end{aligned} \tag{23}$$

In Equation 23, for the case of plane stress,

$$\begin{aligned} W &= \int_0^{\varepsilon_{ij}} \sigma_{ij} d\varepsilon_{ij} \\ &= \left(\frac{E}{1-\nu^2} \right) \left(\frac{\varepsilon_{11}^2}{2} + \nu \varepsilon_{11} \varepsilon_{22} \right) + \left(\frac{E}{1+\nu} \right) \varepsilon_{12}^2 \\ &+ \left(\frac{E}{1-\nu^2} \right) \left(\frac{\varepsilon_{22}^2}{2} + \nu \varepsilon_{11} \varepsilon_{22} \right) \end{aligned} \tag{24}$$

where E is the Young's modulus, ν is the Poisson's ratio for the material of the body, and ε_{ij} is the strain field given by

$$\varepsilon_{ij} = \frac{1}{2} \left(\frac{\partial z_i}{\partial x_j} + \frac{\partial z_j}{\partial x_i} \right), \quad i, j = 1,2 \tag{25}$$

Sensitivity of *J*-integral

For two-dimensional plane stress problem, Equation 23 can be expressed as

$$J = \int_A g dA \tag{26}$$

where

$$g = g_1 + g_2 + g_3 + g_4 - g_5 - g_6 \tag{27}$$

with

$$g_1 = \frac{E}{1-\nu^2} \frac{\varepsilon_{11}^2}{2} \frac{\partial q}{\partial x_1}, \tag{28}$$

$$g_2 = \frac{E}{1+\nu} \varepsilon_{21} \frac{\partial z_2}{\partial x_1} \frac{\partial q}{\partial x_1}, \tag{29}$$

$$g_3 = \frac{E}{1+\nu} \varepsilon_{12} \varepsilon_{11} \frac{\partial q}{\partial x_2}, \tag{30}$$

$$g_4 = \frac{E}{1-\nu^2} (\varepsilon_{22} + \nu \varepsilon_{11}) \frac{\partial z_2}{\partial x_1} \frac{\partial q}{\partial x_2}, \tag{31}$$

$$g_5 = \frac{E}{1+\nu}\varepsilon_{12}^2\frac{\partial q}{\partial x_1}, \quad (32)$$

and

$$g_6 = \frac{E}{1-\nu^2}\left(\frac{\varepsilon_{22}^2}{2} + \nu\varepsilon_{11}\varepsilon_{22})\right)\frac{\partial q}{\partial x_1}. \quad (33)$$

Hence, in relation to Equation 12, the material derivative of J-integral is

$$J' = \int_A [g' + \mathrm{div}(g\boldsymbol{V})]dA \quad (34)$$

where,

$$g' = g_1' + g_2' + g_3' + g_4' - g_5' - g_6' \quad (35)$$

and $\boldsymbol{V} = \{V_1, V_2\}^T$. If the crack length, a is the design variable, consider a change in crack length in the x_1 direction (mode-I) only. Hence, $\boldsymbol{V} = \{V_1, 0\}^T$,

$$g\boldsymbol{V} = \{gV_1, 0\}^T, \quad (36)$$

and

$$\mathrm{div}(g\boldsymbol{V}) = \frac{\partial(gV_1)}{\partial x_1} + \frac{\partial(gV_2)}{\partial x_2} = \frac{\partial(gV_1)}{\partial x_1}. \quad (37)$$

Applying Equation 27,

$$\begin{aligned}\mathrm{div}(g\boldsymbol{V}) &= \frac{\partial(g_1V_1)}{\partial x_1} + \frac{\partial(g_2V_1)}{\partial x_1} + \frac{\partial(g_3V_1)}{\partial x_1} \\ &+ \frac{\partial(g_4V_1)}{\partial x_1} - \frac{\partial(g_5V_1)}{\partial x_1} - \frac{\partial(g_6V_1)}{\partial x_1}\end{aligned} \quad (38)$$

Hence, Equation 34 becomes

$$J' = \int_A (G_1 + G_2 + G_3 + G_4 - G_5 - G_6)dA \quad (39)$$

where

$$G_i = g_i' + \frac{\partial(g_iV_1)}{\partial x_1}, \; i = 1, \cdots, 6. \quad (40)$$

Consider the term G_1, which can be further expanded as

$$\begin{aligned} G_1 &= g_1' + \frac{\partial(g_1V_1)}{\partial x_1} \\ &= \frac{E}{1-\nu^2}\left(\frac{\varepsilon_{11}^2}{2}\frac{\partial q}{\partial x_1}\right)' + \frac{E}{1-\nu^2}\frac{\partial}{\partial x_1}\left(\frac{\varepsilon_{11}^2}{2}\frac{\partial q}{\partial x_1}V_1\right) \\ &= \frac{E}{1-\nu^2}\left[\frac{\partial q}{\partial x_1}\varepsilon_{11}(z)\varepsilon_{11}(z') + \frac{\varepsilon_{11}^2}{2}\left(\frac{\partial q}{\partial x_1}\right)'\right] \\ &+ \frac{E}{1-\nu^2}\left[\varepsilon_{11}\frac{\partial\varepsilon_{11}}{\partial x_1}\frac{\partial q}{\partial x_1}V_1 + \frac{\varepsilon_{11}^2}{2}\frac{\partial^2 q}{\partial x_1^2}V_1 + \frac{\varepsilon_{11}^2}{2}\frac{\partial q}{\partial x_1}\frac{\partial V_1}{\partial x_1}\right]\end{aligned} \quad (41)$$

Since q is an explicit function of x_1 and x_2, so is $\partial q/\partial x_1$. Therefore,

$$\left(\frac{\partial q}{\partial x_1}\right)' = 0. \quad (42)$$

Using Equations 9 and 42 and the following relations,

$$\varepsilon_{11}(z) = \frac{\partial z_1}{\partial x_1} \quad (43)$$

and,

$$\nabla z^T\boldsymbol{V} = \begin{bmatrix}\frac{\partial z_1}{\partial x_1} & \frac{\partial z_1}{\partial x_2} \\ \frac{\partial z_2}{\partial x_1} & \frac{\partial z_2}{\partial x_2}\end{bmatrix}\begin{Bmatrix}V_1 \\ 0\end{Bmatrix} = \begin{Bmatrix}\frac{\partial z_1}{\partial x_1}V_1 \\ \frac{\partial z_2}{\partial x_1}V_1\end{Bmatrix}, \quad (44)$$

Equation 41 becomes,

$$\begin{aligned} G_1 &= \frac{E}{1-\nu^2}\frac{\partial q}{\partial x_1}\frac{\partial z_1}{\partial x_1}\frac{\partial(\dot{z} - \nabla z^T\boldsymbol{V})_1}{\partial x_1} \\ &+ \frac{E}{1-\nu^2}\left[\frac{\partial q}{\partial x_1}\frac{\partial z_1}{\partial x_1}\frac{\partial^2 z_1}{\partial x_1^2}V_1 + \frac{1}{2}\left(\frac{\partial z_1}{\partial x_1}\right)^2\frac{\partial^2 q}{\partial x_1^2}V_1\right] \\ &+ \frac{1}{2}\frac{E}{1-\nu^2}\frac{\partial q}{\partial x_1}\left(\frac{\partial z_1}{\partial x_1}\right)^2\frac{\partial V_1}{\partial x_1} \\ &= \frac{E}{1-\nu^2}\frac{\partial q}{\partial x_1}\frac{\partial z_1}{\partial x_1}\left[\frac{\partial \dot{z}_1}{\partial x_1} - \frac{\partial}{\partial x_1}\left(\frac{\partial z_1}{\partial x_1}V_1)\right)\right] \\ &+ \frac{E}{1-\nu^2}\left[\frac{\partial q}{\partial x_1}\frac{\partial z_1}{\partial x_1}\frac{\partial^2 z_1}{\partial x_1^2}V_1 + \frac{1}{2}\left(\frac{\partial z_1}{\partial x_1}\right)^2\frac{\partial^2 q}{\partial x_1^2}V_1\right] \\ &+ \frac{1}{2}\frac{E}{1-\nu^2}\frac{\partial q}{\partial x_1}\left(\frac{\partial z_1}{\partial x_1}\right)^2\frac{\partial V_1}{\partial x_1}\end{aligned} \quad (45)$$

Since,

$$\frac{\partial}{\partial x_1}\left(\frac{\partial z_1}{\partial x_1}V_1\right)=\frac{\partial^2 z_1}{\partial x_1^2}V_1+\frac{\partial z_1}{\partial x_1}\frac{\partial V_1}{\partial x_1} \tag{46}$$

Equation 41 further simplifies to

$$\begin{aligned} G_1 &= \frac{E}{1-\nu^2}\frac{\partial q}{\partial x_1}\frac{\partial z_1}{\partial x_1}\left[\frac{\partial \dot{z}_1}{\partial x_1}-\left(\frac{\partial^2 z_1}{\partial x_1^2}V_1+\frac{\partial z_1}{\partial x_1}\frac{\partial V_1}{\partial x_1}\right)\right] \\ &\quad +\frac{E}{1-\nu^2}\frac{\partial q}{\partial x_1}\frac{\partial z_1}{\partial x_1}\frac{\partial^2 z_1}{\partial x_1^2}V_1 \\ &\quad +\frac{1}{2}\frac{E}{1-\nu^2}\frac{\partial q}{\partial x_1}\left[\left(\frac{\partial z_1}{\partial x_1}\right)^2\frac{\partial^2 q}{\partial x_1^2}V_1+\left(\frac{\partial z_1}{\partial x_1}\right)^2\frac{\partial V_1}{\partial x_1}\right] \\ &= \frac{E}{1-\nu^2}\frac{\partial q}{\partial x_1}\left[\frac{\partial z_1}{\partial x_1}\frac{\partial \dot{z}_1}{\partial x_1}-\frac{1}{2}\left(\frac{\partial z_1}{\partial x_1}\right)^2\frac{\partial V_1}{\partial x_1}\right] \\ &\quad +\frac{1}{2}\frac{E}{1-\nu^2}\frac{\partial q}{\partial x_1}\left(\frac{\partial z_1}{\partial x_1}\right)^2\frac{\partial^2 q}{\partial x_1^2}V_1 \\ &= \frac{E}{1-\nu^2}\frac{\partial q}{\partial x_1}\left[\varepsilon_{11}\frac{\partial \dot{z}_1}{\partial x_1}-\frac{1}{2}\varepsilon_{11}^2\frac{\partial V_1}{\partial x_1}+\frac{1}{2}\varepsilon_{11}^2\frac{\partial^2 q}{\partial x_1^2}V_1\right] \end{aligned} \tag{47}$$

Following similar considerations,

$$\begin{aligned} G_2 &= \frac{E}{1+\nu}\frac{\partial q}{\partial x_1}\frac{\partial z_2}{\partial x_1}\left(\frac{\partial \dot{z}_2}{\partial x_1}+\frac{1}{2}\frac{\partial \dot{z}_1}{\partial x_2}\right) \\ &\quad -\frac{1}{2}\frac{E}{1+\nu}\frac{\partial q}{\partial x_1}\frac{\partial z_2}{\partial x_1}\left(\varepsilon_{11}\frac{\partial V_1}{\partial x_2}+\frac{\partial z_2}{\partial x_1}\frac{\partial V_1}{\partial x_1}\right), \\ &\quad +\frac{E}{1+\nu}\frac{\partial q}{\partial x_1}\left(\frac{1}{2}\frac{\partial z_1}{\partial x_2}\frac{\partial \dot{z}_2}{\partial x_1}\right) \end{aligned} \tag{48}$$

$$\begin{aligned} G_3 &= \frac{1}{2}\frac{E}{1+\nu}\left(\varepsilon_{11}\varepsilon_{12}\frac{\partial^2 q}{\partial x_1 \partial x_2}V_1\right) \\ &\quad +\frac{1}{2}\frac{E}{1+\nu}\frac{\partial q}{\partial x_2}\left[\varepsilon_{11}\left(\frac{\partial \dot{z}_2}{\partial x_1}+\frac{\partial \dot{z}_1}{\partial x_2}\right)+2\varepsilon_{12}\frac{\partial \dot{z}_1}{\partial x_1}\right] \\ &\quad -\frac{1}{2}\frac{E}{1+\nu}\frac{\partial q}{\partial x_2}\left(\varepsilon_{11}\varepsilon_{22}\frac{\partial V_1}{\partial x_1}-\varepsilon_{11}^2\frac{\partial V_1}{\partial x_2}-\varepsilon_{11}\varepsilon_{12}\frac{\partial V_1}{\partial x_1}\right) \end{aligned} , \tag{49}$$

$$\begin{aligned} G_4 &= \frac{E}{1-\nu^2}\frac{\partial q}{\partial x_2}\frac{\partial z_2}{\partial x_1}\left(\frac{\partial \dot{z}_1}{\partial x_1}+\nu\frac{\partial \dot{z}_2}{\partial x_1}\right) \\ &\quad -\frac{E}{1-\nu^2}\frac{\partial q}{\partial x_2}\left[\left(\frac{\partial z_2}{\partial x_1}\right)^2\frac{\partial V_1}{\partial x_2}+\nu\varepsilon_{11}\frac{\partial z_2}{\partial x_1}\frac{\partial V_1}{\partial x_1}\right] \\ &\quad +\frac{E}{1-\nu^2}(\varepsilon_{22}+\nu\varepsilon_{11})\left[\frac{\partial q}{\partial x_2}\frac{\partial \dot{z}_2}{\partial x_1}+\frac{\partial z_2}{\partial x_1}\frac{\partial^2 q}{\partial x_1 \partial x_2}V_1\right] \end{aligned} , \tag{50}$$

$$\begin{aligned} G_5 &= \frac{E}{1+\nu}\frac{\partial q}{\partial x_1}\varepsilon_{11}\left[\frac{\partial \dot{z}_1}{\partial x_2}+\frac{\partial \dot{z}_2}{\partial x_1}-\frac{\partial z_2}{\partial x_1}\frac{\partial V_1}{\partial x_1}\right] \\ &\quad +\frac{E}{1+\nu}\frac{\partial q}{\partial x_1}\varepsilon_{11}\left[-\varepsilon_{11}\frac{\partial V_1}{\partial x_2}+\varepsilon_{12}\frac{\partial V_1}{\partial x_1}\right] \end{aligned} \tag{51}$$

and

$$G_6=\frac{E}{1-\nu^2}\frac{\partial q}{\partial x_1}\left[(\varepsilon_{22}+\nu\varepsilon_{11})\frac{\partial \dot{z}_2}{\partial x_2}+\nu\varepsilon_{22}\frac{\partial \dot{z}_1}{\partial x_1}+\frac{\varepsilon_{22}^2}{2}\frac{\partial V_1}{\partial x_1}\right]. \tag{52}$$

Equations 47-52 provide explicit expressions of G_i, i = 1,...,6, which can be inserted in Equation 39 to yield first-order sensitivity of J with respect to crack size. The integral in Equation 39 is independent of domain size A and can be calculated numerically using standard Gaussian quadrature. A 2 × 2 or higher integration rule is recommended for calculating J'.

NUMERICAL EXAMPLES

Example 1: Center-Cracked Plate

Consider a center-cracked panel (CCP) with width, $2W$ = 20 units, length, $2L$ = 20 units and a crack length, $2a$. For the CCP specimen, shown in Figure 3(a), two crack sizes with normalized crack lengths, a/W = 0.05 and 0.1 were considered. The plate is subjected to far-field remote tensile stress, σ^∞ = 1 units. For material properties, the elastic modulus E = 26 units, Poisson's ratio, ν = 0.3. A plane stress condition was assumed.

Figure 3(b) shows a finite element mesh for 1/4 model due to the symmetry of this problem. A total of 270 elements and 843 nodes were used in this mesh. Second-order elements from ABAQUS element library were used. The element type was CPS8R – the reduced integration, eight-noded quadrilateral element. Focused elements with collapsed noded were employed in the vicinity of crack tip. A 2 × 2 Gaussian integration was used.

Table 1 shows the numerical results of J and dJ/da for this CCP specimen for a/W = 0.05 and 0.1. For dJ/da, two sets of results are shown. One is based on the proposed method described in this paper. The other is based on the finite difference method by perturbing the crack-tip location. A one-percent perturbation was used in the finite difference calculations. The results in Table 1 shows that very accurate results of dJ/da can be obtained using shape sensitivity analysis when compared with the corresponding results of the finite difference method. Unlike the virtual crack extension techniques, no mesh perturbation is needed in the proposed method. The difference between the results of the proposed method and the finite-difference method is less than 2 percent.

Example 2: Single-Edged-Notched Tension Specimen

Consider a single-edged-notched tension (SENT) specimen with width, $W = 10$ units, length, $L = 10$ units, and a crack length, a. It is shown in Figure 4(a). Two crack sizes with normalized crack lengths, $a/W = 0.25$ and 0.5 were considered. The plate is subjected to far-field remote tensile stress, $\sigma^{\infty} = 1$ unit. For material properties, the elastic modulus $E = 30$ units, Poisson's ratio, $\nu = 0.25$. A plane stress condition was assumed.

Figure 4(b) shows a finite element mesh for 1/2 model due to the symmetry of this problem. A total of 270 elements and 843 nodes were used in this mesh as well. Similar to Example 1, second-order eight-noded quadrilateral elements were employed. Also, focused elements with collapsed nodes were used in the vicinity of crack tip. For calculating J', a 2×2 Gaussian quadrature was used.

Table 2 shows the similar comparisons of J and dJ/da for the SENT specimen calculated by the proposed method and the finite difference method using one-percent perturbation of the crack length. Once again, very accurate results were achieved by the proposed method. The maximum error in calculating the sensitivity of J-integral was less than two percent.

CONCLUSIONS

A new and efficient method was developed for shape sensitivity analysis of J-integral for a crack in a homogeneous, isotropic, and linear-elastic structure subject to model-I loading condition. Based on this study, the following conclusions can be made:

- In the proposed method, the material derivative of continuum mechanics is utilized to conduct continuum-based shape sensitivity analysis. For the sensitivity equation, the domain integral representation of J-integral is considered, and the material derivative of this variational form is then obtained using the design velocity field concept to describe shape design changes. The direct differentiation method was employed to develop sensitivity coefficients. Hence, the proposed method avoids the need for mesh perturbation – a fundamental requirement of all virtual crack extension techniques.

- The proposed shape sensitivity analysis is based on the domain integral form of J-integral. Hence, only first-order sensitivities of stress and strain are required to calculate the sensitivity of J. Existing methods of shape sensitivity analysis, which are based on expressing J as a rate of potential energy, requires second-order sensitivity of potential energy to produce first-order sensitivity of J. Hence, the proposed method is simpler and more efficient than the existing methods.

- Since the governing variational equation is differentiated before the process of discretization, the resulting sensitivity equations are independent of any approximate numerical techniques, such as the finite element method, boundary element method, meshless method, and others.

- The numerical results from two examples based on a center-cracked panel and a single-edged-notched tension specimen suggest that the proposed method yields very accurate sensitivity of J. The maximum error in calculating the sensitivity of J is less than two percent.

Work is on progress in extending the shape sensitivity method for solving problems in mixed-mode and elastic-plastic fracture mechanics.

ACKNOWLEDGMENTS

This work was supported by the Faculty Early Career Development Program of the U.S. National Science Foundation (Grant No. CMS-9733058). The program directors were Drs. Sunil Saigal and Ken Chong.

REFERENCES

1. Madsen, H. O., Krenk, S., and Lind, N. C., Methods of Structural Safety, Prentice-Hall, Inc., Englewood Cliffs, New Jersey, 1986.

2. Bergman, L. A., Bucher, C. G., Dasgupta, G., Deodatis, G., Ghanem, R. G., Grigoriu, M., Hoshiya, M., Johnson, E. A., Naess, A., Pradlwarter, H. J., Schueller, G. I., Shinozuka, M., Sobczyk, K., Spanos, P. D., Spencer, B. F., Sutoh, A., Takada, T., Wedig, W. V., Wojtkiewicz, S. F., Yoshida, I., Zeldin, B. A., and Zhang, R., "A State-of-the-Art Report on Computational Stochastic Mechanics," *Probabilistic Engineering Mechanics*, edited by G. I. Schueller, Vol. 12, No. 4, pp. 197-321, 1997.

3. Provan, James, W., Probabilistic Fracture Mechanics and Reliability, Martinus Nijhoff Publishers, Dordrecht, The Netherlands, 1987.

4. Rahman, S., "A Stochastic Model for Elastic-Plastic Fracture Analysis of Circumferential Through-Wall-Cracked Pipes Subject to Bending," *Engineering Fracture Mechanics*, Vol. 52, No. 2, September 1995.

5. Rahman, S. and Kim, J-S., "Probabilistic Fracture Mechanics for Nonlinear Structures," submitted for review and possible publication in *International Journal of Fracture*, 1999.

6. Lin, S. C. and Abel, J., "Variational Approach for a New Direct-Integration Form of the Virtual Crack Extension Method," *International Journal of Fracture*, Vol. 38, pp. 217-235, 1988.

7. deLorenzi, H. G., "On the Energy Release Rate and the J-integral for 3-D Crack Configurations," *International Journal of Fracture*, Vol. 19, pp. 183-193, 1982.

8. deLorenzi, H. G., "Energy Release Rate Calculations by the Finite Element Method," *Engineering Fracture Mechanics*, Vol. 21, pp. 129-143, 1985.

9. Haber, R. B. and Koh, H. M., "Explicit Expressions for Energy Release Rates using Virtual Crack Extensions," *International Journal of Numerical Methods in Engineering*, Vol. 21, pp. 301-315, 1985.

10. Barbero, E. J. and Reddy, J. N., "The Jacobian Derivative Method for Three-Dimensional Fracture Mechanics," *Communications in Applied Numerical Methods*, Vol. 6, pp. 507-518, 1990.

11. Hwang, C. G., Wawrzynek, P. A., Tayebi, A. K., and Ingraffea, A. R., "On the Virtual Crack Extension Method for Calculation of the Rates of Energy Release Rate," *Engineering Fracture Mechanics*, Vol. 59, pp. 521-542, 1998.

12. Feijóo, R. A., Padra, C., Saliba, R., Taroco, E., and Vénere, M. J., "Shape Sensitivity Analysis for Energy Release Rate Evaluations and Its application to the Study of Three-Dimensional Cracked Bodies," accepted in *Computational Methods in Applied Mechanical Engineering*, 2000.

13. Haug, E. J., Choi, K. K., and Komkov, V., Design Sensitivity Analysis of Structural Systems, Academic Press, New York, NY, 1986.

14. Taroco, E., "Shape Sensitivity Analysis in Linear Elastic Fracture Mechanics," accepted in *Computational Methods in Applied Mechanical Engineering,* 2000.

15. Cèa, J., "Problems of Shape Optimal Design," *Optimization of Distributed Parameter Structures*, eds. E. J. Haug & J. Cèa. Sijthoff and Noordhoff, Alphen aan den Rijn, The Netherlands, pp. 1005-1048, 1981.

16. ABAQUS, User's Guide and Theoretical Manual, Version 5.8, Hibbitt, Karlsson, and Sorenson, Inc., Pawtucket, RI, 1999.

17. Rice, J. R., "A Path Independent Integral and the Approximate Analysis of Strain Concentration by Notches and Cracks," *Journal of Applied Mechanics*, 35, pp. 379-386, 1968.

18. Shih, C. F., Moran, B., and Nakamura, T., "Energy Release Rate Along a Three-Dimensional Crack Front in a Thermally Stressed Body," *International Journal of Fracture*, Vol. 30, pp. 79-102, 1986.

19. Moran, B. and Shih, C. F., "A General Treatment of Crack Tip Contour Integrals," *International Journal of Fracture*, Vol. 35, pp. 295-310, 1987.

20. Li, F. Z., Shih, C. F., and Needleman A., "A comparison of methods for calculating Energy Release Rates," *Engineering Fracture Mechanics*, Vol. 21, No. 2, pp 405-421, 1985.

21. Anderson, T. L., Fracture Mechanics: Fundamentals and Applications, Second Edition, CRC Press Inc., Boca Raton, Florida, 1995.

Table 1. Sensitivity of J for CCP specimen by the proposed and finite difference methods (Example 1)

		Sensitivity of J-integral (dJ/da)		
a/W	J-integral	Proposed Method	Finite Difference	Error[(a)] (percent)
0.05	6.01×10^{-5}	1.20×10^{-4}	1.22×10^{-4}	1.63
0.1	1.23×10^{-4}	1.27×10^{-4}	1.29×10^{-4}	1.55

(a) Error = (dJ/da by finite difference method - dJ/da by proposed method)×100/ dJ/da by finite difference method

Table 2. Sensitivity of J for SENT specimen by the proposed and finite difference methods (Example 2)

		Sensitivity of J-integral (dJ/da)		
a/W	J-integral	Proposed Method	Finite Difference	Error[(a)] (percent)
0.25	5.80×10^{-7}	4.63×10^{-7}	4.59×10^{-7}	- 0.87
0.5	4.70×10^{-6}	3.48×10^{-7}	3.54×10^{-7}	1.69

(a) Error = (dJ/da by finite difference method - dJ/da by proposed method)×100/ dJ/da by finite difference method

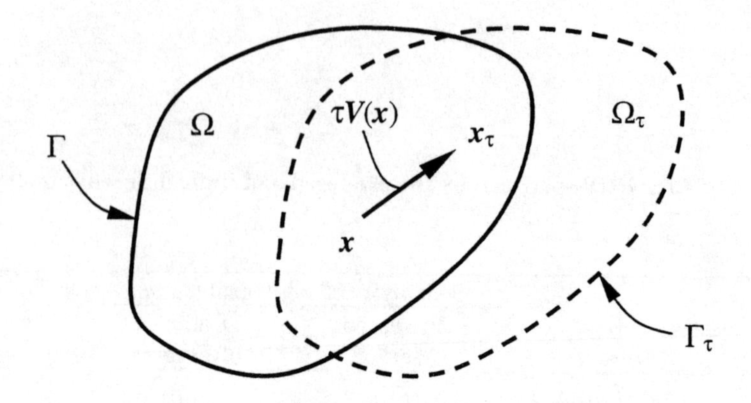

Figure 1. Variation of domain

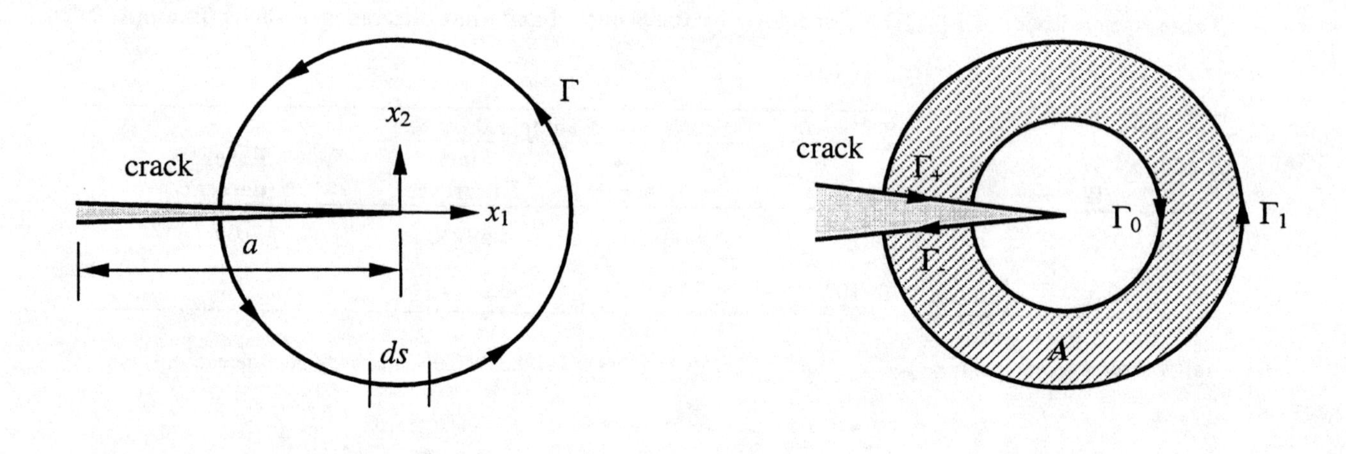

(a) Arbitrary contour around a crack tip **(b) Inner and outer contours enclosing A**

Figure 2. J-integral fracture parameter

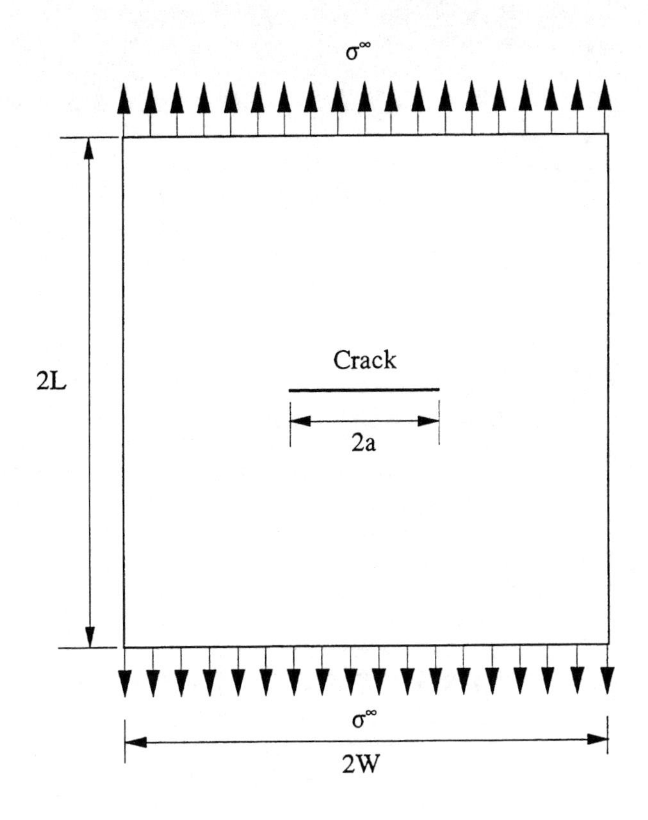

(a) geometry and loads

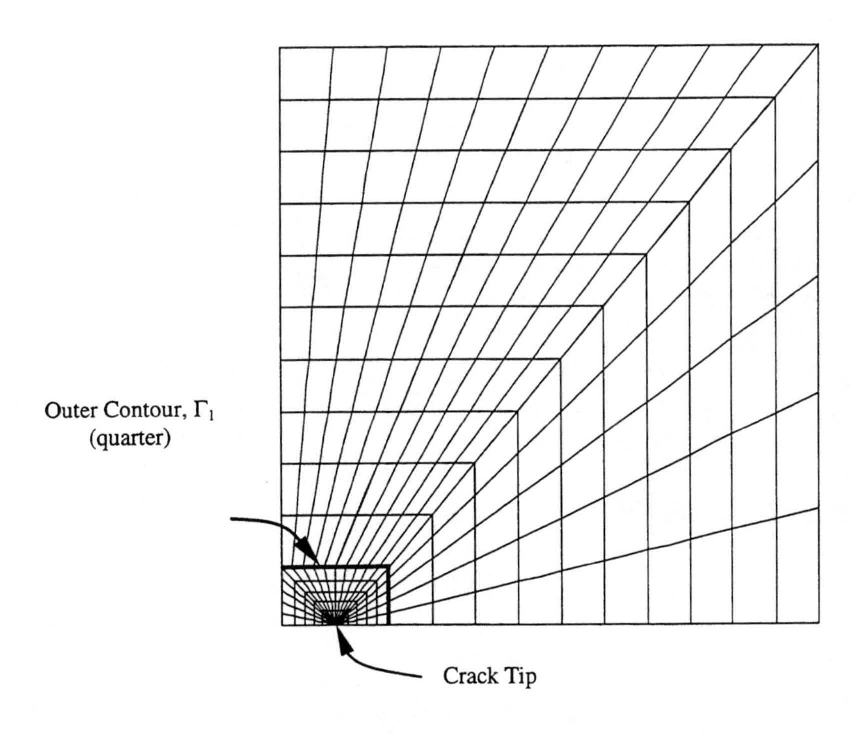

(b) finite element mesh (1/4 model)

Figure 3. Center-cracked panel under mode-I tension loading

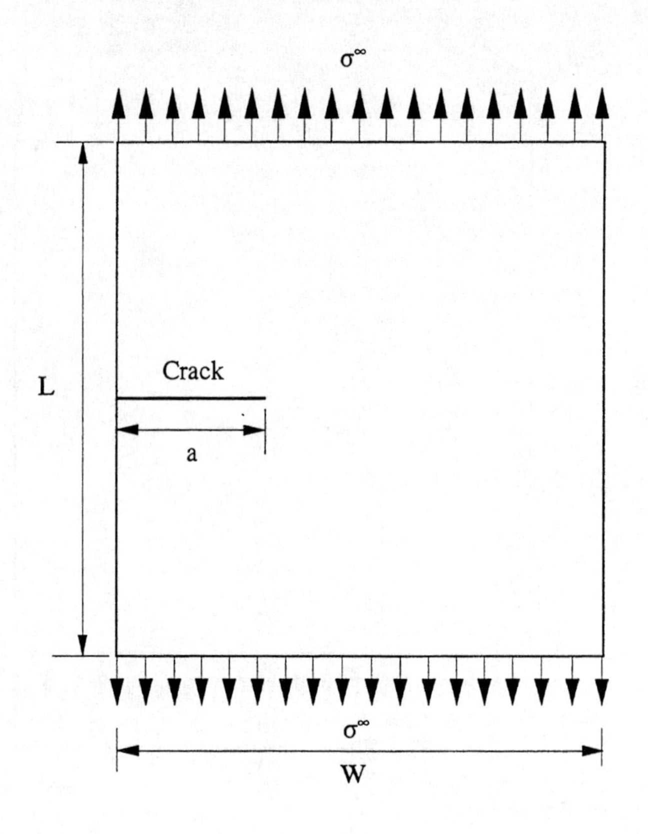

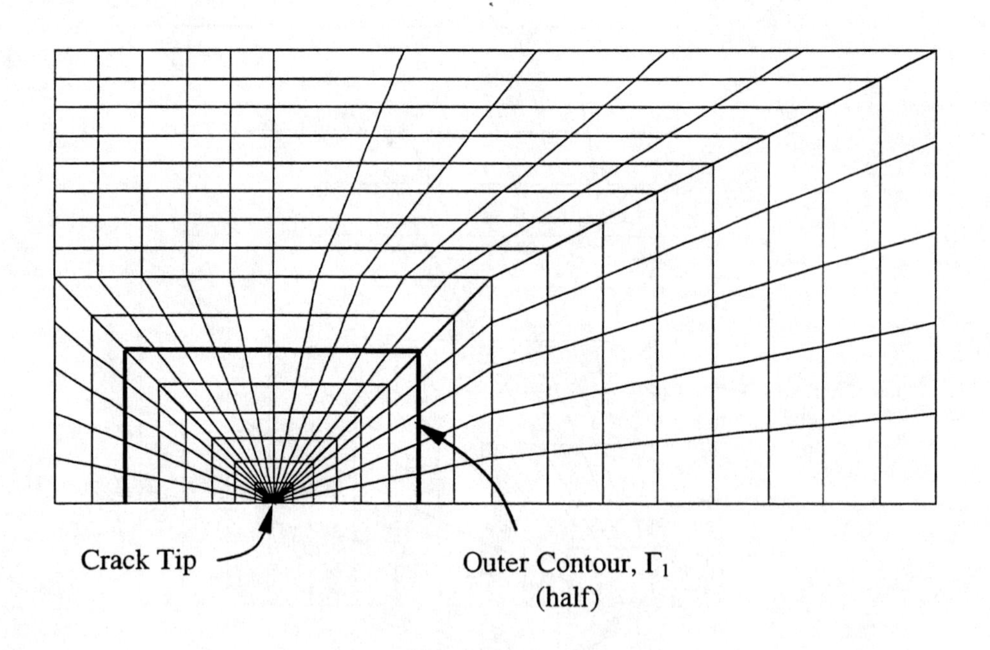

(b) finite element mesh (1/2 model)

Figure 4. Single-edged-notched tension specimen under mode-I loading

PVP-Vol. 410-1, Assessment Methodologies for Preventing Failure:
Deterministic and Probabilistic Aspects and Weld Residual Stress
Volume 1
ASME 2000

INTEGRATION OF NDI RELIABILITY INTO FAD-BASED PROBABILISTIC ANALYSIS OF PRESSURE VESSEL INTEGRITY

Dragos D. Cioclov
Fraunhofer-Institute
For Nondestructive Testing IZFP
University, Building 37
D-66123 Saarbrücken, Germany
Tel/Fax: +49.681.9302.3973/-5930
cioclov@izfp.fhg.de

Michael Kröning
Fraunhofer-Institute
For Nondestructive Testing IZFP
University, Building 37
D-66123 Saarbrücken, Germany
Tel/Fax: +49.681.9302.3800/-5934
kroening@izfp.fhg.de

ABSTRACT

The paper presents a methodology that aims towards the integration of nondestructive inspection (NDI) quality as quantified by the probability of detection (POD), into a probabilistic failure assessment diagram (FAD) for analysis of pressure vessel integrity. The first part of the paper introduces the common models for POD estimation (e.g., Marshall, log-odds) along with a new POD model developed by the authors. The FAD procedure coupled with POD models is developed into a computer code called FADSIM. The statistic of input includes flaw size and material characteristics such as fracture toughness, UTS, and yield strength.

The developed rationale enables the estimation of the gain in reliability against fracture failure when applying a specific NDI procedure characterized by its POD – flaw size dependence in the context of deterministic or statistically variable FAD input parameters. Extended probabilistic sensitivity analyses are thus enabled. Along this line, the accuracy in the underlying statistical sampling simulation methodology was checked against a compact analytic probabilistic model. Acceptable accuracy was demonstrated for a number of simulations (10^5 and greater) that can be performed in a reasonable time with current computing hardware (PC). For the purpose of illustration, representative parametric results have been produced for the case of a cylindrical pressure vessel model having an axial or transverse semi-elliptic crack at the inner surface. The beneficial role of NDI has been quantified in terms of the decrease of failure probability as function of applied stress or a FAD defined safety index.

1. INTRODUCTION

In a previous paper (Cioclov and Kröning 1999), a rationale for obtaining analytic solution for the probability of failure of structural elements under static loading and brittle circumstance has been developed for several geometrical cases. However, only simple geometry of structural elements containing cracks is amenable to such a treatment. This is the case when the crack size is distributed exponentially in large components (compared to the crack size) and the material fracture toughness is deterministic or distributed according to a Rayleigh distribution. Within this framework, the influence of the quality of nondestructive inspection (NDI), as expressed by the probability of detection (POD) was implemented in one limited case (crack size exponentially distributed, material fracture toughness assuming deterministic values and the POD expressed by an asymptotic exponential rule) and a compact solution of the probability of failure has been developed.

Analytical solutions enable rapid point assessments and sensitivity studies of the failure probabilities trends in relationship with loading intensity and other parameters governing the crack size distribution and the quality of NDI. However, many real cases do not fall into the territory of the outlined solution approach due to the complex structural geometry and due to the fact that elastic-plastic failure behavior may prevail. Moreover, in the outlined analytic solution a mean correction factor in the stress intensity factor (SIF) is involved. The sense given to the mean value of this factor for a population of crack size is by no means obvious.

In order to account on the elastic-plastic behavior in a static failure model, the concept of deterministic failure assessment diagram (FAD) model has recently gained in popularity. The basis of this methodology has been described in detail elsewhere (Milne et al., 1988) together with its limitations and hidden conservatism (e.g. Chell et al., 1995). In recent years, there is a trend to adapt FAD procedure to probabilistic treatment of structural failures. Two ways of thinking have emerged along this line:

The first approach can be quantified as purely statistical. Thus, the key FAD governing parameters that are prone to statistical scatter, such as crack size, material strength and toughness characteristics are simulated by Monte-Carlo algorithms and the analysis is performed by repeated numeric (synthetic) sampling with FAD assessment performed in each sampling sequence. The probability of failure simply results as the ratio of the number of cases when representative FAD points fall in the "failure space" to the total number of samples (simulations).

The second approach (e.g. Warke et al., 1999 – with an overview on probabilistic FAD) is based on computing the probability of failure according to First Order Reliability Method (FORM) and mapping the results as iso-probable limit curves in FAD space. The user needs to perform deterministic assessments and according to the position of the assessment point on iso-probable FAD contours, the probability of failure results.

In the present paper, a rationale based on statistical approach will be presented with emphasis on pressure vessels technology applications.

2. THE NDI RELIABILITY

The management of failure risk in pressure vessels is closely related to the capability of inspection systems to find bility of inspection systems to find small acute flaws (cracks) that might trigger fracture. When applied at the lower extreme size, not all flaws will be detected. Even the repeated inspection of a small flaw will not necessarily improve the NDI results. Because of inherent uncertainties in NDI, its reliability is characterized in terms of probability of detection (POD) or probability of non-detection (PND) as function of the flaw size *a*. Obviously, POD(*a*) + PND(*a*) = 1. Statistically planned reliability experiments on specimens containing flaws of known size, enables to infer for the POD curves analytical functions (e.g., Berens and Hovey 1981). Many extensive round-robin tests have resulted in various analytical pro posals for POD – laws distributions. An exponential distribution with asymptotic truncation on POD at values lower than unity have been proposed such as in the early stages of European PISC programs devoted to nuclear power plant NDI (PISC, 1979). The asymptotic value of POD smaller than unity for large flaws size indicates merely a real-life fact that NDI reliability cannot be "perfect" i.e., always POD<1.

Other experimental NDI programs have favored POD(*a*) analytical description in terms of error function (Lo et al. 1989), Rayleigh distribution of PND (Swift and Conolly, 1985) or log-logistic (log odds) function (Berens and Hovey 1983). Recently, the authors of the present paper have used a POD(*a*) distribution based on special "inverse-log power - ILP" functions (developed previously for fatigue reliability assessments – Cioclov, 1975).

For the purpose of the present analysis of NDI influence on the static fracture reliability of pressure vessels, only the exponential-asymptotic (EA) and inverse-log exponential (ILP) POD distributions will be considered.

The analytical form of EA – POD distribution is:

$$POD(a) = A\left[1 - \exp\left(-a/a1\right)\right] \qquad (1)$$

where *a* is regarded as the depth extent of a crack-type defect, while *a*1 and *A* are empirical parameters. Figure 1a shows a principle sketch of Equation (1). As outlined, *A* represents an asymptotic value of POD for large crack size. Based on initial PISC programs, Marshall (1982) proposed *A* = 0,995 and *a*1 = 8,85 mm. These values will be used further for exemplification purpose.

A modified form of Equation (1) can be introduced in order to account for the observed inferior threshold of crack size a_o under which there is no chance to detect the flaw i.e. POD = 0 for $a \le a_o$. The modified Equation (1) is:

$$POD(a) = A\{1 - \exp\left[-\left(\frac{a_o - a_o}{a1 - a_o}\right)\right]\} \qquad (2)$$

where *A* and *a*1 as in Equation 1.

The ILP analytical form of POD has a good versatility to describe steep variation of POD associated with asymptotic trends both at upper threshold POD and inferior crack size threshold a_o. The analytic form of ILP-POD distribution is:

$$POD(a) = POD*\left(\frac{POD1}{POD*}\right)^{\frac{\ln(a1/a_o)}{\ln(a/a_o)}} \quad (3)$$

For $a<a_o$ the probability of detection is vanishing. POD1 is the probability of detection corresponding to the crack size $a1$ and POD* is the asymptotic POD value for large crack size. Figure 1b illustrates the significance of the outlined parameters. It is apparent, that the analytical form expressed by Equation 3 can be easily adjusted to experimental data with a straightforward significance of the fitting parameters.

Figure 2 and 3 illustrates the fitting of experimental data with POD-distributions expressed by Equations 1 and 3 respectively. In Figure 2, the POD data obtained by ultrasonic testing of structural steel welded specimens with lack of fusion flaws are fitted according to Equation 2. Details on the NDI technique and POD experimental data processing have been reported in detail by Froeli (e.g. Froeli, 1997). Figure 3 illustrates POD data obtained by ultrasonic testing of 2219-T87 aluminum alloy specimens containing fatigue cracks of varying size and aspect ratio (depth/length). In Figure 3, lower-envelope POD curves have been fitted. Sources of this data and detailed description of the experimental technique can be find in Rummel et al., 1994.

In the following, for the purpose of sensitivity studies related to failure risk in pressure vessels conjointly with the influence of NDI reliability, the POD curves as given by Eq. 1 and 3 will be retained.

3. FAD APPROACH IN A DETERMINISTIC AND PROBABILISTIC FORMAT

The probabilistic fracture analysis presented in this paper is based on Failure Assessment Diagram (FAD) or R6 methodology (for a detailed description, see Milne et al. 1988). The analysis is two-parametric. In the simplest form, one parameter is defined as the ratio of elastic stress-intensity factor (SIF), K_I to material fracture toughness K_c, i.e. $Kr = K_I/K_c$. This parameter is related to fracture under prevailing elastic state of stress, hence with brittle fracture circumstance. The second parameter S_r, is defined as the ratio of the loading intensity P vs. the plastic collapse load P_L in a ductile fracture scenario of a flawed structural element i.e. $S_r = P/P_L$. Alternatively, S_r may be expressed in terms of the ratio of net stress S_{NET} acting in the flawed cross-section to the flaw stress $\sigma_{flow} = (R_m + R_e)/2$, with R_m the ultimate tensile stress and R_e the yield strength of the material.

According to the FAD philosophy, failure occurs when in the (K_r, S_r) presentation the state point falls outside the region bound by a limiting curve $K_r = f_{LIM}(S_r)$ – the "failure space".

In the present analysis ("Option 1"), FAD route has been employed. This simple assessment route is based only on the knowledge of 0,2 % proof stress ($R_{e,\,0.2}$), ultimate tensile strength (R_m) and material fracture toughness (K_c). Details on other possible options may be found in Milne et al., 1988. Under Option 1, the limiting at curve at failure is defined as:

$$K_r = \left(1-0.14S_r^2\right)\left[0.3+0.7\exp\left(-0.65S_r^6\right)\right]$$
$$for S_r \le L_r^{max} and \quad (4)$$
$$K_r = 0 for\ S_r > L_r^{max}$$

where L_r^{max} is the material dependent cut-off of the f_{LIM} curve:

$$L_r^{max} = R_m / R_{e,0.2}$$

These are the basics of deterministic FAD-methodology that ends for a given flawed structural element under specified loading circumstance with a statement of Yes or No, as related to the failure occurrence.

However, in real-life cases various degrees of uncertainty are encountered in specifying numerical values of key FAD parameters. This is manifested as statistical scatter in experimentally assessed material characteristics such as K_c, (J_c), R_m, R_e, ... or in sizing the flaws dimensions by a NDI procedure of a given reliability described by POD. Without approaching the nature and sources of uncertainty involved in FAD analysis, which is beyond the scope of the present paper, we will quantify it in terms of random variables. This is a natural way to associate a probabilistic format to FAD procedure.

From a computational standpoint, probabilistic FAD assessment can be constructed in a statistical way by sequential Monte-Carlo simulated sampling of sets of FAD involved parameters $\{a, K_c, R_m, R_e, \ldots\}_i$. Furthermore, every parameter set generation sequence is regarded as a possible occurrence. In a sequence, the FAD state point $(K_r, S_r)_i$ is computed and the failure analysis is performed on the Yes/No basis as related to deterministic FAD limit boundary, see Equation (4). With a sufficiently great number of sampling sequences, an estimate of the failure probability P_f results as the ratio of the number of cases when state points $(S_r, K_r)_i$ fall outside of the FAD-limiting curve, i.e. in the failure space to the total number of simulations. Obviously, the convergence of P_f estimate to the limit theoretical values is improved as the number of simulations increase, typically for any numerical procedure based on Monte-Carlo methodology.

Within this rationale, the reliability of NDI is modeled in every sampling sequence by simulating a "targeting game". Accordingly, the NDI crack detection success in a FAD sequence is retained when for the sampled crack size *a*, the considered POD (*a*) is greater than a sampling outcome from a uniform random variable distributed over the interval (0,1).

Based on this principles, a comprehensive software (FADSIM) for probabilistic FAD analysis has been developed including the influence of POD, with the main aim of enabling probabilistic failure risk assessments beyond the very limited set of available compact analytical solutions.

4. NUMERICAL PROBABILISTIC FAD SIMULATION VS. ANALYTIC SOLUTION

As described in a previous paper (Cioclov and Kröning 1999), a compact solution of failure probability P_f under pure brittle circumstance, i.e. on K_r – FAD axis ($S_r = 0$) was obtained. The solution pertains to the case of a large structural element containing a through-wall crack, the element being subject to a remote uniform traction σ. The crack size is considered to be exponentially distributed with the mean size value of $\bar{a}$ and material fracture toughness assumes a deterministic value K_o. The probability of fracture results in a simple analytic reparations function of exponential form:

$$P_f = \exp\left[-\left(\frac{K_o}{\bar{K}}\right)^2\right] \tag{5}$$

where $\bar{K} = Y\sigma\sqrt{\pi \bar{a}}$ is the stress intensity factor related to the mean crack size $\bar{a}$ and Y is SIF correction factor. Figure 4 illustrates the P_f variation as against the applied stress according to Equation (5). The mean crack $\bar{a}$ = 6.25 mm, a value that originates in PISC-Programs and was used for sensitivity analysis in our outlined previous paper. This configuration will be use as basis for comparison with the output of FADSIM numeric algorithm.

Two series (A and B) of 10^6 simulations have been performed for the structural case exemplified in Figure 4 at an applied stress of σ=250 MPa (representation by the solid point in Figure 4). Figure 5 illustrates the convergence of simulation towards the analytical value P_f = $1.54.10^{-3}$. It is apparent that a good approximation can be attained at this level of P_f in some 10^5 simulations or more, a matter of five minutes of computation on a commercial PC.

When real-life structural elements with finite geometry are considered, the analytical approach may become tedious due to complicated convolution integrals involved in fracture probability computation. Figure 6 illustrates the case of a rectangular plate of a finite width H with an edge crack. The crack size is assumed to be a random variable following an exponential distribution with the mean crack size value a = 6.25 mm. The overall geometry is set at $H/\bar{a}$ = 10. The plate is subject to a uniform traction stress σ (parametric in the analysis). Material characteristics are chosen for an aluminum alloy of class 2219-T87 with strength and toughness characteristics as follows: UTS = 430 MPa; YP = 310 MPa; Kc = 100 MPa $\sqrt{m}$ (deterministic). The computation of fracture probability by FADSIM algorithm has been performed for the circumstance of no – NDI and with NDI, according to a procedure which reliability is defined by an ILP-POD distribution, given by Equation (3). The POD-distribution parameters are as in Figure 3: POD* = 0.995; a_o = 0.55 mm; POD1 = 0.87 and a1 = 8.4 mm, corresponding to a lower conservative envelope.

Figure 6 shows the results of several set of simulation. One set is represented in P_f – σ coordinates by a distinct point, each point resulting from 10^5 simulations. It is apparent that an acceptable convergence can be obtained at this computation extent.

The computation results shown in Figure 6 quantify the beneficial role of NDI for reducing the failure risk. For the considered NDI-procedure, which POD-function and defining parameters are in line with current practice for aluminum alloys, it results in a decrease of fracture probability that may be an order of magnitude lower in comparison with the case when NDI is not applied. For comparison purposes, Figure 6 also shows the case when the plate has a very large width (infinite), where the other input data concerning crack size distribution and material characteristics remaining unchanged. The influence of finite width on P_f is apparent the increase of P_f being significant especially in the lower range of applied stress.

5. A SENSITIVITY PROBABILISTIC FAD ANALYSIS OF CYLINDRICAL PRESSURE VESSELS

Case description and FADSIM-algorithm input options

The objective of the analysis is to evince (in the framework of a probabilistic FAD approach) the influence of NDI quality, as expressed by POD distributions, on the probability of fracture for two configurations of inner cracks in pressurized cylindrical vessels. One crack configuration pertains to a semi-elliptical crack placed in axial position (SE-AX) on the inner face of the vessel while the other case considers the same crack shape but placed in transverse position (SE-TR) on the inner face. Figures 7a and b illustrate the overall geometry of the considered pressure vessels with SE-AX and SE-TR crack orientation, respectively.

The geometry of the considered PV is defined by the wall thickness of h = 25 mm and the inner radius to wall thickness ratio R_i/h = 50. The mean crack depth $\bar{b}$ and semi-length $\bar{a}$ have been allowed to have a parametric variation $\bar{b}/h = 0.25\ and\ 0.50$ while the crack aspect ratio have been maintained constant in mean sense, i.e. $\bar{a}/\bar{b} = 2.$

Uniform inner pressure has been considered as a variable parameter in the analysis.

In the framework of Option I-FAD methodology, the material mean values have been choose to represent a low-alloyed carbon steel with:

$$R_m = 600 MPa;\ R_{e,0.2} = 460\ MPa\ and$$
$$K_c = 125\ MPa\sqrt{m}.$$

Analytical stress intensity factors analytical solutions implemented in FADSIM algorithm are those developed by Newman and Raju (1980) for SE-AX crack configuration and Zahoor (1985) for SE-TR.

The probabilistic features in the analysis stems from the uncertainty involved in the crack size knowledge and NDI reliability. The crack size has been considered as normally distributed. For illustration purposes, the uncertainty that may appear in the knowledge of material properties have been superimposed by considering R_m, $R_{e,\ 0.2}$ and K_c as random normal variables.

The NDI reliability has been quantified by an exponential type POD-distribution with asymptotic trend at POD values close to unity, Eq (1). The defining parameters of the POD-distribution have been chosen in line with PISC-results for carbon steels i.e.: A = 0.995 and a1 = 8,85 mm (Marshall 1982).

In the probabilistic FAD analysis, 10^5 simulations have been performed for every estimated fracture probability P_f. For low P_f-values in the range of 10^{-4} to 10^{-5}, 10^6 simulation have been performed per analysis point.

Figure 8a and b shows an example of FADSIM-algorithm simulation output as it appears on the computer screen. The "clouds" of FAD-representative points associated with the crack depth *b* (red points) and the crack semi-length *a* on the inner face (light blue points) result from the crack size scatter imposed in the simulation according to a normal distribution with $\bar{b} = 6.25$, coefficient of variation CV = 0.2 % and $\bar{a} = 12.5$, CV = 0.2 %. Under internal pressure of 50 bar and the overall geometry of the vessel (outlined in previous section, Figure 8a) illustrates the case when no NDI is applied and P_f = 7.10^{-4} results while Figure 8b illustrates the case when NDI is applied. In this latter case the POD distribution is as given by Equation (1) with A = 0.995; a1 = 8.85 mm. A probability of fracture of 9.10^{-5} is resulting, clearly smaller in comparison with the case when no NDI is applied.

Discussion of the results

In an attempt to link the probability of failure with some global index reflecting in FAD representation the degree of safety, an index defined as in Figure 9 is usually used. This safety index (SI) is smaller than unity for FAD points in the failure space and greater than unity for FAD points inside FAD-limiting curve. Figures 10a, 10b give for two crack size populations $\left(\bar{b}/h = 0.25 \text{ and } \bar{b}/h = 0.5\right)$. a plot of simulated P_f values versus SI, comparatively for SE-AX and SE-TR crack configurations. The influence of applying a NDI with POD-function as given by Eq (1) with A = 0.995 and a1 = 8.85 mm is also evinced. The beneficial role of NDI on the mitigation of failure risk is obvious. It appears that NDI is more influencing for a distribution of deep cracks.

$$(\bar{b}/H = 0{,}5, \ Fig.\ 10b)$$

in comparison with

$$(\bar{b}/H = 0{,}25, \ Fig.\ 10a)$$

In the former case a higher proportion of large cracks exists, hence with an "imperfect" NDI that implies a finite probability of non-detection, higher is the chance to escape critical cracks. An increase in the probability of failure results.

The results in Figures 11a and 11b are the same as presented in Figure 10a, while the Figure 11c shows the results of Figure 10b plotted in a presentation of P_f vs. applied stress S. For the SE-AX crack orientation, the hoop stress is taken as a stress parameter and for the SE-TR orientation the axial stress generated by pressure (closed ends cylinders) is corresponding. The beneficial influence of NDI is clearly demonstrated in this presentation. When the comparison is made on applied stress bases, it is interesting to note that for deep cracks (Figure 11c, $\bar{b}/H = 0.5$) at the same distribution of crack size, the transverse inner face semi-elliptic cracks in cylindrical PV with closed ends (under uniform pressure) are much more deleterious in comparison to cracks in axial orientation.

Under the FAD-Option 1 condition, this trend results mainly from the interaction of K_r parameters, i.e. from the inherent character of employed SIF analytical solutions (Zahoor, 1985, vs. Newman and Raju, 1980).

6. SUMMARY AND CONCLUSIONS

This paper describes a rationale for the integrating of nondestructive inspection quality, as quantified by POD, into probabilistic failure risk assessments based on FAD-Option 1 methodology. For this purpose, a common exponential model for POD estimation, as proposed in the framework of the PISC-research program, has been used along with a new POD-model based on log-inverse power (LIP) functions. Both POD models, as well as log-normal and log-logistic POD models are integrated in the FADSIM algorithm and code developed for probabilistic failure risk assessment based of FAD procedure. The algorithm of probabilistic FAD assessment is constructed on the statistical base, e.g., by sequential Monte-Carlo simulated sampling of key FAD parameters, such as crack size, material fracture toughness, and yield and ultimate tensile strength. One feature of the developed FADSIM algorithm is that once the pertinent stress intensity factor solutions (or J-solutions) for the considered geometry is implemented, the point and sensitivity (parametric) estimations of failure probability may be easily performed.

For the purpose of validation, probabilistic results obtained with the FADSIM have been compared to analytical solutions developed in previous work. The equivalence of the two approaches has been demonstrated in the limit case of large plates having small crack. As the number of simulated scenarios increases, a good convergence of simulated failure probabilities towards an analytical solution has been substantiated.

The application of the developed probabilistic FAD rationale for the case of cylindrical pressure vessels has resulted in a sensitivity (parametric) study for two orientations of an inner semi-elliptic crack in axial and transverse orientation. The uncertainty in crack size knowledge has been modeled by normal distribution while the reliability of NDI by a POD following an exponential rule. The beneficial influence of NDI has been quantified in terms of the decrease of failure probability as function of applied stress or a FAD defined safety index. The peculiarities of a deep crack in axial or transverse position in a cylindrical pressure vessel have been demonstrated. Finally, the authors consider this contribution as a first step in addressing the prominent problem of integrating quantitative NDI methods into probabilistic fracture mechanics models. It was evinced, quantitatively, the influence of POD on the pressure vessel failure risk by fracture. By this way an engineering rationale have been developed in order to cope with the problem of NDI optimization and structural integrity assessment.

REFERENCES

Berens, A. P. and Hovey P. W., 1983: "Statistical Methods for Estimating Crack Detection Probabilities", *Probabilistic Fracture Mechanics and Fatigue Methods: Application for Structural Design and Maintenance,* ASTM STP 798, pp 79 – 94.

Chell, G.G., Millwater, H. R., Mauney, D. A. and Houdak, S. J. Jr.: "Combining Probabilistic Failure Assessment and Decision Analysis for the Safe and Economic Management of Flawed Petrochemical Components". *Proc.1995 Pressure Vessels and Piping Conference,* PVP - Vol.315, *Fitness-for-Service and Decisions for Petroleum and Chemical Equipment,* ASME.

Cioclov, D. D. and Kröning, M., 1999: "Probabilistic Fracture Mechanics Approach to Pressure Vessel Reliability Evaluation", PVP - Vol. 386, *Probabilistic and Environmental Aspects of Fracture and Fatigue*, ASME.

Cioclov, D. D., 1975: "Strength and Reliability under Variable Loading" (in Roumanian) Editura Facla, Timisoara, Romania.

Froeli O., 1997: "How to Develop Acceptance Criteria for Pipeline Girth Weld Defects", *European – American Workshop Determination of Reliability and Validation Methods of NDE*, Berlin, 18-20 June, 1997.

Lo, T.Y. et al., 1984: "Probability of Pipe Failure in the Reactor Coolant Loop of Combustion Engineering PWR plants", *Rep. NUREG / CR - 3663, UCRL – 53500.*

Marshall, W., 1982: "Assessment of the Integrity of PWR Pressure Vessels", *Summary Report, UKEA Authority, UK,* June 1982.

Milne, I., Ainsworth, R. A., Dowling A. R, and Stewart, A. T., 1988: "Assessment of the Integrity of Structures Containing Defects", *Int. J. Pres. Ves. Z. Piping,* Vol. 32, No. 3, pp 3-104.

Newman, Jr., J.C. and Raju, I.S., (1980): "Stress-Intensity Factors for Internal Surface Cracks in Cylindrical Pressure Vessels", *Trans. ASME,* Ser.J., Vol. 102, pp 342-346.

PISC, 1979, PISC: Plate Inspection Steering Committee, CEC. Rep. EUR 637 / Vol.I to V.

Rummel, W. D., Hardy, G. L. and Cooper, T. D.: "Application of NDE Reliability to Systems", in *Nondestructive Evaluation and Quality Control* - ASM - International, Vol 17, pp 664-688, 1994.

Swift, R. and Connolly, M. P. I, 1985: Paper No.7A presented at *20th Annual British Conference on Non-Destructive Testing,* 1985.

Warke, R.W., Ferregut, C., Carrasco, C. J., Wang, Y. Y. and Morsley, D. J., 1999: "A FAD-Based Method for Probabilistic Flaw Assessment of Strength-Mismatched Girth Welds", PVP - Vol. 386, *Probabilistic and Environmental Aspects of Fracture and Fatigue* – ASME, pp 89-99.

Zahoor, A., 1985: "Closed Form Expressions for Fracture Mechanics Analysis of Cracked Pipes", *Journal of Pressure Vessel Technology,* Vol. 107, pp 203-205.

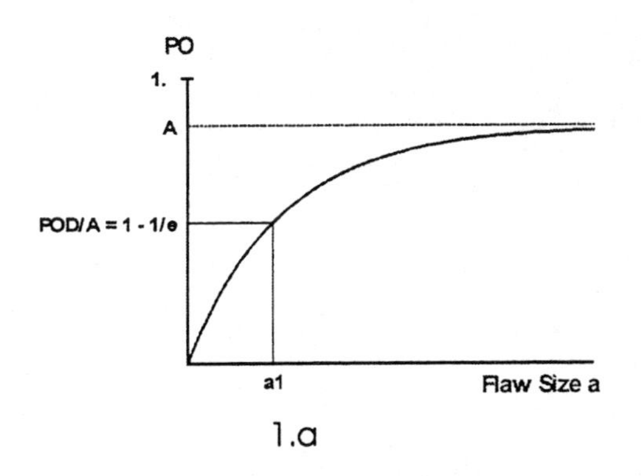

1.a

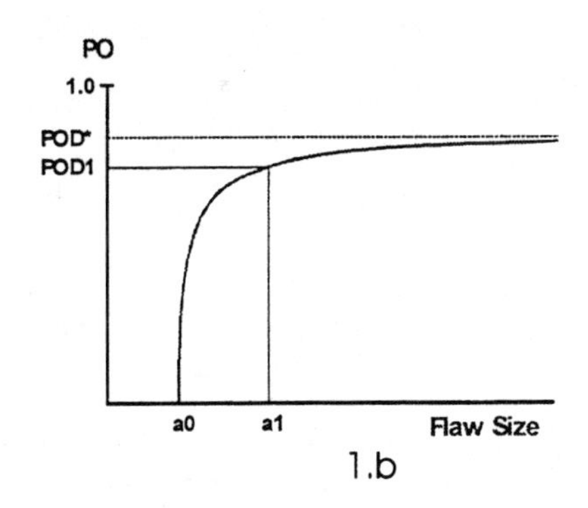

1.b

Fig. 1: Schematic representation of POD vs. Flaws size
a) Exponential – asymptotic (EA); b) Inverse – log. Power (ILP) distribution

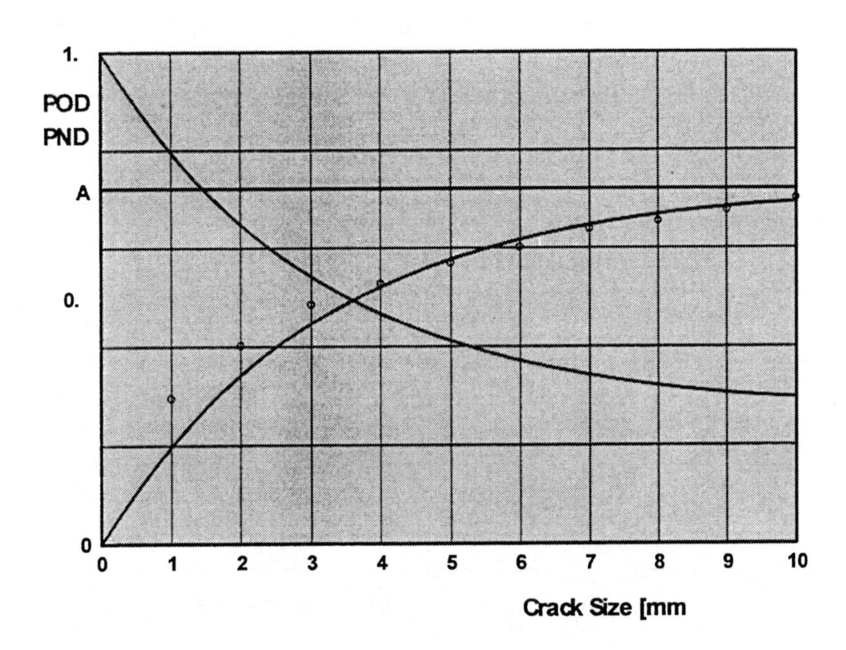

Fig. 2: POD vs. flaws size. Experimental data on welded structural steel specimens with lack of fusion flaws. Plate thickness <15 mm. Ultrasonic testing (source: Froeli 1997). Data fitting according to EA distribution Eq (1): A = 0.995; a1 = 3.06 mm.

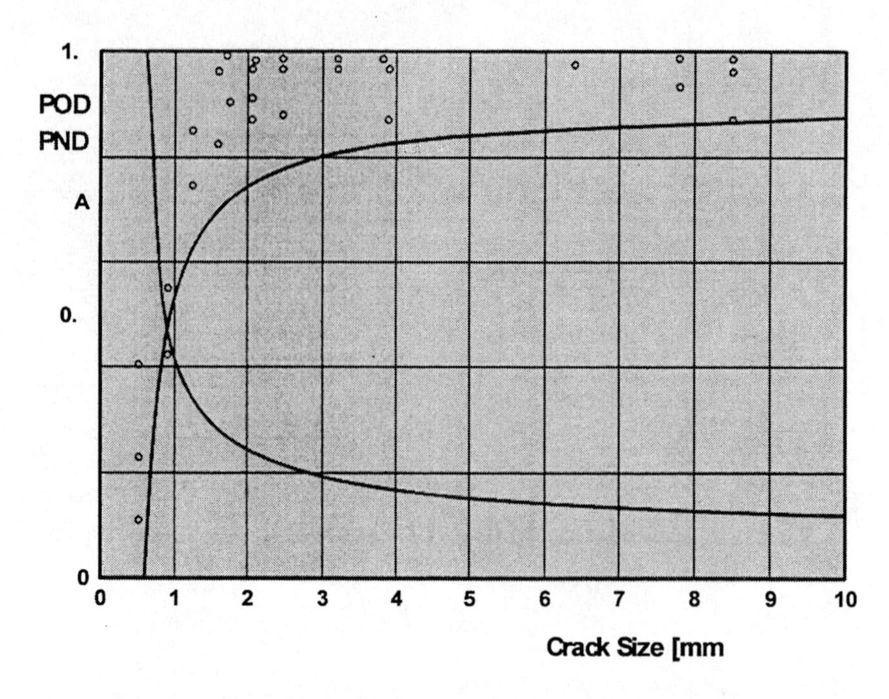

Fig. 3: POD vs. flaws size. Experimental data on AL 2219 – T87 alloy. Specimens with fatigue cracks. Ultrasonic testing (source: Rummel et.al., 1994). Data fitting according to ILP distribution, Eq. (3), lower envelope: POD* = 0.995; POD1 = 0.87; a1 = 8. 4 mm; a_o = 0.55 mm.

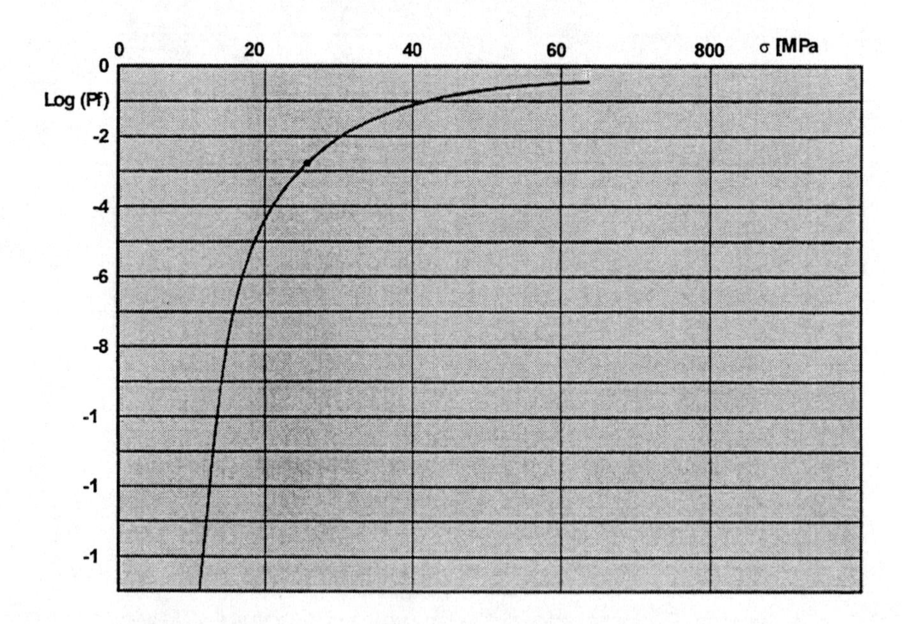

Fig. 4: Probability of failure vs. applied traction stress. Large plate with a side through crack. Crack size exponentially di stributed with mean value $\bar{a}$ = 6.25 mm. Material fracture toughness K_o = 100 MPa √m – deterministic.

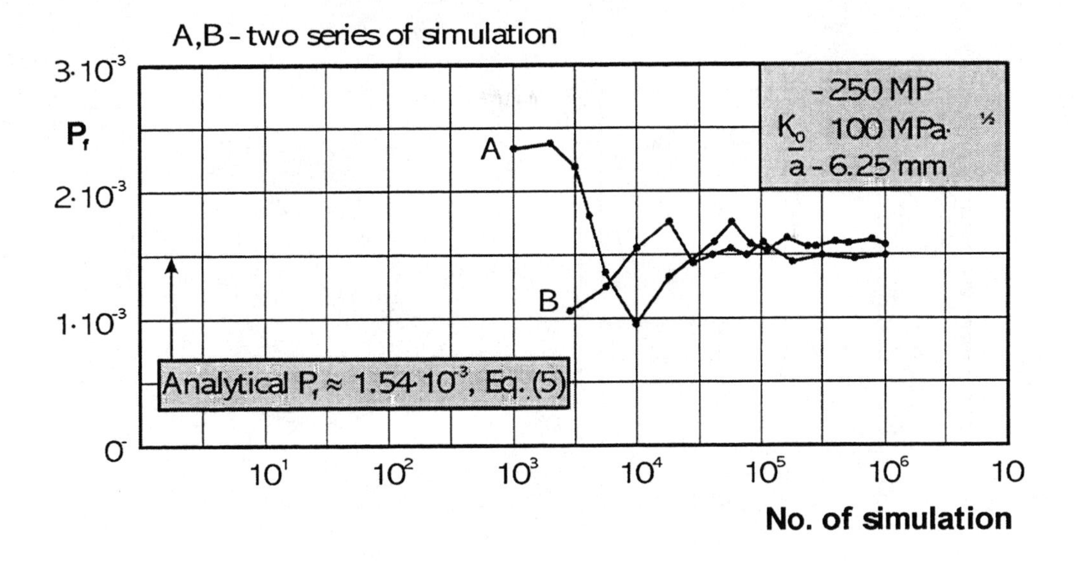

Fig. 5: Convergence of simulated prob ability of fracture according to FADSIM alg orithm vs. number of simulations. Simulation parameters as in Fig. 4. Applied stress σ = 250 MPa.

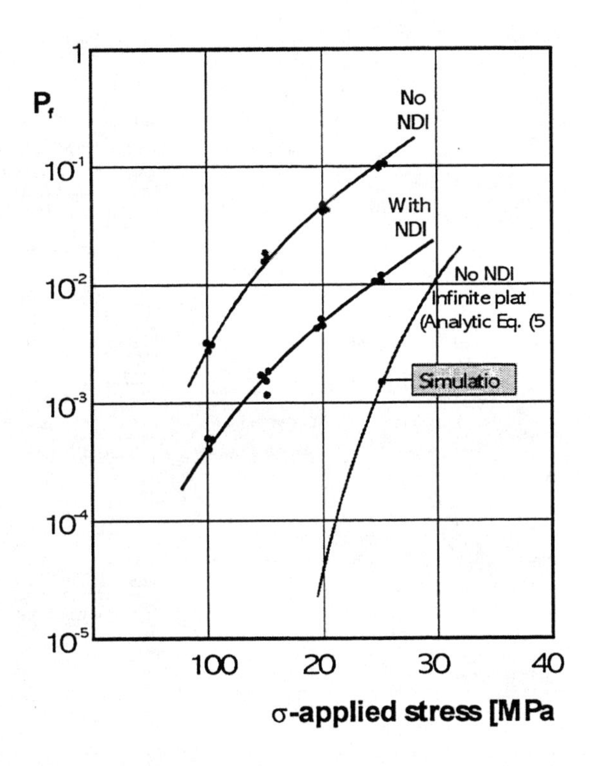

Fig. 6: Probability of failure vs. applied traction stress. Finite rectangular plate with side through crack. Crack size exponentially distributed with mean value $\bar{a}$ = 6.25 mm. Plate width / $\bar{a}$. = 10. Material fracture toughness K_o = 100 MPa$\sqrt{m}$ – deterministic. Influence of NDI with POD according to ILP – distribution Eq. 3. POD parameters as in Fig. 3. For no-NDI a comparison with infinite plate case is shown.

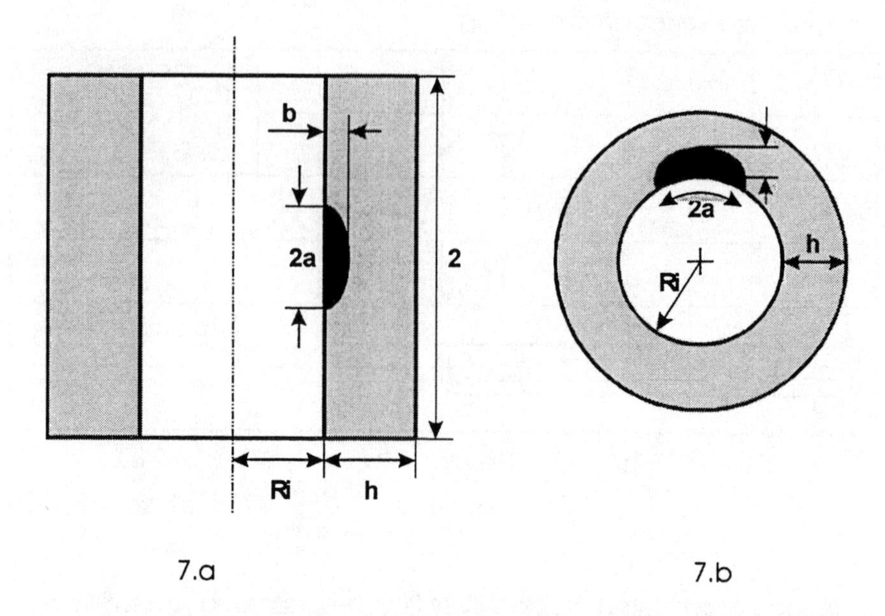

7.a 7.b

Fig. 7: Schematic representation of the geometry of cylindrical pressure vessel with crack

a) inner semi-elliptic axial crack (SE-AX) ; b) inner semi- elliptic transverse crack (SE-TR)

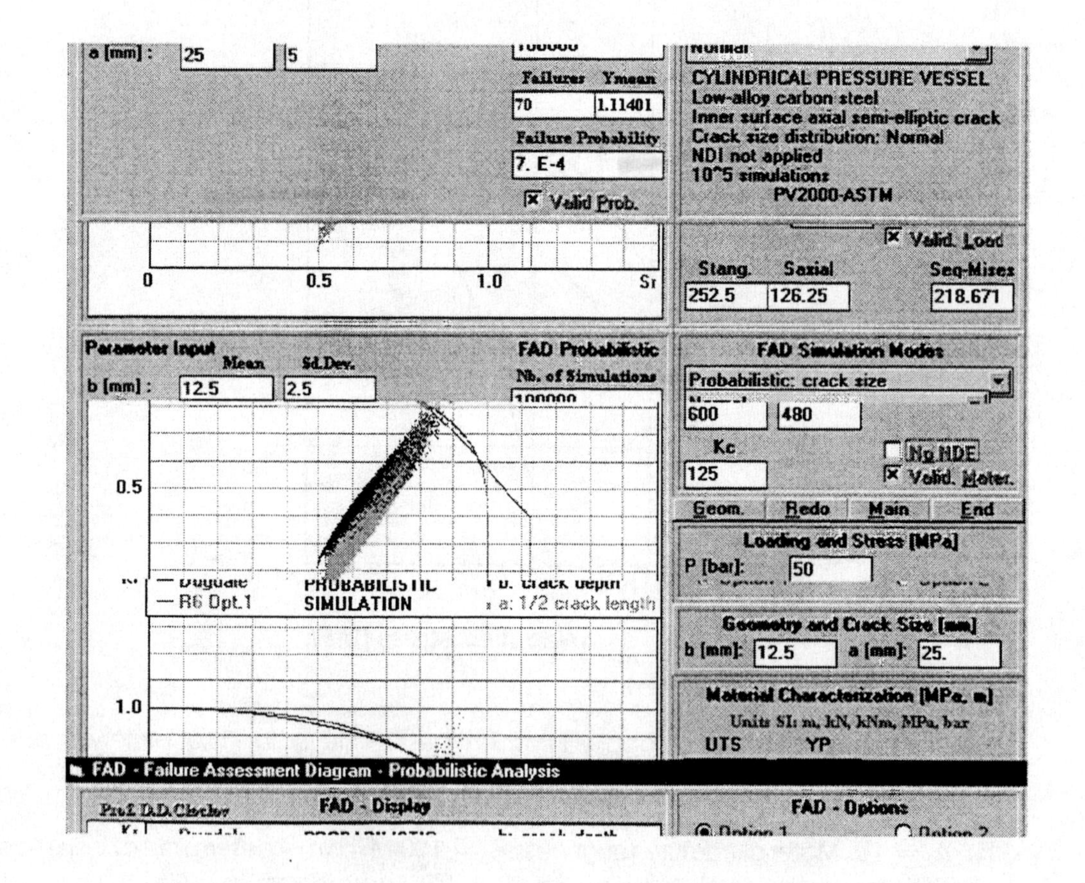

8.a

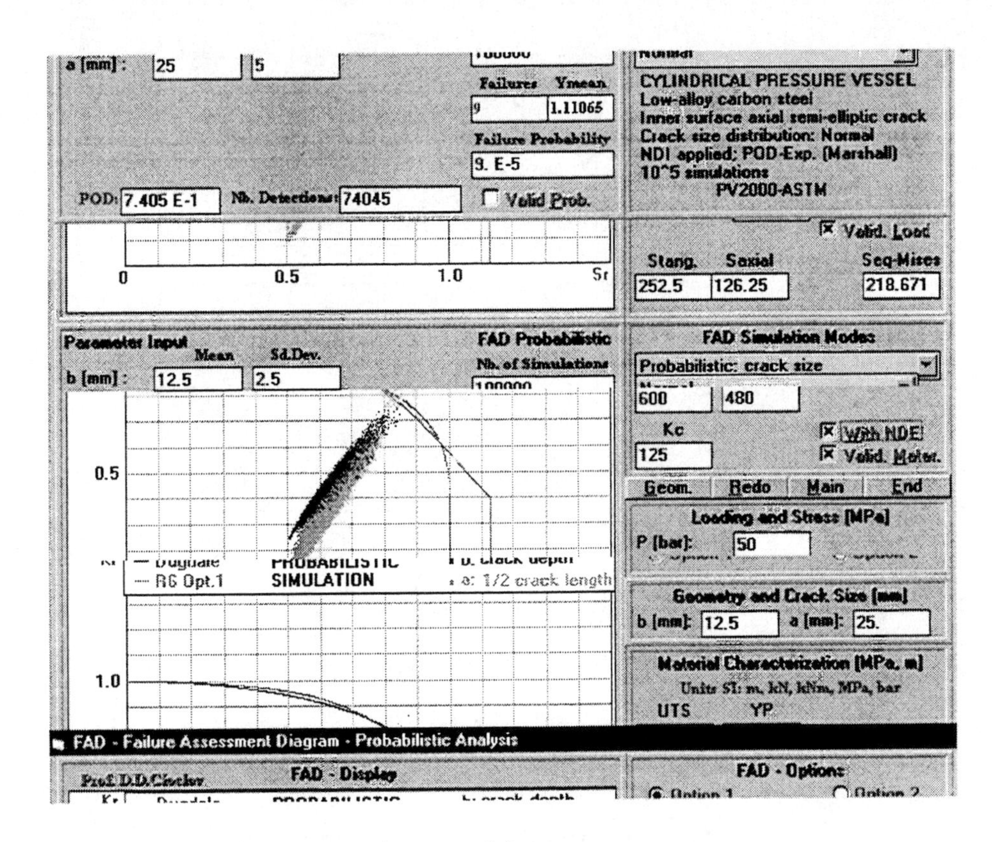

8.b

Fig. 8: FADSIM presentation of probability of failure simulation of a cylindrical PV (Ri/H=50). Inner surface semi-elliptic crack (b -crack depth and a – ½ crack length).

Crack size follows Normal distribution: $\bar{b}$ = 12.5 mm, SD = 2.5 mm and $\bar{a}$ = 25 mm, SD = 5 mm. a) No NDI; b) NDI applied; POD distribution: exponential, (Eq. 1); A = 0.995; a1 = 8.85 mm.

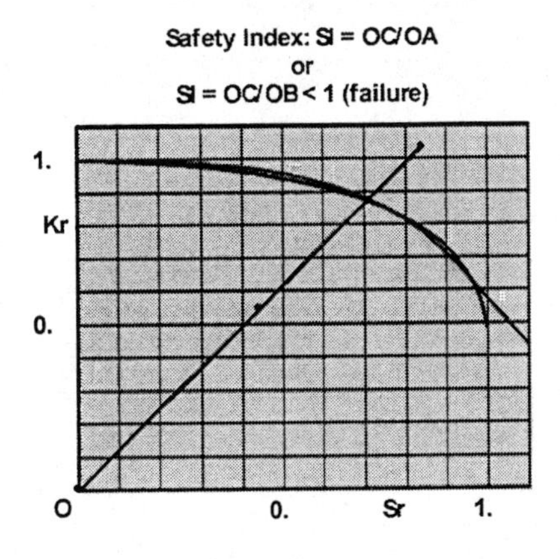

Fig. 9: FAD Safety Index (SI) definition.

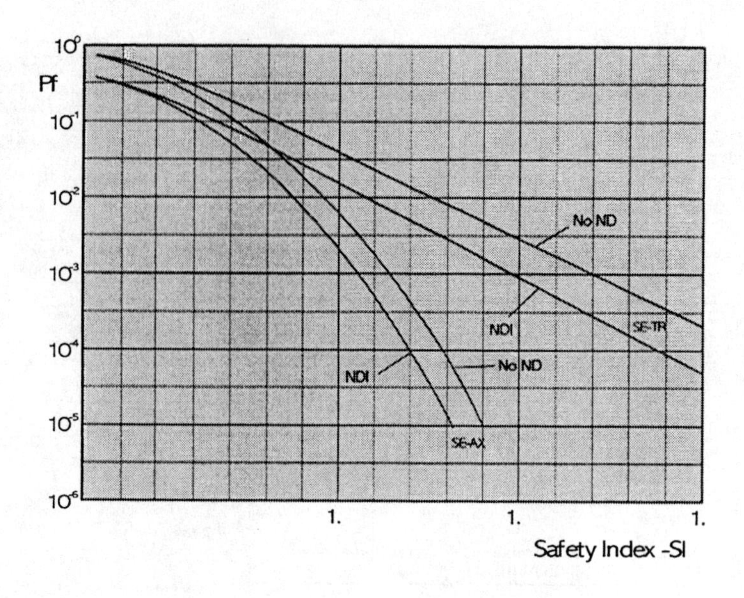

10.a

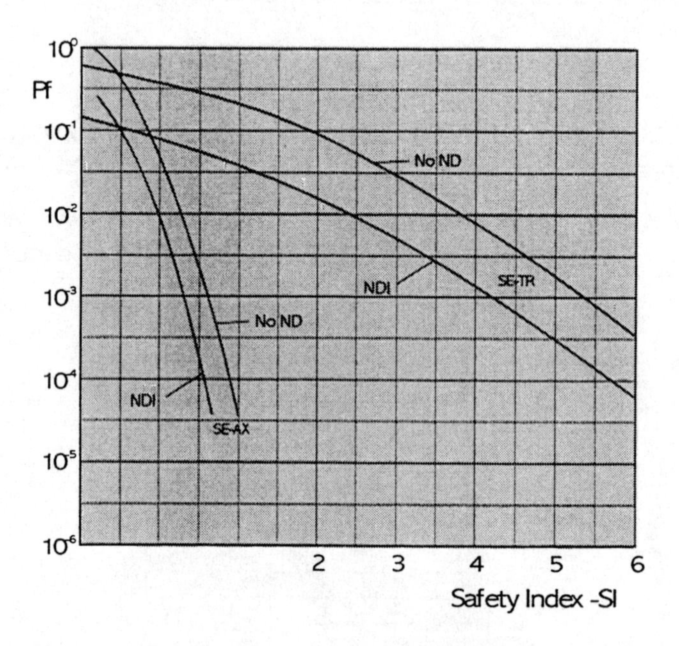

10.b

Fig. 10: Probability of failure P_f vs. Safety Index SI. Cylindrical PV (Ri/H=50). Inner face semi-elliptic crack. Crack size follows Normal Distribution. NDI as in Fig. 8b.

a) mean crack depth $\bar{b}$ = 0,25 H; CV = 10 %;

b) mean crack depth $\bar{b}$ = 0,5 H (deep crack); CV = 10 %;

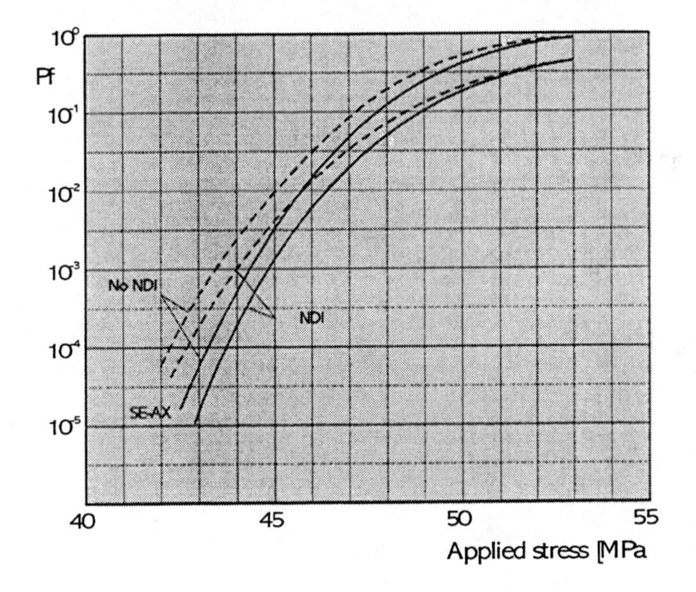
Pf
No NDI
NDI
SE-AX
40
45
50
55
Applied stress [MPa

11.a

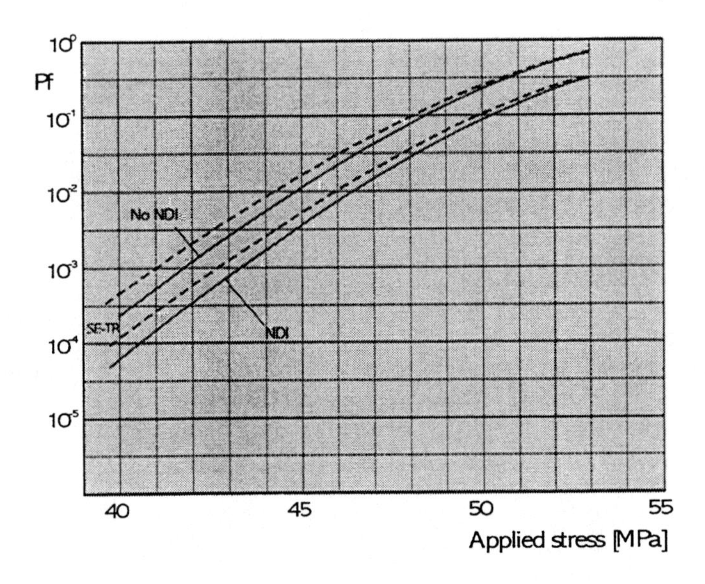
Pf
No NDI
SE-TR
NDI
40
45
50
55
Applied stress [MPa]

11.b

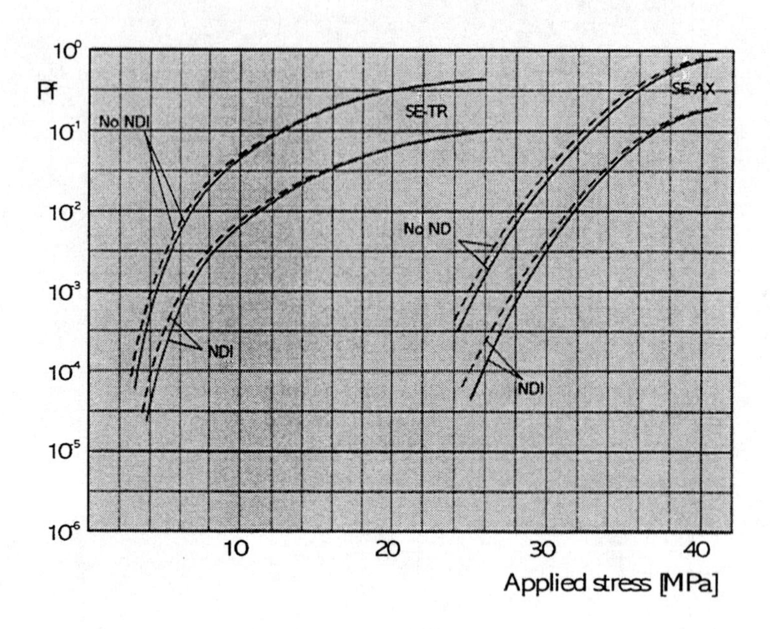

11.c

Fig. 11: Probability of failure P_f vs. applied stress. Crack size distribution as in Fig. 10 – solid lines. Dashed lines pertains to probabilistic variation of K_c, YP, UTS; CV = 5 % (all)

a) mean crack depth $\bar{b}$ = 0.25 H; CV = 10 % - transverse semi-elliptic crack (SE-TR);

a) mean crack depth $\bar{b}$ = 0.25 H; CV = 10 % - axial semi-elliptic crack (SE-AX);

c) mean crack depth $\bar{b}$ = 0.5 H (deep cracks); CV = 10 %, SE-AX and SE- TR crack orientation. NDI as in Fig. 8b.

PVP-Vol. 410-1, Assessment Methodologies for Preventing Failure:
Deterministic and Probabilistic Aspects and Weld Residual Stress
Volume 1
ASME 2000

LIFETIME ASSESSMENT OF A PIPELINE WITH MULTIPLE LONGITUDINAL CRACKS SUBJECTED TO CYCLIC LOADING AND LOCAL CORROSION

Margarita G. Malyukova **Sviatoslav A. Timashev**

Science & Engineering Center "Reliability & Resource of Large Machine Systems",
Ural Branch, Russian Academy of Sciences
54a, Studencheskaya Str, Ekaterinburg, 620049, Russia
Ph./Fax: (3432) 741-682. E-mail: WEKT@DIALUP.MPLIK.RU

ABSTRACT

Methods are presented of deriving the deterministic and the probabilistic assessment of longevity of a pipeline with multiple longitudinal cracks in the form of semi-elliptical flaws on the inner surface of the pipe subjected to cyclic loading and local corrosion. It is an extended and deployed generalization of the methods described in (Malyukova and Timashev, 1998, 1999). The deterministic approach is based on non-linear fracture mechanics for the case when the prefracture zone (PEZ) at the tip of the crack is in the form of two bands of plastic flow which are at an angle and symmetrical to the plane of crack propagation. In this NLFM setting the following cases are considered: local corrosion is concentrated inside the crack, and is uniform or orthotropic (along the longitudinal x and hoop y axes). Moreover, cluster of two longitudinal cracks is considered, with local corrosion also having different rates along the x, y axes.

The influence of the random character of most important parameters, including the velocity of internal corrosion is investigated. For this, the maximum entropy method in conjunction with the Monte Carlo simulation method is used. As a result, the least predetermined PDF of pipeline remaining lifetime is constructed and its first four moments are calculated.

INTRODUCTION

The overwhelming majority of known papers on this topic (Charles et al, 1995, Yunovich et al, 1997, Garud, 1990) are restricted to the case of external electrochemical corrosion of pipelines.

At the same time internal corrosion is of substantial influence. According to (Medvedev, 1995) main factors that generate internal corrosion in oil collection pipelines are high water content of transported oil, presence of suspension of mechanical admixtures and precipitation of Fe-salts and Ca-salts on the internal surface of the pipeline. These factors may induce not only uniform corrosion, but also local corrosion. In the latter case, corrosion attacks, in the first place, internal crack-like flaws. To the best of the authors' knowledge, there are no papers that have dealt with singular cracks subjected to a combination of cyclic loading and local corrosion.

The results of inspections of performing oil pipelines show that they have considerable amounts of singular crack-like flaws and also clusters of cracks. Under corrosion and cyclic loading each of the cracks in a cluster is growing, and in some cases several cracks merge, jump-forming one large crack, which in many cases dramatically diminishes the pipeline capacity and may lead to a through-wall crack. Therefore the problem of remaining lifetime assessment of pipelines with single and multiple longitudinal cracks subjected to internal local corrosion and cyclic loading is of considerable practical interest.

MAIN EQUATIONS

The kinetics of growth of a flat semielliptical crack in a pipe (Fig. 1) is defined by the change of the lengths of its semiaxes a(N) and b(N) and is described (Malyukova and Timashev, 1998) by the following system of two nonlinear differential equations (DE):

$$\frac{da}{dN} = \Phi^{-1}[\lambda(\alpha)], \text{ when } \alpha = 0$$
$$\frac{db}{dN} = \Phi^{-1}[\lambda(\alpha)], \text{ when } \alpha = \pi/2 \qquad (1)$$

at following initial conditions (IC)

$$a(N)\big|_{N=0} = a_0; \quad b(N)\big|_{N=0} = b_0 \qquad (1a)$$

Here

$$\Phi[\lambda(\alpha)] = A\left[\left(\frac{\lambda_0}{\lambda_0 - \lambda}\right)^m - 1\right]$$

is the characteristic function of fatigue fracture, which is connected with the fatigue crack growth FCG velocity v by the relation

$$v = \Phi^{-1}[\lambda(\alpha)] \tag{1b}$$

$$\lambda = 1 - \sqrt{\delta_{max}(\alpha)/\delta_{fc}}\ ; \quad \lambda_0 = 1 - \sqrt{\delta_{th}/\delta_{fc}}\ ; \tag{1c}$$

$\delta_{max}(\alpha)$ is the maximal value of crack opening between its edges at its tip (is obtained for maximal value of load amplitude p_{max}); δ_{th} is the threshold value of crack opening (is determined experimentally); δ_{fc} is the critical value of crack opening at cyclic loading (may be calculated as shown in (McMeeking, 1977)); A and m are pipe material characteristics, which are defined numerically (from the condition of minimal quadratic deviation of design and experimental values) using the kinetic diagram of fatigue fracture (constructed using results of experimental studies of specimens with cracks).

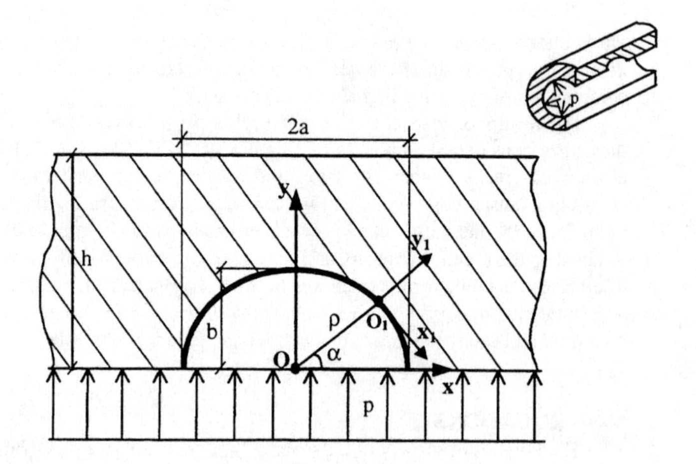

Fig. 1 Longitudinal cross-section of the pipe in the crack plane

FRACTURE CRITERIA

The longevity assessment of a pipe $N=N^*$ is defined from the condition that at least one of the following equations holds true:

$$\delta_{max}\left[a(N^*), b(N^*)\right] = \delta_{fc} \quad \text{when } \alpha = 0 \text{ or } \alpha = \pi/2 \tag{2}$$

or

$$b(N^*) = h, \tag{3}$$

where h is pipeline wall thickness.

The equation (2) relates to the beginning of the spontaneous pipe fracture due to unrestricted growth of the crack semiaxes a(N) and b(N).

The equation (3) holds true when the crack as it grows becomes a through-wall crack. In this case (when a(N) grows infinitely) for $N>N^*$ one cannot exclude avalanche type longitudinal fracture of the tube.

ALGORITHM FOR THE DETERMINISTIC CASE

It is assumed that internal corrosion in the pipe takes place at a uniform rate V_c in all directions.

For this case the basic algorithm (Malyukova, Timashev, 1998) is modified in the following manner.

In step 1, when solving the DE system by the 4-th order Runge-Kutta method (Korn, 1984) when $N_1 = \Delta N$, one is using two types of IC, which correspond to two possible methods of accounting for corrosion damage Δx_1 during the time from $N_0 = 0$ to $N_1 = \Delta N$: addition of Δx_1 either to a_0 at the beginning of the interval at $N = 0$ or to $a(N_1)$ at the end of the interval, at $N_1 = \Delta N$ (Fig. 2).

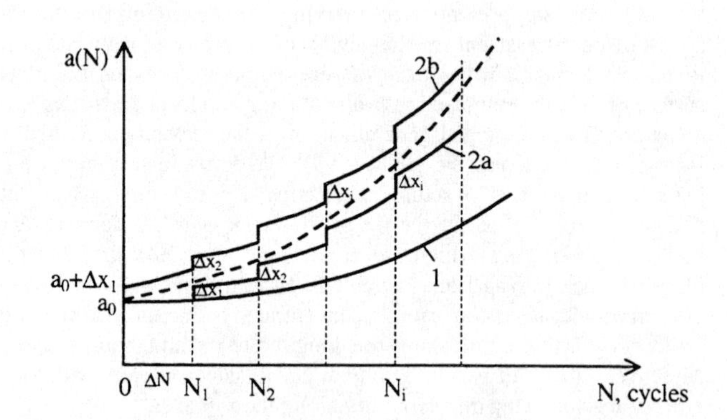

Fig. 2 Semiaxis a(N) growth: 1) without corrosion; 2a and 2b – accounting for corrosion (IC type a and IC type b correspondingly)

LIFETIME ASSESSMENT OF A PIPELINE SEGMENT WITH LONGITUDINAL CRACK UNDER CONDITIONS OF CYCLIC LOADING AND LOCAL CORROSION

Results obtained in (Malyukova and Timashev, 1999) allow solving of the problem of pipeline lifetime assessment when subjected to cyclic loading and local corrosion. In many practical cases this type of pitting corrosion is the most dangerous. Solution of this problem is also a prerequisite for solving the problem of multiple cracks, subjected to combined cyclic loading and local corrosion.

- Changes in the algorithm

In the considered case pipe inner radius R_1=const.

In step 1 the initial conditions may be written as:

type a $$a(N)\big|_{N=0} = a_0; \quad b(N)\big|_{N=0} = b_0 \tag{4}$$

type b $$a(N)\big|_{N=0} = a_0 + \Delta x_1; \quad b(N)\big|_{N=0} = b_0 + \Delta x_1 \tag{5}$$

Corrosion damage Δx_k is defined by expression

$$\Delta x_k = V_{lc} \cdot \frac{N_k - N_{k-1}}{f}, \; k = 1,2,3,...,$$

where v_{lc} is the rate of local corrosion; N_k and N_{k-1} are the total number of cyclic loading on the k-th and (k-1)–st step accordingly; f is the load frequency.

In all consequent steps $N_k=N_{k-1}+\Delta N$ the following IC are considered:

for type a
$$a(N)\big|_{N=N_{k-1}} = a(N_{k-1}) + \Delta x_{k-1}; \quad b(N)\big|_{N=N_{k-1}} = b(N_{k-1}) + \Delta x_{k-1}; \tag{6}$$

for type b
$$a(N)\big|_{N=N_{k-1}} = a(N_{k-1}) + \Delta x_k; \quad b(N)\big|_{N=N_{k-1}} = b(N_{k-1}) + \Delta x_k; \tag{7}$$

The deterministic problem is solved for a pipeline segment, for which geometry and material characteristics are given in Table 1.

The results of calculations for two different initial crack sizes with semiaxes lengths a_0 and b_0 for different values of local corrosion v_{lc} rates are given in Table 2 and in Fig. 3 – Fig. 6.

According to calculations:

- as in the case of general corrosion of the pipe with a crack the $b(N^*)=h$ fracture criteria (the LBB case) is realized;
- with growing value of local corrosion v_{lc} rate the pipeline longevity substantially decreases. For the case of a deep crack ($b_0=4.5\cdot10^{-3}$ m) the increase of v_{lc} from 0 to $3\cdot10^{-3}$ m/yr decreases the pipeline longevity from 6.5 years to 0.74 year (a 8.8 times decrease). And a 40-fold increase of v_{lc} (from $0.2\cdot10^{-3}$ m/yr to $8\cdot10^{-3}$ m/yr) induces a 14.1 fold decrease of pipeline lifetime (from 4.24 yrs to 0.30 yr);
- with equal corrosion rates v_c and v_{lc} of uniform general and local corrosion, the lifetime of a pipeline with a crack in the case of local corrosion is larger than in the case of general uniform corrosion. When solving the problem the selected increment ΔN of cyclic loading ($\Delta N=1000$ cycles for $v_{lc}=0$ and $\Delta N=100$ cycles for $v_{lc}\neq0$) yields practically the same values of lifetime of a pipeline with crack for both IC (type a and type b);
- in every case of crack propagation the ratio b(N)/a(N) is not larger than 0.75, which means that the crack form is far from being circular;
- pipelines with deeper cracks in conditions of local corrosion have a shorter lifetime than pipelines with shallow cracks of the same length.

Table 1. Initial data for a pipeline segment with an inner surface crack subjected to CO_2-type corrosion and cyclic loading

Parameters	Connotation	Numerical value	Measure-ment unit
R_1	Pipe inner radius	0.206	m
R_2	Pipe outer radius	0.213	m
p_{max}	Maximal inner pressure pulse	3.9	MPa
f	Pulsations frequency	6300	cycle/yr
σ_y	Yield stress limit of pipe material	330	MPa
E	Young's modulus	$2.1\cdot10^5$	MPa
μ	Poisson's ratio	0.26	-
A	Material fracture toughness parameters	$8.4\cdot10^5$	cycle/m
m		2.09	-
δ_{fc}		$1.5\cdot10^{-5}$	m
λ_0		0.89	-
$(a_0;b_0)$	Initial sizes of the crack ellipse (large and small semi-axes)	$(8\cdot10^{-3};3.5\cdot10^{-3})$ $(7\cdot10^{-3};4.5\cdot10^{-3})$	m
v_{lc}	Corrosion rate	$(0;0.2;1.8)\cdot10^{-3}$ $(3;5;8)\cdot10^{-3}$	m/yr

Table 2. Deterministic lifetime of a pipeline with internal crack subjected to cyclic loading and local uniform internal corrosion

Crack parameters		Local corrosion rate v_{lc}, 10^{-3}m/yr	IC type in the corrosion problem	Lifetime	
$a_0,10^{-3}$m	$b_0,10^{-3}$m			N^*, cycles	T^*, yrs
7	4.5	0	-	40942	6.50
		0.2	a	26960	4.28
			b	26461	4.20
		1.8	a	7200	1.14
			b	7200	1.14
		3.0	a	4700	0.75
			b	4600	0.73
		5.0	a	3000	0.48
			b	2900	0.46
		8.0	a	1900	0.30
			b	1800	0.29
8	3.5	0	-	47904	7.60
		0.2	a	31936	5.07
			b	31437	4.99
		1.8	a	9400	1.49
			b	9400	1.49
		3.0	a	6300	1.00
			b	6200	0.98

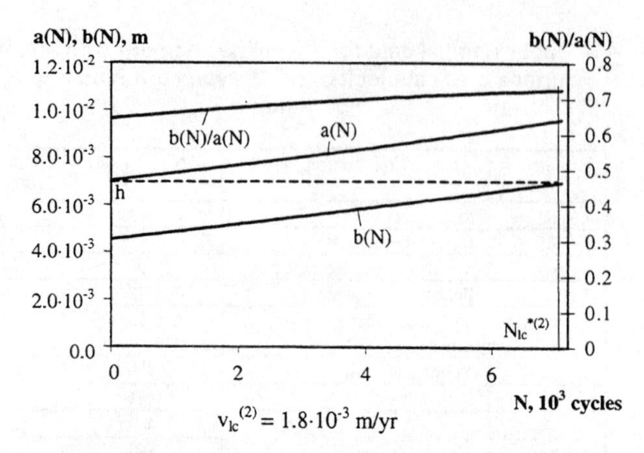

Fig. 3 Kinetics of internal crack growth (a_0=7·10^{-3} m; b_0=4.5·10^{-3} m) under combined influence of internal uniform local corrosion and cyclic loading for local corrosion and IC type b. Dashed line indicates wall thickness

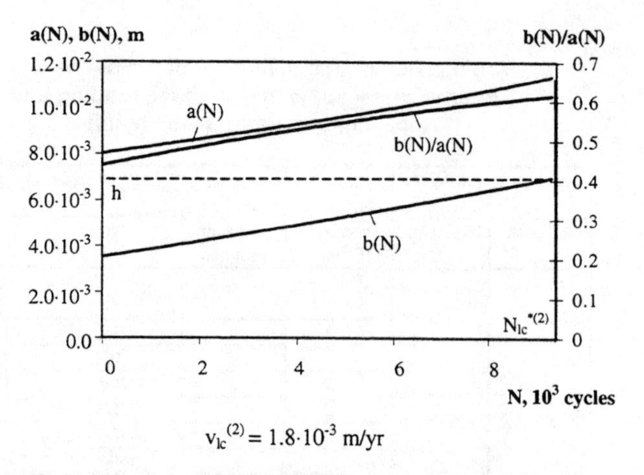

Fig. 4 Kinetics of internal crack growth (a_0=8·10^{-3} m; b_0=3.5·10^{-3} m) in a pipeline under combined influence of cyclic loading and local uniform corrosion for IC type b. Dashed line indicates wall thickness

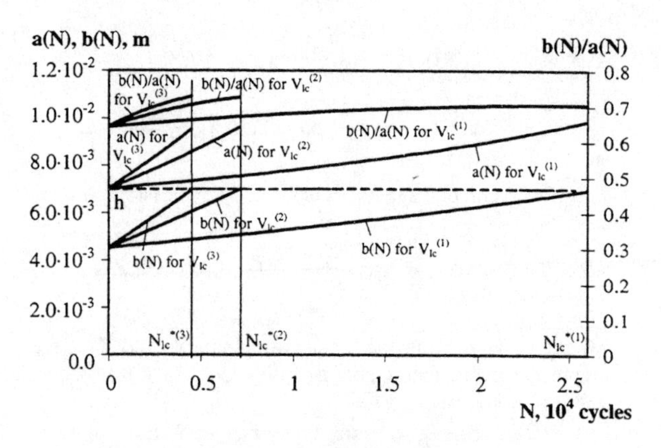

Fig. 5 Kinetics of internal crack growth (a_0=7·10^{-3} m; b_0=4.5·10^{-3} m) under combined influence of cyclic loading and local uniform corrosion of different rate ($v_{lc}^{(1)}$=0.2·10^{-3} m/yr; $v_{lc}^{(2)}$=1.8·10^{-3} m/yr; $v_{lc}^{(3)}$=3·10^{-3} m/yr) for IC type b. Dashed line indicates wall thickness

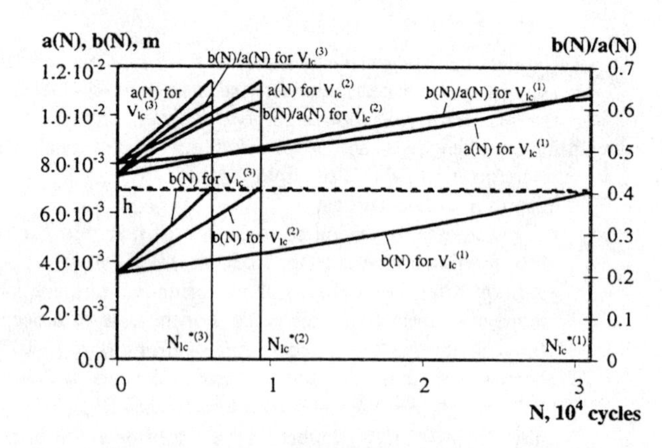

Fig. 6 Kinetics of internal initial crack (a_0=8·10^{-3} m; b_0=3.5·10^{-3} m) in a pipeline under combined influence of cyclic loading and local uniform corrosion of different rate ($v_{lc}^{(1)}$=0.2·10^{-3} m/yr; $v_{lc}^{(2)}$=1.8·10^{-3} m/yr; $v_{lc}^{(3)}$=3·10^{-3} m/yr) for IC type b. Dashed line indicates wall thickness

The probabilistic problem is solved for a pipeline with the most dangerous (among considered) crack ($a_0=7\cdot10^{-3}$ m; $b_0=4.5\cdot10^{-3}$ m). Parameters b_0, m, v_{lc} and p_{max} are assumed to be RVs. Their characteristics as well as variants of the problem solved are given in Tables 3 and 4. (M stands for mathematical expectation, σ stands for standard deviation.) The rate of v_{lc} is assumed to be the same as for v_c.

Simulation of RV's was performed using the Monte Carlo method. The size of simulated sets was assigned to be 100. For each simulated value of the above parameters the system (1) of DE with IC (8) and (10) type b (yielding a conservative assessment of pipeline lifetime) was solved.

According to the above results sets of pipeline lifetime assessments $\{N^*_{b_0}\}$, $\{N^*_m\}$, $\{N^*_{V_{lc}}\}$, $\{N^*_p\}$ were obtained. For each of them the maximal, minimal and average values were selected, histograms built and its first four moments calculated (Fig. 7).

The maximal scatter (96%) of longevity assessments around the average value is due to the random value of the crack depth b_0. For the RV v_{lc} this scatter is less – 41%.

According to the methods described in detail in (Malyukova and Timashev, 1999) the PDFs of pipeline lifetime N* (Fig. 7) and $N^i_{res}(T)$ residual lifetime (in cycles) of pipeline (Fig. 8) are obtained. In each scenario the mathematical expectations of $N^{(i)}$ and its variance σ_{N^i} are taken as certain fractions of the mean and variance of the relevant PDFs of N*.

Table 3. Simulation of random variables (RV)

RV	PDF	Method of simulating RV realizations	RV dimen-tion	PDF parameters value
b_0	Lognormal	$b_{0,k}=\exp\{M_b+\sigma_b\cdot S_k\}$ where $S_k=\frac{1}{\sqrt{k/12}}\cdot\left(\sum_{i=1}^{k}\eta_i-\frac{k}{2}\right)$ $k\geq 10$; η_i – an evenly distributed on [0,1] RV	m	$M_b = -5.43$ $\sigma_b = 0.17$
v_{lc}	Lognormal	$v_{lc,k}=\exp\{M_{V_{lc}}+\sigma_{V_{lc}}\cdot S_k\}$ (S_k described above)	m/yr	$M_{V_{lc}} = -6.33$ $\sigma_{V_{lc}} = 0.17$
m	Weibull	$m_j=\sigma_m\cdot\{-\ln\eta_j\}^{1/\alpha_m}$ (η_i described above)	-	$\sigma_m = 2.23$; $\alpha_m = 7.1$
p_{max}	Raleigh	$p_{max,j}=\sigma_p\cdot\sqrt{-2\ln\eta_j}$ (η_i described above)	MPa	$\sigma_p = 3.12$

Table 4. Design parameters for the probabilistic problem

b_0, m	m	V_{lc}, m/yr	p_{max}, MPa
Lognormal RV (see Table 3)	2.09	$1.8\cdot10^{-3}$	3.9
$4.5\cdot10^{-3}$	Weibull RV (see Table 3)	$1.8\cdot10^{-3}$	3.9
$4.5\cdot10^{-3}$	2.09	Lognormal RV (see Table 3)	3.9
$4.5\cdot10^{-3}$	2.09	$1.8\cdot10^{-3}$	Raleigh RV (see Table 3)

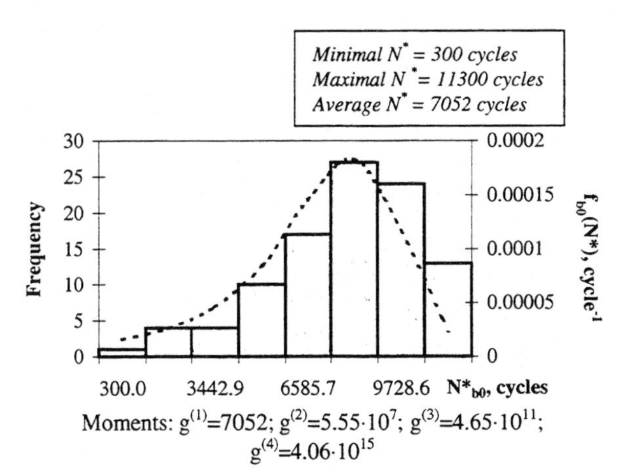

Moments: $g^{(1)}=7052$; $g^{(2)}=5.55\cdot10^{7}$; $g^{(3)}=4.65\cdot10^{11}$; $g^{(4)}=4.06\cdot10^{15}$

Fig. 7 Histogram and PDF of pipeline lifetime for Lognormal b_0, IC type b

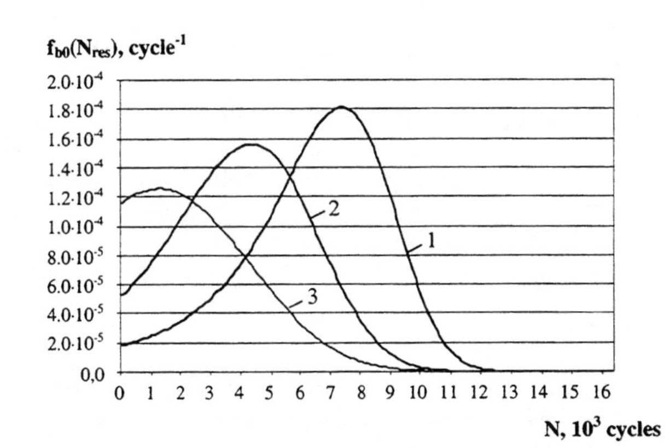

Fig. 8 Future residual lifetime PDFof a pipe segment subjected to local corrosion and cyclic loading (IC type b) calculated for the cases when the randomness is allocated to crack depth b_0. 1: $MN^i=g_i^{(1)}/10$; $\sigma^2_{N^i}=g_i^{(2)}/10$; **2:** $MN^i=g_i^{(1)}/2$; $\sigma^2_{N^i}=g_i^{(2)}/2$; **3:** $MN^i=9g_i^{(1)}/10$; $\sigma^2_{N^i}=9g_i^{(2)}/10$

LONGEVITY OF PIPELINE WITH A CLUSTER OF TWO LONGITUDINAL INTERNAL CRACKS

The deterministic problem. The problem of longevity assessment of a pipeline with two longitudinal cracks on its internal surface of semi-elliptical form is solved here. The sizes of the large and small semi-axes of the first and second crack are defined as a_{01}, b_{01} and a_{02}, b_{02} (Fig. 9). They satisfy the ratio b_{0i}/a_{0i}=0,3÷0,84, b_{0i}/h=0,15÷1. In this section the algorithm previously described in (Malyukova and Timashev, 1998, 1999) is utilized, which gives the solution of the longevity assessment of a pipeline with a single crack whose initial parameters also satisfy the above restrictions on b_0/a_0 and b_0/h.

A case when no corrosion is present is considered, as well as the case of local orthotropic internal corrosion, with rates v_{lc} and $v^t_{lc}=k_c \cdot v_{lc}$ (v_{lc} is the corrosion rate along the longitudinal direction of the pipe, v^t_{lc} is the corrosion rate in the transverse direction).

The first step is to consider the case when no corrosion is present. From the condition (see Fig. 9)

$$l(N_j) = l_0 + a_{01} + a_{02} - a_1(N_j) - a_2(N_j) = 0 \tag{8}$$

the $N_j=N_m$ is found which corresponds to the moment of merger of two cracks during cyclic loading. Here l_0 and $l(N_j)$ are the initial and current distance between the cracks, $a_1(N_j)$ and $a_2(N_j)$ are the relevant solutions of the deterministic problem for the pipeline with one crack with the parameters (a_{01},b_{01}) and (a_{02},b_{02}).

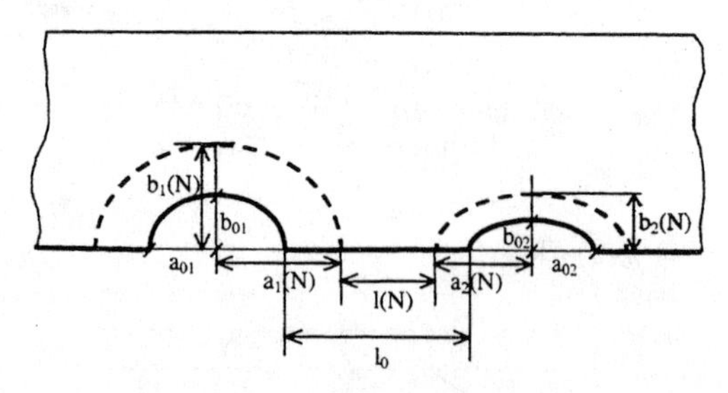

Fig. 9 Design scheme of a pipeline with a cluster of two longitudinal cracks. a_{0i}, b_{0i} are initial parameters of cracks, $a_i(N)$, $b_i(N)$ are current parameters of cracks, l_0 is initial distance between cracks, $l(N)$ is current distance between cracks, i=1,2

After merger the parameters of the new crack (assuming it has a semi-elliptical form) are found using following formulas

$$a_m = a_1(N_j) + a_2(N_j)$$
$$b_m = \max\{ b_1(N_j), b_2(N_j)\} \tag{9}$$

When $N>N_j$ the deterministic problem is solved for the pipe with one crack (a_m,b_m) and the number of cycles N* corresponding to the pipe longevity is found. For example, in step $N_{j+1}=N_j+\Delta N$, IC of type

$$a(N)\big|_{N=N_j} = a_m$$
$$b(N)\big|_{N=N_j} = b_m \tag{10}$$

are considered.

Solution of the problem with accounting for the local corrosion. From the condition

$$l(N_{c,j}) = l_0 + a_{01} + a_{02} - a_1(N_{c,j}) - a_2(N_{c,j}) = 0 \tag{11}$$

one finds $N_m=N_{c,j}$ which corresponds to the moment of merger of two cracks under conditions of cyclic loading and local corrosion. Here $a_1(N_{c,j})$ and $a_2(N_{c,j})$ are the relevant solutions of the deterministic problem accounting for local corrosion for a pipeline with a single crack with sizes (a_{01},b_{01}) and (a_{02},b_{02}) accordingly.

The newly formed crack of semi-elliptical form has the following parameters

$$a_m = a_1(N_{c,j}) + a_2(N_{c,j})$$
$$b_m = \max\{ b_1(N_{c,j}), b_2(N_{c,j})\} \tag{12}$$

For $N>N_{c,j}$ one solves the deterministic problem for the pipe with one crack (of size a_{cm}, b_{cm}) under conditions of local corrosion. In step $N_{c,j+1}=N_{c,j}+\Delta N$, the IC type b have the form

$$a_c(N)\big|_{N=N_{c,j}} = a_{c,m} + \Delta x_{j+1}$$
$$b_c(N)\big|_{N=N_{c,j}} = b_{c,m} + k_c \Delta x_{j+1} \tag{13}$$

On all other steps $N_{c,j+k} = N_{c,j+k-1} + \Delta N$

$$a_c(N)\big|_{N=N_{c,j+k-1}} = a_c(N_{c,j+k-1}) + \Delta x_{j+k}$$
$$b_c(N)\big|_{N=N_{c,j+k-1}} = b_c(N_{c,j+k-1}) + k_c \Delta x_{j+k} \tag{14}$$

Results of calculations. Calculations are performed for initial data given in Table 5. The results of calculation are presented in Fig. 10 – Fig. 12 and Table 6.

According to these results in all cases the criteria LBB, b(N*)=h is realized. When local corrosion is absent, after N_m=32128 cycles of loading, merger of two cracks occurs (crack one with initial sizes $7 \cdot 10^{-3}$ m; $4.5 \cdot 10^{-3}$ m and crack two with initial sizes $4 \cdot 10^{-3}$ m; $1.2 \cdot 10^{-3}$ m). The initial distance between cracks is $2 \cdot 10^{-3}$ m. The single crack which is formed during the merger of the two initial cracks has the following parameters: $1.3 \cdot 10^{-2}$ m; $6.1 \cdot 10^{-3}$ m. The FCG combined with the thinning of pipe wall due to local corrosion leads to a through-wall crack at N*=35712 cycles (Table 6, Fig. 10).

In the case of local corrosion the crack merger takes place much faster: for corrosion rate $v_{lc}=0.2 \cdot 10^{-3}$ m at N_m=16384 cycles, for $v_{lc}=1.8 \cdot 10^{-3}$ m at N_m=3200 cycles. With this the relevant longevity assessments dramatically drop as compared to the case when corrosion is absent: for the first case N*=20608 cycles (1.73 times less), for the second case N*=5376 cycles (6.64 times less).

In all observed cases the curves that define the change of semi-axes sizes a(N) and b(N) are linear, both before and after the two cracks merge. It should be noted that the above graphs (see Fig. 4) of parameters growth of a single crack ($a=7\cdot10^{-3}$ m, $b=4.5\cdot10^{-3}$ m) were non-linear. But in this case non-linearity developed when the number of cycles $N>N_m$.

The probabilistic problem. Parameters m, v_{lc}, p_{max} are taken as RV's. Their parameters and variants of the solution of the problem are the same as in the case of local corrosion of the pipe with a single internal crack. In calculations the initial data was taken from Tables 3 and 4, with following substitutions $M_{v_{lc}} = -8.53$, $v_{lc}=0.2\cdot10^{-3}$ m/yr.

The realizations of RV were simulated by the Monte-Carlo method. The size of simulated sets $\{m\}$, $\{v_{lc}\}$ is equal to 1000 and of the set $\{p_{max}\}$ is 600.

According to the solution of the basis system of DE with given initial conditions [Eqs. (6), (7)], for each of modeled values of the above parameters sets of pipeline longevities $\{N^*_m\}$, $\{N^*_{v_{lc}}\}$, $\{N^*_p\}$ are obtained. Statistical analysis of the obtained sets included calculation of the maximal, minimal and average longevities, construction of histograms and assessment of the first four PDF moments (Figs. 13).

According to numerical results, the maximum scatter of longevity assessment around the average value (131.8%) is observed for the random loading parameter p_{max}. The least scatter is observed for the longevity assessment, obtained for the random local corrosion rate v_{lc} (17.7%). The scatter of longevity assessment due to cyclic crack toughness parameter m is 84.5%.

Also, the PDFs of pipeline lifetime N* (Fig. 13) and $N^i_{res}(T)$ residual lifetime (in cycles) of pipeline (Fig. 14) are obtained using the methods described in (Malyukova and Timashev, 1999). Here in each of the scenarios the mathematical expectations of $N^{(i)}$ and its variance σ_{N^i} are taken as certain fractions of the mean and variance of the relevant PDFs of N*.

Table 5. Initial data for a pipeline segment with two cracks

Pipeline segment	Parameters	Numerical value	Measure-ment unit
	R_1	0.206	m
	R_2	0.213	m
	p_{max}	3.9	MPa
	f	6300	cycle/yr
	v_{lc}	0 $0.2\cdot10^{-3}$ $1.8\cdot10^{-3}$	m/yr
	k_c	1.2	-
Two different cracks	$(a_{01}; b_{01})$	$(7\cdot10^{-3}; 4.5\cdot10^{-3})$	m
	$(a_{02}; b_{02})$	$(4\cdot10^{-3}; 1.2\cdot10^{-3})$	m
	l_0	$2\cdot10^{-3}$	m

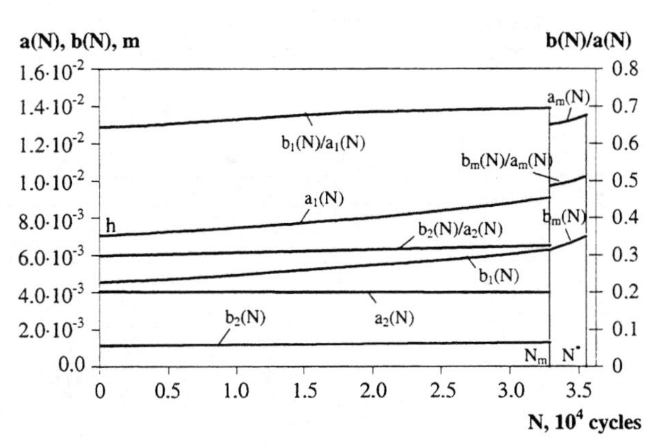

Fig. 10 Pipeline with two different cracks (without corrosion)

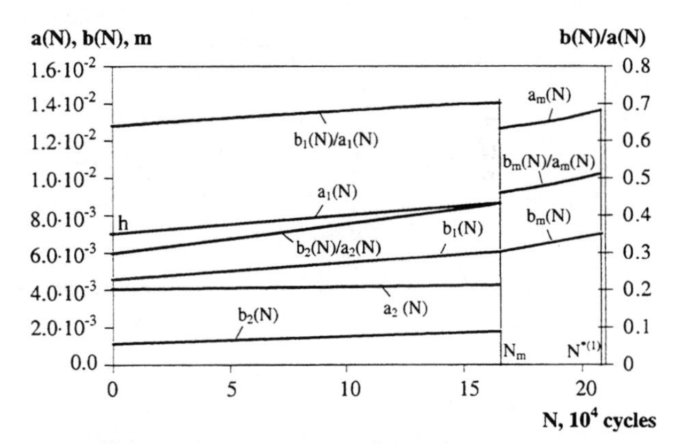

Fig. 11 Pipeline with two different cracks ($v_{lc}^{(1)}=0.2\cdot10^{-3}$ m/yr)

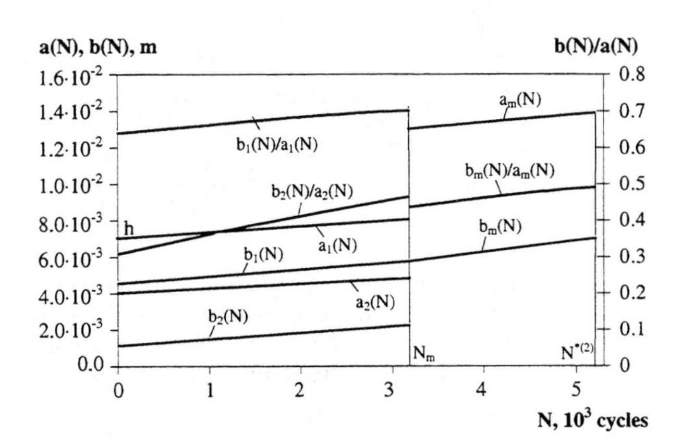

Fig. 12 Pipeline with two different cracks ($v_{lc}^{(2)}=1.8\cdot10^{-3}$ m/yr)

Table 6. Pipeline longevity assessment (case of two longitudinal cracks on the pipe inner surface)

Presence and corrosion type	Local corrosion rate v_{lc}, m/yr	Number of load cycles at merger of two cracks N_m, cycles	Pipeline lifetime assessment N*, cycles
No corrosion	0	32128	35712
Local corrosion	$0.2 \cdot 10^{-3}$	16384	20608
	$1.8 \cdot 10^{-3}$	3200	5376

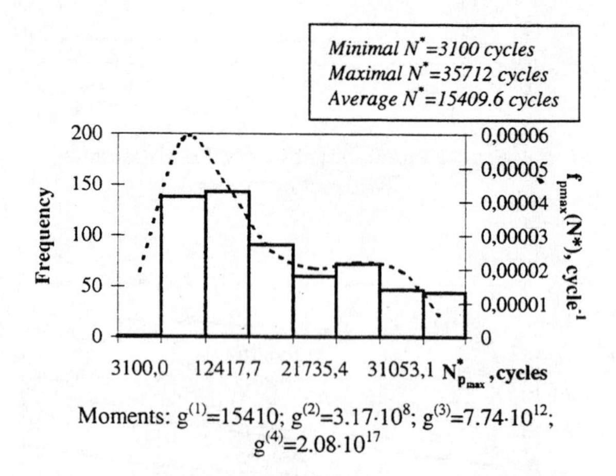

Moments: $g^{(1)}=15410$; $g^{(2)}=3.17 \cdot 10^{8}$; $g^{(3)}=7.74 \cdot 10^{12}$; $g^{(4)}=2.08 \cdot 10^{17}$

Fig. 13 Histogram and PDF of pipeline lifetime for Raleigh type PDF for p_{max}, IC type b

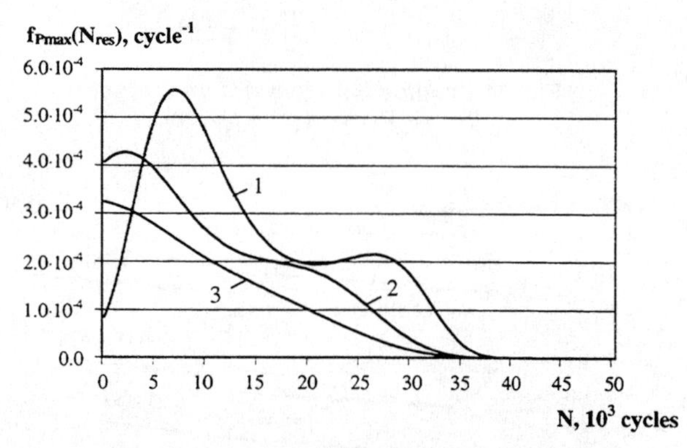

Fig. 14 Future residual lifetime PDF of a pipe segment with two cracks subjected to local corrosion and cyclic loading (IC type b) calculated for the cases when the randomness is allocated to parameter p_{max}.

1: $MN^{i} = g_i^{(1)}/10; \sigma^2_{N^i} = g_i^{(2)}/10$, **2:** $MN^{i} = g_i^{(1)}/2; \sigma^2_{N^i} = g_i^{(2)}/2$

3: $MN^{i} = 9g_i^{(1)}/10; \sigma^2_{N^i} = 9g_i^{(2)}/10$

CONCLUSION AND SUMMARY

1. The algorithm outlined here based in non-linear fracture mechanics, is a further development of computational assessment of pipelines residual resource. It accounts for real-life conditions of performance, including the corrosion activity of the environment and the presence of multiple crack-like defects on the inner surface of the pipe.

2. By using the Monte-Carlo simulation method together with the maximum entropy principle for the case of one and two cracks on the inner surface of pipelines, subjected to internal cyclic pressure and corrosion, the PDFs of pipeline longevity are obtained, and safe margins of pipeline residual resource can be appointed.

3. Systematic usage of these methods allows one to estimate time length of safe operation, depending on the expected dimensions of initial cracks, real properties of the pipeline material, corrosion rate and pumping technology. It provides for implementation of the Fit-for-Purpose (FFP) principle and Risk-Based Inspection (RBI) in pipeline maintenance.

4. The method outlined is part of an expert system for structural safety and reliability monitoring of main pipelines and is implemented in one of the largest Russian oil transportation companies.

ACKNOWLEDGEMENT

Authors express their gratitude to I.V. Makarov and M.G. Shalin for their programming expertise and help in obtaining numerical results.

REFERENCES

Charles, E. A., and Parkins, R. N., 1995, "Generation of Stress Corrosion Cracking Environments at Pipeline Surface," J. Corrosion. – Vol. 51, #7, pp. 518-527.

Garud, Y. S., 1990, "Incremental Damage Formulation and its Application to Assess IGSCC Growth of Circumferential Cracks in a Tube," J. Corrosion. – Vol. 47, # 7, pp. 523-527.

Korn, G., and Korn, T., 1984, "Reference Book on Mathematics," Moscow, Nauka, 831 pp.

Malyukova, M. G., and Timashev, S. A., 1998, "Residual Life of Pipelines with Longitudinal Initial Cracks," Fatigue, Fracture, and Residual Stresses, Advanced Fracture Mechanics, Pergamon Press, NY, 19, Joint ASME/JSME PVP Conference, pp. 99-103.

Malyukova, M. G., and Timashev, S. A., 1999, "Probabilistic Longevity of an Oil Pipeline with Crack Subjected to Internal Corrosion",ASME PVP Conference (to be published).

McMeeking, R. M., 1977, "Finite Deformation Analysis of Crack Tip Opening in Elastic Material and Implications for Fracture," J. of the Mech. and Physics of Solids, - 25, Vol. 2, pp. 357-382.

Medvedev, A. P., and Markin, A. N., 1995, "On Intensified Corrosion of Reservoir Oil Collecting Pipeline Systems", J. Oil Industry, #11, pp. 56-59 (in Russian).

Yunovich, M., Xia, Z., and Szklarska – Smialowska, Z., 1997, "Factors Influencing Stress Corrosion Cracking of Carbon Steel in Diluted Bicarbonate Environments," J. Corrosion. – Vol. 54, #2, pp. 155-161.

PVP-Vol. 410-1, Assessment Methodologies for Preventing Failure:
Deterministic and Probabilistic Aspects and Weld Residual Stress
Volume 1
ASME 2000

Probabilistic Analysis of Fatigue Crack Initiation and Growth in Reactor Pressure Boundary Components[1]

F.A. Simonen
Pacific Northwest National Laboratory[2]
P.O. Box 999
Richland, Washington 99352
509-375-2087
509-375-6497 (fax)
fredric.simonen@pnl.gov

M.A. Khaleel
Pacific Northwest National Laboratory
P.O. Box 999
Richland, Washington 99352
509-375-2438
509-375-2426 (fax)
moe.khaleel@pnl.gov

D.O. Harris
Engineering Mechanics Technology
Suite 166
4340 Stevens Creek Boulevard
San Jose, California 95129
408-247-9274
408-247-9272 (fax)
dharris@emtinc.com

D. Dedhia
Engineering Mechanics Technology
Suite 166
4340 Stevens Creek Boulevard
San Jose, California 95129
408-247-9275
408-247-9272 (fax)
ddedhia@emtinc.com

ABSTRACT

Procedures for the probabilistic analysis of the initiation of fatigue cracks in pressure boundary components of commercial light water reactors (LWRs) are described. This analysis is then combined with a probabilistic fracture mechanics analysis to predict the probability of such initiated cracks growing to become through-wall cracks, thereby causing leaks or catastrophic breaks. The crack initiation procedures are based on data analyses performed by Argonne National Laboratory (ANL) using data from fatigue tests of pressure boundary materials in LWR environments. The ANL work provided relations giving the probability of crack initiation as a function of the number of cycles for a given cyclic stress. The ANL correlations included the influence of the strain rate, sulfur content, oxygen content of the reactor water and temperature. The probabilistic fracture mechanics model of the pcPRAISE code was modified to account for crack initiation and the possible linking of fatigue cracks at multiple initiation sites. Example calculations are presented to show the effects of crack initiation and crack linking on the probabilities of through-wall cracks and the relative probabilities of large versus small leaks. The results obtained are of use in assessments of risk contributions of components with high cyclic usage and the risks posed by extending the lifetime of these components beyond their initial design life.

INTRODUCTION

Reactor pressure boundaries of many older plants were designed to codes that did not require an explicit component fatigue analysis. Subsequently, the ASME Code Section III did require a fatigue evaluation of components of the reactor pressure boundary, but aspects of the code methodology have come under review. Specifically, recent test data indicate that the effects of LWR environments can significantly reduce the fatigue resistance of materials thereby possibly leading to lack of conservatism in the ASME design fatigue curves. The purpose of this paper is to describe the procedures developed for reactor pressure boundary components that analyze the probability of fatigue crack initiation and subsequent growth to become through-wall cracks. These procedures can be used to estimate the failure probability of components subject to high fatigue usage. Such results can serve as inputs to risk assessments that address the risk contributions of these components. These risk assessments are based on the recent fatigue test data and are not dependent on the previous ASME design curves.

[1] Work supported by the U.S. Nuclear Regulatory Commission under Contract DE-AC066RLO 1830; NRC JCN W6671-6; S.K. Shaukat and D.N. Kalinousky Program Monitors
[2] Pacific Northwest National Laboratory is operated for the U.S. Department of Energy by Battelle Memorial Institute under Contract DE-AC06-76RLO 1830

ANL has developed revised fatigue curves based on test data from small polished specimens which were cycled to failure (or a significant load drop) in water simulating LWR conditions (Keissler et al., 1995a, 1995b, 1996). ANL also developed statistical models for estimating the effects of various materials, loading, and environmental conditions on the fatigue life of carbon steels, low alloy steels, and austenitic stainless steels. These statistical models are used here to estimate the probability of crack initiation, where crack initiation corresponds to a significant load drop in the fatigue test and corresponds to a crack of about 3mm depth (Keissler et al., 1995).

The Idaho Engineering Laboratory (INEL) (Ware et al., 1995) investigated the significance of the interim fatigue curves in Majumdar et al (1993) by performing deterministic fatigue evaluations for a sample of LWR components. This investigation involved the gathering of information on the cyclic stresses and number of cycles for a large number of components, with an emphasis on components with high fatigue usage. The cyclic stresses and cycles from these efforts were used as inputs to the efforts reported herein and in Simonen et al. (1999) to provide a probabilistic analysis of crack initiation and growth to failure. These calculations were performed to evaluate the risks posed by these components for their originally intended lifetime of 40 years and an extended lifetime of 60 years. The failure probability results for the many components considered are described in Simonen et al. (1999) along with estimated contributions to core damage frequencies. The present paper focuses on the procedures developed for the probabilistic analyses of crack initiation and crack growth, and provides example results for one component.

REVIEW OF ANL STATISTICAL INITIATION MODEL

The ANL statistical model of crack initiation described in Keissler et al. (1995a, 1995b and 1996) employs the following form of the strain-life relation

$$\varepsilon_a = \frac{B}{N^b} + A, \tag{1}$$

where ε_a is the strain amplitude (one half the peak-to-peak value) and N is the number of stress cycles. A, B and n are curve fitting constants. This can also be written as

$$\ln(N) = \frac{1}{b}\ln(B) + \frac{1}{b}\ln(\varepsilon_a - A). \tag{2}$$

This alternative form shows that A is the strain endurance limit, i.e. the number of cycles approaches infinity as ε_a decreases towards A. The ANL statistical model is equivalent to considering A and $\ln(B)$ to be normally distributed, with parameters as given in Table 1. The particular constants given in Table 1 are those employed in the present efforts and in Simonen et al. (1999), and differ somewhat from values reported in the Keissler references. The modifications are based on communications between ANL and PNNL personnel. The $-\ln(4)$ term at the end of the expressions for the mean of $[\ln(B)]/b$ is intended by ANL to account for size effects, geometry and surface finish. As discussed below, this was modified in the present efforts in order to predict a size effect with the probabilistic fracture mechanics analysis that depended on the specific size of the component involved.

The number of cycles in Equations 1 and 2 is actually the cycles to a noticeable load drop during the fatigue test, such as a 25% drop. This is taken to correspond to an initiated crack depth of 3 mm (Keissler et al., 1995a, 1996). Tests were performed on smooth specimens (i.e. un-notched) with fully reversed loading.

Figure 1 shows the data for carbon steel in air at 550°F as given in Figure 4 of Keissler et al. (1995b). Also shown in Figure 1 are various quantiles of the distribution of the strain range as functions of the number of cycles as calculated for a strain rate of greater than 1%/sec. At this high strain rate, the sulfur content has no effect. The constants used to calculate the various quantiles are as reported in Keissler et al. (1995b), and do not include the $-\ln(4)$ term in the mean of $[\ln(B)]/b$. The lowest quantile shown is 10^{-4}, which is just below the lower limit of extrapolation of the probabilistic curve fits of 2×10^{-4} mentioned in Keissler et al. (1996). Extrapolations to smaller quantiles can lead to unreasonable strain-life behavior, such as (in extreme cases) negative endurance limits.

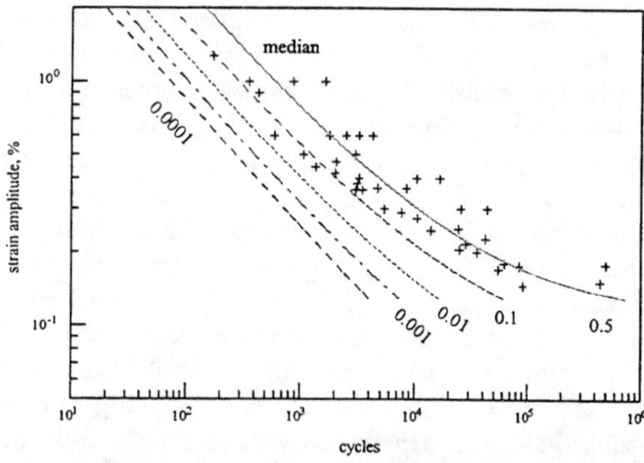

Figure 1: Fatigue Data for Carbon Steel in Air at 550°F with Various Quantiles of the Statistical Distributions from the ANL Model (data and fit from Keissler et al., 1995b).

Table 1
Summary of ANL Strain-Life Relations

	Low Alloy and Carbon Steel in Air and Water	304 and 316 Stainless Steel in Water	304 and 316 Stainless Steel in Air
1/b	$1.813+0.219I_s$	2.172	2.172
Parameters of A			
Mean	$0.080+0.014I_s$	0.108	0.108
Standard Deviation	0.026	0.026	0.026
Parameters of [ln(B)]/b			
Mean	$(6.857-0.766I_w)$ $-(0.275-0.382I_w)I_s$ $-0.00133T(1-I_w)$ $+0.1097S^*T^*O^*\varepsilon^*$ $-\ln(4)$	$5.841+T^*O^*\varepsilon^*$ $-\ln(4)$	$6.776+T^*\varepsilon^*$ $-\ln(4)$
Standard Deviation	0.52	0.52	0.52
Indicator Functions			
I_w	0 for air 1 for water	--	--
I_s	0 for low alloy steel 1 for carbon steel	--	--
S*	S, $S<0.015$ 0.015, $S>0.015$	--	--
T*	0, $T<150$ T-150, $T>150$	0, $T<200$ 1, $T>200$	0, $T<250$ $[(T-250)/525]^{0.84}$ $T>250$
O*	0, $O<0.05$ O, $0.05<O<0.5$ 0.5, $0.5<O$	0.260, $O<0.05$ 0.172, $O>0.05$	--
ε*	0, $\dot{\varepsilon}>1$ $\ln(\dot{\varepsilon})$, $0.001<\dot{\varepsilon}<1$ ln(0.001), $\dot{\varepsilon}<0.001$	0, $\dot{\varepsilon}>0.4$ $\ln(\dot{\varepsilon})$, $0.0004<\dot{\varepsilon}<0.4$ ln(0.0004/0.4), $\dot{\varepsilon}<0.0004$	

strains in %
strain rates in %/second
T, temperatures in °C
S, sulfur content in wt. %
O, dissolved oxygen contents in ppm

MODIFICATIONS TO ANL MODEL

Some modifications were made to the ANL curve fits based on discussions between ANL and PNNL personnel. The constants in Table 1 were used in the efforts reported here and Simonen et al. (1999). Some of the constants in Table 1 are different than those appearing in the Keissler et al. (1995a, 1995b, and 1996) references based on recent work by ANL.

Another adjustment was made to have a size, geometry and surface finish effect that depended on the size of the specific component. The –ln(4) term at the end of the expressions for the mean of $[\ln(B)]/b$ was intended to provide the adjustment for size, geometry and surface finish. The –ln(4) term is equivalent to dividing the number of cycles to initiation by 4. A size-dependent adjustment was believed to be desirable, because the model was to be applied to components of widely varying size, all of which are much larger than a typical laboratory fatigue specimen. The size adjustment consisted of retaining the –ln(4) term, but then multiplying the initiation cycles by a factor that calibrated the predicted number of cycles to initiation to results of the 9 inch diameter pressure vessel fatigue tests reported in the ANL documents. This resulted in multiplying the number of cycles by 3, and considering the relation to then be suitable for a fatigue test specimen of 2-inch length. Each component was then considered to consist of a number of 2 inch long "specimens." For instance, the circumference of a girth weld in a pipe was broken up into 2-inch lengths. Allowing for multiple initiation sites allows stress gradients along the surface to be accounted for. Stresses can vary from specimen to specimen. The fatigue crack initiation in each "specimen" was taken to be independent of the other specimens. The other extreme of considering the initiations to be dependent leads to simultaneous initiation all the way around a girth weld, which is not the expected behavior.

PROBABILISTIC ANALYSIS OF CRACK INITIATION

The ANL relations for crack initiation are for a constant strain amplitude, whereas reactor components typically see several different transient types, each with a given strain amplitude and frequency. The history is usually defined in terms of a stress range (peak-to-peak stresses) and the number of occurrences in 40 years, such as in the middle two columns of Table 2.

Table 2
Typical Definition of Stress History

ID	Stress Amplitude	Cycles per 40 Year Life	Cycles per Year
1	$\sigma_{a,1}$	n_1	$\dot{n}_1$
2	$\sigma_{a,2}$	n_2	$\dot{n}_2$
--	--	--	
M	$\sigma_{a,M}$	n_M	$\dot{n}_M$

The cycles per year (fourth column) is equal to the cycles per 40 year life divided by 40, that is, the cycling rate is considered to be constant. A description of each transient is usually provided. The fatigue damage as a function of time is expressed as

$$D(t)=\sum_{j=1}^{M}\frac{\dot{n}_j t}{N_I(\sigma_{a,j})}=t\sum_{j=1}^{M}\frac{\dot{n}_j}{N_I(\sigma_{a,j})}=t\dot{D} \tag{3}$$

$N_I(\sigma_{a,j})$ is the cycles to initiation for the cyclic stress, $\sigma_{a,j}$, and is a random variable defined by the ANL relations summarized in Table 1. The time to initiation, t_I, is the time for D to reach the value of unity. Hence, t_I is equal to $1/\dot{D}$.

Previous efforts (Khaleel and Simonen, 1998) considered D to be normally distributed, because (as a result of the central limit theorem) the sum of random variables ($\dot{D}$ in this case) will tend to be normal independently of the type of distribution of the random variables themselves. This is true for the sum of a large number of variates (large number of transient types) which is often not the case for the components of interest. Some sensitivity studies on representative problems showed that the assumption of D being normal was not very good. Hence, this assumption is not made, and Monte Carlo simulation is used to compute the distribution of initiation time. The Monte Carlo simulation consists of: 1. Sampling from a uniform unit variate, 2. Calculating the cycles to initiation for this quantile for each cyclic stress, and 3. Calculating $t_I=1/\dot{D}$. This provides a histogram of t_I, from which the probability of crack initiation as a function of time follows. This capability was incorporated into the software that already existed for the probabilistic analysis of the growth of cracks, the pcPRAISE code (Harris et al. 1981, 1992), but some additional features were necessary before the software would be applicable to the problem at hand. This included defining the length of initiated cracks (the depth is 3 mm), the linking of multiple cracks and defining the stress gradients – primarily the through-thickness gradient.

TREATMENT OF INITIATED CRACKS

The ANL initiation relations are for the number of cycles to a crack depth of 3 mm. It is assumed here that the initiated cracks in real components will be semi-elliptical surface cracks, which have both a depth and length. The depth is most important in the growth calculations because it has more influence on the stress intensity factor than the surface length. However, the surface length, $2b$, is also important because it also has an influence on the stress intensity factor, and because it effects crack linking and, therefore, the frequency of leaks relative to catastrophic failures. The small initiated cracks may be nearly semi-circular in shape, and a value of b at initiation (denoted as b_o) of 3 mm is quite likely. A value of b_o less than 3 mm is however unlikely. Therefore b_o-a_o was selected to be random with a minimum value of 0. It is also unlikely that the initiated value of b will be greater than the length of a "specimen", meaning that the probability of b_o exceeding 1 inch should be very small. The random variable

b_o-a_o was assumed to be lognormally distributed with a median value of b_o of 0.30 inches (7.6 mm). This corresponds to a median b_o/a_o of 2.5, which is well above 1 and, therefore, believed to be conservative. It is still necessary to define the second parameter of b_o-a_o, which was selected as 0.682, which results in one crack in 100 having a total initiated length greater than the length of a "specimen." This is also believed to be a conservative assumption.

Since multiple cracks can initiate, it is necessary to have a set of criteria defining when growing cracks in adjacent specimens will link together. These criteria were already in pcPRAISE (Harris et al., 1986) and were based on the ASME Boiler and Pressure Code crack linking procedures as they existed in 1986.

Stress histories, such as exemplified in Table 2, is typically in terms of peak stresses in the component. When these stresses are high, they generally have steep spatial gradients away from the high stress location. The high stresses often have very steep through-wall gradients resulting from extreme and rapid temperature excursions. There can also be stress gradients along the surface away from the high stress location, such as can occur near nozzles. pcPRAISE can account for the through-thickness gradients, and it is unrealistic to assume that the peak stresses are uniform through the thickness. Hence, it is desirable to estimate through-thickness gradients, even though only surface stress information is available. A representative through-thickness stress gradient is given by the expression

$$\sigma(\xi) = \sigma_s (1 - 3\xi + 1.5\xi^2) \qquad (4)$$

where $\xi = x / h$, x is the distance into the wall and h is the wall thickness. σ_S is the stress at the inner wall. Equation (4) describes the stress gradient that is present shortly after a temperature excursion occurs in which the temperature at the inner surface of a pipe is changing linearly with time (ramp temperature), and the outer surface is insulated (h/R<<1). This stress gradient can also be used to estimate the maximum stress intensity factors resulting from a step change in temperature. Equation (4) is the only second order polynomial variation of stress that is self-equilibrating and has a zero slope at the outside surface (a consequence of the insulated outer surface). Hence, this generic gradient was used to represent through-thickness gradients that are self-equilibrating (such as required for radial gradient thermal stresses). This gradient is believed to be conservative when applied to geometric stress concentrations, such as at counterbores in pipes, because such stress concentrations may be more localized than described by this gradient.

Surface stress gradients can also be appreciable, such as at nozzles. It is difficult to define a generic gradient in this case, but the results of finite element calculations reported in Cohen et al. (1977) are useful in this regard. Surface gradients are handled in pcPRAISE by allowing the maximum stress in each 2-inch specimen to vary from specimen to specimen in a given component.

The stress history information from Ware et al. (1995) consisted only of the peak stress range and number of stress cycles in 40 years, i.e. the center two columns of Table 2. It is necessary to define the spatial gradients of these stresses in order to perform the fatigue crack growth portion of the calculations. The following set of rules was devised to conservatively estimate radial stress gradients:

1. Cyclic stresses associated with seismic loads were treated as uniform,

2. Stresses with ranges less than 45 ksi were taken to be uniform (with the exception of item 3 below). Nonseismic stresses with ranges greater than 45 ksi were treated as 45 ksi of uniform stress and the remainder as the generic radial gradient of Equation 4. The 45 ksi limit on uniform stress is based on the ASME Code allowables on primary and secondary membrane being $3S_m$, which is typically close to 45 ksi for the materials considered here.

3. Transients with more than 1000 cycles in a 40 year life were assumed to have equal amounts of uniform and the generic radial gradient components, unless the uniform component exceeded 10 ksi, in which case the uniform component was taken to be 10 ksi and the remainder of the generic radial gradient category. This assumption was made because if there are a large number of cycles for a transient, they are usually due to thermal transients; which would have appreciable radial gradients.

EXAMPLE

The crack growth portion of the analysis is performed using the pcPRAISE software (Harris et al., 1981, 1992). In addition to the random variables discussed above, the random variables of the probabilistic fracture mechanics analysis are the fatigue crack growth rate for a given ΔK and the flow stress used with the critical net section stress failure criterion employed for the stainless steel components. The tearing instability failure criterion based on the J-integral was also employed using deterministic material properties. pcPRAISE can compute leak rates through cracks, and can evaluate the probability of leaks of various magnitudes. A leak is taken to be the existence of a through-wall crack, a big leak is the existence of a through-wall crack that is large enough to result in a user-defined system disabling leak, and a double-ended-pipe-break (DEPB) is a sudden and complete pipe severance. A DEPB can occur when the crack is long enough at the time it breaks through the pipe wall that it will grow circumferentially suddenly and catastrophically to fail the pipe.

Simonen et al. (1999) reports results for some 47 components in plants of newer and older vintages for the four US commercial reactor vendors. Results are included for both 40 and 60 year lifetimes. The example presented here is based on the makeup/HPI nozzle safe end for a PWR. The following information was used

thickness = 0.50 inches
inner diameter = 6 inches
material 304 stainless steel
sulfur content 0.015 wt%
water with dissolved oxygen 0.010 ppm
temperature 590°F
strain rate – zero

The stress history consisted of the three transient types summarized in Table 3.

Table 3
Summary of Stress Transients in Example Component

Stress Range ksi	Number in 40 Years	Description
221.24	33	HPI actuation A/B
169.31	7	test/null
11.98	200	heat up/cool down

Calculations for this component assumed a uniform stress around the circumference. The stresses were decomposed into uniform and generic radial gradient by the criteria given above. Table 4 summarizes the results for crack initiation and leak probabilities for the three cases of: multiple cracks with uniform stress, multiple cracks with radial stress gradients, and a single crack with radial stress gradients. A big leak was defined as a leak greater than 500 gallons per minute and a small leak the occurrence of any through-wall crack. Results were generated with 10^6 or 10^7 Monte Carlo trials, as indicated in the table, and considered operating times up to 60 years. Results are also given for big leaks with multiple initiation sites. No big leaks occurred within the number of trials for the other two cases, and no double ended pipe breaks occurred with the number of trials for any of the three cases. Thus in all instances there was a strong predicted tendency for leak before break, and although the leak probabilities were not necessarily low, the probabilities for the big leak (500 gpm) and the double ended pipe break probabilities were very low.

Figure 2 provides a plot of the initiation and leak results from Table 4. These results show that the case of multiple cracks gave consistently higher crack initiation and leak probabilities, with the probabilities being roughly in proportion to the number of initiation sites (a factor of 9). A comparison of the "multiple leak uniform" and "multiple leak" results show a very large influence of the stress gradient on the leak probabilities. Hence, the assumption of a uniform stress through the wall is very conservative.

The results in Table 4 and Figure 2 show that the failure probabilities do not increase markedly in the period from 40 to 60 years, thereby indicating that the 40 year period is not at the useful end-of-life. The very low probabilities of big leaks or double ended pipe breaks relative to the leak probabilities indicate that leak-before-break will still occur and will provide warning of problems due to initiation and growth of fatigue cracks even beyond 40 years.

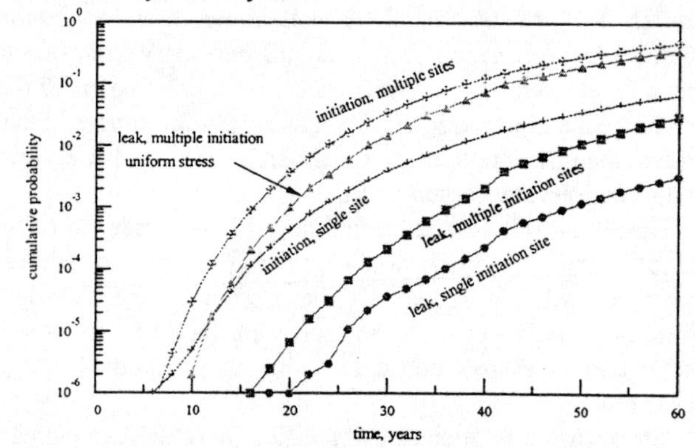

Figure 2: Crack Initiation and Leak Probabilities as Functions of Time for the Example Problem.

Simonen et al. (1999) includes results for initiation and leak probabilities for nearly 50 components from five PWR and two BWR plants. The results indicate that some components do have high failure probabilities, with through-wall crack frequencies approaching 0.05 per year. In these relatively high failure probability components, there is little or no increase in the failure frequency in the period from 40 to 60 years. Simonen et al. (1999) carry the process one step further to address the contributions of the high-fatigue usage components to the core damage frequency, and find that the calculated core damage frequencies of the most critical components with the highest failure frequencies show essentially no increase from 40 to 60 years.

COMBINING INITIATING AND PRE-EXISTING CRACKS

The failure probability for components subject to cyclic stresses will be governed by both initiating cracks and pre-existing fabrication cracks. In some instances both contributors may be significant, and the following discusses how to account for both. In the case that initiating cracks dominate, the procedures discussed above can be used. In the case that pre-existing cracks dominate, pcPRAISE can also be employed. In fact, pcPRAISE gives the failure probability given the presence of one crack, $P(t_f<t|1) \equiv P_1$, and the probability of having a crack initially present is accounted for outside of pcPRAISE. The contributions of both pre-existing and initiating cracks can be accounted for by making two assumptions:

1. the number of cracks in the weld is Poisson distributed with mean value λ, and
2. the failure probability with the presence of N cracks ($N \geq 1$) is given by the expression

$$P(t_f < t|N) \equiv P_N = 1-(1-P_1)^N \tag{5}$$

Table 4
Summary of Cumulative Probability Results for Example Problem

cracks	single		multiple (9)			
stress	gradient		gradient			uniform
trials	10^6		10^7			10^6
time, years	initiation	leak	initiation	leak	big leak	leak
4	$<10^{-6}$	$<10^{-6}$	$<10^{-7}$	$<10^{-7}$	$<10^{-7}$	$<10^{-6}$
8	2.0×10^{-6}	$<10^{-6}$	4.40×10^{-6}	$<10^{-7}$	$<10^{-7}$	$<10^{-6}$
12	1.6×10^{-5}	$<10^{-6}$	1.19×10^{-4}	$<10^{-7}$	$<10^{-7}$	1.7×10^{-5}
16	1.16×10^{-4}	1.0×10^{-6}	9.14×10^{-4}	1.0×10^{-6}	$<10^{-7}$	2.12×10^{-4}
20	4.43×10^{-4}	1.0×10^{-6}	3.75×10^{-3}	6.8×10^{-6}	$<10^{-7}$	9.97×10^{-4}
24	1.25×10^{-3}	3.0×10^{-6}	1.08×10^{-2}	3.06×10^{-5}	$<10^{-7}$	3.50×10^{-3}
28	2.80×10^{-3}	2.2×10^{-5}	2.46×10^{-2}	1.32×10^{-4}	$<10^{-7}$	1.04×10^{-2}
32	5.43×10^{-3}	5.0×10^{-5}	4.75×10^{-2}	3.70×10^{-4}	$<10^{-7}$	2.19×10^{-2}
36	9.33×10^{-3}	1.05×10^{-4}	8.15×10^{-2}	9.41×10^{-4}	$<10^{-7}$	4.15×10^{-2}
40	1.49×10^{-2}	2.31×10^{-4}	1.27×10^{-1}	2.10×10^{-3}	$<10^{-7}$	7.50×10^{-2}
44	2.23×10^{-2}	5.92×10^{-4}	1.83×10^{-1}	5.37×10^{-3}	2.0×10^{-7}	1.22×10^{-1}
48	3.14×10^{-2}	9.53×10^{-4}	2.48×10^{-1}	8.89×10^{-3}	2.0×10^{-7}	1.66×10^{-1}
52	4.23×10^{-2}	1.49×10^{-3}	3.21×10^{-1}	1.40×10^{-2}	3.0×10^{-7}	2.24×10^{-1}
56	5.47×10^{-2}	2.27×10^{-3}	3.97×10^{-1}	2.14×10^{-2}	5.0×10^{-7}	2.97×10^{-1}
60	6.89×10^{-2}	3.32×10^{-3}	4.73×10^{-1}	3.10×10^{-2}	7.0×10^{-7}	3.66×10^{-1}

The second assumption is equivalent to assuming that the N cracks are independent of one another. This accounts for the increase in the failure probability if more than one crack is present, but does not explicitly account for crack interactions. The first assumption means that the probability of having N cracks in the weld is given by the expression

$$p(N) = e^{-\lambda} \frac{\lambda^N}{N!} \tag{6}$$

Then the probability of failure accounting for the number of pre-existing cracks is given by

$$\begin{aligned} P(t_f < t) &= \sum_{n=0}^{\infty} p(n) P_n = e^{-\lambda} P_0 + e^{-\lambda} \sum_{n=1}^{\infty} \frac{\lambda^n}{n!} [1 - (1 - P_1)^n] \\ &= e^{-\lambda} P_0 + e^{-\lambda} \sum_{n=1}^{\infty} \frac{\lambda^n}{n!} - e^{-\lambda} \sum_{n=1}^{\infty} \frac{\lambda^n}{n!} (1 - P_1)^n \\ &= e^{-\lambda} P_0 + e^{-\lambda} \sum_{n=1}^{\infty} \frac{\lambda^n}{n!} - e^{-\lambda} \sum_{n=1}^{\infty} \frac{[\lambda(1 - P_1)]^n}{n!} \end{aligned} \tag{7}$$

P_0 is the failure probability given no initial crack. Making use of the infinite series for the exponential function, this equation can be simplified to the following:

$$\begin{aligned} P(t_f < t) &= e^{-\lambda} P_0 + e^{-\lambda} (e^{\lambda} - 1) - e^{-\lambda} (e^{\lambda(1-P_1)} - 1) \\ &= e^{-\lambda} P_0 + 1 - e^{-\lambda P_1} \end{aligned} \tag{8}$$

Equation 8 provides a means of combining the failure probabilities for crack initiation and growth (making use of the above procedures) with conventional pcPRAISE results for one pre-existing crack. If there are no initial cracks, then $\lambda=0$ and the failure probability is simply the probability obtained by use of the above initiation procedures. If the initiation probability is zero, then $P_0=0$ and the failure probability accounting for the probability of a crack being present is $1 - e^{-\lambda P_1}$. Khaleel et al. (1999) provides information on values of λ for LWR piping welds.

The above example problem with uniform stress was run for pre-existing cracks using pcPRAISE, with the results in Table 5 being obtained for the leak probability. The default values of the crack size distribution from Khaleel et al. (1999) were employed, with a lognormal initial crack depth distribution with median 0.10 inches and shape parameter of 0.250. The initial aspect ratio (half surface length divided by crack depth) was also lognormal with median 1.34 and shape parameter of 0.538. The value of λ (mean pre-existing cracks per weld) was 0.0558. Other inputs were as described above for the initiation example problem.

Table 5
Summary of Cumulative Leak Probability for Example Problem with Pre-Existing Cracks (conditional on crack being present)

Time, Years	Cumulative Probability	Time, Years	Cumulative Probability
4	2.59×10^{-3}	36	2.56×10^{-1}
8	8.22×10^{-3}	40	2.93×10^{-1}
12	2.72×10^{-2}	44	3.14×10^{-1}
16	4.40×10^{-2}	48	3.47×10^{-1}
20	7.27×10^{-2}	52	3.79×10^{-1}
24	1.26×10^{-1}	56	4.17×10^{-1}
28	1.77×10^{-1}	60	4.39×10^{-1}
32	2.11×10^{-1}		

Figure 3 provides plots of the cumulative leak results on both log-linear and lognormal scales. Results are included for initiated cracks only (right column of Table 4), pre-existing cracks only (compensated for the probability of a crack being initially present) and for the combined result.
The lower portion of Figure 3 shows that the initiated crack and pre-existing crack results are each nearly straight lines on lognormal probability scales. This means that the lifetimes for each of these conditions are approximately lognormally distributed. This is not surprising because the dominant random variables are all lognormally distributed, and the lifetimes are approximately proportional to products of powers of the random variables. This is not exactly true and the lifetime equations can not be explicitly written. The combined result is not lognormal, but quickly transitions from one lognormal (for pre-existing cracks) to another lognormal (for initiating cracks).

Figure 3 shows that the pre-existing cracks dominate early in life because they have a "head start" on the initiating cracks. As time progresses, the initiating cracks take over because they may be present with a relatively high probability that can exceed the probability of having a crack initially present. (The leak probability for pre-existing cracks can never be larger than $1\text{-}e^{-\lambda}$, which is 0.0542 for this example). This shows that the pre-existing cracks dominate for cases where stress cycling is not excessive, but that crack initiation becomes an important aspect at long times and for high cyclic stresses and frequencies.

DISCUSSION AND CONCLUDING REMARKS

The methodology of this paper corrects a significant shortcoming that has existed for many years in existing codes for probabilistic fracture mechanics that address structures and components subject to fatigue. Implementation of the approach into an enhanced version of the pc-PRAISE code has demonstrateed the ability to simulate the initiation and growth of fatigue cracks in addition to the simulation of failures due to preexisting fabrication flaws. Trial calculations show large differences between probabilities of crack initiation and probabilities of through-wall cracks associated with these initiated cracks. Other calculations show that initiated fatigue cracks can make important contributions to structural failure probabilities. For conditions of high fatigue usage the calculations indicate that intiated cracks, rather than preexisting cracks, are the dominate contributor to pipe leaks due to cyclic fatigue stresses. The model has found application to the resolution of regulatory issues related to the fatigue of nuclear power plant components for extended operation to 60-year plant life (Simonen et al. 1999). Future applications are expected address the ability of inservice inspections to maintain piping failure probabilities for fatigue sensitive components at acceptable levels (Gosselin and Simonen 1999).

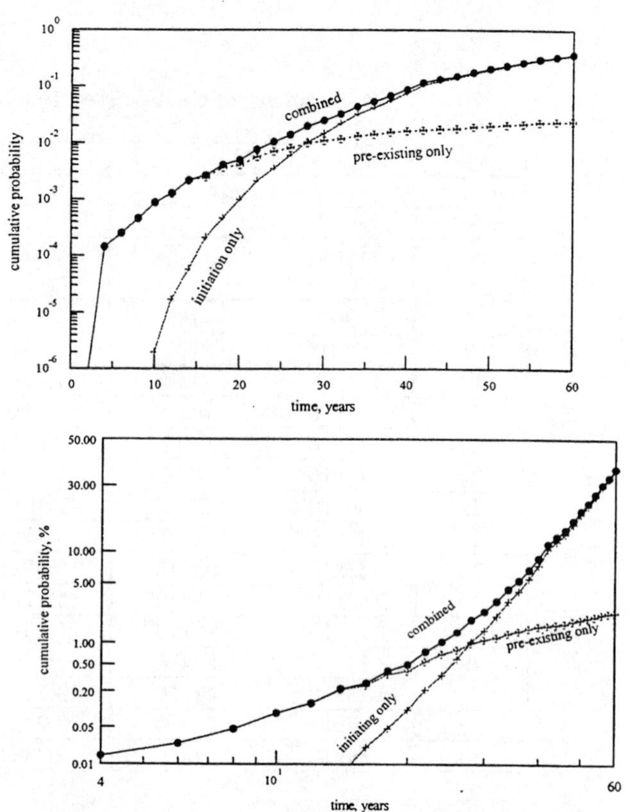

Figure 3: Cumulative Leak Probabilities as Functions of Time for the Example Problem with Uniform Stress Including the Contribution of Pre-Existing Cracks.

REFERENCES

Cohen, L.M., J.L. McLean, G. Moy and P.M. Besuner 1977. *Improved Evaluation of Nozzle Corner Cracking*, Report EPRI NP-399, Electric Power Research Institute, Palo Alto, California.

Gosselin, S.R. and F.A. Simonen 1999. "A Probabilistic Basis for Damage Tolerance Assessments and Component Fatigue Life Extension," *Pressure Vessel and Piping Codes and Standards – 1999,* ASME PVP-Vol. 383, pp. 105-110.

Harris, D.O., E.Y. Lim and D.D. Dedhia 1981. *Probability of Pipe Fracture in the Primary Coolant Loop of a PWR Plant, Vol. 5: Probabilistic Fracture Mechanics Analysis*, U.S. Nuclear Regulatory Commission, Washington, D.C., Report NUREG/CR–2189, Vol. 5.

Harris, D.O., D. Dedhia, E.D. Eason and S.P. Patterson 1986. *Probability of Failure in BWR Reactor Coolant Piping*, U.S. Nuclear Regulatory Commission, Washington, D.C., report NUREG/CR–4792, Vol. 3.

Harris, D.O., D. Dedhia and S.C. Lu 1992. *Theoretical and User's Manual for pc-PRAISE*, U.S. Nuclear Regulatory Commission, Washington, D.C., Report NUREG/CR-5864.

Keissler, J., O.K. Chopra and W.J. Shack 1995a. *Statistical Analysis of Fatigue Strain-Life Data for Carbon and Low-Alloy Steels, Austenitic Steels and Alloy 600 in LWR Environments*, U.S. Nuclear Regulatory Commission, Washington, D.C., Report NUREG/CR-6335.

Keissler, J. and O.K. Chopra 1995b. "Statistical Analysis of Fatigue Strain-Life Data for Carbon and Low-Alloy Steels", *Risk and Safety Assessment: Where is the Balance?"* ASME PVP-Vol 296/SERA Vol. 3, pp. 355-366.

Keissler, J. O.K. Chopra and W.J. Shack 1996. "Statistical Models for Estimating Fatigue Strain-Life Behavior of Pressure Boundary Materials in Light Water Reactor Environments" *Nuclear Engineering and Design,* Vol. 167, pp. 129-154.

Khaleel, M.A., O.J.V. Chapman, D.O. Harris and F.A. Simonen 1999. "Flaw Size Distribution and Flaw Existence Frequencies in Nuclear Piping", *Probabilistic and Environmental Aspects of Fracture and Fatigue*, ASME PVP-Vol 386, pp. 127-144.

Khaleel, M.A. and F.A. Simonen 1998. "A Probabilistic Model for Fatigue Crack Initiation and Propagation," *Fatigue, Fracture, and Residual Stress,* ASME PVP-Vol 373, pp. 27-34.

Majumdar, S., O.K. Chopra and W.J. Shack 1993. *Interim Fatigue Design Curves for Carbon, Low-Alloy and Austenitic Stainless Steels in LWR Environments*, U.S. Nuclear Regulatory Commission, Washington, D.C., Report NUREG/CR-5999.

Simonen, F.A., M.A. Khaleel, H.K. Phan, D.O. Harris, D. Dedhia, D.N. Kalinousky and S.K. Shaukat 1999. "Evaluation of Environmental Effects on Fatigue Life of Piping," Transactions of the Twenty Seventh Water Reactor Safety Information Meeting, NUREG/CP-0168, Vol. 1, pp. 29-30, U.S. Nuclear Regulatory Commission.

Ware, A.G., D.K. Morton and M.E. Nitzel 1995. *Application of NUREG/CR-5999 Interim Fatigue Curves to Selected Nuclear Power Plant Components*, U.S. Nuclear Regulatory Commission, Washington, D.C., Report NUREG/CR-6260.

PVP-Vol. 410-1, Assessment Methodologies for Preventing Failure:
Deterministic and Probabilistic Aspects and Weld Residual Stress
Volume 1
ASME 2000

FAILURE OF A HIGH-TEMPERATURE WATER PIPE IN A FOSSIL POWER PLANT PIPEWORK: A PROBABILISTIC ANALYSIS

Dragos D. Cioclov
Fraunhofer-Institute
for Nondestructive Testing IZFP
University, Building 37
D-66123 Saarbrücken, Germany
Tel/Fax: +49 (681) 9302-3973/5930
cioclov@izfp.fhg.de

Helmut Wiedemann
SGS-TÜV Saarland GmbH
Anlagentechnik TAB
Saarbrücker Str. 8
D-66274 Sulzbach/Saar, Germany
Tel/Fax: +49 (6897) 506-134/198
SGS-TUV@t-online.de

Abstract

A horizontal pipe in the water level control system of a power plant boiler failed in service at a temperature of T ≅ 300°C and a pressure of 172 bar (registered). The fractographic analysis revealed a four stage time-dependent failure scenario: I) corrosion pit formation on the inside surface of the pipe; II) stable corrosion fatigue crack growth originating at the corrosion pits; III) cracks coalescence into a single crack; and IV), sudden pipe burst from the single coalesced crack. Probabilistic FAD analysis of stages III and IV evinced a high probability of failure under operating conditions. Probabilistic FAD algorithm (FADSIM software) accounts for the uncertainty encountered in the estimation of crack size, material fracture toughness, and strength characteristics. Normal distributions has been assumed throughout. The analysis gives due account to thermo-mechanical aging and adjacent crack interaction effects at the onset of multiple cracks coalescence (Stage III). The probabilistic FAD analysis demonstrated that the pipe failure was a natural final event after a long process of material damage where corrosion, mechanical and thermal fatigue all combined with global material aging were synergistically involved in the failure.

In a broader sense, the outlined failure analysis offers an example of the increasing insight that can be achieved when fracture mechanics principles combined with probabilistic methods are used for risk evaluation in pressure vessels and piping systems.

1. Case description

In the pipe-work of a power plant boiler, section water level control, a horizontal pipe of 300 mm ID and 20 mm wall thickness manufactured of W36 steel (15NiCUMoNb5 – Material Nr 1.6368) failed abruptly with an apparent bulging and axially oriented fractured surfaces. Decompression of the system resulted. The failure was located on a straight portion of the pipe just before a ¼ circular loop of the 1500 mm radius. The support elements adjacent to the failure location were also damaged, resulting in plastic deformation and cracking. At the time of failure, the pipe was filled with water at a pressure of 172 bar. No information on the pipe temperature at the failure location was available. Under normal thermal isolation, the pipe wall temperature presumably was in the range of 300 - 320°C (water saturation temperature of 355°C corresponds 172bar). After the pipe cool-down (34 hours after the failure) a secondary spontaneous pipe fracture occurred adjacent to the primary fracture, but oriented in a transverse direction. The entire operating time of the boiler block at failure was 107,850 hours and the number of restarts amounted to 1,123 cycles. The pipe had been designed for a maximum operating pressure and temperature of $P_{operating}$ = 250bar and $T_{operating}$ = 420°C, respectively.

By design, the pipe should be filled with water only intermittently. However, between control maneuvers, when the water surplus was directed to the condenser, remnant water was always left behind at the bottom of the pipe. During control activities, hot water from the boiler was mixed with cold stagnant water by intermittently opening the control valve. Under this circumstance, repeated loading by pressure induced fluctuating stresses and transient thermal stresses occurred.

2. Fractured surface morphology

Figure 1 shows an overall macroscopic view of the fractured surface in the bulged area. Some main macroscopic features are obvious. In the region where the primary fracture initiated, well-defined, multiple, semi-elliptic cracks with a "smooth" colored appearance were identified together with some traces of macroscopic striations. They prove that a stable crack growth occurred from multiple sites a long time before the actual failure. A closer examination revealed that the central point the of semi-elliptic crack's originated at corrosion pits on the inside surface of the pipe. Outside the elliptic cracks as results from Fig. 1, distinct "chevron" patterns displayed a convergent orientation toward the cracks' initiation points, typical for a sudden brittle crack extension. The interaction of chevron patterns emanating from semi-elliptic crack proves the cracks' coalescence process continued into an overall deep penetrating crack. A scaled sketch of the macroscopic fracture morphology is shown in Figure 2. Once a single deep dominant crack resulted from coalescence, the remaining material ligament in the region was sufficiently thinned such that the crack to enters unstable plastic deformation followed by local bulging and pipe rupture.

A micro-structural investigation confirmed corrosion pitting at the inside pipe surface as the crack initiation location. Figures 3a and 3b illustrate this fact. It was found that corrosion pits were concentrated in a band limited to the 5 and 7 o'clock positions, coinciding with the region where water stagnation occurred over most of the operating time.

3. A tentative description of the pipe damage process

Based on macro- and micro-fractographic observations, the material damage process resulting in the pipe failure can by described in their main stages as follows (see Figure 4):

I) Multiple corrosion pits developed on the inside surface of the pipe, predominantly in the bottom region of water stagnation (around the 6 o'clock position).

II) Corrosion pits acting as local stress concentrators developed into multiple cracks that grew in a steady manner as a result of synergistic fatigue (repeated pressure loading and thermal transients) and corrosion effects. In this stage, steady crack growth does not extend over the entire pipe wall thickness, thus leaking prior to breaking did not occur.

III) A stage is reached in the steady crack growth when the driving force for crack extension exceeds the local material resistance to initiation of unstable crack growth. This occurs first at the depth of the most advanced crack. The process is triggered when the driving force for crack extension expressed by the stress intensity factor (SIF) at the crack tip, K_I, exceeds the local material fracture toughness K_{CC}. After the first unstable crack extension occurrence of the most advanced crack, an intemperate coalescence of cracks into a single dominant crack follows.

IV) The stage of final fracture was governed by the single dominant crack and nearly plane-stress fracture toughness K_C. In this stage, pipe bulging had fully developed.

For the purpose of quantitative failure assessment, two circumstances will be considered in this analysis: Stage III and Stage IV.

The assessment of the time-dependent corrosion pits generation and crack growth under mechanical, thermal, and corrosion fatigue falls beyond the scope of the present paper.

4. Material Characterization

The pipe was manufactured of W36 steel (15NiCuMo5) of a composition given in Table 1.

Table 1: Chemical composition (% weight)

Source	C	Si	Mn	P	S
Vd TÜV Code Blatt 377/2-9.86	0.08 0.19	0.20 0.56	0.75 1.30	max. 0.035	max. 0.030
Pooled data MIF & ALIANZ	0.17 0.18	0.34 0.37	1.04 1.06	0.009 0.014	0.012 0.017

Source	Al	Cu	Cr	Ni
Vd TÜV Code Blatt 377/2-9.86	max. 0.055	0.45 0.85	Max. 0.35	0.95 1.35
Pooled data MIF & ALIANZ	0.010 0.011	0.63 0.68	0.19 0.20	1.17 1.24

Source	Mo	Nb	N
Vd TÜV Code Blatt 377/2-9.86	0.22 0.54	0.010 0.050	-
Pooled data MIF & ALIANZ	0.34 0.35	0.022 0.025	0.01 0.012

Source; ALIANZ, 1995; MFI, 1996

The tensile strength characteristics of the W36 steel coupons taken in the neighboring area of the failure are shown in Table 2. Four coupons have been tasted by ALIANZ and three coupons by MFI.

Table 2: Tensile strength characteristics DIN EN 10002 transverse specimens

	Room Temperature		
	R_m (MPa)	R_{eH} (MPa)	Z (%)
Vd TÜV Code	610-780	>440	-

Blatt 377/2-9.86			
MFI	752-753	631-633	53
ALIANZ	765-815	639-680	51-55
	350°C		
	Rm (MPa)	Rp 0.2 (MPa)	Z (%)
Vd TÜV Code Blatt 377/2-9.86	>510	>373	-
ALIANZ	69.2-743	5 % - 582	45-49
	400°C		
	Rm (MPa)	Rp 0.2 (MPa)	Z (%)
Vd TÜV Code Blatt 377/2-9.86	>500	>343	-
MFI	642-649	488-490	51-54

*) Source: ALIANZ, 1995; MFI, 1996

Charpy-V impact test results at various temperatures are shown in Figure 5. The specimens have been taken in the hoop direction from three coupons removed from a neighboring area of the failure. One set of results (from coupon I) displays a rather low KV-toughness at room temperature (20, 22, 23 J). A repeated set of tests on an annealed coupon (5h, 930° C/Air) revealed a restoration of material fracture toughness at KV-values of 58-64 J. It is obvious that material aging occurred due to the long exposure under stress in the range of 300°C (thermo-mechanical aging).

5. Backgrounds of Applied Fracture Mechanics Methodology

The fracture mechanics analysis is based on the Failure Assessment Diagram (FAD) or R6 methodology (Milne et al., 1988).

The algorithm of the FAD analysis is derived from a modified variant of the Dugdale (1960) model of elastic-plastic failure of a wide plate containing a through-wall crack. The analysis considers two-parameter (Figure 6). One parameter K_r is defined as the ratio of elastic stress-intensity factor (SIF), K_I, to material fracture toughness K_C, i.e. $K_r = K_I/K_C$. This parameter is related to fracture under the prevailing elastic state of stress, hence under brittle fracture condition. In the presented analysis, due consideration is given to neighboring crack interaction. The second parameter S_r, is defined as the ratio of the loading intensity to the loading for plastic collapse in a ductile fracture scenario of the flawed structure. The applied load is related to the nominal (net) hoop stress σ_{NET} in the cylinder wall resulting from water hydrostatic pressure, while the plastic yield load of the flawed structure is identified as the flow stress $\sigma_f = (R_m + R_e)/2$, supposedly acting in the remaining ligament at the crack tip when the plastic collapse occurs.

According to the FAD philosophy, failure occurs when the state point related to the loading, crack size, and material fracture characteristics in the (K_r, S_r) presentation fall outside the region boundary of the limiting curve $K_r = f_{LIM}(S_r)$, illustrated in Figure 6.

In the present analysis, "Option 1" recommended by R6 methodology has been employed. Option 1 implies the knowledge of 0.2 % proof stress (R_e), ultimate tensile strength (R_m), and material fracture toughness (K_C). Details on other possible options can be found in Milne et al. (1988). Under Option 1, the limiting curve at failure is defined as:

$$K_r = \left(1 - 0.14\, S_r^2\right)\left[0.3 + 0.7 \exp\left(-0.65\, S_r^6\right)\right] \quad for\ S_r \le L_r^{max} \quad (1)$$
$$K_r = 0 \quad for\ S_r > L_r^{max}$$

L_r^{max} is the material dependent cut-off of the f_{LIM} curve, as illustrated in Figure 6, being defined by:

$$L_r^{max} = \sigma_f / R_{e(0.2\%)} \quad (2)$$

In the present analysis, the required material characteristics involved in Option 1 of the FAD approach have been evaluated on a limited number of tests performed on specimens taken from the area of the failure initiation.

For the K_r parameter calculation, semi-elliptical surface crack models have been considered with the SIF solution as given by Raju and Newman (1980), together with required corrections for adjacent crack interaction (Miyoshi et al., 1985).

6. Material and crack geometry evaluation for probabilistic FAD analysis

Material fracture toughness evaluation

At the time of failure analysis, the only available material characteristics pertain to standard tensile (Table 2) and Charpy-V impact tests (Figure 5).

In order to apply FAD principles, material fracture toughness K_c (or J integral based J_c toughness) must be ascertained. Moreover, K_c data must reflect the actual material fracture toughness according to the failure phase under analysis (Section 3). As outlined in this report, only the circumstance of quasi-static widespread crack extension will be evaluated.

This occurs at:

a) The transition from Stage II of steady crack growth to Stage III of widespread crack extension and coalescence into a final, dominant crack. The time coordinate t_c corresponds to this stage (Figure 4).

b) The onset of ultimate extension of the single coalesced crack to the final failure (Stage IV).

In the former case a), the fracture toughness K_{cc} at time t_c results after a process of steady crack growth due to mechanical and thermal fatigue damage combined with corrosion effects. Point C in Figure 7 portrays in log-log crack growth rate vs. maximum applied SIF, K_{Imax}, the fracture toughness at time t_c. For $K_{Imax} > K_{cc}$ the crack extension at the level of individual cracks is intemperate and Stage III of rapid crack coalescence is triggered.

In the absence of K_{cc} data for the very specific circumstance of failure, a compilation of literature data have been made for carbon steels under corrosion fatigue (Vosikovsky 1975, Misawa et al. 1976, ASM 1996), elevated temperature operations in the range of 300°C (Mc Henry and Pense, 1983), and very low loading frequencies (Vosikovsky 1975). From these data it was obvious that for any combination of fatigue, corrosion, temperature, and frequency effects at high ultimate crack propagation rates of 0.1 to 0.5 mm / cycle, K_{cc} ($R \cong 0$) values of 60 to 80 MPa$\sqrt{m}$ correspond. However, the values in the analyzed case are affected by long exposure thermo-mechanical aging.

This effect is responsible for some decreasing of K_{cc} that can be estimated on the base of comparison of Charpy-V toughness data obtained from material a coupon taken after the failure and a coupon of restored material by annealing (see sect. 3 and Fig. 5). When transposing these data into K_{cc} values by correlation (Equation 3) a correction factor of 0.9 for aging results. Finally, the outlined estimated K_{cc} values from literature corrected for thermo-mechanical aging yield a mean value of 0.9 x 70 = 63 MPa√m. The standard deviation (SD) was computed with the "3-sigma" rule applied to the corrected K_{cc} two-side range (54; 72 MPa√m). SD = 3 MPa√m results in the case when normal distribution is assumed. Of course, this was a coarse "rule of thumbs" estimation, but it is believed to be rather robust since the outlined data sources demonstrate stable results obtained from extensive testing.

In the case b) at the beginning of Stage IV, i.e. at ultimate extension to failure of the coalesced crack, the material fracture toughness K_c was estimated by correlation procedures as related to conventional material characteristics. Since in this stage the material was not influenced by corrosion at the tip of the deep coalesced crack, the correlation procedure proposed by Rolfe and Novak, 1970, for the upper-shelf domain of Charpy KV toughness vs. temperature curve (see Fig. 5) seems appropriate. This correlation has the form (units: psi√in, psi, ft·lb):

$$(K_c / R_e)^2 = 5\left(KV / R_e - 0.05\right) \quad (3)$$

where KV and yield strength were considered at 300°C. With the interplay of R_e data given in Table 2 and KV data (58, 62, 66 J) illustrated in Figure 5, a set of nine computed K_c values were obtained. Table 3 provides ordered K_c values with their associated probability of occurrence estimated by the methods of order statistics (binomial distribution).

Table 3: - K_c - estimation (Rolfe-Novak correlation) and associated order statistics probabilities (binomial distribution).

Temperature 300°C

Rank	1	2	3	4
Kc (MPa√m)	143	144	148	149
Prob.	0.074	0.181	0.287	0.394

Rank	5	6	7	8	9
Kc (MPa√m)	149	152	153	153	158
Prob.	0.500	0.606	0.713	0.819	0.926

The empirical point estimates of K_c data presented in Table 3 are: median K_c = 149 MPa√m mean K_c = 150 MPa√m and standard deviation SD = 5 MPa√m. Note that normal distribution assumption is supported in this case by nearly identical empirical median and mean value. These estimations will be retained for the probabilistic FAD analysis of Stage IV of failure.

For the purpose of quantifying the statistical scatter of ultimate tensile strength (UTS) and yield point YP, the normal distribution combined with point estimations resulting from the data given in Table 2 will be assumed (300 °C):

Mean UTS = 715 MP; SD = 25 MPa
Mean YP = 571 MPa; SD = 22 MPa

Stress intensity factors evaluation

a) Stage III

At the start of the multiple crack coalescence, adjacent cracks have been linked as indicated by fractographic patterns shown in Figure 1. Schematically, a crack geometry configuration as shown in Figure 8 is assumed. For this case, a maximum SIF in point A (Figure 8) was calculated by compounding Newman and Raju, 1980, solution for the inside axial semi-elliptic crack in a cylindrical pressure vessel and Miyoshi et al. 1985, finite elements analysis for overlapping twin semi-elliptic cracks emerging at the surface of a finite width plate. On these bases, the SIF computation formula implemented in FADSIF algorithm is:

$$K_{IA} = \sigma_{HOOP} Y_{NR}\left(\frac{b}{H},\frac{b}{a};\frac{R_i}{h},\varphi\right) \cdot Y_{INTR}\left(\frac{d}{a};\frac{b}{a};\frac{b}{H}\right) \quad (4)$$

where Y_{NR} (...) and Y_{INTR} (...) are Newman – Raju and Miyoshi et. al. geometry correction factors, R_i is the inside pipe radius; H the wall thickness, (φ)– angle index on the semi-elliptic crack contour (*b*), and *a* and *d* parameters related to the crack geometry as defined in Figure 8. The details for calculation correction factors are given in Appendix (Eqs A1 to A9). In the FADSIM analysis, the pressure vessel geometry parameters R_i = 150 mm and H = 20 mm were considered as deterministic values, while the size of the most advanced crack was assumed to have a normal distribution with mean values and standard deviations resulting from direct measurements on the fracture surface. Mean *b* = 4.6 mm; SD (*b*) ≅ 0.2 mm; mean *a* = 12.5 mm; SD (*a*) = 2 mm.

The dual crack overlapping extension has been measured as *d* = 3.2 mm. According to Myoshi et. al., 1985, the finite-element numerical data a correction factor Y_{INTR} = 2.8 due to this effect can be approximated by interpolation. In the probabilistic FADSIM algorithm, the *d* value has been considered a deterministic quantity, hence Y_{INTR} assumes a constant value throughout.

b) Stage IV

In this stage, a single coalesced deep and nearly semi-elliptic crack corresponds to a geometry estimated by direct observation on the fractured surface (contour B in Figure 2). The point estimates of the crack size are: mean *b* = 16.2 mm; SD = 0.2 mm and mean *a* = 130 mm; SD = 2 mm.

Table 4: Input data for probabilistic FADSIM analysis (Normal distribution)

Failure Stage	b mm		a mm	
	Mean	SD	Mean	SD
III-Onset of multiple crack link-up*	4.6	0.2	12.5	2
IV-final Failure from single coalesced crack	16.5	0.2	130	2

Failure Stage	Fracture Toughness		UTS MPa		YP MPa	
	Mean	SD	Mean	SD	Mean	SD
III-Onset of multiple cracks link-up*	63	3				

			715	25	571	22
IV-final Failure from single coalesced crack	150	5				

*) The most advanced crack at the start of Stage III.

7. Results of FADSIM probabilistic analysis

Table 4 summarizes the input data used in the probabilistic FADSIM analysis of the pipe failure.

At the beginning of Stage III, i.e. the triggering of cracks' coalescence, following the previous steady crack growth by mechanical, thermal and corrosion effects, 10^5 probabilistic FADSIM Monte-Carlo iteration resulted in a simulated probability of cracks coalescence of ≈ 30%. The parameters of the distributions have been evaluated as outlined in Section 6 and are summarized in Table 4. Figure 9a illustrates the output of the simulation according to FADSIM algorithm as it appeared on the computer screen.

The evaluation of the failure probability of the ultimate Stage IV of pipe rupture has been performed on the same basis as for Stage III, but for a single coalesced crack. The input parameters are given in Table 4 and the simulation result, after a 10^5 iterations, yields in a probability of failure in the range of 42.5 %. Figure 9b illustrates the simulation output of this case.

8. Discussion and Conclusions

The outlined probabilistic FAD analysis of a pipe rupture have been performed after a previously made thorough failure analysis in a classical sense, i.e. without using fracture mechanics and probabilistic approach to parameters scatter or encountered uncertainties. However, the classical approach to this case of failure was not able to explain the fact that failure occurred at a pressure of 172 bar, unambiguously registered at the time of failure. By applying fracture mechanics principles in accordance with the Failure Assessment Diagram rationale, it was possible to account for strong influencing factors on failure, such as: a) the increase of SIF factors due to adjacent crack interaction and the decrease in material fracture toughness after a long exposure to water corrosion; and b), mechanical and thermal fatigue together with global thermo-mechanical aging of the pipe material after long exposure under stress at temperatures of approximately 300°C. By estimating the scatter or uncertainty related to crack size, material strength, and fracture toughness characteristics, the probability of crack coalescence after steady growth as well as the final pipe failure probability have been simulated with the aid of FADSIM algorithm. The output of a 30% probability of the crack coalescence into a dominant final crack and the further 42.5% probability of pipe rupture (both in the area of immediate occurrence) revealed unambiguously that the pipe failure was a logical consequence of a natural long-term material damage process that developed under normal operating conditions.

Note that in the present post – failure analysis the probability of failure in the Stage IV has been associated with an independent event i.e. not regarded as a conditional probability on the occurrence of Stage III as required by Boolean rules in an *a priori* predictive analysis. This follows from the fact that in this *a posteriori* assessment we know that Stage III - event did occurred.

As concerns analysis sensitivity related to uncertainty in the estimation of mean and standard deviation of material characteristics, exploratory simulations showed that within the normal range of coefficients of variations (3 - 5 %) a high probability of failure (greater than 10^{-1}) corresponds in Stage 3 and 4. Mean values given in Table 4 are fixed on the base of available experimental data.

One main highlight of the present analysis was to clarify that an initially suspected overpressure (not observed on pressure records at the time of failure) is not required by probabilistic fracture mechanics to explain the failure.

In a broader sense, the outlined failure analysis offers an example of the increasing insight that can be achieved when fracture mechanics principles combined with probabilistic methods are used for risk evaluation in pressure vessels and piping systems.

APPENDIX

Raju and Newman (1980) stress intensity factor solutions for a semi-elliptical surface crack on the inside surface of a cylindrical vessel are described in this appendix.

This solution was obtained based on finite elements analysis. It is claimed that its accuracy is better than 10%. For the geometry of the cylindrical vessel with an inside semi-elliptical crack (illustrated in Figure A) and hoop stress σ_{Hoop}, the expression of the stress intensity factor is:

$$K_I = \sigma_{Hoop}\sqrt{\frac{\pi b}{Q}}\; Y_{NR}\left(\frac{b}{h}, \frac{b}{a}, \frac{R_i}{h}, \varphi\right) \quad \text{(A1)}$$

where the shape factor Q is given by:

$$Q = 1 + 1.464\,(b/a)^{1.65} \quad \text{(A2)}$$

and the correction Y_{NR} was obtained by fitting-double-series polynomials in terms of b/a, a/h and φ

$$Y_{NR} = 0.97\left[M_1 + M_2\left(\frac{b}{h}\right)^2 + M_3\left(\frac{b}{h}\right)^4\right] g \cdot f_\varphi \cdot f_c \quad \text{(A3)}$$

where

$$M_1 = 1.13 - 0.09\frac{b}{a}; \quad \text{(A4)}$$

$$M_2 = -0.54 + 0.89/\left(0.2 + \frac{b}{a}\right) \quad \text{(A5)}$$

$$M_3 = 0.15 - \left(0.65 + \frac{b}{a}\right)^{-1} + 14\left(1 - \frac{b}{a}\right)^{24} \quad \text{(A6)}$$

$$g = 1 + \left[0.1 + 0.35\left(\frac{b}{h}\right)^2\right]\;(1 - \sin\varphi)^2 \quad \text{(A7)}$$

$$f_\varphi = \left[\sin^2\varphi + \left(\frac{b}{a}\right)^2 \cos^2\varphi\right]^{1/4} \quad \text{(A8)}$$

$$f_c = \left[\left(R_o^2 + R_i^2\right)/\left(R_o^2 - R_i^2\right) + 1 - 0.5\sqrt{\frac{b}{n}}\right]\frac{h}{R_i} \quad \text{(A9)}$$

REFERENCES

Alianz; 1995, *Internal Report Nr. ATVE 95. 12.15/01*, Ismaning, Germany

ASM, 1996, Fracture Mechanics Properties of Carbon and Alloy Steels, in ASM-Handbook, Vol. 19, *Fatigue and Fracture*, pp 614-654

Cioclov, D.D., 1996, Bruchmechanische Bewertung von Endgewaltbruch an der Schnellablaßleitung RK 10. *Internal Report. TÜV Saarland, Germany*

Dugdale, D.S., 1960, *Yielding of Steel Sheets Containing Slits,* J. of Mech. Phys. Solids, Vol. 8, pp. 100-104

Mc Henry, H.J. and Pense, A.W., 1973, Fatigue Crack Propagation in Steel Alloys at Elevated Temperatures, in: *ASTM STP 520,* pp 345-354

MFI, 1996, *Mannesmann Forschungsinstitut Internal Report Nr. 4/96,* Duisburg, Germany

Milne, I., Ainsworth, R.A., Dowling, A.R. Stewart, A.T., 1988, *Assessment of the Integrity of Structures Containing Defects,* Int. J. of Pressure Vessels and Piping, Vol. 32, pp. 3-104

Misawa, T., Ringshall, N. and Knott, J.F., 1976, Fatigue Crack Propagation in Low Alloy Steel in a De-aerated Distiled Water Environment, *Corrosion Science,* Vol. 16 (No 11), pp 805-818

Miyoshi, T., Shiratori, M., Tanabe, O. (1985), Stress Intensity Factors for Surface Cracks with Arbitrary Shapes in Plates and Shells, in: *ASTM STP 868,* M.F. Kanninen and A.T. Hopper Eds., pp 521-536

Vosikovsky, O., 1975, Fatigue-Crack Growth in an X-65 Line-Pipe Steel at Low Cycle Frequencies in Aqneos Environments, *Trans. ASME,* Series H, Vol. 97, pp 298-305

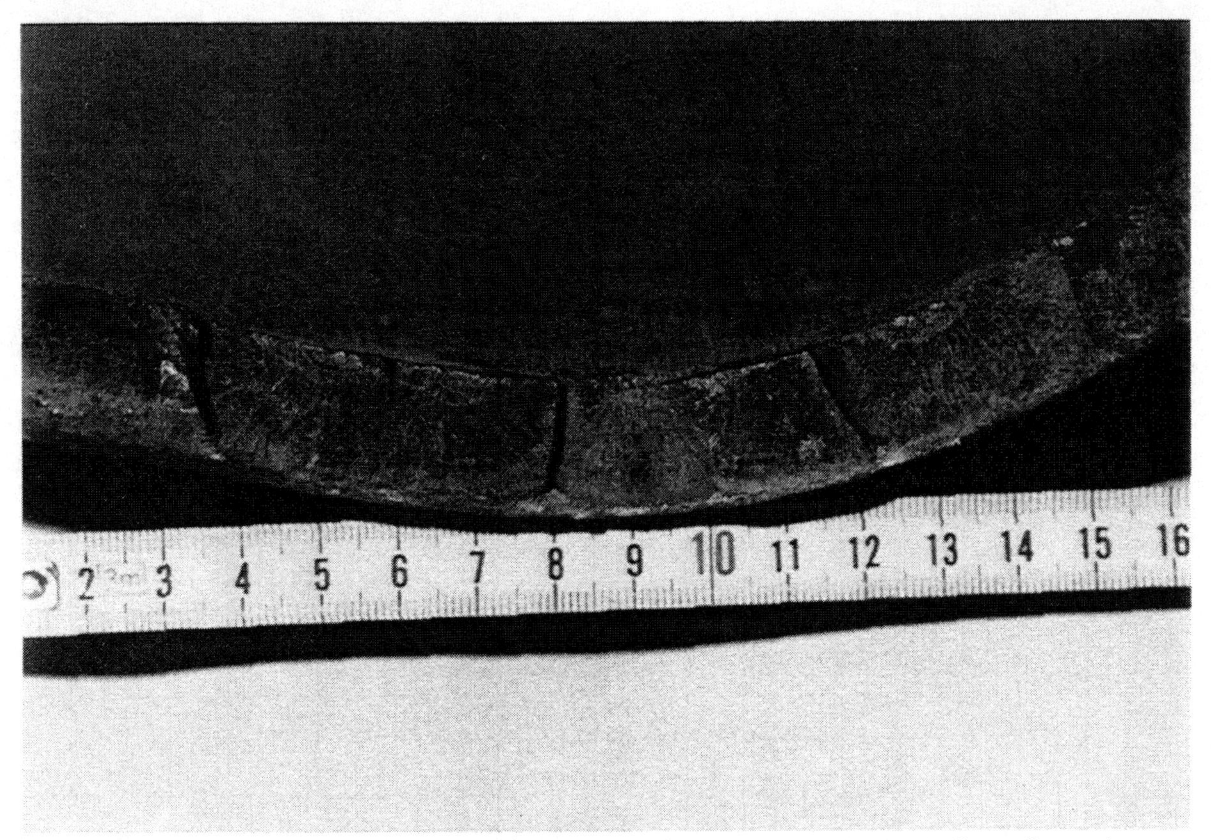

Figure 1: A macroscopic view of fractured surface in axial orientation. Curvature resulted from bulging before burst. Multiple cracks and "chevron" patterns related to cracks coalescence process are visible

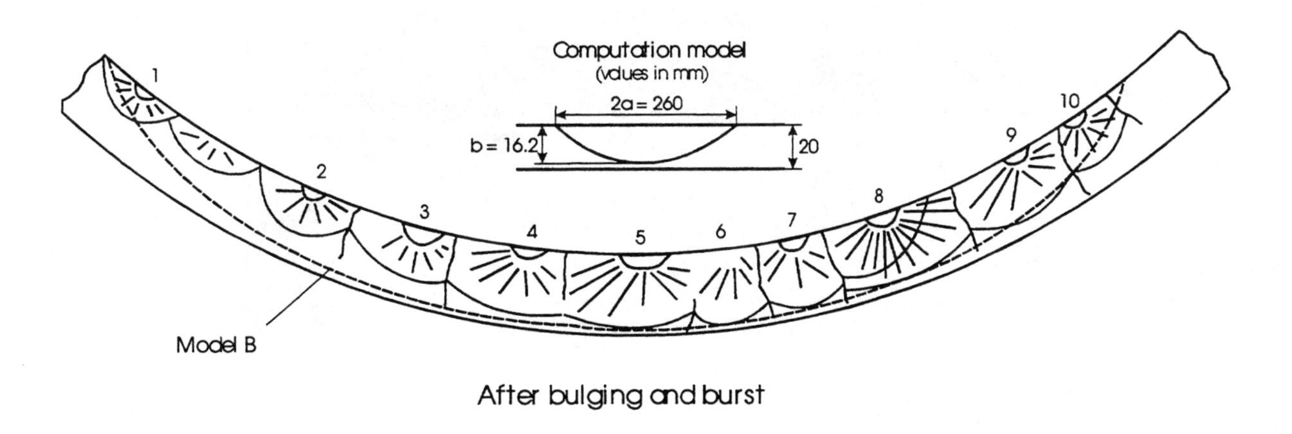

Figure 2: A scaled sketch of macroscopic fracture morphology. Contour B corresponds to the approximated single coalesced semi-elliptic crack that triggered the final fracture

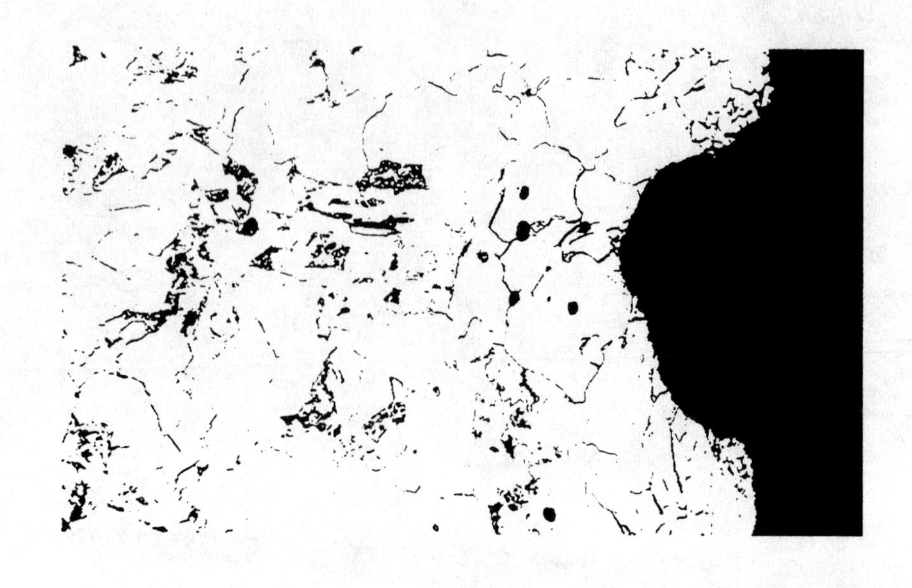

Figure 3: Cracks initiating sites.
a) Initial corrosion pits

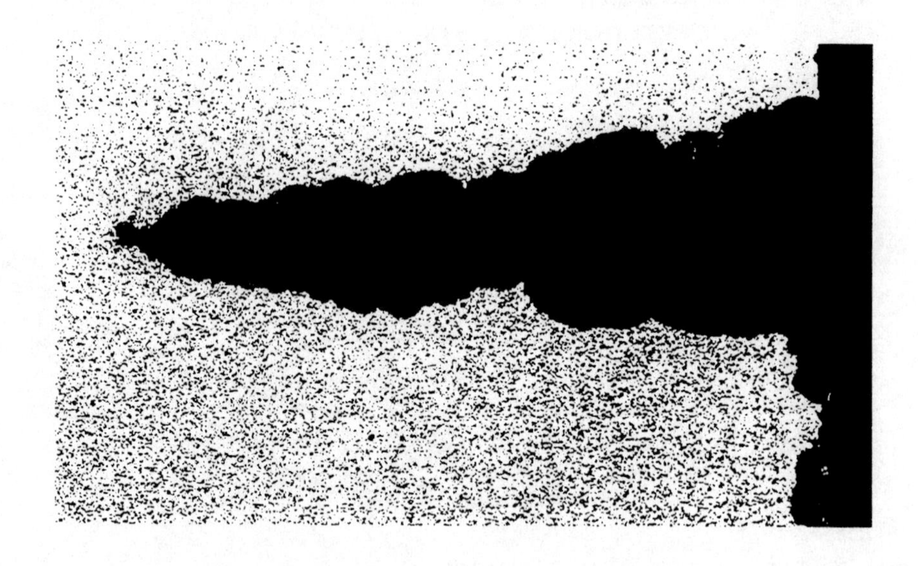

b) Crack initiated at a corrosion pit

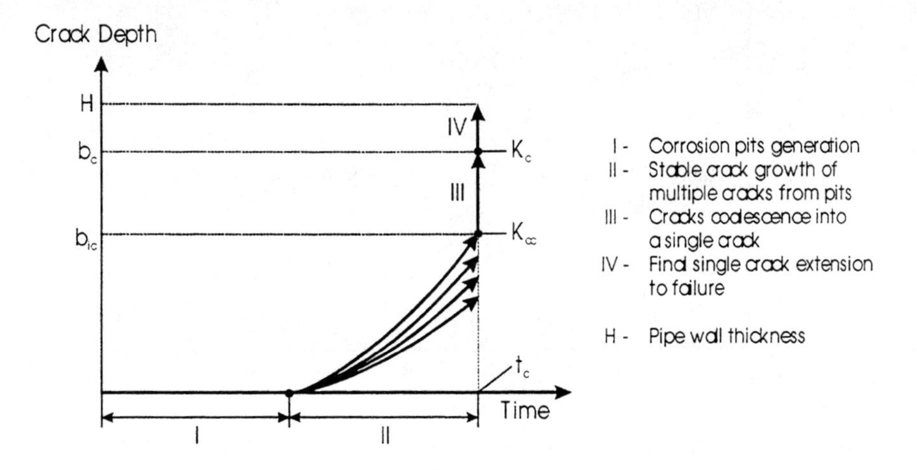

Figure 4: Phases of material damage process

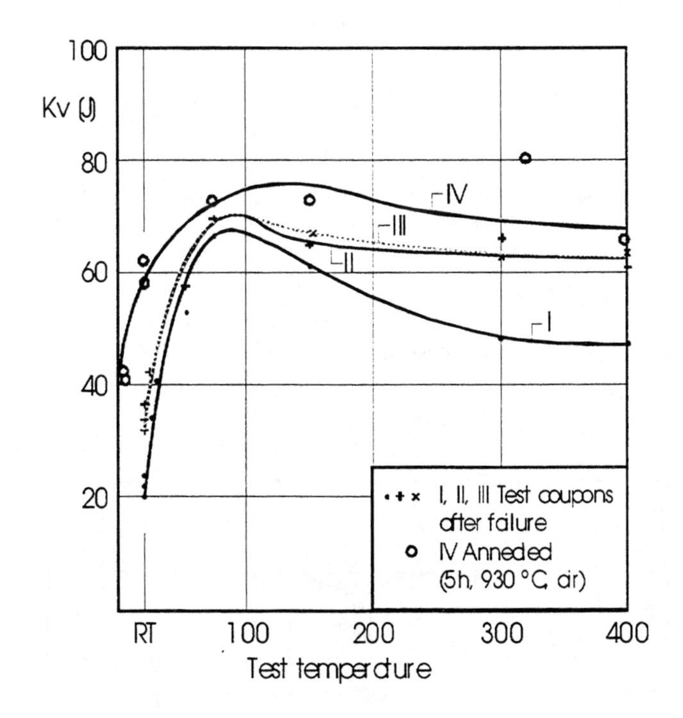

Figure 5: KV-Charpy V impact fracture vs. test temperature. Material as resulted after failure
(coupons I, II, III) and restored by annealing (coupon IV)

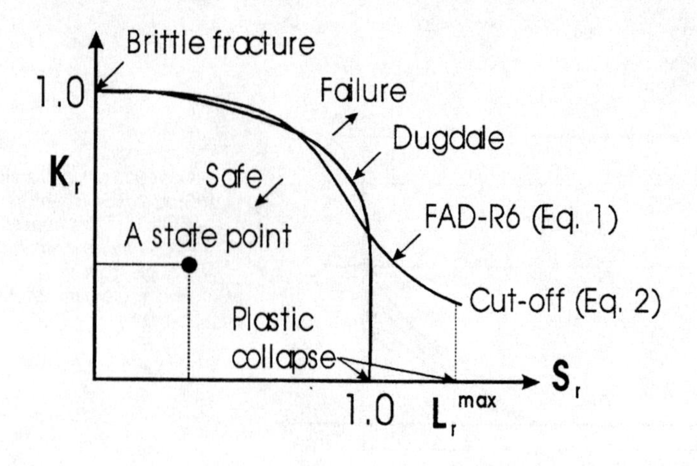

Figure 6: The principle of FAD analysis

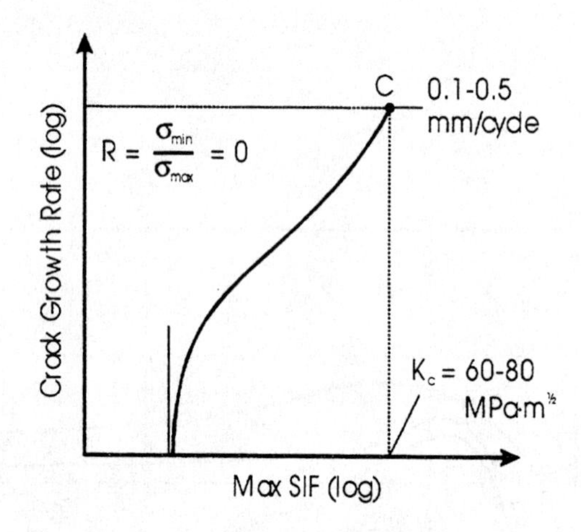

Figure 7: Phase II stable crack growth rate vs. maximum SIF - schematically

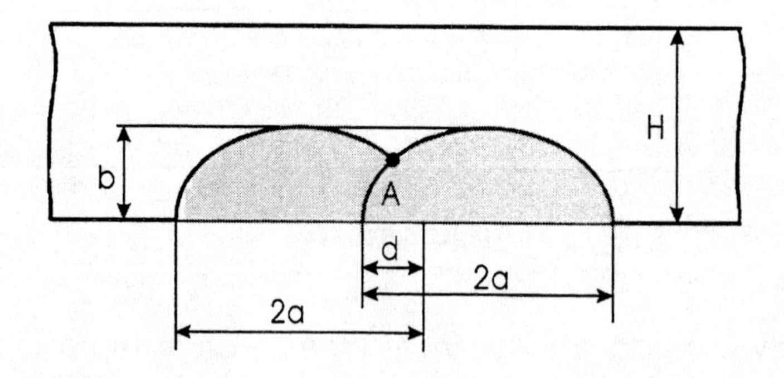

Figure 8: The geometry of twin overlapping cracks

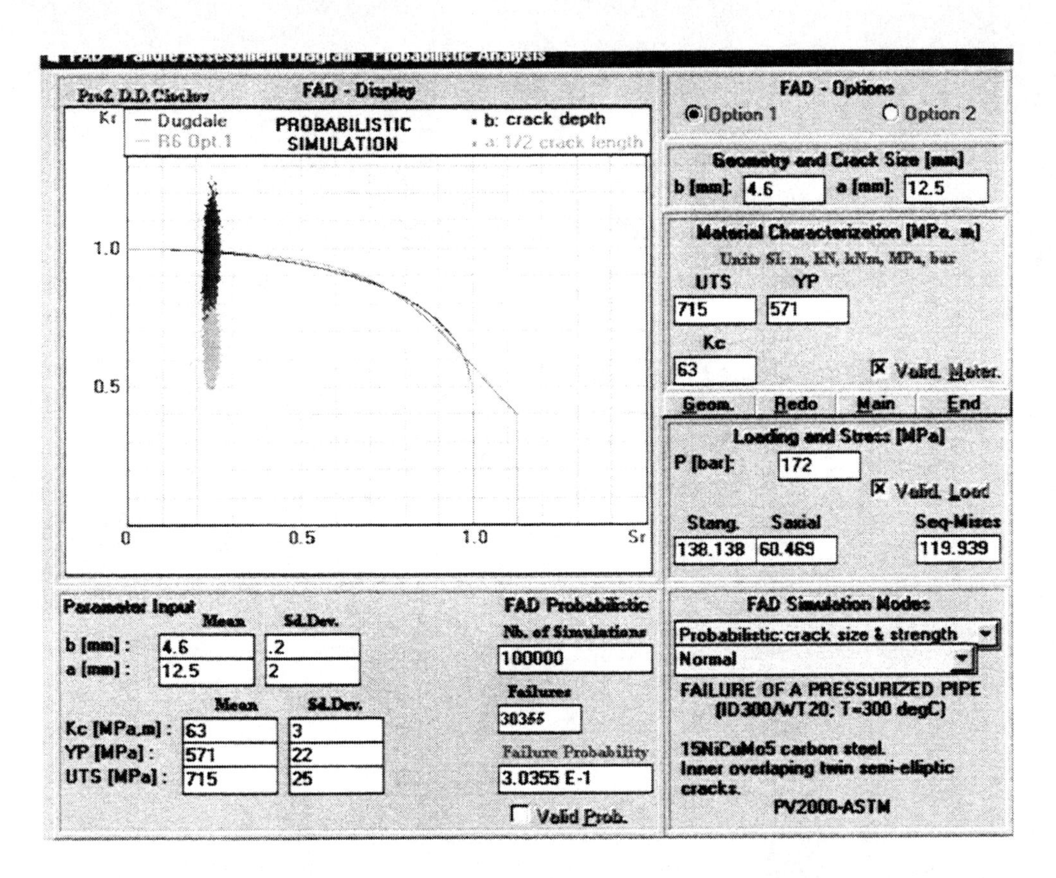

Figure 9: PC screen view of FADSIM failure probability simulation
a) onset of rapid crack extension from multiple adjacent cracks (phase III), $P_f \cong 30$ %

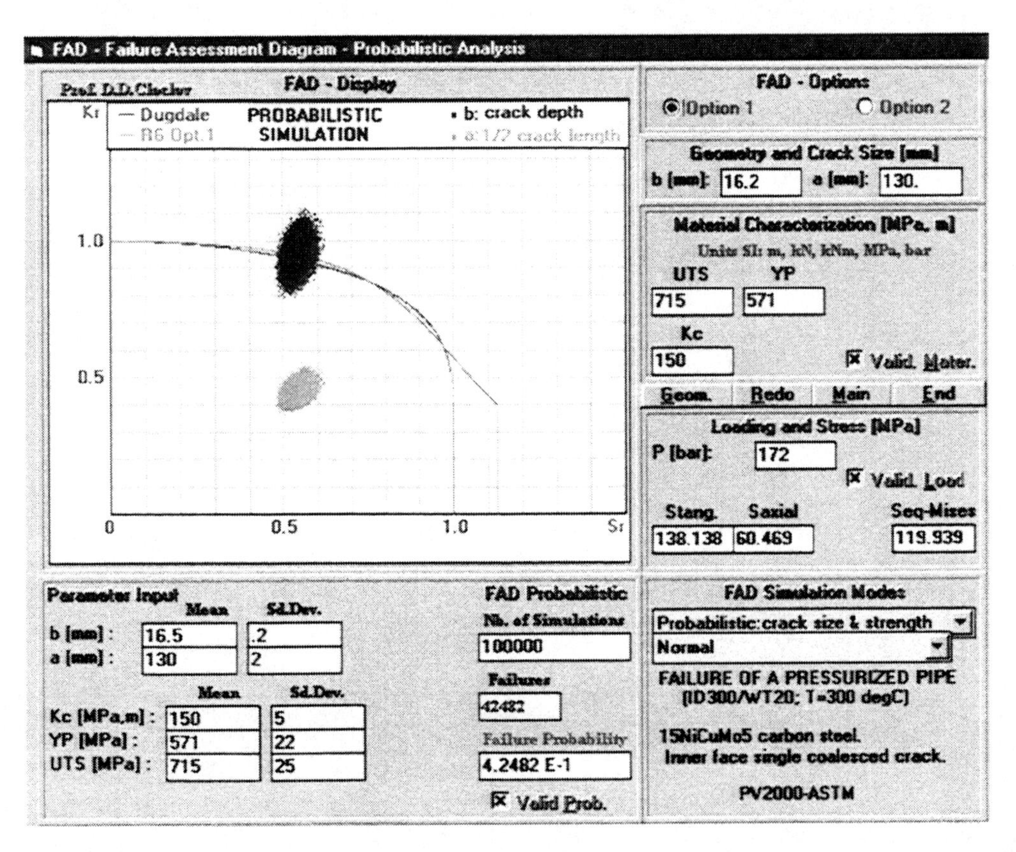

b) final failure from single coalesced crack (phase IV); $P_f = 42{,}5$ %

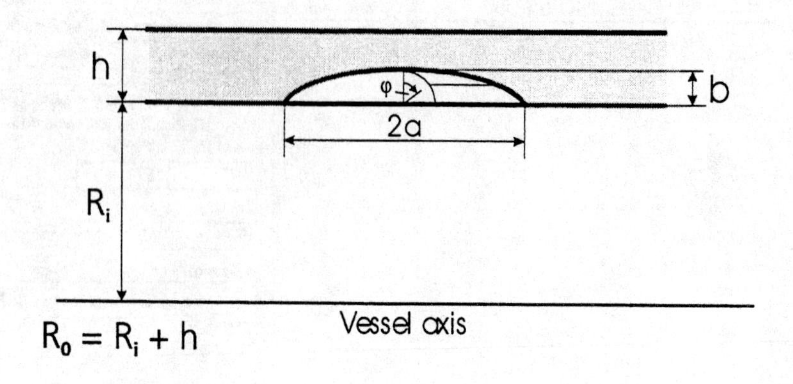

Figure A: Geometry of a cylindrical pressure vessel with an inner surface semi-elliptical crack in longitudinal orientation

PVP-Vol. 410-1, Assessment Methodologies for Preventing Failure:
Deterministic and Probabilistic Aspects and Weld Residual Stress
Volume 1
ASME 2000

RELIABILITY ANALYSIS OF STRUCTURES CONTAINING DEFECTS WITH A PROBABILISTIC FAILURE ASSESSMENT DIAGRAM

Liu, Changjun
Institute of Process Equipment and Pressure Vessel Technology,
East China University of Science and Technology,
200237, Shanghai, P R China
changjunliu@263.net

Wang, Weiqiang
College of Environmental and Chemical Engineering,
Shandong University of Technology,
250061, Jinan, P R China
wqwang@dms.sdut.edu.cn

ABSTRACT

In order to develop the reliability analysis method of structures containing defects based on the R6 FAD method, the factors affecting structure failure and the probabilistic models are discussed and built respectively in the paper.

Firstly, the different factors affecting structure failure are analyzed by fault tree analysis method, and the relationships among the factors are declared and analyzed. Secondly, the probabilistic models are built by proposing a random safety factor and by simplifying three options of R6 FACs. The option 3 curve was built by *J*-integral which is calculated by the multi-piecewise-power-law superposition engineering evaluation approach based on EPRI fully plastic solutions and especially simplified to multi-piecewise-power-law curves by reference stress. Finally, through some assumptions to the stochastic variables and a so called "assumption of independence" discussed in the paper, a very similar diagram to R6 FAD, probabilistic failure assessment diagram can be established.

The framework and procedure discussed in the paper would be very useful to the reliability analysis of structures containing defects and easy to use by practicing engineers.

INTRODUCTION

Much better quantitative analysis results of structures containing defects can be obtained by reliability analysis as the parameters and properties adopted in reliability analysis are more scientific than those adopted in deterministic analysis to describe the different factors affecting structure failure. Within the deterministic analysis methods in present, the failure assessment diagram (FAD) technology of R6 (Milne, et al, 1986) is extensively used in the safety assessment of structures containing defects because of its advantages of rigour in theory and simplicity in usage, etc., and is also widely adopted in many country's codes (for example, BS PD6493-96 (British Standard Institute, 1996) and SA/FOU-Report91/01 of Sweden (Bergman, et al, 1991), etc.). In recent years the idea of R6 FAD is again widely introduced into the reliability analysis of structures containing defects (Gate, 1985; Kaiser, 1986; Zhao, et al, 1997; Wang, et al, 1999; Warke, et al, 1999).

Based on this, the procedure how to utilize R6 option 3 curve and EPRI fully plastic solution method establishing a probabilistic failure assessment diagram (PFAD) will be approached emphatically in this paper. Furthermore, how the simplification should be made to the solution of the complex *J*-integral of common materials will be expounded in order for easy to establish a PFAD.

THE INFLUENCE FACTORS OF RELIABILITY ANALYSIS BASED ON R6 FAD METHOD

When the reliability analysis is carried out for structures containing defects, the factors affecting structure failure should be learned first. In order for easy to describe the failure causes and the mutual relation among the factors reflecting the structure failure causes, the method of the fault tree analysis (FTA) diagram is adopted, which provides a graphical description of all the causes of failure in a construction. The fault tree of structure failure with respect to R6 option 3 (Milne, et al, 1986) is shown in Fig.1. Some of the different factors that influence a failure (of fracture) process are:

10 Geometry
11 crack length (a)
12 crack shape
13 crack growth (Δa)
14 structure type
15 stress condition (γ)
  for plane stress γ=1
  for plane strain γ=2

20 Material
21 yield strength (σ_{ys})
22 hardening exponent (N or N_i, i=1,2,..., k, where k+1 is the pieces of stress-strain curve divided into piecewise curve)
23 Poisson ratio (ν)
24 dimensionless material constant (α or α_i, i=1,2, ..., k)
25 Young's modulus (E)
26 the stress and strain values at the intersection point of two different piece stress-strain curves (σ_i, ε_i, i=1,2,..., k)

30 Load (P)
31 controlled load/displacement
32 residual stresses
33 thermal stresses
34 loading rate

40 Model
41 calculation methods
42 testing methods

50 Environment
51 temperature
52 radiation
53 corrosion
54 age

These factors are used in Fig.1 with the normal FTA symbols. That means that a rectangle constitutes a top or intermediate stochastic event, a circle means a basic stochastic event which can be determined from experiments or by calculations, and a rhomb indicates a stochastic event of uncertain influence which often means a factor has no effect in theory but has real effect in practice. Fig. 1 does not show a purest form of a fault tree, however, since it is not built of events but of parameters that influence the top event—whether the structure fails or not. Instead the figure shows what has to be known to make a reasonable estimation of the random safety factor. The basic event then indicates the basic level at which each parameter is determined from experiments or by calculations.

After making the determination of the different factors affecting structure failure and the mutual relation among them and the statistical property description of the factors, the structure reliability analysis can thus be in progress. In many methods representing and describing the relation among the different factors affecting structure failure, the method of FAC in revision 3 R6 (Milne, et al, 1986) may after all be accepted as a simple and scientific method.

If R6 option 1 FAC is selected in reliability analysis, a convenient and practical PFAD (Wang, et al, 1999) can be easy to build up as the FAC is a only L_r determined event. On the other hand, if R6 option 2 is selected to carry out the structure reliability analysis will also be comparatively simple as the FAC will varies stochastically only with the stochastic changes of the material constitutive relation.

But as only a definition is given in R6 when the R6 option 3 FAC is selected:

$$f_3(L_r)=\sqrt{\frac{J_e}{J}} \qquad L_r \le L_r^{\max} \qquad (1)$$

the problem is getting more complex than the other. When determining

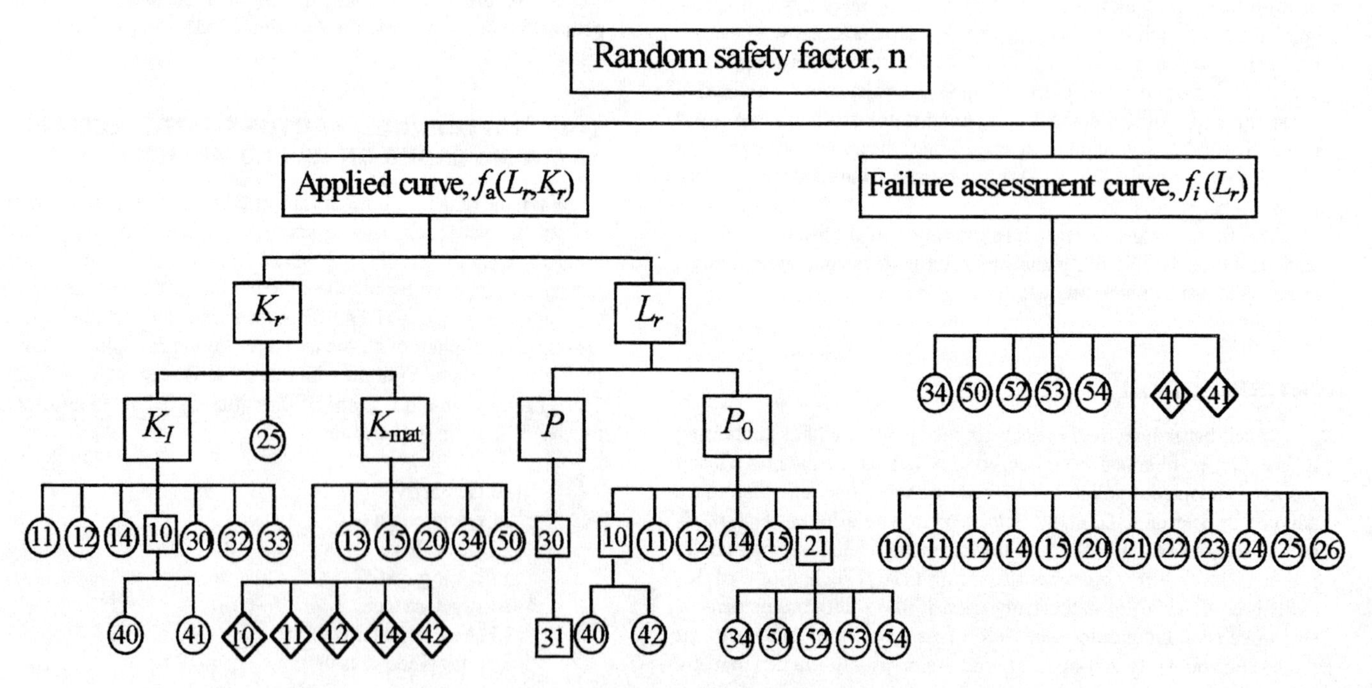

Fig. 1 The fault tree analysis diagram of different factors affecting structure failure

FAC, the first thing is to calculate J-integral and the precision of the FAC is determined by the accuracy of J-integral calculating results. At present the much better developed methods to calculate J-integral are finite element method, EPRI engineering evaluation approach (Zahoor, 1991), and multi-piecewise-power-law superposition (MPPLS) engineering evaluation approach (Wang, et al, 1994; Lei, et al, 1996a; Lei, et al, 1996b). Although to adopt EPRI and MPPLS engineering evaluation approaches is much simpler than to adopt finite element method, the calculation is still very over elaborate and the approaches are not convenient to general engineering staffs to conduct the engineering assessment. For this reason, the reference stress method is introduced to J-integral EPRI and MPPLS engineering evaluation approaches, and thus a more succinct FAC is derived based on the two approaches (detailed in Appendix A).

Hence, the option 3 FAC established convenient to engineering application lays a foundation for further carrying out reliability analysis.

RELIABILITY ANALYSIS MODEL AND ASSESSMENT APPROACH

Reliability Analysis Model

Since the complexity in reliability analysis process, the Monte Carlo (MC) simulation method is adopted here to carry out the numerical analysis of reliability. To establish an effective reliability analysis model is the key to carrying out the reliability analysis of structures. It can be seen from Fig. 1 that the factors affecting structure failure may be divided into two parts, the applied curve, $f_a(L_r, K_r)$, of assessment and failure assessment curve, $f_i(L_r)$. As shown in Fig.2, the probabilistic density function, $P_a(L_r, K_r)$, of an assessed point appears at (L_r, K_r) which results from the variations of material fracture toughness (K_{mat}) and yield strength (σ_{ys}) under a given crack length or depth (a), is a double-parameter probabilistic density distribution function. The R6 FAD representation is then taken as the basic model of the reliability analysis.

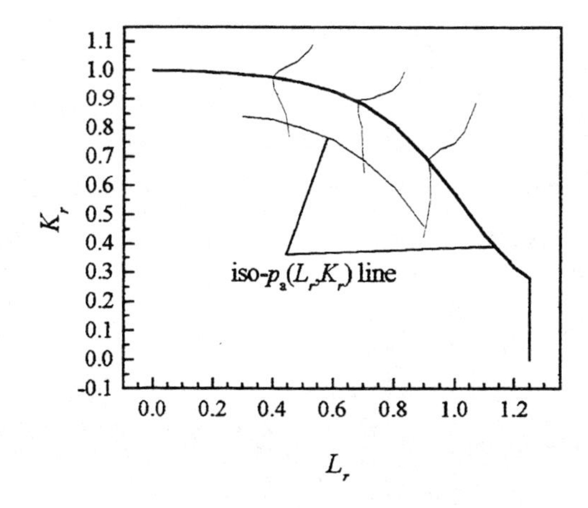

Fig. 2 Principle diagram of calculating the structure failure probability

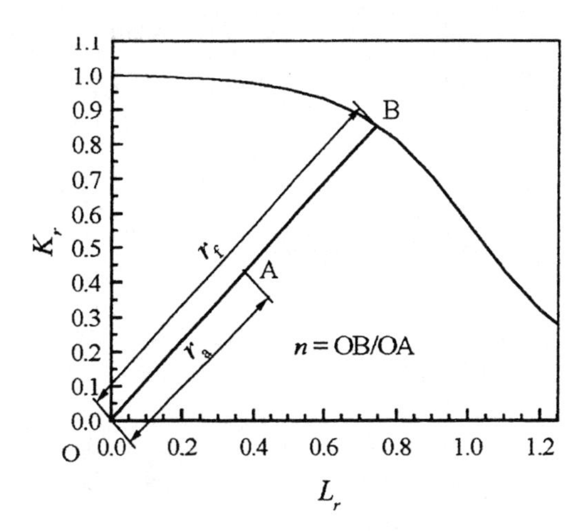

Fig. 3 The definition of random safety factor, n

In order to raise the efficiency during MC simulating and convenience the computation programming and statistical analyzing, the Cartesian coordinates (L_r, K_r) is transformed into the polar coordinates (r, θ) in FAD as shown in Fig. 3. Thus the radius relating to the assessed point A, r_a is equal to OA, and the intersecting point of the elongated OA with the assessment curve is B. For the convenience of structure reliability analysis, a random safety factor, n=OB/OA, is introduced here. To introduce the random safety factor, n, may mainly has the advantages as follows:

a. The information of each assessed point in MC simulations may fully be reflected.

The random safety factor, n, can reflect the extent of each assessed point approaching failure more accurate than that the conventional method, the ratio of K_r to $f_i(L_r)$, reflects (see Fig. 4). The relative distribution of n can be obtained after a certain number of simulating tests. A higher extent of precision of failure probability, P_f, can thus be obtained by using method of least squares to fit the tail part of data of n approaching 1 or $n \leq 1$ to be concerned about. The random safety factor, n, especially reveals its superiority when the curve is selected above R6 option 1 in the analysis because the failure assessment curves will be different for each simulating test.

When only σ^P exists, the random safety factor, n, is equivalent to the load reserving coefficient, F^L, of R6; when σ^S exists along with σ^P, the extent of an assessed point approaching to failure represented by n is less than that represented by F^L, and their values are also different from each other. But when F^L is approaching 1, that is, which is nearby the very concerned about tail data of F^L, the difference between n and F^L will decrease along with the approaching. And when F^L is equal to 1, n will be equal to F^L. After a certain number of simulations, the distribution of n in substance is transforming the information of assessed point (L_r, K_r) from two-dimensional into one-dimensional problem. The inadequacy of the information of assessed point (L_r, K_r) represented by n than that by F^L may almost be neglected. The most important point is that the calculation of n is much simpler than that of F^L.

b. It needs not draw the FAD when solving the problem and is convenient to analyze.

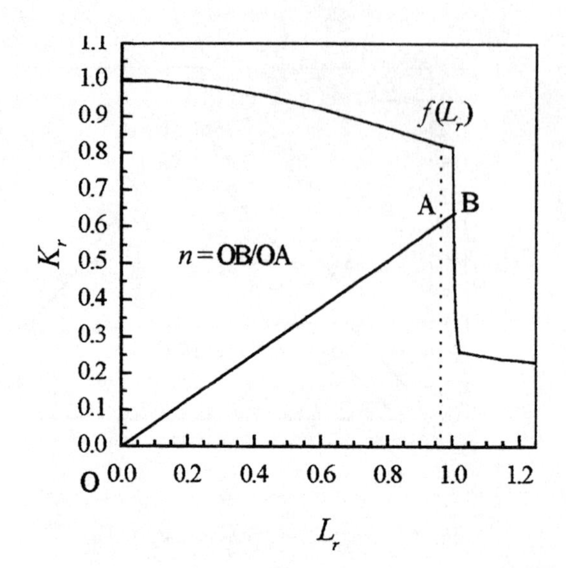

Fig. 4 The comparison of n with the conventional method

When adopting the MC method carries out the multiple-simulation, it is not only wasting time but also no need for a new assessment curve to be built each time followed the changes of assessing condition. As n simplifies the problem from two-dimensional to one-dimensional, the changes of FAC will be transformed to the representation of n. When assessing, however, it is no need to draw a complete FAD, and it is only need to find the relative radius OB as shown in Fig. 3 by analytical or numerical method. Hence, a very simple basic model can be obtained so as to be convenient to be used in MC method analysis.

When the Cartesian coordinates (L_r, K_r) is change into the polar coordinates (r, θ), the relative relations of each parameter are changed as:

$$K_r = r \cdot \cos\theta \qquad L_r = r \cdot \sin\theta \qquad \theta = \text{arctg}\left(\frac{L_r}{K_r}\right) \tag{2}$$

radius of assessed point, OA: $$r_a = \left(L_r^2 + K_r^2\right)^{\frac{1}{2}} \tag{3}$$

radius of related failure point at FAC, OB: $$r_f = G_i(\theta) \tag{4}$$

Where $G_i(\theta)$ represents the r-θ curve function of the option i FAC. The random safety factor:

$$n = \frac{r_f}{r_a} \tag{5}$$

Reliability Analysis Method

After the mathematical description of the statistical properties of the factors affecting structure failure and the sampled simulation to the factors have been made, the distribution function, $F(n)$, of the random safety factor, n, can be obtained. Therefore, the relative reliability is:

$$P_r = \int_{-\infty}^{1} F(n)dn \tag{6}$$

and the failure probability is:

$$P_f = 1 - \int_{-\infty}^{1} F(n)dn \tag{7}$$

The reliability analysis procedure is shown in Fig. 5 and the related block diagram of analysis programming is shown in Fig. 6.

PROBABILISTIC FAILURE ASSESSMENT DIAGRAM

Based on the assumptions that: (1) the load is a deterministic variable; (2) the fracture toughness and yield strength satisfy the mean and variance determined distribution such as normal or Γ distribution, etc, and their variational coefficients are constant c_1 and c_2 respectively; (3) the initial crack length, l_0, remains unchanged; (4) the crack is a penetrated one in an infinite plate and the probabilistic density function of the crack length follows:

$$f_l(x) = \begin{cases} 0.1979\ \exp(-0.1979\ l) & l > l_0 \\ 1 - \exp(-0.1979\ l_0 & l = l_0 \\ 0 & l < l_0 \end{cases} \tag{8}$$

the independence of L_r and K_r on the distributions of load, fracture toughness, yield strength, and initial crack length can be demonstrated as follows:

(a) the independence of L_r

If the probabilistic density function of the material yield strength, σ_{ys}, is known as $f_{\sigma_{ys}}(x)$, the expected value, variance, and variational coefficient of σ_{ys} can be written as:

$$\begin{aligned} E(\sigma_{ys}) &= \int_0^{\infty} x f_{\sigma_{ys}}(x)dx \\ D(\sigma_{ys}) &= \int_0^{\infty} x^2 f_{\sigma_{ys}}(x)dx - [E(\sigma_{ys})]^2 \\ C(\sigma_{ys}) &= \sqrt{D(\sigma_{ys})}/E(\sigma_{ys}) = c_2 \end{aligned} \tag{9}$$

Because the stress, σ_L, induced by applied load is also a deterministic variable, the distribution type of σ_{ys}/σ_L is the same as that of σ_{ys} and its probabilistic density function is $\sigma_L f_{\sigma_{ys}}(x/\sigma_L)$. The expected value, variance, and variational coefficient of σ_{ys}/σ_L can also be written as:

$$\begin{aligned} E(\sigma_{ys}/\sigma_L) &= \sigma_L \int_0^{\infty} x f_{\sigma_{ys}}(x/\sigma_L)dx \\ &= E(\sigma_{ys})/\sigma_L \\ D(\sigma_{ys}/\sigma_L) &= \sigma_L \int_0^{\infty} x^2 f_{\sigma_{ys}}(x/\sigma_L)dx - [E(\sigma_{ys}/\sigma_L)]^2 \\ &= D(\sigma_{ys})/\sigma_L^2 \\ C(\sigma_{ys}/\sigma_L) &= \sqrt{D(\sigma_{ys}/\sigma_L)}/E(\sigma_{ys}/\sigma_L) = c_2 \end{aligned} \tag{10}$$

$1/L_r=\sigma_{ys}/\sigma_L$ thus follows the same distribution type as that of σ_{ys} and its variational coefficient keeps unchanged. Furthermore, when the variational coefficient of σ_{ys} is determined, the distribution function of $1/L_r$ is merely related to the mean, μ_{L_r}. Therefore, the position of failure probabilistic curve in FAD is merely relative to variational coefficient, c_2, and the mean of L_r and irrelative to the choice of mean of σ_{ys}.

(b) the independence of K_r

With regard to $K_r = \frac{\sigma_K\sqrt{\pi l/2}}{K_{mat}}$, it can also be demonstrated with the same reasoning as mentioned above that the distribution function of K_{mat}/σ_K is only related to the mean, $\mu_{\frac{K_{mat}}{\sigma_K}}$ when the variational coefficient of K_{mat} remains unchanged. In addition, the distribution of $\sqrt{\pi d/2}$ remains unchanged when l_0 is fixed. Therefore, the distribution

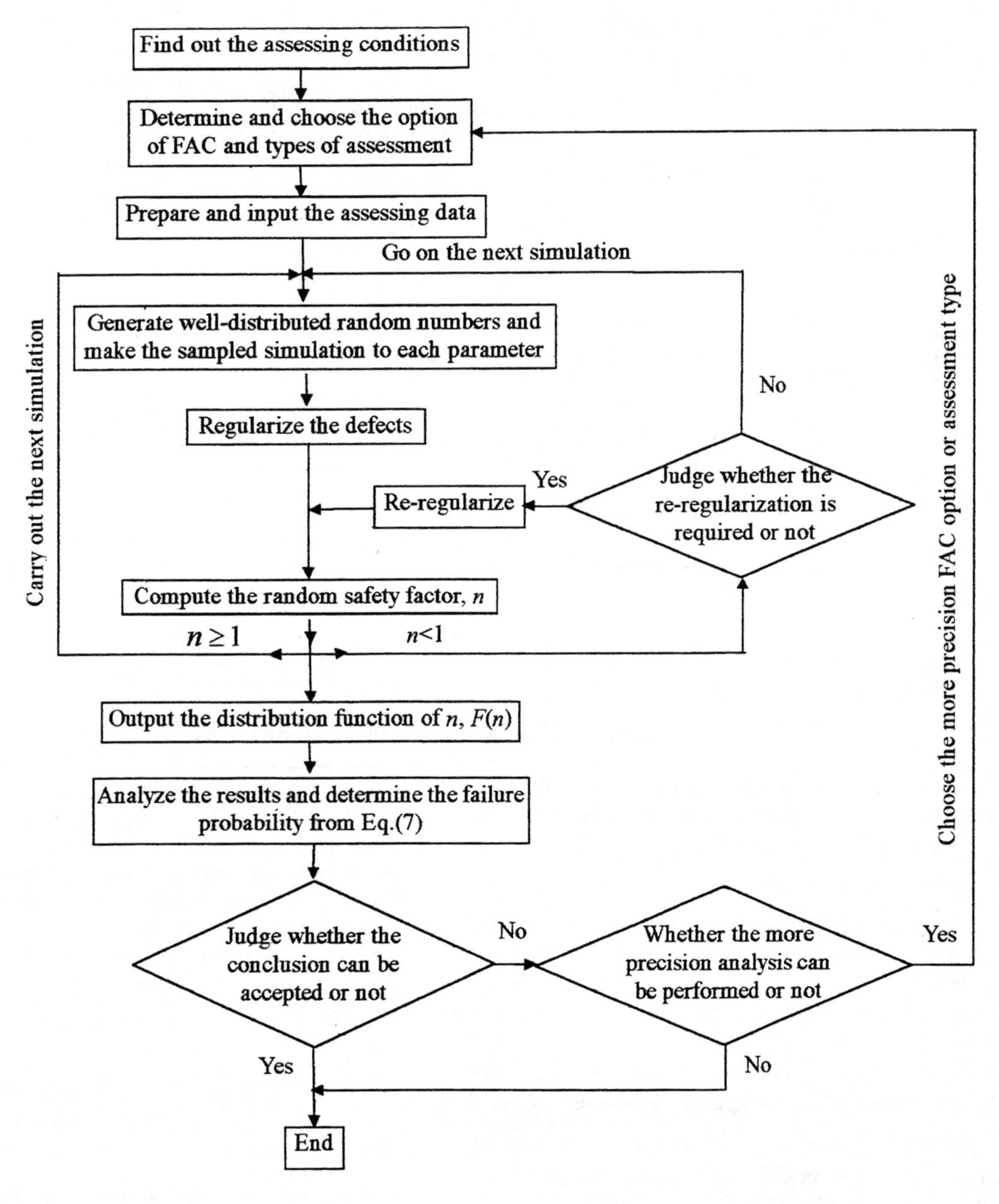

Fig. 5 The procedure of reliability analysis based on R6 FAD method

function of K_r is only relative to its mean, μ_{K_r}, when the conditions are fitted above.

When assessing, the value of assessed mean point, $P(L_r', K_r')$, can be

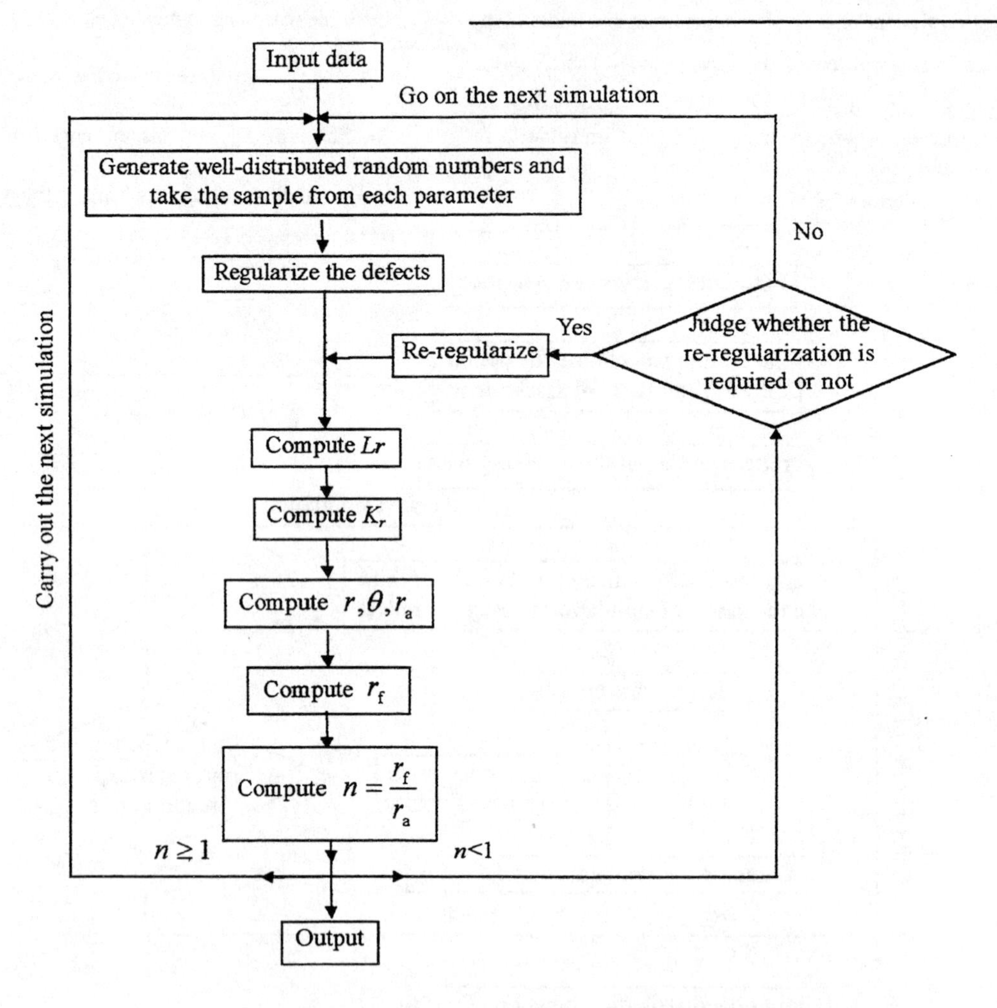

Fig.6 The block diagram of reliability assessment programming

As a consequence, "an assumption of independence" of L_r and K_r on variables load, fracture toughness, etc can be proposed and the distribution function of assessed point (L_r, K_r), $P(L_r, K_r)$ is only relative to the mean values, μ_{L_r} and μ_{K_r}. That means the distribution function and failure probability are same respectively for the points having the same mean values. Hence, the distribution function of n is also related to its mean, μ_n.

Thereby, for a given initial crack length and fixed variational coefficients of fracture toughness and yield strength, the values of distribution function of assessed points can be obtained for different $P(L_r, K_r)$ after a series of calculations. Furthermore, the means of $P(L_r, K_r)$ with the corresponding failure probabilities on FAD are plotted and the iso-failure-probability points are linked to form a series of iso-failure-probabilistic curves and a probabilistic failure assessment diagram (PFAD) (Wang, et al, 1999) as shown in Fig.7.

calculated according to the deterministic method by taking the mean of each random variable so long as the variational coefficients of fracture toughness and yield strength are not bigger than that adopted in the establishment of the probabilistic failure assessment curves. Consequently, the failure probability on the safe side can be obtained by marking the assessed mean point on PFAD, and then the elementary reliability analysis to the structures containing defects could be carried out. The assessed mean point $P(L_r', K_r')$, has a slight different to the mean of $P(L_r, K_r)$ which is discussed by Wang, et al (1999).

CONCLUSIONS

1. The fault tree of the factors affecting structure failure is given, and relationships among the factors are declared and analyzed;

2. The probabilistic analysis models were built by proposing a random safety factor and by simplifying three kinds of options of R6 FACs. The option 3 curve can be built by J-integral which is calculated by MPPLS engineering evaluation approach and especially by simplified multi-piecewise-power-law curves by reference stress;
3. Based on the proposed assumptions to the basic stochastic variables, load, yield strength, fracture toughness, initial crack length, etc., the independence of L_r and K_r on them has been demonstrated. As a consequence, a very similar diagram to R6 FAD, PFAD could be derived. The PFAD may be varied with the different assumptions to the distributions of the basic stochastic variables, load, yield strength, fracture toughness, initial crack length, etc.

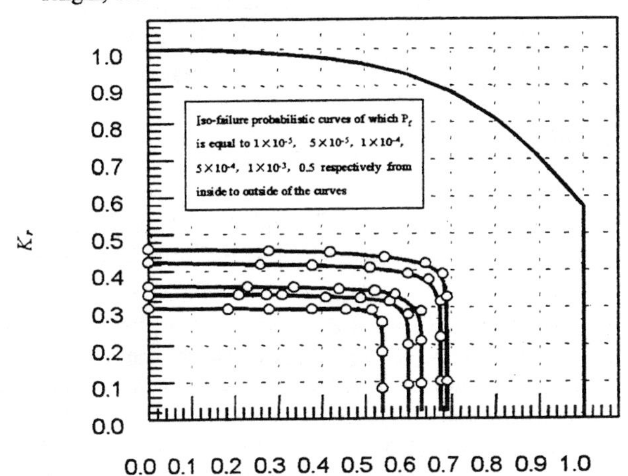

(a) Probabilistic failure assessment diagram by taking the maximum values of the variational coefficients

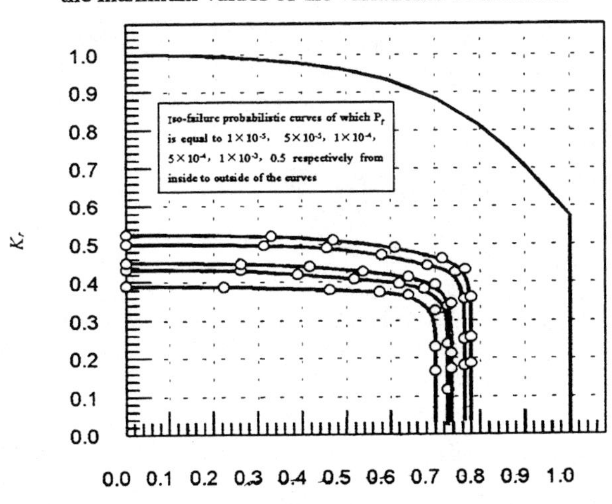

(b) Probabilistic failure assessment diagram by taking the recommended adoptive values of variational coefficients

Fig. 7 Probabilistic failure assessment diagram by taking the different variational coefficients of fracture toughness and yield strength

REFERENCES

Ainsworth, R A, 1984, "The assessment of defects in structure of strain hardening material," Eng Fract Mech, Vol.19, p633

Bergman, M, Brickstad, B, Dahlberg, L, Nilsson, F, and Sttari-Far, I, 1991, "A procedure for safety assessment of components with cracks—Handbook," SA/FOU-Report 91/01, Sweden: The Swedish Plant Inspectiorate

British Standard Institute, 1996, "Guidance on some methods for the derivation of acceptance levels for defects in fusion weld joints," Published Document PD 6493

Gate, R S, 1985, " Probabilistic elastic-plastic fracture mechanics analysis based on R6 methodology," Int J Pres Ves & Piping, Vol.18, pp1~13

Kaiser, S, 1986, "A deterministic and probabilistic model for ductile fracture," Int J Pres Ves & Piping, Vol.22, pp1~21

Lei, Y B, Li, P N, Wang, Z W, Wu, X H, Chen, Z F, 1996a, "The verification and improvement for J-integral estimation scheme of 3-section piecewise exponential function, Part I," Pressure Vessel Technology, Vol.13, No.4, pp16~22(in Chinese)

Lei, Y B, Li, P N, Wang, Z W, Wu, X H, Chen, Z F, 1996b, "The verification and improvement for J-integral estimation scheme of 3-section piecewise exponential function, Part II," Pressure Vessel Technology, Vol.13, No.5, pp35~41(in Chinese)

Milne, I, Ainsworth, R A, Dowling, A R, Stewart, A J, 1986, "Assessment of the integrity of structures containing defects," CEGB Report R/H/R6-Revision 3

Wang, W Q, Liu, C J, Zhou, S J, 1999, "On the probabilistic failure assessment diagram," Int J Pres Ves & Piping, Vol.76, pp653~662

Wang, W Q, Li, A J And Zhang, X D, 1994, "The engineering approach of J-integral evaluation by means of fully plastic solutions for pressure vessel steels in common use in China," Proc 1994 ASME PVP Conf, PVP-Vol.287/MD-Vol.47, Fracture Mechanics Applications, ASME, pp11~18

Warke, R W, Wang, Y Y, Ferregut, C, Carrasco, C J, Horsley, D J, 1999, "A FAD-based method for probabilistic flaw assessment of strength-mismatched girth welds," Proc 1999 ASME PVP Conf, PVP-Vol 386, Probabilistic and Environmental Aspects of Fracture and Fatigue, ASME

Zahoor, A, 1991, "Ductile Fracture Handbook," Vol.1~3, New York: Novetech, EPRI

Zhao, J P, Huang, W L, Dai, S H, 1997, "A new concept: probabilistic failure assessment diagram," Int Pres Ves & Piping, Vol.71, pp165~168

AKNOWLEDGEMENT

THE AUTHOR GRATEFULLY ACKNOWLEDGES THE SUPPORT OF K. C. WONG EDUCATION FOUNDATION, HONG KONG.

APPENDIX A

A1 The Multi-Piecewise-Power-Law Superposition Engineering Evaluation Approach of *J*-integral

The EPRI engineering evaluation approach can be used to calculate *J*-integral values of many kinds of structures, but it only suite to calculate the material satisfying the Ramberg-Osgood rule (ROR). In fact the constitutive relation of practical materials cannot be described only by the ROR. In order to solve this problem, Wang et al (1994) and Lei et al (1996a, 1996b) have proposed an engineering evaluation approach to calculate *J*-integral values by using multi-piecewise-power-law superposition method based on EPRI fully plastic solutions (it is called the multi-piecewise-power-law superposition engineering evaluation approach, or abbreviated as MPPLS engineering evaluation approach) for the material constitutive relation may be described by multi-piecewise RORs.

When the material constitutive relation cannot be described by only one piece of ROR, it may be described by multi-piecewise RORs (see Fig. A1 and Eq. (A1)).

Piece 1: $$\frac{\varepsilon}{\varepsilon_0}=\frac{\sigma}{\sigma_0}+\alpha_1\left(\frac{\sigma}{\sigma_0}\right)^{N_1}\qquad 0\le\sigma\le\sigma_1$$

Piece 2: $$\frac{\varepsilon}{\varepsilon_0}=\frac{\sigma}{\sigma_0}+\alpha_2\left(\frac{\sigma}{\sigma_0}\right)^{N_2}\qquad \sigma_1\le\sigma\le\sigma_2\qquad\text{(A1)}$$

Piece *k*: $$\frac{\varepsilon}{\varepsilon_0}=\frac{\sigma}{\sigma_0}+\alpha_k\left(\frac{\sigma}{\sigma_0}\right)^{N_k}\qquad \sigma_{k-1}\le\sigma\le\sigma_k$$

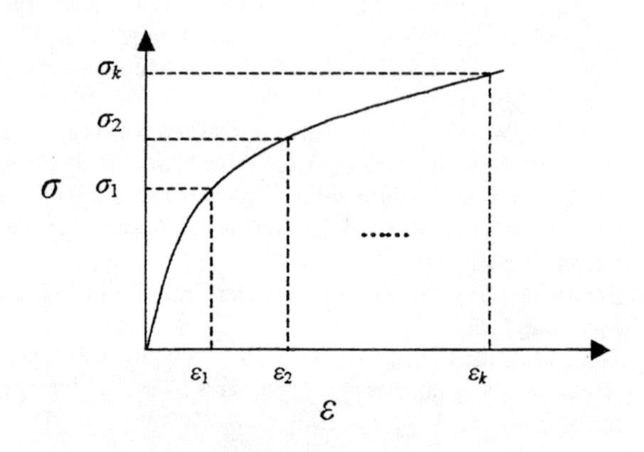

Fig. A1 The diagram of the material constitutive relation fitted by multi-piecewise RORs

where: σ_0 can take σ_{ys} or other proper value as its value;

σ_k is the stress value at the intersection of the *k*th and (*k*-1)th sections of ROR curves.

According to the MPPLS engineering evaluation approach:

$$J=J_e(a_e)+J_p\qquad\text{(A2)}$$

$$J_e(a_e)=\frac{K_I^{\,2}(a_e)}{E'}=\frac{Y^2(a_e)\pi a_e P^2}{E'}\qquad\text{(A3)}$$

Where: $$a_e=a_e^{k-1}+\phi\gamma_y\qquad\text{(A4) (Wang, et al, 1994)}$$

$$a_e^0=a$$

$$\gamma_y=\frac{1}{\beta\pi}\frac{(N_k-1)}{(N_k+1)}\left(\frac{K_I(a_e^{k-1})}{\sigma_0}\right)^2$$

$$\phi=\frac{1}{[1+(P/P_0)^2]}$$

$$J_p=\bar{J}_{p(k)}[(\frac{P}{P_0})^{N_k+1}-(\frac{P_{0(k-1)}}{P_0})^{N_k+1}]+\sum_{i=1}^{k-1}\bar{J}_{p(i)}[(\frac{P_{0(i)}}{P_0})^{N_i+1}-(\frac{P_{0(i-1)}}{P_0})^{N_i+1}]\qquad\text{(A5)}$$

(Lei, et al, 1996a; 1996b)

where: $$P_{0(k)}=\frac{\sigma_k}{\sigma_0}P_0$$

$$\bar{J}_{p(k)}=\alpha_k\sigma_0\varepsilon_0 cg_1\left(\frac{a}{b}\right)h_1(N=N_k)\qquad\text{(A6)}$$

A2 The FACs Established By *J*-integral Which Is Calculated By MPPLS Engineering Evaluation Approach

For the practical materials which constitutive relations usually cannot be represented by only one-piece ROR, the establishments of their FAC are suggested to adopt the MPPLS engineering evaluation approach calculating *J*-integral and forming a series of piecewise-FAC.

The failure assessment curve defined by *J*-integral can strictly be written as:

$$f(L_r)=\sqrt{\frac{J_e}{J}}=\sqrt{\frac{J_e}{J_e(a_e)+J_p}}\qquad L_r\le L_r^{\max}\qquad\text{(A7)}$$

or written as: $$f(L_r)=[\frac{J_e(a_e)}{J_e(a)}+\frac{J_p}{J_e(a)}]^{-1/2}\qquad\text{(A8)}$$

When introducing the reference stress

$$\sigma_{ref}=\frac{P}{P_0}\sigma_0\qquad\text{(A9)}$$

and adopting the similar derivation procedure concerning to the one-piece-power-law relation material given by Ainsworth (1984), the following results can be obtained.

A2.1 When $0\le\sigma_{ref}\le\sigma_1$

In Eq.(A8):

$$J_p=\frac{cg_1\left(\frac{a}{b}\right)h_1(N_1)\sigma_{ref}^{\;2}}{E}(\frac{E\varepsilon_{ref}}{\sigma_{ref}}-1)\qquad\text{(A10)}$$

$$\frac{J_e(a_e)}{J_e(a)}=1+\frac{L_r^2}{2(1+L_r^2)} \quad \text{(A11)}$$

Where, it is conservatively taking $\frac{N_k-1}{N_k+1}\approx 1$ and $\beta = 2$ under plane stress condition in Eq.(A4) when calculating $J_e(a_e)$ (the same below).

$$\frac{J_p}{J_e(a)}=\frac{h_1(N_1)}{h_1(N=1)}\left(\frac{E\varepsilon_{ref}}{\sigma_{ref}}-1\right) \quad \text{(A12)}$$

As the fully plastic solution, J_p, can be believed as the elastic solution when $N=1$ and $\alpha = 1$ in Eq.(A1), and Poisson ratio, $\nu = 1/2$, $J_e(a)$ is thus equal to $(\mu/E)ch_1(N=1)\sigma_{ref}$, where $\mu = 1$ for plane stress condition and $\mu=4/3\cdot E/E'$ for plane strain condition. For conservative, μ takes the value under plane stress condition when calculating $J_e(a)$ in Eq.(A12) (the same below).

A2.2 When $\sigma_1 \leq \sigma_{ref} \leq \sigma_2$

The reference stress, $\sigma_{ref}=\frac{P}{P_0}\sigma_0$, is introduced again. And then in Eq.(A8):

$$J_p=\frac{cg_1\left(\frac{a}{b}\right)h_1(N=N_2)}{E}\left[\sigma_{ref}^2\left(\frac{E\varepsilon_{ref}}{\sigma_{ref}}-1\right)-\sigma_1^2\left(\frac{E\varepsilon_1}{\sigma_1}-1\right)\right]+\frac{cg_1\left(\frac{a}{b}\right)h_1(N=N_1)}{E}\left[\sigma_1^2\left(\frac{E\varepsilon_1}{\sigma_1}-1\right)-\sigma_0^2\left(\frac{E\varepsilon_0}{\sigma_0}-1\right)\right] \quad \text{(A13)}$$

$$\frac{J_e(a_e)}{J_e(a)}=1+\frac{L_r^2}{2(1+L_r^2)} \quad \text{(A11)}$$

$$\frac{J_p}{J_e(a)}=\frac{1}{h_1(N=1)\sigma_{ref}^2}\left\{h_1(N=N_2)\left[\sigma_{ref}^2\left(\frac{E\varepsilon_{ref}}{\sigma_{ref}}-1\right)-\sigma_1^2\left(\frac{E\varepsilon_1}{\sigma_1}-1\right)\right]+h_1(N=N_1)\left[\sigma_1^2\left(\frac{E\varepsilon_1}{\sigma_1}-1\right)-\sigma_0^2\left(\frac{E\varepsilon_0}{\sigma_0}-1\right)\right]\right\} \quad \text{(A14)}$$

Where $\sigma_0^2\left(\frac{E\varepsilon_0}{\sigma_0}-1\right)$ is added in order to unify the equation below when $k>2$. In fact as $\sigma_0^2\left(\frac{E\varepsilon_0}{\sigma_0}-1\right)$ is equal to 0, it does not affect the result.

A2.3 When $\sigma_{k-1} \leq \sigma_{ref} \leq \sigma_k$

The same reasoning as mentioned above when $k>2$:

$$\frac{J_e(a_e)}{J_e(a)}=1+\frac{L_r^2}{2(1+L_r^2)} \quad \text{(A11)}$$

with regard to $\frac{J_p}{J_e(a)}$, and as

$$J_p=cg_1\left(\frac{a}{b}\right)h_1(N=N_k)\left[\sigma_{ref}^2\left(\frac{E\varepsilon_{ref}}{\sigma_{ref}}-1\right)-\sigma_{k-1}^2\left(\frac{E\varepsilon_{k-1}}{\sigma_{k-1}}-1\right)\right]+\sum_{i=1}^{k-1}cg_1\left(\frac{a}{b}\right)h_1(N=N_i)\left[\sigma_i^2\left(\frac{E\varepsilon_i}{\sigma_i}-1\right)-\sigma_{i-1}^2\left(\frac{E\varepsilon_{i-1}}{\sigma_{i-1}}-1\right)\right] \quad \text{(A15)}$$

hence,

$$\frac{J_p}{J_e}=\frac{1}{h_1(N=1)\sigma_{ref}^2}\left\{h_1(N=N_k)\left[\sigma_{ref}^2\left(\frac{E\varepsilon_{ref}}{\sigma_{ref}}-1\right)-\sigma_{k-1}^2\left(\frac{E\varepsilon_{k-1}}{\sigma_{k-1}}-1\right)\right]+\sum_{i=1}^{k-1}h_1(N=N_i)\left[\sigma_i^2\left(\frac{E\varepsilon_i}{\sigma_i}-1\right)-\sigma_{i-1}^2\left(\frac{E\varepsilon_{i-1}}{\sigma_{i-1}}-1\right)\right]\right\} \quad \text{(A16)}$$

Making a summary as mentioned above and replacing $\frac{L_r^2}{2[1+L_r^2]}$ by $\frac{L_r^3\sigma_0}{2E\varepsilon_{ref}}$ which makes the FAC much smoother (Ainsworth, 1984), it can be obtained that:

$$f(L_r)=\left(\frac{1}{h_1(N=1)\sigma_{ref}^2}\left\{h_1(N=N_k)\left[\sigma_{ref}^2\left(\frac{E\varepsilon_{ref}}{\sigma_{ref}}-1\right)-\sigma_{k-1}^2\left(\frac{E\varepsilon_{k-1}}{\sigma_{k-1}}-1\right)\right]+\sum_{i=1}^{k-1}h_1(N=N_i)\left[\sigma_i^2\left(\frac{E\varepsilon_i}{\sigma_i}-1\right)-\sigma_{i-1}^2\left(\frac{E\varepsilon_{i-1}}{\sigma_{i-1}}-1\right)\right]\right\}+1+\frac{L_r^3\sigma_0}{2E\varepsilon_{ref}}\right)^{-\frac{1}{2}} \quad \text{(A17)}$$

The derivation procedure described above is carried out based on the strictly verified EPRI and MPPLS engineering evaluation approaches of J-integral. In the derivation procedure, only the plastic effect modification item concerned in $\frac{J_e(a_e)}{J_e(a)}$ is simplified. As the plastic effect modification item $\frac{J_e(a_e)}{J_e(a)}$ is less important compared to the plastic effect item $\frac{J_p}{J_e(a)}$, and the result under plane stress condition has been taken on the safe side in deriving the item $\frac{J_p}{J_e(a)}$, the FAC used in super assessment derived in this paper is on the safe side and rational and has no need to be verified. To use the FAC derived in this paper as a FAC used in super assessment for structures containing defects is simpler than the FAC is computed and established after calculation of J-integral in the past. Because the material sort and structure type are not limited in the derivation procedure of FAC in the paper, the FAC (Eq.(A17)) is available to every kind of structures and materials.

PVP-Vol. 410-1, Assessment Methodologies for Preventing Failure:
Deterministic and Probabilistic Aspects and Weld Residual Stress
Volume 1
ASME 2000

RESIDUAL STRESS MEASUREMENT IN THICK SECTION COMPONENTS

D. George [1] E. Kingston [1]

D.J. Smith [1]

[1] Department of Mechanical Engineering, University of Bristol, BS8 1TR, UK.

ABSTRACT

This paper describes the application of the deep-hole technique to measure residual stresses in thick welded components. Residual stresses were evaluated in three specimens with a wall thickness ranging from 2.36 inch to 4.33 inch. The three components were a fillet welded ferritic steel T-section, a stainless steel "Stub-beam" welded to a thick section cylinder and a repair welded ferritic steel thick section plate. In all components the deep-hole technique was used to determine the distribution of residual stresses through the wall thickness. The measurements are compared with finite element results for the "Stub-beam" and the repair weld. There is limited agreement between the experiment and predictions.

INTRODUCTION

Welding is of primary importance in most engineering applications. It is used particularly in the offshore, automotive and power industries. A critical issue about welding is the effect it has on the structural integrity of a component and its lifetime. All welding processes generate residual stresses within the component and knowledge of these stresses may be of primary importance for the correct use of the component. There are different techniques to measure residual stresses and they are usually subdivided into three categories. V. De Angelis and C. Sampietri [6], J. Liu [11] and S.K. Bate et al [2] have given reviews of these techniques. Non-destructive techniques such as X-ray and neutron diffraction enable several measurements to be taken at the same place. The semi-destructive techniques such as surface hole drilling, deep-hole drilling and ring coring damage only a small part of the specimen and enable other measurements at other locations. Finally, techniques such as layering can be used but they completely destroy the specimen. In this paper, we describe the application of the deep-hole technique to three welded specimens of complex shape. This technique measures the residual stresses through the wall at sections of interest, particularly where other techniques could not be applied easily. Comparisons are made with other experimental measurements and finite element analysis predictions. The results demonstrate that the deep-hole technique can be applied to a wide range of component shapes.

NOMENCLATURE

E	Young's modulus
$[M]$	Compliance matrix
$[M]^*$	Pseudo-Inverse of the matrix [M]
$f[\theta,z]$, $g[\theta,z]$, $h[\theta,z]$	Angular and through thickness functions
x, y	Co-ordinates in the plane normal to the axis of the reference hole
z	Position through thickness
ε_θ	Reference hole hoop strain
ε_z	Reference hole through thickness strain
$\overline{\varepsilon}$	Vector of measured reference hole strains
ν	Poisson's ration
σ_{xx}, σ_{yy}	Residual stresses on the plane normal to the axis of the reference hole
σ_{zz}	Residual stresses parallel to the axis of the reference hole
$\overline{\sigma}$	Vector of residual stresses

WELDED COMPONENTS

Measurements were carried out on three different components: a fillet welded "T" ferritic steel specimen, a stainless steel "Stub-beam" specimen and a ferritic steel plate containing a large repair weld.

Fillet Welded T–Section

The fillet welded T-section was cut from a large fillet welded T-section ferritic steel plate. The fillet welds were created using the manual metal arc process. The plate thickness was 3.94 inch and the overall height of the T-section was 7.87 inch. The cut specimen length was 3.94 inch (see Figure 1). Young's modulus for the steel was 29 MSI and the yield strength was 43.51 kSI. One deep-hole measurement was carried out through the weld heat-affected zone (HAZ), with the axis, 'z', of the reference-hole perpendicular to the surface of the specimen.

Stainless Steel Stub-beam

The stub-beam specimen was made from two components, an AISI Type 316H stainless steel pipe section of outside diameter 16.85 inch and thickness 2.52 inch and a Type 316L stainless steel rectangular section. The weld preparation had the form of a 'J' shape of 1.73 inch depth on the stub-beam. The square section was welded on all four sides to the pipe using a weaved manual metal arc process with electrodes of diameter between 0.16 inch and 0.197 inch. Young's modulus for the weld and parent component materials was 28.28 MSI. The yield strengths were 34.39 kSI for the parent materials and 61.79 kSI for the weld metal. Three measurements were carried out in the welded specimen. In this paper, results from a measurement in the HAZ of the final welded side of the joint are presented. The dimensions of the specimen and measurement location are shown in Figure 2. The axis, *z*, of the reference hole is parallel to the fusion boundary of the weld.

Repair Welded Plate

The final specimen was a large low alloy ferritic CrMoV steel plate containing a repair weld. Prior to the introduction of the repair, seven strong-backs were welded onto one side of the plate. Young's modulus and the yield strength for the parent steel were 29 MSI and 92.24 kSI respectively. The weld repair was a lower strength C-Mn steel having a yield strength of 59.175 kSI. The original weld yield strength was 71.07 kSI. A measurement was carried out in the HAZ of the repair weld, at an angle of 25° from the normal relative to the top face of the plate. The arrangement of the specimen containing the weld repair, and with strong-backs is shown in Figure 3.

RESIDUAL STRESS MEASUREMENT

Basic principles and procedures

The deep-hole (DH) method relies on the measurements of the distortion of a reference hole drilled through the thickness of the specimen. A schematic of the operation is shown in Figure 4. The method is divided into 5 steps. Small steel blocks or bushes are glued onto the specimen surface to act as a reference. A $1/8^{th}$ inch diameter hole is gun-drilled through the bushes and specimen. Accurate measurement of its diameter is undertaken each 0.0079 inch mm in depth and at 18 different angles around the hole. A core containing the reference hole as its axis is then trepanned out of the specimen releasing residual stresses and modifying the diameter of the reference hole. Measurement of the length of the core during trepanning is also undertaken. The hole is then re-measured at the same locations as before trepanning and distortions of the diameter converted into strain and stresses. Leggatt et al. [10] and Bonner et al. [5] describe further details of the process.

Measurement of the hole diameter is done using an air probe which converts variation of air pressure into diameter by using an air-electric transducer. The trepanning operation is carried out using an Electro Discharge Machining (EDM) system. It is also assumed that the reference hole axis is a principal stress direction and plane stress analysis is undertaken in the plane normal to the hole axis.

Analysis and results

The trepanning process releases residual stresses present in the trepanned column. The experimental method measures the change in diameter of the reference hole and variation of length of the trepanned column. These displacements are used to determine the residual stresses in the component. Full details of the analysis can be found in Bonner [4]. Some more recent developments can be found in Garcia-Granada et al. [7] and Smith et al. [12].

A summary of the analysis is given thereafter. In the DH method, the radial displacements, u_r, are measured as a function of angle, θ, of the reference hole, (of diameter, d_o), and through-thickness position, z. The change in thickness is measured also when residual stress is released as a result of trepanning. The relaxed normalised displacements, $2u_r/d_o$ are therefore related to the residual stresses, σ_{xx}, σ_{yy}, τ_{xy}, where :

$$\left.\begin{aligned} &\varepsilon_\theta[\theta,z].E = E\frac{2u_r}{d_0} \\ &= -\{\sigma_{xx}.f[\theta,z] + \sigma_{yy}.g[\theta,z] + \tau_{xy}.h[\theta,z] - \upsilon.\sigma_{zz}\} \\ &E.\varepsilon_{zz} = -\sigma_{zz} - \nu(\sigma_{xx} + \sigma_{yy}) \end{aligned}\right\} \quad (1)$$

The functions f [θ,z], g [θ,z], and h [θ,z] are given by

$$\begin{aligned} f[\theta,z] &= A[z]\{1 + B[z]2\cos(2\theta)\} \\ g[\theta,z] &= A[z]\{1 - B[z]2\cos(2\theta)\} \\ h[\theta,z] &= 4.A[z].B[z].\sin(2\theta) \end{aligned} \quad (2)$$

The parameters A[z] and B[z] represent uniform expansion and eccentricity of the hole, respectively (Garcia-Granada et al, [7]).

If measurements are obtained at n angles a least squares fit to the diametral strains can be used to determine the stresses. For a given through thickness position z_1 equation (1) is rewritten in matrix form as :

$$\overline{\varepsilon} = -[M].\overline{\sigma} \quad (3)$$

where the measured strain and residual stress vectors are

$$\begin{aligned} \varepsilon &= [\varepsilon_\theta[\theta_1,z_1],\varepsilon_\theta[\theta_2,z_1],...,\varepsilon_\theta[\theta_n,z_1],\varepsilon_{zz}]^T \\ \sigma &= [\sigma_{xx},\sigma_{yy},\tau_{xy},\sigma_{zz}]^T \end{aligned} \quad (4)$$

The compliance matrix M is given by

$$[M]=\begin{bmatrix} f[\theta_1,z_1] & g[\theta_1,z_1] & h[\theta_1,z_1] & -\upsilon \\ f[\theta_2,z_1] & g[\theta_2,z_1] & h[\theta_2,z_1] & -\upsilon \\ . & . & . & . \\ . & . & . & . \\ f[\theta_n,z_1] & g[\theta_n,z_1] & h[\theta_n,z_1] & -\upsilon \\ -\upsilon & -\upsilon & 0 & 1 \end{bmatrix} \qquad (5)$$

The optimum residual stress can be obtained by using Pseudo-Inverse or Moore-Penrose inverse matrices, where

$$\overline{\sigma} = -[M]^*.\overline{\varepsilon} \qquad (6)$$

where $[M]^* = (M^T.M)^{-1}.M^T$ is the Pseudo-Inverse of the matrix [M] and $\overline{\sigma}$ is the optimum stress vector $[\sigma_{xx}, \sigma_{yy}, \tau_{xy}, \sigma_{zz}]^T$ that best fits the measured strains ε. The residual stresses for the three components were obtained by inserting the measured strains into equation 6.

Experimental results obtained from the three components are shown in Figure 5 to 7. In each specimen, experimental measurements through the front bushes and the components are shown. The residual stresses in the front bushes are zero as expected. The distance from the outer surface is along the axis, z, of the reference hole through the specimen. In all specimens, results were obtained for the residual stresses on planes normal to the axis of the reference hole. Although measurements in the changes in length of the core can be obtained, for these specimens no measurements were taken.

Figure 5 shows the residual stress measurements obtained through the 3.937 inch thick T-section starting at the toe of the weld. The residual stresses σ_{xx}, σ_{yy}, on the plane normal the axis of the reference hole are shown. For this specimen σ_{xx} represents the residual stress transverse to the axis of the fillet weld, and σ_{yy} represents the residual stress parallel to the weld axis. The in-plane residual stresses were essentially equi-biaxial for depths, z, up to about 0.787 inch. A peak tensile residual stress of 36.26 kSI occurred at the toe of the weld on the surface of the specimen. This peak value decreased rapidly into compression within 0.59 inch from the surface. A peak compressive residual stress of –26.11 kSI occurred at about 0.98 inch to 1.18 inch from the weld surface. Both the σ_{xx} and σ_{yy} residual stresses were in tension on the opposite face to the weld. Force and moment equilibrium across the section is exhibited by the measured transverse residual stresses. The shear residual stresses were relatively small. This indicates that the in-plane residual stresses σ_{xx} and σ_{yy} were close to the principal stress direction.

The results obtained from the stainless steel Stub-beam component are shown in Figure 6. The measured residual stresses were mostly in tension through the specimen thickness. Peak values of transverse stress σ_{xx} and longitudinal stress σ_{yy} were 43.5 kSI and 58 kSI respectively at about 0.59 inch from the outer surface. Both stress components dropped to about zero at the inner surface. It is noticeable that the measured residual stress distributions across the section thickness are not in force and moment equilibrium.

Results obtained from the HAZ of the repair weld are shown in Figure 7. Again, as with the other specimens, the in-plane residual stresses were tensile at the top surface of the specimen. The peak residual stress was about 600MPa in the region adjacent to the repair weld. σ_{yy} represents the residual stress parallel to the weld repair axis. The longitudinal component of residual stress was essentially tensile through the plate thickness. The measured σ_{xx} residual stress did not provide complete force and moment equilibrium across the section, but this is not surprising because of the presence of the strong-backs.

DISCUSSION

Finite element (FE) analyses were carried out to simulate welding of the Stub-beam and repair welded specimens. Further details of these types of analyses can be found in Anderson and Little [1] and Bouchard et al. [3]. Here the residual stress distributions predicted by the analyses are shown for these two specimens in Figures 6 and 7. For the 'Stub-beam' the FE analyses overestimated the longitudinal σ_{yy} residual stresses, particularly close to the outer surface of the specimen. There is, however, good agreement between the measured and the predicted σ_{xx} residual stress. It is anticipated that the higher predicted stresses arise because the FE model for the Stub-beam was much more highly constrained than the real component.

Results for the thick repair weld (see Figure 7) show that there is a reasonably good correlation in the first 0.79 inch of the wall thickness between the FE and the experimental measurements. The experimental measurements reached a peak tensile residual stress of about 87 kSI and the predicted residual stress is about 29 kSI lower. For greater depths (in the original weld) the FE analysis predicted compressive stresses of higher magnitude. Finally, the FE analysis indicates that the σ_{xx} returns to tension on the opposite surface. The experimental results do not agree with these predictions. However, the measurements are consistent with similar studies carried out on repairs in stainless steel welds (George et al, [8]), which showed that the repair weld re-distributed the residual stresses associated with the original weld.

Measurements of residual stresses in a smaller and geometrically similar T-section have been made using neutron diffraction. The thickness t of the T-section (see Figure 1) was 0.985 inch. The transverse residual stress σ_{xx} from the neutron diffraction measurements are shown in Figure 8. The distance 'z' through the thickness is normalised with respect to the wall thickness t. Therefore the neutron diffraction measurements for t = 0.985 inch can be compared with the deep-hole measurements for t = 3.937 inch. The distributions are similar, although the neutron diffraction measurements do not reveal tensile residual stresses at the opposite face to the weld.

CONCLUSIONS

- It has been demonstrated that the deep-hole technique can be used to measure residual stress in thick sections of complex shape.
- There is limited agreement between experimental measurements and finite element predictions.
- The deep-hole measurement method provides a means of validating finite element residual stress simulations so that further refinements to models can be made.

ACKNOWLEDGMENTS

The authors would like to thank British Energy (UK) for their financial help and providing the stub-beam specimen, Mitsui Babcock (UK) for the repair welded plate and Imperial College (London) for the T-specimen and their neutron diffraction measurements. Acknowledgement is also given to Batelle (USA) and British Energy for providing the FE analysis results for the different components.

REFERENCES

1. Anderson, G.S. and Little, W.J.G., "Repair Options for Defects Found in Boiler Shells", I Mech E Seminar Publication 1999-14, Boiler Shell Weld Repair Sizewell A Nuclear Power Station, 33-53, Prof. Eng. Publishing Ltd., London, 1999.
2. Bate, S.K., Green, D., and Buttle , D.J., "A review of residual stress distributions in welded joints for the defect assessment of offshore structures", OHT 482, HMSO, 1997.
3. Bouchard, J., Holt, P., and Smith, D.J., "Prediction and Measurement of Residual Stresses in a Thick Section Stainless Steel Weld", Proceedings of ASME-PVP Conference, vol. 347, pp. 77-82, 1997.
4. Bonner, N.W., "Measurement of Residual Stresses in Thick Section Steel Welds", PhD Thesis, University of Bristol, 1996.
5. Bonner, N.W., and Smith, D.J., Measurement of Residual Stresses Using the Deep Hole Method", PVP, vol 327, pp 53-65, 1996.
6. De Angelis, V., and Sampietri, C.. "Report on the state of the art regarding the problem of residual stresses in welds of LMFBR components", CISE Technology Innovative, Report No. 5495, 1990.
7. Garcia-Granada, A.A., George, D., and Smith, D.J., "Assessment of Distortions in the Deep Hole Technique for Measuring Residual Stresses", Proceedings of 11th Conf. Exp. Mech., pp 1302-1306, 1998.
8. George, D., Bouchard, P.J., and Smith, D.J., "Evaluation of Through Wall Residual Stresses in Stainless Steel Welds Repairs", Proceedings Of the 5th European Conference on Residual Stresses, Noordwijkerhout, The Netherlands, 2000.
9. Lacarac, V.D., Pavier, M.J., Smith, D.J., Keltjenz, J. and Lemmens, D., "Measurement and Prediction of Residual Stresses in Autofrettage High Pressure Tubing Incorporating Temperature and Bauschinger Effect", PVP conference, Seattle, USA, 2000.
10. Leggatt, R.H., Smith, D.J., Smith, S.D. and Faure, F., "Development and Experimental Validation of the Deep Hole Method for Residual Stress Measurement", Journal of Strain Analysis, vol 31, no 3, 1996.
11. Liu, J., "Handbook of measurement of residual stresses", Society for Experimental Mechanics Inc., Fairmont Press Inc, 1997.
12. Smith, D.J., Bouchard, P.J. and George, D., "Measurement And Prediction Of Through Thickness Residual Stresses In Thick Section Welds", Special Issue of Journal of Strain Analysis on Residual Stresses, to be published, 2000.

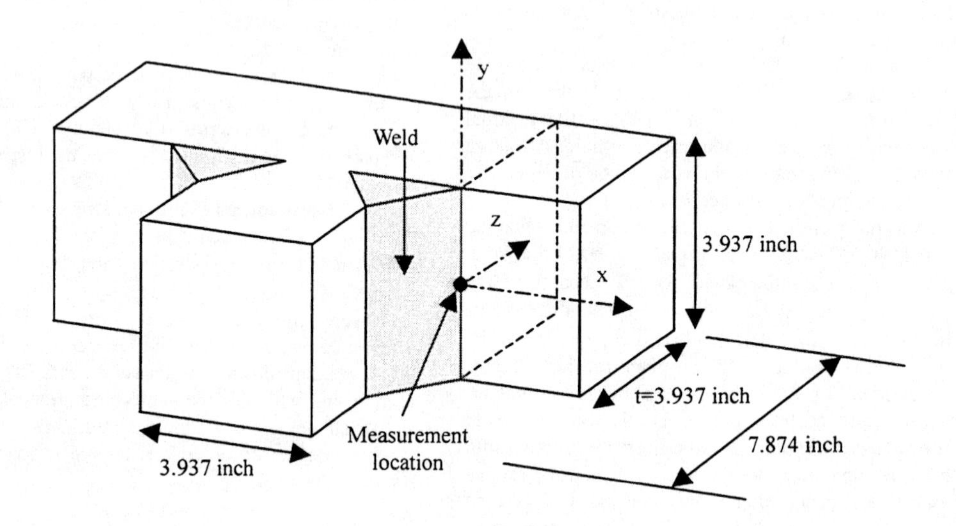

Figure 1 : Fillet welded T-Section.

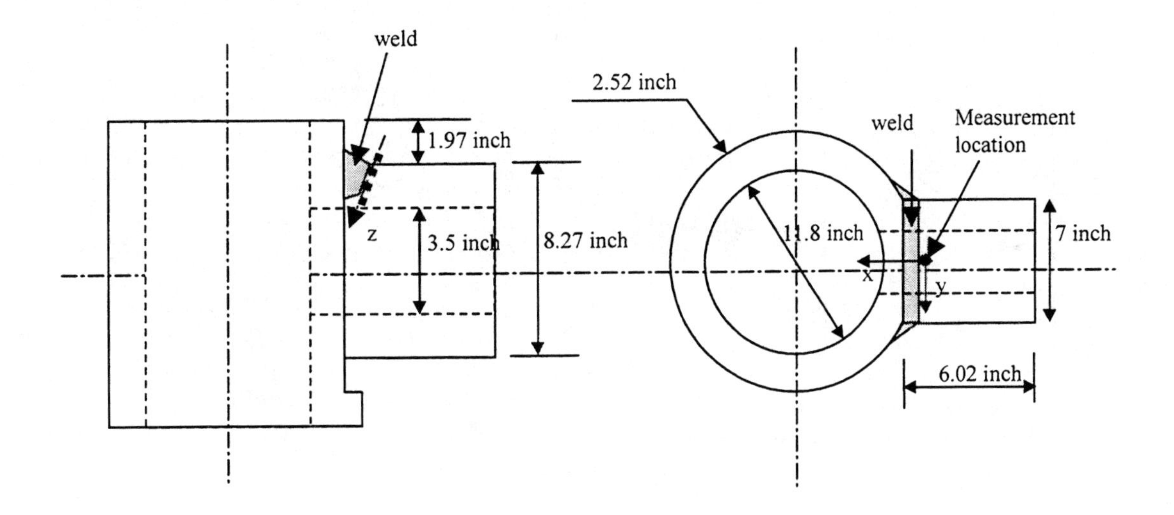

Figure 2 : Stainless steel stub-beam component

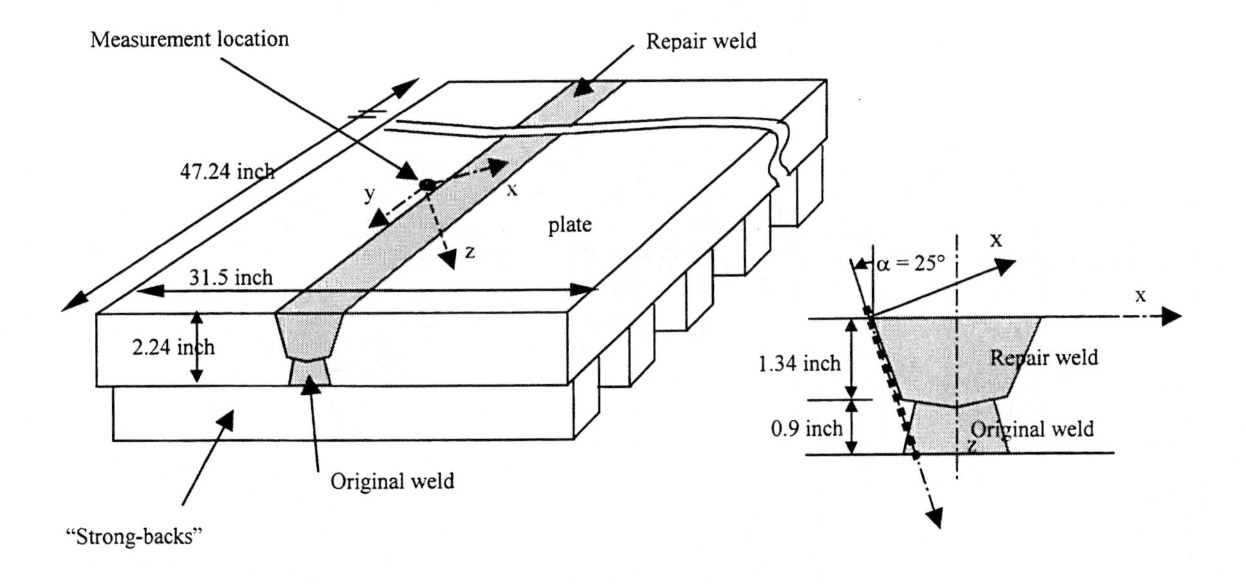

Figure 3 : Thick section plate containing repair weld.

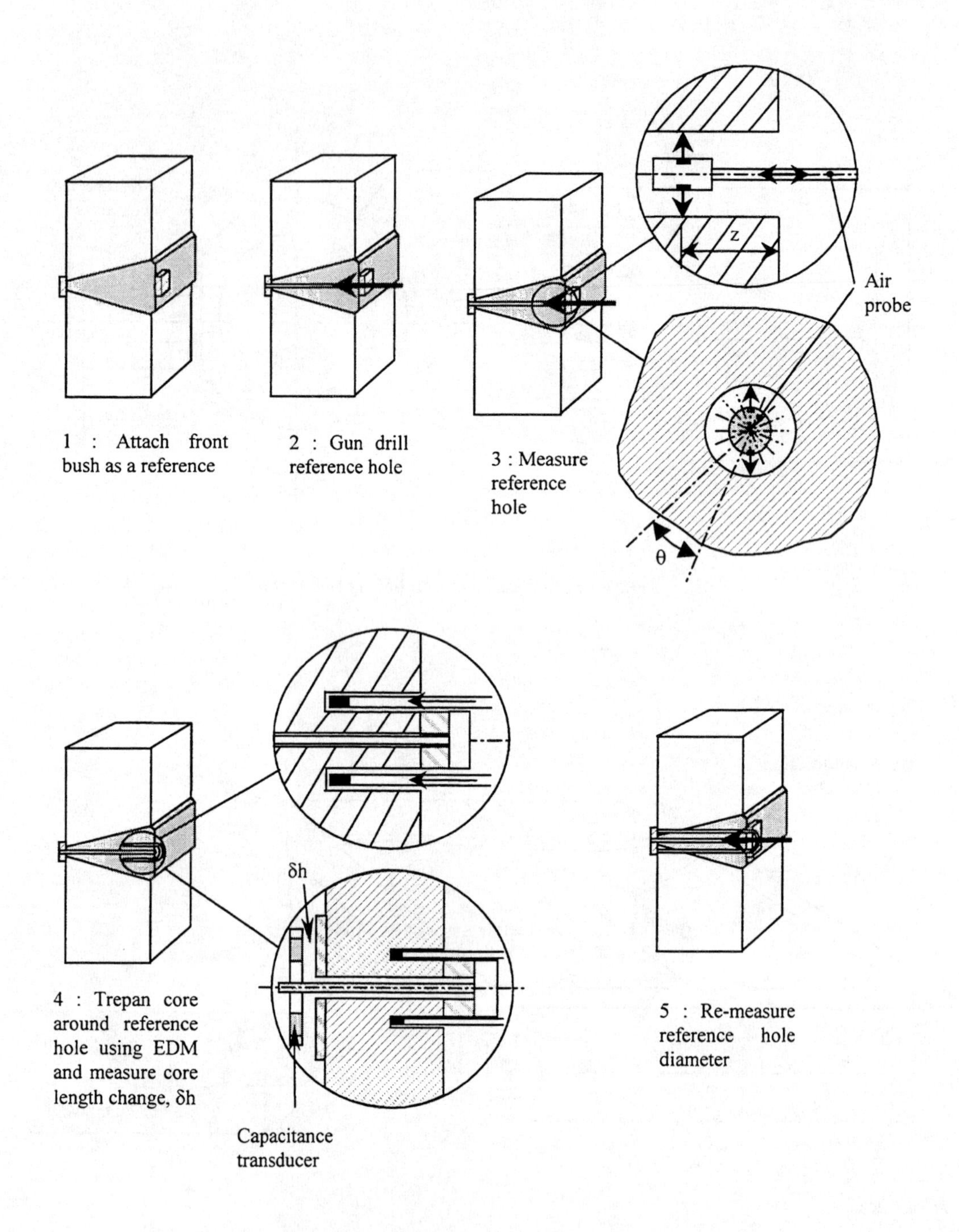

Figure 4 : Schematic of the steps for the deep hole technique.

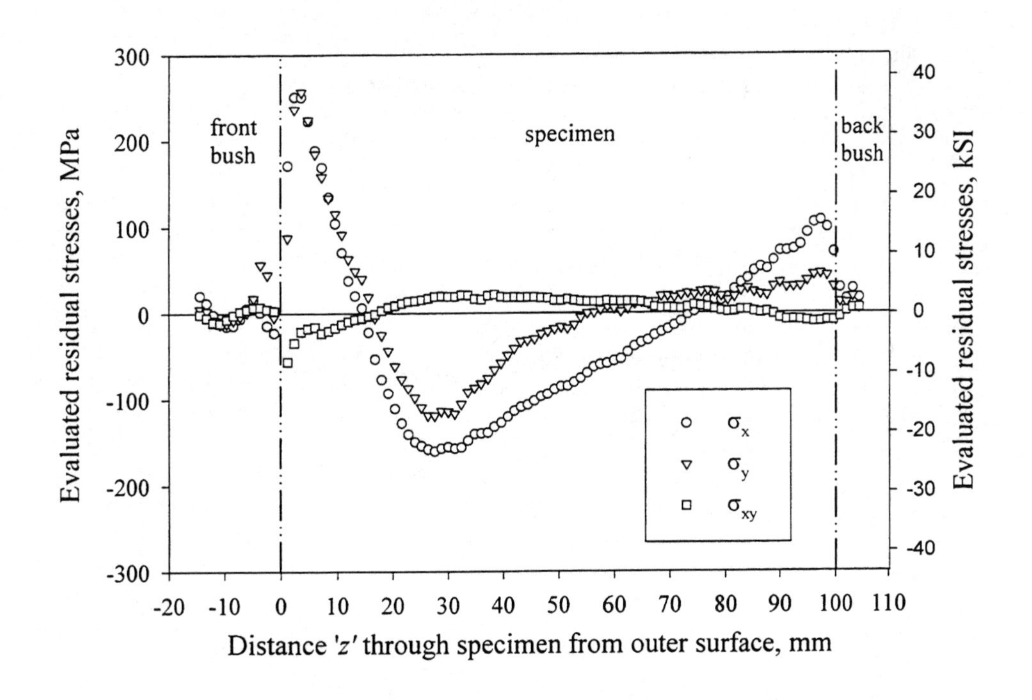

Figure 5 : Experimental results for the T-section.

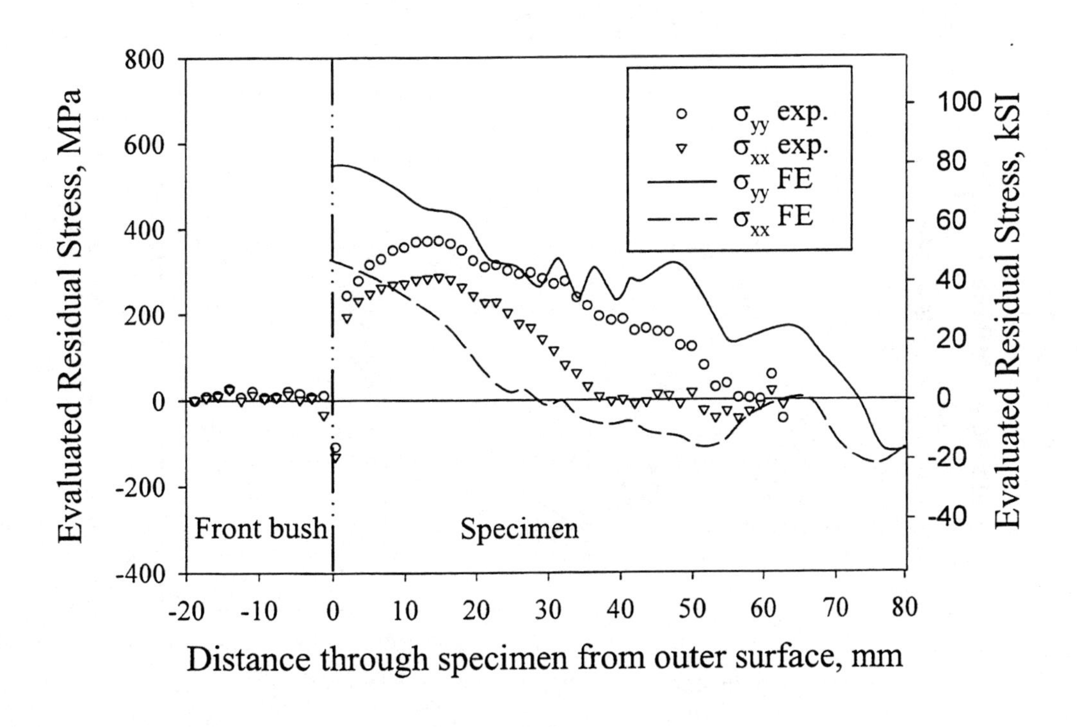

Figure 6 : Comparison of FE and experimental results for the stub-beam specimen

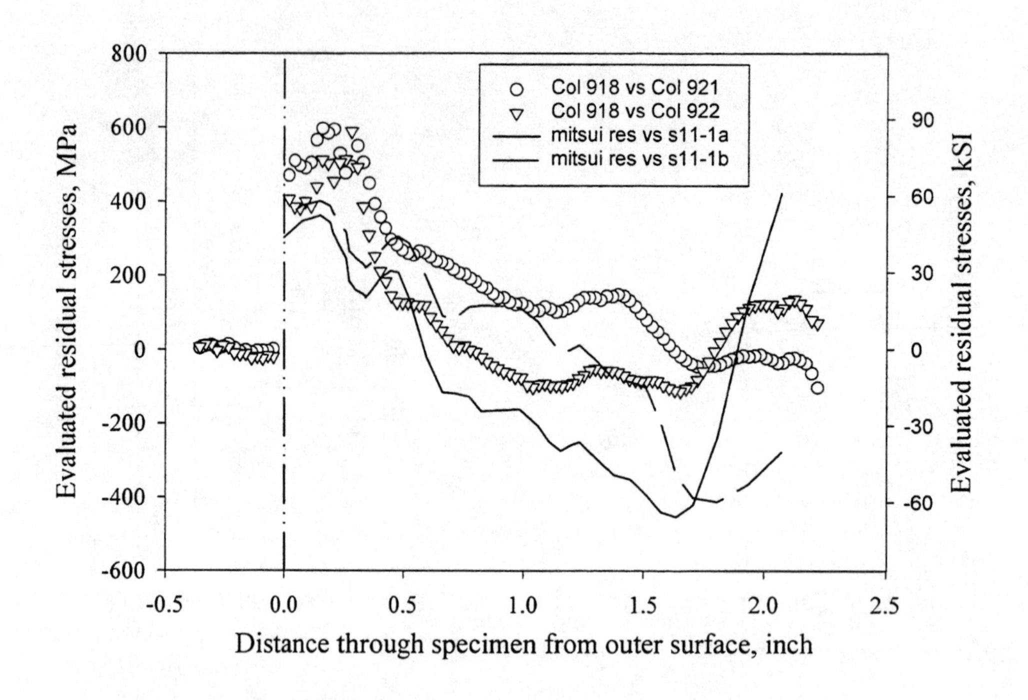

Figure 7 : Comparison of FE and experimental results for a repair welded thick plate specimen

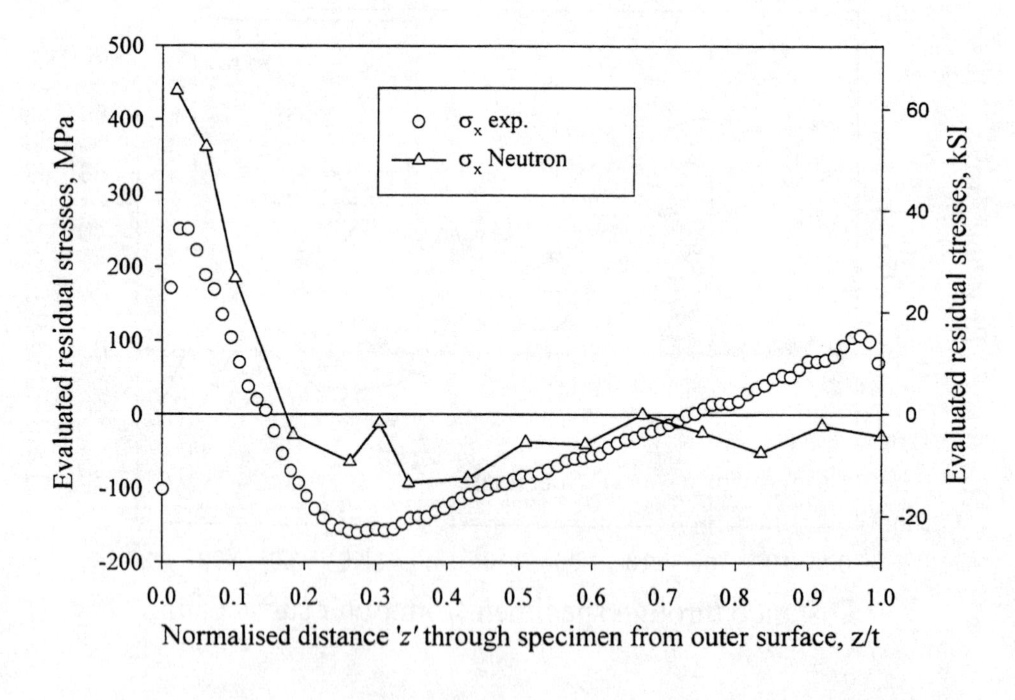

Figure 8 : Comparison of Deep-hole and Neutron diffraction measurements for "T-specimens"

AUTHOR INDEX

PVP-Vol. 410
Assessment Methodologies for Preventing Failure: Volume 1 — Deterministic and Probabilistic Aspects and Weld Residual Stress

Book Number: H01178